Wetlands: Biodiversity, Health and their Role in Ecosystem Services

Wetlands: Biodiversity, Health and their Role in Ecosystem Services

Edited by

Inderjeet Tyagi
Centre for DNA Taxonomy, Molecular Systematics Division, Zoological Survey of India, Kolkata, West Bengal, India

Vikas Kumar
Centre for DNA Taxonomy, Molecular Systematics Division, Zoological Survey of India, Kolkata, West Bengal, India

Kaomud Tyagi
Centre for DNA Taxonomy, Molecular Systematics Division, Zoological Survey of India, Kolkata, West Bengal, India

ELSEVIER

Elsevier
Radarweg 29, PO Box 211, 1000 AE Amsterdam, Netherlands
125 London Wall, London EC2Y 5AS, United Kingdom
50 Hampshire Street, 5th Floor, Cambridge, MA 02139, United States

ISBN: 978-0-443-34222-6

For Information on all Elsevier publications
visit our website at https://www.elsevier.com/books-and-journals

Publisher: Peter Llewellyn
Acquisitions Editor: Maria Elekidou
Editorial Project Manager: Pratishtha Gupta
Production Project Manager: Vijayaraj Purushothaman
Cover Designer: Mark Rogers

Typeset by MPS Limited, Chennai, India

Contents

List of contributors

Neha Anees Department of Zoology, Wildlife and Fisheries, University of Agriculture, Faisalabad, Pakistan

Khadija Batool Department of Zoology, Wildlife and Fisheries, University of Agriculture, Faisalabad, Pakistan

Shumaila Batool Department of Zoology, Wildlife and Fisheries, University of Agriculture, Faisalabad, Pakistan

Bhaskar Behera P.G. Department of Zoology, Fakir Mohan University, Balasore, Odisha, India; P.G. Department of Bioscience and Biotechnology, Fakir Mohan University, Balasore, Odisha, India

Sufie Bhaskara Manggala Agni Kalimantan Daops VI/ Tanah Laut, Balai Pengendalian Kebakaran Hutan Wilayah Kalimantan, Pelaihari, South Kalimantan, Indonesia

Rakesh Bhutiani Department of Environmental Science, Gurukul Kangri (Deemed to Be University), Hardwar, Uttrakhand, India

Sadia BiBi College of Life Sciences and Technology, Tarim University, Xinjiang, P.R. China

Harish Chandra Singh Bisht Department of Zoology, D.S.B Campus, Kumaun University, Nainital, Uttarakhand, India

Satpal Singh Bisht Vice Chancellor, Soban Singh Jeena University, Almora, Uttarakhand, India

Arghya Chakrabarty Wetlands International South Asia, New Delhi, India; ICFRE—Rain Forest Research Institute, Jorhat, Assam, India

Surendra Kumar Chandniha BRSM College of Agricultural Engineering and Technology & Research Station, IGKV, Mungeli, Chhattisgarh, India

Arnab Chattopadhyay Zoological Survey of India, Prani Vigyan Bhavan, New Alipore, Kolkata, West Bengal, India

Sushmita Chettri GB Pant National Institute of Himalayan Environment, Sikkim Regional Centre, Gangtok, Sikkim, India

Prakash Chhetri GB Pant National Institute of Himalayan Environment, Sikkim Regional Centre, Gangtok, Sikkim, India

Girija D.S. Division of Ecotechnology, M. S. Swaminathan Research Foundation, Chennai, Tamil Nadu, India

Rudi Windra Darisman Manggala Agni Kalimantan Daops X/Ketapang, Balai Pengendalian Kebakaran Hutan Wilayah Kalimantan, Pelaihari, South Kalimantan, Indonesia

Mihir Tanay Das School of Environmental Sciences, Fakir Mohan University, Balasore, Odisha, India

Soumya Ranjan Dash School of Biological Sciences, AIPH University, Bhubaneswar, Odisha, India

Ashis Debnath Zoological Survey of India, Prani Vigyan Bhavan, New Alipore, Kolkata, West Bengal, India

Seeta Dewali Department of Zoology, D.S.B Campus, Kumaun University, Nainital, Uttarakhand, India

Rajni Dhalaria Department of Biosciences, Minerva P.G. College of Arts, Sciences and Commerce, Kangra, Himachal Pradesh, India

Kailash Singh Gaira GB Pant National Institute of Himalayan Environment, Sikkim Regional Centre, Gangtok, Sikkim, India

Kantharajan Ganesan ICAR-National Bureau of Fish Genetic Resources, Lucknow, Uttar Pradesh, India

Amit Kumar Ghosh Centre for DNA Taxonomy, Molecular Systematics Division, Zoological Survey of India, Kolkata, West Bengal, India

Rohit Goutam School of Environmental Sciences, Jawaharlal Nehru University, New Delhi, India

Devanshu Gupta Zoological Survey of India, Prani Vigyan Bhavan, New Alipore, Kolkata, West Bengal, India

Shantabala Devi Gurumayum Arunachal Pradesh Regional Centre (APRC), Zoological Survey of India, Itanagar, Arunachal Pradesh, India

Raghib Hussain Centre for DNA Taxonomy, Molecular Systematics Division, Zoological Survey of India, Kolkata, West Bengal, India

Sk Md Abu Imam Saadi Department of Biological Sciences, Aliah University, Kolkata, West Bengal, India

Sk Saruk Islam Department of Natural & Applied Sciences, Raja Narendra Lal Khan Women's College (Autonomous), Midnapore, West Bengal, India

Rajesh Joshi GB Pant National Institute of Himalayan Environment, Sikkim Regional Centre, Gangtok, Sikkim, India

Sivaganga Kandoth GB Pant National Institute of Himalayan Environment, Sikkim Regional Centre, Gangtok, Sikkim, India

Richa Kothari Department of Environmental Sciences, Central University of Jammu, Jammu, Jammu and Kashmir, India

Ankit Kumar Department of Pharmaceutical Sciences, Sir J.C. Bose Technical Campus, Bhimtal, Uttarakhand, India

Ritesh Kumar Wetlands International South Asia, New Delhi, India

Vikas Kumar Centre for DNA Taxonomy, Molecular Systematics Division, Zoological Survey of India, Kolkata, West Bengal, India

Krishma Kumari Department of Environmental Sciences, Central University of Jammu, Jammu, Jammu and Kashmir, India

Sananda Kundu Department of Geography, Manipur University, Imphal, Manipur, India

Saravanaraman M. Division of Ecotechnology, M. S. Swaminathan Research Foundation, Chennai, Tamil Nadu, India

Sumana Mahato Department of Natural & Applied Sciences, Raja Narendra Lal Khan Women's College (Autonomous), Midnapore, West Bengal, India

Gopinathan Maheswaran Zoological Survey of India, Kolkata, West Bengal, India

Sujoy Midya Department of Zoology, Raja Narendra Lal Khan Women's College (Autonomous), Midnapore, West Bengal, India

Rashi Miglani Department of Forensic Science, D.S.B. Campus, Kumaun University, Nainital, Uttarakhand, India

Arun Kumar Mohanta Indira Gandhi Institute of Technology, Sarang, Utkal University, Bhubaneswar, Odisha, India

Arun Mondal Department of Geography, University of Allahabad (A Central University), Prayagraj, Uttar Pradesh, India

Diwakar Naidu SV College of Agricultural Engineering and Technology & Research Station, IGKV, Raipur, Chhattisgarh, India

Bhaskar Dutt Nainwal Department of Chemistry, Doon University, Dehradun, Uttarakhand, India

Yogesh Chandra Nainwal Department of Chemistry, Government Post Graduate College Maldevta (Raipur), Dehradun, Uttarakhand, India

Manikandan P. Division of Ecotechnology, M. S. Swaminathan Research Foundation, Chennai, Tamil Nadu, India

Anupam Pandey Department of Geography, University of Allahabad, Prayagraj, Uttar Pradesh, India

Nagma Parveen Department of Forensic Science, D.S.B. Campus, Kumaun University, Nainital, Uttarakhand, India

Deepak Pathania Department of Environmental Sciences, Central University of Jammu, Jammu, Jammu and Kashmir, India

Sarita Pathania Department of Zoology, Sardar Patel University, Mandi, Himachal Pradesh, India

Biswajit Patra PG Department of Botany, Fakir Mohan University, Balasore, Odisha, India

Surya Narayan Pradhan School of Life Sciences, Sambalpur University, Sambalpur, Odisha, India

Rengalakshmi R. Division of Ecotechnology, M. S. Swaminathan Research Foundation, Chennai, Tamil Nadu, India

Annisa Utami Rauf Department of Health Behavior, Environment and Social Medicine, Faculty of Medicine, Public Health and Nursing, Universitas Gadjah Mada, Sleman, Indonesia

Gaurav Rawat Department of Zoology, D.S.B Campus, Kumaun University, Nainital, Uttarakhand, India

Sushree Swati Rout P.G. Department of Zoology, Fakir Mohan University, Balasore, Odisha, India

Uttam Kumar Sarkar ICAR-National Bureau of Fish Genetic Resources, Lucknow, Uttar Pradesh, India

Rishabh Sen Centre for DNA Taxonomy, Molecular Systematics Division, Zoological Survey of India, Kolkata, West Bengal, India

Deepa Senapathi Department of Sustainable Land Management, University of Reading, Reading, Berkshire, United Kingdom

Anam Shehzadi Department of Zoology, Wildlife and Fisheries, University of Agriculture, Faisalabad, Pakistan

Urwah Sheikh Department of Zoology, Wildlife and Fisheries, University of Agriculture, Faisalabad, Pakistan

Kavyanjali Shukla Indian Science Communication Society, Lucknow, Uttar Pradesh, India

Uzma Siddiqui Department of Zoology, D.S.B Campus, Kumaun University, Nainital, Uttarakhand, India

Leena Singh Department of Applied Sciences, Galgotias College of Engineering and Technology, Greater Noida, Uttar Pradesh, India

Santosh Singh Centre for DNA Taxonomy, Molecular Systematics Division, Zoological Survey of India, Kolkata, West Bengal, India

Shipra Sonali Zoological Survey of India, Prani Vigyan Bhavan, New Alipore, Kolkata, West Bengal, India; University of Calcutta, Kolkata, West Bengal, India

Jullen Subba GB Pant National Institute of Himalayan Environment, Sikkim Regional Centre, Gangtok, Sikkim, India

Abhiseka Swain Centre for DNA Taxonomy, Molecular Systematics Division, Zoological Survey of India, Kolkata, West Bengal, India

Lakpa Tamang GB Pant National Institute of Himalayan Environment, Sikkim Regional Centre, Gangtok, Sikkim, India

Kranti Thakur Department of Botany, Shoolini Institute of Life Sciences and Business Management, Solan, Himachal Pradesh

Binsar Oktavianus Togatorop Manggala Agni Kalimantan Daops III/Pangkalan Bun, Balai Pengendalian Kebakaran Hutan Wilayah Kalimantan, Pelaihari, South Kalimantan, Indonesia

Inderjeet Tyagi Centre for DNA Taxonomy, Molecular Systematics Division, Zoological Survey of India, Kolkata, West Bengal, India

Kaomud Tyagi Centre for DNA Taxonomy, Molecular Systematics Division, Zoological Survey of India, Kolkata, West Bengal, India

Harithalekshmi V. Project Scientist, Institute of Climate Change Studies, Kottayam, Kerala, India

Jagdeep Verma Department of Botany, Sardar Patel University, Mandi, Himachal Pradesh, India

Bryony K. Willcox Centre for Agri-Environmental Research, School of Agriculture, Policy and Development, University of Reading, Reading, Berkshire, United Kingdom

Chapter 1

Wetlands: formation, ecological significance, emerging threats, and pathways for sustainable conservation

Santosh Singh[1], Abhiseka Swain[1], Vikas Kumar[1], Kaomud Tyagi[1], Shantabala Devi Gurumayum[2], Rakesh Bhutiani[3], and Inderjeet Tyagi[1]

[1]*Centre for DNA Taxonomy, Molecular Systematics Division, Zoological Survey of India, Kolkata, West Bengal, India,* [2]*Arunachal Pradesh Regional Centre (APRC), Zoological Survey of India, Itanagar, Arunachal Pradesh, India,* [3]*Department of Environmental Science, Gurukul Kangri (Deemed to Be University), Hardwar, Uttrakhand, India*

1.1 Introduction

According to the Ramsar Convention (2018), wetlands are defined as "areas of marsh, fen, peatland or water, whether natural or artificial, permanent or temporary, with water that is static or flowing, fresh, brackish or salt, including areas of marine water the depth of which at low tide does not exceed six meters." Given the diverse ecological functions, ecosystem services, and regulatory frameworks associated with wetlands, their definition often varies depending on the disciplinary and management contexts. As a result, wetland scientists increasingly endorse a comprehensive, integrative definition that emphasizes core ecological attributes, namely, biological diversity, environmental conditions, and the provisioning of ecosystem services as central to the identification and classification of wetland systems. An et al. (2024) defined wetlands as areas where the water table is typically close to the surface and shallow water covers the land, thereby acting as a transition between terrestrial and aquatic ecosystems. Further, they must possess one of the three mentioned characteristics: (1) the substrate is primarily undrained hydric soil; (2) the land supports primarily hydrophytes at least occasionally; and (3) the substrate is saturated with water or covered by shallow water at some point throughout each year's growing season.

Wetlands are vital ecosystems for a wide range of plants and animals and play a crucial role in watershed management. These ecosystems are also referred to as "Kidneys of Earth" or "biological supermarket," facilitating 47.6% of the total global ecosystem services with a cover area of 12.1 million km^2 (Xu et al., 2019; Xu et al., 2020). In terms of global ecosystem services, wetlands perform Earth's ecosystem functioning, such as carbon sequestration, nutrient metabolism, water and wastewater remediation, and biodiversity hotspots (Collen et al., 2014; Mitsch & Gosselink, 2015; Orihuela-Torres et al., 2025). Moreover, they also play a significant role in human benefit, providing services such as food and fuel, flood mitigation and prevention, coastal abatement and protection, groundwater recharge and streamflow maintenance, and by supporting a wide range of cultural and economic activities (Gupta et al., 2020; Ramsar Convention, 2008; Turner et al., 2008). Despite their potential and crucial role, global wetland areas is declining rapidly andthe precise number of wetland losses worldwide is unknown (Thorslund et al., 2017). Studies indicate the world's wetland cover has decreased approximately by 87% since 1700 in the regions having some data. The majority of the degradation was mainly occurred during 20th and early 21st centuries due to anthropogenic and agricultural activities (Davidson, 2014; Mitsch & Gosselink, 2015; Ramsar Convention, 2015; Xu et al., 2020). Furthermore, 50% of Europe's wetlands have disappeared in the last three centuries as a result of drainage and transformation for agriculture (Fluet-Chouinard et al., 2023; Zedler & Kercher, 2005). Additionally, the natural wetlands experienced reduced hydroperiods due to groundwater abstraction and decreased precipitation from climate change (Díaz-Delgado et al., 2016; Green et al., 2024; Le Landes et al., 2014; Orihuela-Torres et al., 2025). Besides

Wetlands: Biodiversity, Health and their Role in Ecosystem Services. DOI: https://doi.org/10.1016/B978-0-443-34222-6.00008-5

agricultural expansion and rapid urbanization, population growth, tourism development, and inappropriate land use practices are some of the driving forces for wetland degradation (Kulik et al., 2024). However, the negative impact on wetland ecosystems was also due to false beliefs of people; previously, they considered wetlands as a heaven for mosquitoes, disease carriers, and a source of mortality rather than acknowledging and disseminating knowledge about the importance and significance of wetlands (Trakala et al, 2023; Xu et al., 2020). Hence, mostly human activities and natural factors (flood, landslide, drought, etc.) have led to a steady decline in the world wetland area and a significant worsening of wetland quality (Finlayson, 2012; Mitsch, 1994).

In comparison to previous decades, wetland loss has slowed significantly in certain regions of the world, including the United States and Europe, but it is still occurring rapidly in many other global regions (Davidson, 2014; Mitsch & Gosselink, 2015; Thorslund et al., 2017). Despite 87% wetland devastation, global wetlands are still providing more than 20% of the ecosystem services valued at $26.4 trillion year^{-1} surpassing the contributions of coral reefs and terrestrial forests (Costanza et al., 2014). The extent of wetlands contribution to human well-being underscores the importance of assessing the most effective ways for the wise use of wetlands as a sustainable, nature-based solution to the current and future problems. Since, most of the wetland ecosystem services have not been promoted in the form of trade in different economic markets as well as forums, causing continuous disregard, undervaluation, and ignorance of the wetland's significance and importance by stakeholders, government, and general public (Jiang et al., 2011; Prasad et al., 2002). Although, it is recognized that wetlands can offer a wide range of environmental services in addition to their value as habitats for plant, animal, and different floral and faunal species, despite their importance, many wetlands are still at risk of deterioration and degradation (Burks et al., 2006; Nabuyanda, 2024; Russi et al., 2013; Xu et al., 2020). The continuous degradation of wetlands led to the formulation of policies and framework focusing mainly on preservation and wise use of wetlands . The first global attempt focusing on wetland conservation and management took place in the form of the Ramsar Convention, signed in 1971 (Finlayson et al., 2018). This environmental agreement was aimed at bringing an international intergovernmental pact on wetland protection and wise use of wetlands by preserving wetland ecosystem functioning, upholding wetland culture, and ultimately achieving sustainable socioeconomic development in terms of local, regional, and national action as well as international cooperation (Cui & Yang, 2006; Erwin, 2009; Kim, 2010). Till 2018, 2303 Ramsar sites (Wetlands of International Importance) from 169 nations have been recognized, accounting for over 229 million hectares, or nearly 19% of the 1210 million hectares of the global wetland area (Davidson et al., 2018; Xu et al., 2020).

1.1.1 Wetland global and Indian extent

Excluding Antarctica, wetland habitats are found all over the world, with Asia and North America accounting for the highest shares, 31.8% and 27.1%, respectively, followed by Latin America, Caribbean (15.8%), Europe (12.5%), Africa (9.9%), and Oceania 2.9% (An et al., 2024). Over 12.1 million Km^2 of wetlands exist worldwide, 54% of the area were permanently occupied by water, while 46% were seasonally occupied. From an Indian perspective, varied topographic and climatic attributes support and sustain unique and diverse wetland habitats (Prasad et al., 2002), including high-altitude lakes, wetlands in the floodplains of river systems, saline and temporary wetlands in arid and semiarid biogeographic regions, coastal wetlands (lagoons, estuaries, mangrove swamps, coral reefs, and marine wetlands). With the exception of bogs, fens, and typical salt marshes, India's wetlands encompass nearly every type of environment. Additionally, India has a high number of constructed wetlands (CWs) for flood control, irrigation, water supply, and fishing (Bassi et al., 2014; Prasad et al., 2002; Ragavan et al., 2021). India harbors a total of about 1307 wetlands, out of which 100 wetlands are listed in Ramsar sites (https://indianwetlands.in/). In terms of geographical area, Gujarat has the largest share (17.5%), and Mizoram has the lowest (0.66%) of the area under wetlands. Among the union territories, Lakshadweep has the largest percentage (~ 96%) while Chandigarh has the lowest proportion (~3%) of geographical area under wetlands. However, due to water dispersion during the postmonsoon season, Indian states like Sikkim, Nagaland, Mizoram, Meghalaya, and Jharkhand account for over 90% of the total wetland area while Uttar Pradesh (28%), Chhattisgarh (29%), Himachal Pradesh (29%), Tripura (29%), Sikkim (30%), Andhra Pradesh (31%), Jharkhand (32.5%), Punjab (33%), Bihar (34%), Gujarat (36%), Karnataka (38.5%), Maharashtra (53.5%), Tamil Nadu (55%), Madhya Pradesh (57%), and Rajasthan (57%) have all shown a notable decrease in the water spread area of wetlands from postmonsoon to premonsoon (Bassi et al., 2014).

1.1.2 Wetland formation

The wetland formation process started about 8100 calendar years ago by the accretion of sediments beyond the ecstatic rise of water level (Turner & Ohimain, 2024; Turner et al., 2018). The entire wetland ecosystem formation is completed into five consecutive steps, such as (1) establishment of floral diversity in shallow water and resulting green toupee, (2) peat accumulation through land surface burial and increased water level, (3) increase in vegetation due to plant roots penetration into sediments, (4) stabilization of benthic sediments microflora, and (5) organic accumulations. Step-by-step changes in river or soil shore subsequently modified rigid nonliving precursor system into a living ecosystem of wetlands (Tolhurst et al., 2002; Turner & Ohimain, 2024). As the sea level rises, these living component becomes a physical agent as it slows and direct the heavier minerals to suspend, settle and accumulate nearby (Wu & Wang, 2006). The pedological changes (aerated soil chemistry and geomorphologies) are governed by floral and faunal components, tributaries, and their embankments, either established or vanished as water flow changed (Turner & Ohimain, 2024; Xin et al., 2022; Xin, Rui, et al., 2022). Underground flows are constrained by immersing channels, soil strength, pressure from overweight, and soil porosity. These small- and large-scale imbalances in time and space allows the wetland ecosystem to synchronize with various confinements and opportunities as the water level increases and also shape delta morphologies (Turner & Ohimain, 2024).

1.1.3 Historical and modern perspectives on wetlands

For centuries, humans have lived close to or around wetlands to survive and flourish since wetlands provide food, water, and other raw materials. Wetlands were gradually incorporated into expanding urban areas and frequently altered to allow people to keep using them for communication and transportation, rainwater storage, and other day to day activities(Falfán & Zambrano, 2024; Ghermandi et al., 2010; Stefanakis et al., 2014). However, the establishment of wetlands to offer a variety of services for residents has been a fairly frequent practice once the city was developed (Falfán & Zambrano, 2024; Stefanakis et al., 2014; White et al., 2020). Besides these tangible benefits, people in the past attached value to wetlands for nonmaterial and cultural services, mainly pilgrimage, spiritual fulfillment, rituals, and healing. Many wetlands, such as Titicaca (Peru), Poyang (China), Victoria (East Africa), Manasarovar (Tibet), and Renuka (India), around the world were revered for their holy, religious, and spiritual value. Mainstream faiths like Daoism, Shinto, Hinduism, and Jainism are rooted in spirituality that views wetlands as indigenous, while the theological principles of other religious faiths, such as Buddhism, Judaism, Christianity, and Islam, regard wetlands as part of God's creation (Verschuuren, 2016). However, past research had also oriented toward frequently used static indicators like population, GDP (gross domestic product) to represent anthropogenic factors and overlooking the dynamic spatial distribution of human activities (Cui et al., 2024; Culhane et al., 2019) but in current perspective, scientists, environmentalist, policymakers, and lawyers have significantly shifted toward sustainable use, management, and conservation of wetlands rather than wetland ecosystem services (Mandishona, 2021). Recently, research on wetlands become a focal point to improve the effectiveness and viability of adopting wetland protection and management in terms of biodiversity conservation and restoration, environmental protection, and carbon sequestration (Salimi et al., 2021; Strassburg et al., 2020). To synchronize the recent research trends on wetlands, the Ramsar Convention on Wetlands (COP 14) has also emphasized the formulation of a similar format for mapping the strategic framework for global wetland conservation and development, along with 2020 global biodiversity framework conservation targets, including sustainable development goals (SDGs) (Yi et al., 2024).

1.1.4 Classification of wetland

On the basis of different criteria like geological, biological, and hydrological conditions, wetlands are classified as inland and coastal, natural and man-made , freshwater and brackish (Junk, 2024). Moreover, based on hydrological status, vegetation type, location, and water salinity, wetlands can be further divided into subtypes. Nonetheless, Ramsar Convention (2018) has classified wetlands into three major groups: (1) coastal or marine ; (2) inland ; (3) man-made ; these categories comprise 42 different types of wetlands. According to Xu et al. (2020), these wetlands are further divided into 9 categories and 28 subcategories, that is, coastal (3 subcategories), estuarine (4 subcategories), lacustrine (lake) (4 subcategories), and riverine (4 subcategories), palustrine (5 subcategories), peatland, and man-made (8–8 subcategories), followed by multiple wetlands, and others. According to the global wetland distribution network, 53% of total inland wetlands are marshes and peatlands, 33% are lakes , 5% are riverine , and 9% are forest wetlands. On the other hand, coastal and marine wetlands include 34% salt marshes and 8% mangroves, 17% coral reefs, 11% seagrass beds, 2% coastal deltas, and 0.2% (Binilkumar, 2010; Davidson et al., 2018; Reithmaier et al., 2024). Thus, the global wetlands distribution network comprises coastal, estuarine, lacustrine, riverine, palustrine, human made, peatland, etc. (Fig. 1.1). The details of each component are provided in the subsections.

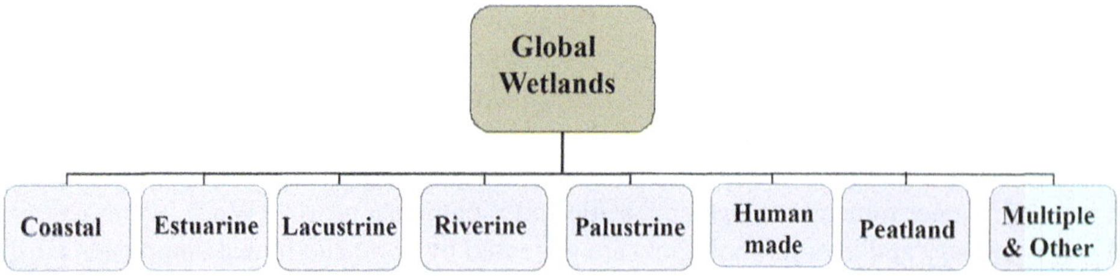

FIGURE 1.1 Global wetlands distribution and its components.

1.1.4.1 Coastal wetland

Coastal wetlands comprise an ecotone between terrestrial and marine ecosystems having high susceptibility to the effects of human activity, severe storms, and sea-level rise (Lin & Yu, 2018). These wetlands represent the most valuable and sensitive ecosystem and rank one of the most productive as well as endangered ecosystems in the world (Navarro & Rodríguez-Santalla, 2023). Lagoons, coral reefs, mangroves, salt marshes, seagrass beds, beaches, and coastal waterbodies that are less than 6 m deep are all considered coastal wetlands. Among the coastal wetlands, mangroves and salt marshes are the most notable types. Mangroves are typically found along muddy tropical coasts, whereas salt marshes are more prevalent along muddy subtropical, temperate, subpolar, and arctic coasts (Scott et al., 2014). Coastal wetlands provide numerous significant benefits, including direct supplies like food and biomaterials, wildlife habitat, carbon sequestration, storm surge protection, and sediment accumulation for land accretion. Additionally, they also offer tourism resorts, water purification, and other services along muddy shores from subtropical, temperate, subpolar, and arctic zones (Ramsar, 2012). Among these ecosystem services, carbon sequestration is one of the most important environmental services provided by coastal wetlands, they absorb atmospheric carbon dioxide (CO_2) and store it in plants and soil to mitigate the effects of climate change (Reithmaier et al., 2023). Coastal wetlands are among the planet's most effective natural carbon sinks, removing up to 10 times as much carbon per unit area as tropical rainforests (Mcleod et al., 2011). However, because of rapid economic development, these wetlands are gradually shrinking with an estimation of 0.7%–1.2% global loss resulting in approximately 63% of world coastal wetland loss (Davidson, 2014). According to Li et al. (2018) over 67% of coastal wetland areas, including 65% of seagrass and 48% of other submerged aquatic vegetation, have been lost during 20th century. Out of which roughly 25%–50% of global coastal wetlands reclaimed were turned into agricultural lands (Kirwan & Megonigal, 2013). The global coastal wetlands, particularly salt marshes, are also functionally degraded due to eutrophication (Chislock et al., 2013).

1.1.4.2 Estuarine wetlands

Estuarine wetlands are ecologically more sensitive and fragile wetlands which comprises transition zone between marine, freshwater, and terrestrial ecosystems (Yang et al., 2025). These wetlands are characterized as semienclosed bodies of water in which mixing of freshwater from rivers or streams with seawater to create brackish (slightly salty) conditions. Freshwater runoff from the land periodically dilutes the ocean water in these wetlands. The salinity of estuaries varies from hyperhaline to oligohaline and rises periodically above sea level as a result of evaporation (Cowardin et al., 1985). Despite making up only 0.2% of the world's ocean area, estuaries have a disproportionately large impact on marine carbon budgets (Regnier et al., 2022). The dominant flora and fauna in offshore areas include red mangroves (*Rhizophora mangle*) and eastern oysters (*Crassostrea virginica*) (Cowardin et al., 1985). Estuarine water regimes and water chemistry are usually affected by factors such as oceanic tides, precipitation, freshwater runoff from land areas, evaporation, and wind. These wetlands include mangrove swamps, deltas, and tidal marshes (Xu et al., 2020).

1.1.4.3 Lacustrine wetlands

Lacustrine wetlands are wetlands in which swamps are connected to a lake's shallow section and harbor different aquatic plants capable of water tolerance. These wetlands can be found near a lakeshore or a channel that leads to the lake and are mainly characterized by vegetation, particularly grass or low bushes, with stagnant water. Along the beach where the ground is saturated with water, the terrestrial portion of lacustrine wetlands forms a bell-shaped belt with tiny ponds scattered throughout. Lake water replenishes both the terrestrial and aquatic portions of lacustrine wetlands, resulting in replaceable and susceptible changes in the lake's water level (https://trenchlesspedia.com/definition/3086/lacustrine-wetland). The lake becomes a terrestrial wetland when more of the lake floor is exposed at low levels, and vice versa.

After prolonged exposure to the water, the flora that once covered the newly formed wetlands may have faded or perhaps died. On the other hand, at greater water levels, more onshore wetlands got submerged by lake water and become aquatic and they become the part of the larger lake if the water level keeps increasing. The line separating piedmont and terrestrial lacustrine wetlands becomes hazy when the lake is adjacent to a mountain. Nevertheless, the two can still be distinguished from one another both topographically and ecologically. At 4230 m, lacustrine wetlands are significantly lower than piedmont marshes (Gao et al., 2013). Compared to the steep surface (10.4%) of piedmont wetlands, they have a softer gradient (only 1%) that is comparable to the lake floor. The kind of vegetation that is present can also be used to define the precise border between the two (Gao et al., 2013; Xu et al., 2020). The extent of lacustrine wetlands is indicated by the presence of hydrophytes, or species that can withstand water, such as *Polygonum* spp., *Glaux maritima sibiricum* Laxm, and *Carex Polygonum sibiricum* Laxm. They also differ from one another in terms of hydrology. Terrestrial lacustrine wetlands, unlike piedmont wetlands, are replenished by lake water or rainwater that flows overland from higher ground. They are therefore more resilient to change and possess a greater water reserve (Gao et al., 2013).

1.1.4.4 Riverine wetland

Riverine wetlands refer to swamps related to river channels. In a plateau setting, these channels often branch and travel freely. The sandwich of long grassland or riverine wetland islands is carried by braiding channels. These waterways typically feature as a tributary or stangnantbranch that is not actively contributing to the discharge of channel flow. It is difficult to distinguish between a wetland and waterbodies in riverine wetlands due to the confluence of such defective channels (An et al., 2024). These wetlands recharge lakes and possess sediments that enrich the floodplain and marshes. The biotic structure and physical characteristics of riverine wetlands are determined and influenced by water velocity. The riverine wetland in low order is characterized by fast water velocity, shallow water, rocky substrate, and saturated dissolved oxygen. On the contrary, high-order river stretches are characterized by floodplains that have frequent inundation with mud/sand and slow water velocity (An et al., 2024; Tiner, 2016). Mostly, the riverine system is primarily characterized by trees, shrubs, emergent mosses, and lichens or is restricted on the landward side by upland as well as by the channel bank (including natural and man-made levees). The tidal, lower perennial, upper perennial, and intermittent subtypes are the four subdivisions of the riverine wetland system. Water permanence, gradient, water velocity, substrate, and degree of floodplain development are used to describe each of these subtypes. These wetland floodplains can improve water quality, recharge groundwater, significantly reduce flood impacts, and enhance the breakdown of organic matter (An et al., 2024; Tiner, 2016). Riverine wetlands adjacent to the channel may gradually transform into terrace wetlands if the channel bed is deeply cut or the river water falls so low that it only reaches the wetlands during flooding (An et al., 2024; Cowardin et al., 1985; Gao et al., 2013; Tiner, 2016). Rivers, streams, and wetlands along the rivers are the main types of riverine wetlands (Xu et al., 2020).

1.1.4.5 Palustrine wetlands

Palustrine wetlands occupy nearly 6% of the Earth's terrestrial surface and are recognized as some of the most ecologically sensitive ecosystems, owing to their critical roles in global biogeochemical cycles (Junk et al., 2013; Mitsch & Gosselink, 2015). These wetlands are defined as nontidal systems with salinity levels below 0.5% and are typically dominated by trees, shrubs, persistent emergent vegetation, mosses, or lichens. They may also encompass areas lacking such vegetation provided they meet specific criteria: (1) a surface area smaller than 8 ha (20 acres); (2) absence of active wave-formed or bedrock shoreline features; (3) a maximum water depth of less than 2 m in the deepest part of the basin under low-water conditions; and (4) salinity below 0.5%.

Palustrine wetlands occur across diverse landscape settings, including margins of lakes, river channels, estuaries, and floodplains, as well as in isolated catchments and on slopes. They may also form as vegetated islands within lakes or rivers (Gao et al., 2013). Marshes, bogs, fens, and swamps constitute the principal wetland types within palustrine systems (Xu et al., 2020).

1.1.4.6 Peatland wetlands

Peatland ecosystems are formed by deposition of organic matter (peat) originating from dead and decomposing plant material in a permanently water-saturated condition. These wetlands occur mostly in northern, colder, and tropical areas with heavy rainfall. The wetland covers around 2%–4% and is characterized by plant species such as Sphagnum Mosses, Sedges, and shrubs (Xu et al., 2018). Peatlands possess high biodiversity conservation values and play a significant role in climate regulation and human welfare. Global Peatlands contain the greatest lithospheric carbon stock, approximately 61,011–71,011 tons on the earth, since decomposition of dead plants and mosses slows down due to waterlogging

(Turetsky et al., 2015). This huge amount of stored carbon indicates the significance of peatland ecosystems in the regulation of the global carbon cycle (Dise, 2009). However, inappropriate and unwise management strategies lead toward extensive degradation of peatlands with critical environmental and social impacts. To acquire multiple benefits such as reducing poverty, combating land degradation, maintaining biodiversity, and mitigating climate change, an integrated management along with proper rehabilitation of the peatland ecosystem must be required (Subcommittee, 2013). Peatlands are further divided into two subtypes that is, bogs and fens. Fens characterized by alkaline-rich water (pH>5.5) usually support herbaceous plants. For fens, groundwater, surface source, and precipitation are the main sources for the derivation of water and nutrients. Whereas bog's water is usually less saline and tends to be acidic (pH<5.5) and dominated by woody shrubs or trees among the herbaceous plants. Nutrient input in these peatlands comes only from precipitation (An et al., 2024).

1.1.4.7 Constructed or human-made wetlands

Human-made wetlands constitute a significant subset of worldwide wetlands, in addition to the ecosystem of natural wetlands already discussed. According to An et al. (2024), the total surface area of man-made wetlands makes up more than 12% of all wetlands worldwide and it is still growing. These wetlands are mainly constructed to acquire a broad range of services such as wastewater treatment, effluent purification, biodiversity conservation, agricultural irrigation, aquaculture, fish farming, water reservoirs, and recreational values (An et al., 2024; Liu, Zhao, et al., 2023). CWs are constructed to replicate the natural wetlands and their natural processes include biodegradation, sorption, phytostabilization, phytoextraction, and rhizofiltration (Hassan et al., 2021). These wetlands are characterized by decreased water flow and shallow water depth. Low water flow promotes extended retention, which makes it easier for sediments to settle. CWs primarily assist in the treatment of contaminated water, including stormwater, textile wastewater, mine drainage, municipal wastewater, industrial wastewater (especially from petroleum refineries and sour-water treatment wastewater products), agricultural wastewater, landfill leachates, and agricultural wastewater (Liu, He et al., 2023; Vymazal et al., 2021). The execution of wastewater treatment by CWs is accomplished by a natural process already mentioned in the paragraph; therefore, it presents an environmentally friendly and cost-effective remediation approach that has the least negative impact on the environment (Dan et al., 2011; Sial et al., 2023). The entire process is accomplished in three stages: physical processes (filtration and sedimentation), chemical processes (adsorption and precipitation), and biological processes (biodegradation and plant assimilation) (Hassan et al., 2021). Rapid construction of CWs throughout the globe seems to continue due to ecological benefits such ecofriendly measures and a wide range of socioeconomic advantages, that is, enhancing wastewater quality by degradation or sorption, preventing flooding by storing runoff and storm rainfall, and recycle nutrients (Christofilopoulos et al., 2019) and wetlands value such as recreational, research and education and cooling urban area (Hassan et al., 2021). The major advantages of CWs include low construction costs, eco-friendliness, low maintenance and operational costs, habitat for wildlife, and the reuse of treated water. However, a few disadvantages are also associated with this approach, such as the requirement for large areas, inconsistent treatment, and lower tolerance to drying conditions (Azubuike et al., 2016). On the basis of design, maintenance, and operation, human-made wetlands are classified into three types, which are summarized in Table 1.1.

1.2 Wetlands' ecological and functional significance

Wetland's significance can be divided into two broad classes: ecological significance and cultural significance. The former type comprises biomaterials and food as direct resources, wildlife habitat, carbon sequestration, storm protection, and land accretion through sedimentation. On the other hand, wetland cultural significance mainly includes tourism, resorts, and other functions.

1.2.1 Integrated water services

Tanks, ponds, lakes, and reservoirs are examples of wetlands that offer multipurpose water services, primarily for household, recreational, and agricultural irrigation; groundwater table recharging; flood control; and silt capture. With 0.12 million, the Southern Indian states of Andhra Pradesh, Karnataka, and Tamil Nadu have the highest concentration of irrigation tanks (Bassi et al., 2014), which accounts for around 60% of the tank-irrigated land in India. Similarly, in the states of Bihar, Orissa, Uttar Pradesh, and West Bengal, over 25% of the net tank-irrigated land is used for traditional tank systems (Pant & Verma, 2010). Tanks are crucial for harvesting surface runoff at the time of the monsoon and then allowing it to be used in water-deprived seasons, either of pre- or postmonsoon. In addition to being used in irrigation, tanks are also associated with brick production, fodder grass collection, domestic needs (domestic and livestock

TABLE 1.1 Types of constructed wetlands (*CWs*) on the basis of design, maintenance, and operation.

Type of constructed wetland	Design	Operation	Maintenance
SF	Easy, needs a lot of land, and relies on gravity flow	High evapotranspiration, low temperature, minimal operating costs, and a straightforward operation process all have an impact on microbial activity	Low maintenance cost, significantly impacted by temperature, flocculation, odor, and insect issues (Mosquitoes)
SSF	Complex, requires pumps, a sedimentation tank, and less space than SF	Higher operating cost, more sorption sites than SF, uniform flow with low solid concentration, and only transpiration	Greater cold resistance, reduced odor, and pests. Clogging problems, higher maintenance costs
Vertical subsurface flow	Complex, requires pumps, a sedimentation pond, and less space than SSF	Greater sorption sites than SF, comparatively higher operating costs, low solid concentration that should be homogeneous, and only transpiration	Greater cold tolerance, reduced odor, and pests. clogging issues, high maintenance expenses

SF, Surface flow; *SSF*, surface subsurface flow.

consumption), and fisheries. Hence, multiple use of wetland water significantly impacts household income, nutrition, and health of the poor (Kumar et al., 2013). In India, recreation, tourism, fishing, irrigation, and domestic water supply services have long been linked to a number of lakes, including Carambolim (Goa), Chilika (Orissa), Dal Jheel (Jammu and Kashmir), Deep or Beel (Assam), Khabar Tal (Bihar), Kolleru (Andhra Pradesh), Loktak (Manipur), Nainital (Uttarakhand), Nalsarovar (Gujarat), and Vembanad (Kerala) (Bassi et al., 2014; Jain et al., 2007). To supply water for municipal, industrial, hydropower, agricultural, recreational, and flood control purposes, more than 4700 large reservoirs with a minimum capacity 2 million cubic meter were constructed in India till 2014 (https://cwc.gov.in/sites/default/files/AR_08_09.pdf). For instance, reservoirs such as Govind Sagar Lake formed by diverting river Satluj (Bhakra Dam, Punjab) and Hirakud reservoir (Sambalpur, Orissa) not only support variety of wildlife but also a focal point for tourist attraction (Bassi et al., 2014).

1.2.2 Biogeochemical cycle functioning

Wetlands play an essential role in regulating biogeochemical cycling between terrestrial and aquatic ecosystems (Jarvie et al., 2020). The biogeochemical cycle is functionally affected by hydrological factors. Wetlands mainly regulate carbon, nitrogen, phosphorus, sulfur, etc. . Biogeochemical cycle of these elements within the wetland is accomplished and regulated due to significant spatial and temporal heterogeneity in chemical properties, temperature, and water saturation. The significant variation in these attributes results in proper transportation and transformation of carbon, nutrients, and redox-active elements in wetlands (Leopold et al., 2010; Reddy et al., 2011). Hence, global wetlands are referred to as "biogeochemical hotspots" for cycling of nutrients, carbon, and trace metals (Cherry, 2011). The carbon cycle is the main driver of coupling biogeochemical cycles together through primary productivity to decomposition. Wetland's carbon fixation intensity is generally regulated and controlled by either N and P or a combination of the two (Güsewell et al., 2003). Anaerobic conditions in wetland sediments drastically effect the decomposition of organic matter by facultative or obligate anaerobes, they usesa suite of electron acceptors instead of oxygen, with interactions involving nitrogen, sulfur, iron, and manganese during the carbon cycle. However, in aerated sediment patches, obligate aerobes such as fungi and bacteria execute organic matter decomposition and regulate carbon cycling. The interaction and involvement of N_2, Mn, Fe, S, Ca, and P cycles in the carbon cycle are summarized in Table 1.2 (Verhoeven, 2009).

1.2.3 Bioproductivity and habitats

In terms of primary production, wetland ecosystems are comparable to other abundant terrestrial ecosystems, despite their comparatively low productivity. For instance, the average primary productivity of mangroves is 2.2 Kg m^{-2} $year^{-1}$, which is higher than that of tropical rainforests, whereas the average primary productivity of algal beds and reefs is 2.5 Kg m^{-2} $year^{-1}$. Additionally, the temperate salt marshes have a greater net primary productivity (1.7 Kg m^{-2} $year^{-1}$) than

TABLE 1.2 Coupling between the nitrogen, manganese, iron, and sulfur cycles and the carbon cycle: breakdown with nonoxygen electron acceptors (Maltby & Barker, 2009; Verhoeven, 2009).

S. no.	Biogeochemical cycle coupling	Reaction of the coupling cycles	Organism in wetland ecosystem
I.	C cycle (fermentation)	$C_6H_{12}O_6 \rightarrow 2CH_3CH_2OCOOH$ (lactic acid)	*Saccharomyces* sp.
		$C_6H_{12}O_6 \rightarrow 2\ CH_3CH_2OH + 2CO_2$ (ethanol and carbon dioxide)	
II.	N_2+C (denitrification nitrate reduction to ammonium)	$C_6H_{12}O_6 + 4NO_3^- \rightarrow 6CO_2 + 6H_2O + 2N_2$	*Pseudomonas denitrificans*, *Bacillus* spp., *Alcaligenes* sp.
		$2NO_3^- + 2CH_2O \rightarrow N_2O + H_2O + 2HCO_3$	
		$N_2O + H_2 \rightarrow N_2 + H_2O$	
	Dissimilatory nitrate reduction to ammonium	$C_6H_{12}O_6 + 4NO_3^- + 4\ H^+ \rightarrow 6CO_2 + 6H_2O + 4NH_4^+$	*Clostridium* sp., *Bacillus* sp.
III.	Mn+C (manganese reduction)	$MnO_2 + 2\ e^- + 4\ H^+ \rightarrow Mn^{2+} + 2\ H2O$	*Pseudomonas* sp.
IV.	C+F (iron reduction)	$Fe(OH)_3 + e^- + 3\ H^+ \rightarrow Fe^{2+} + 3\ H_2O$	*Pseudomonas* sp.
V.	C+S (sulfate reduction)	$2CH_3CHOHCOO^- + SO_4^{2-} \rightarrow 2CH_3COO^- + 2CO^{2+}\ 2H_2O + HS^- + H^+$	*Desulfovibrio* sp., *Desulfotomaculum* sp.
		$CH_3COO^- + 2H^+ + SO_4{}^{2-} \rightarrow 2CO_2 + 2H_2O + HS^-$	
VI.	Methanogenesis	$CO_2 + 8H^+ + 8e^- \rightarrow CH_4 + 2H_2O$	*Methanobacillus* sp., *Methanobacterium* sp.
		$CH_3COO^- + H^+ \rightarrow CH_4 + CO_2$	
VII.	C+S coupling cycle	$6CO_2 + 6H_2S +$ light $\rightarrow C_6H_{12}O_6 + 6S$	Purple or green sulfur bacteria
VIII.	N+S coupling cycle	$6NO_3^- + 5\ S + 2H_2O \rightarrow 3N_2 + 5SO_4^{2-} + 4\ H^+$	*Thiobacillus*
		$8NO_3^- + 5HS^- + 3\ H^+ \rightarrow 4N_2 + 5\ SO_4^{2-} + 4\ H_2O$	
IX.	Fe+S coupling cycle	$H_2S + Fe^{2+} \rightarrow FeS\downarrow + H_2$	–
		$2FeOOH + 2\ H_2S + 2H+ \rightarrow FeS_2\downarrow + 4H_2O + Fe^{2+}$	

temperate forests (1.2–1.3 kg m^{-2} $year^{-1}$). Apart from higher productivity, wetlands, particularly coastal wetlands, provide shelter and a food source for various wild animals (particularly for millions of waterbirds), resulting in significant biodiversity and a distinct food web structure. More than 90% of marine fish are obtained from coastal ecosystems, and nearly two-thirds of marine animals, including fish, shrimps, crabs, mollusks, and turtles, spend some of their life history in coastal wetlands (Li et al., 2018). Similarly, urban wetlands in China also provide habitat for a variety of endangered aquatic plant species such as *Isoetes sinensis* (Wang et al., 2022). Due to significant advantage in terms of biodiversity homing and food resources construction of wetland protected area (Wetland Nature Reserves, Wetland Parks) gradually accelerated in China to protect certain animal species including *Echinotriton chinhaiensis*, *Protohynobius puxiongensis*, and *Onychodactylus zhaoermii* in Caudata; *Glandirana minima*, *Oreolalax chuanbeiensis*, and *Limnonectes fragilis* in Anura; *Cuora trifasciata*, *Mauremys nigricans*, and *Sacalia quadriocellata* in Testudines; *Balaenoptera borealis* in Cetacea; and Falconiformes, Strigiformes, Pelecaniformes, Ciconiiformes as well as Coraciiformes (Guo et al., 2019). For instance, Caohai Lake wetland (China) covers an area of 120 km^2 and harbors 17 orders, 53 families, and 246 bird species. Out of which 132 migratory bird species, including 46, are threatened species. According to reports, in May 2022, a significant number of migratory overwintering birds, nearly 90,000, were also seen in the lake (Liu et al., 2022). In a similar vein, the North African wetlands also act as replenishing sites for various birds during autumn and spring before they travel across the Sahara Desert to the Mediterranean Sea. Apart from providing shelter to varied wild animals, some wetlands are also known for their strategic location, which favors migratory birds. Wetlands in Northeastern Algeria are important for their strategic placement between the Saharan Atlas and Tell Atlas ranges, which creates a special climate and habitat for migratory birds. Between 2008 and 2021, 68 species of birds belonging to 16 families and 10 orders used 21 wetlands as wintering grounds (Bezzalla et al., 2025).

1.2.4 Carbon sequestration

Wetlands are a special ecosystem that is essential for reducing the negative impact of climate change. In the context of climate change, one recent concern is blue carbon. To reduce blue carbon, carbon sequestration in wetland refers to the process by which the wetland captures and stores carbon from the atmosphere. The carbon sequestration in soil and sediments of different ecosystem mainly achieved at a high rate (>100 g cm^{-2} $year^{-1}$) by salt marshes, mangroves, and seagrasses, nearly 20 times greater than that of forest ecosystems (<5 g cm^{-2} $year^{-1}$) (Li, Piao, et al., 2018; Mcleod et al., 2011). Because of this, "blue carbon" has emerged as one of the most pressing issues of the past 10 years (Howard et al., 2017). However, freshwater wetlands are frequently thought of as a source of greenhouse gas emissions, but sulfate-reducing bacteria of coastal wetlands inhibit the methanogens by outcompeting them for energy sources, particularly in polyhaline tidal marshes (salinity>18), and hardly contribute to greenhouse gas emissions (Poffenbarger et al., 2011). Mangroves, salt marshes, and seagrass beds contribute between 10% and 13% of the global ecosystem's CO_2 intake or carbon sequestration (which is about 0.23 and 0.77 Pg CO_2 $year^{-1}$). This amounts to between 3% and 19% of the CO_2 emissions brought on by deforestation worldwide (Pendleton et al., 2012). The majority of carbon is stored underground in coastal wetlands, just like the organic carbon in terrestrial ecosystems. In the mangrove environment (salt marshes), soil and sediments store 90% of the organic carbon (Donato et al., 2011; Li et al., 2018). In comparison to the forest ecosystem, the estimated stored carbon in the top 1 m of sediments is approximately 1800–40,000 gCm^{-2} for seagrasses, 9000–54,000 gCm^{-2} for salt marshes, and 28,900–55,100 gCm^{-2} for mangroves (Lavery et al., 2013; Ninan, 2014).

1.2.5 Coastal protection

Wetlands particularly mangroves, salt marshes and salt water swamps play a critical role in protecting coastline from erosion and storm surges by acting an buffer and dissipating as well as neutralizing wave energy. The significant role of mangrove wetland in protecting the coastline is evident during the Indian Ocean tsunami that occurred in December 2014 (Dahdouh-Guebas et al., 2005). Compared to settlements with few or no mangroves, those protected by mangroves saw less mortality. While the cryptically deteriorated area occupied by companion species was removed, the area dominated by real mangroves remained mostly intact (Das & Vincent, 2009). The waves are weakened and eventually neutralized by Mangroves before penetrating into the coast (Li et al., 2018; Massel et al., 1999) similarly 90% impact of tsunami pressure is also reduced within 100 m in dense mangrove stands (Tanaka et al., 2007). In certain cases, a 7-Km wide mangrove strip was also found lowering of 70% wave height during hurricanes (Zhang et al., 2012). The salt marshes exponentially reduce wave height with increasing landward distance from the marsh edge (Yang et al., 2012). Apart from wave height, during storm surge conditions, marsh vegetation is also capable to minimize 60% wave energy (Möller et al., 2014). However, the impact of wave attenuation depends significantly on species characteristics and the spatial distribution of marshes. Unlike more rigid and towering plant canopies, which are prone to stem folding and breaking, flexible low-growing plant canopies are very resilient to storm surge conditions (Rupprecht et al., 2017). The survival and persistence of salt marshes are supported by large amounts of sediments found in the vegetated zone, causing land to accrete both horizontally and vertically (Mudd et al., 2010).

1.2.6 Mitigation of extreme flood and severe drought

Wetlands also play a crucial role in mitigating the impact of both the costliest hydrological event: flood and severe drought, and offer a natural system to manage water systems (Wu et al., 2023a; Wu et al., 2023b). Flood and drought severely affected lives of thousands of people (more than 2 billion and 1.5 billion respectively) globally during 1998–2017, causing considerable economic loss. Risk of increased floods and severe droughts globally intensifies due to high frequency and severity of hydrological extremes (Wu et al., 2023b). Wetlands' potent hydrologic regulatory role makes them a potential natural remedy for bolstering downstream hydrologic infrastructures and enhancing basin resilience to hydrologic extremes (Wu et al., 2020a, 2020b). Wetlands regulate hydrological services and reduce flood impact by minimizing flood volume and intense flow through storing surface runoff at mountainous slopes and excessive water from river channels; relieving as well as reassuring drought risk by supporting base flow (2023) and maintaining water table sustainability via replenishing groundwater (Thorslund et al., 2017). Among these crucial services, the flood reducing capacity is critical and pivotal for the attenuation of broad-scale flood events (Wu et al., 2023a; Zhang and Song, 2014) and base flow support function significantly during drought alleviation periods (Golden et al., 2021). Capabilities of a wetland flood control can be precisely reflected by the HEC-RAS model in CW (Kim, 2010). In this model, three conditions are required: (1) *The boundary conditions are used as input data to evaluate unsteady flow in HEC-RAS model*; (2) *Wetland construction on the basis of storage area provided in by HEC-RAS model. The storage volume capacity of the wetlands is used as input with respect to the elevation level*; (3) *finally the river is connected to CW* (Kim, 2010). Furthermore, the reservoir runoff computation method is used to compute the water balance in the SWAT model. The surface area of the water body is computed with respect to the reservoir capacity based on water level and surface area of the water body of an emergency spillway and the principal spillway as follows (Kim, 2010).

$$SA = \beta_{sa} \cdot V^{expsa} \tag{1.1}$$

$$\text{Exp sa} = \log_{10}(S_{em)} - \log_{10}\left(SA_{pr}\right)$$
$$\text{Log}_{10}(V_{em)} - \log_{10}\left(V_{pr}\right) \tag{1.2}$$

$$B = SA_{em}{}^{expsa}$$
$$V_{em} \tag{1.3}$$

where SA is the surface area of the reservoir (ha); V the volume of storage (m^3), expsa the exponent; SA_{em} the surface area of the reservoir when the water is filled in the emergency spillway; SA_{pr} the surface area of the reservoir when the principal spillway is filled; Here, SA is the surface area of the reservoir (ha); V the volume of storage (m^3), expsa the exponent; SA_{em} the surface area of the reservoir when the water is filled in the emergency spillway; SA_{pr} the surface area of the reservoir when the principal spillway is filled; and V_{em} and V_{pr} refer to the volume of storage when the emergency spillway and the principal spillway are filled, respectively.

The water balance that computes the reservoir storage is calculated in the SWAT model as follows

$$V = V_{stored} + V_{flowin} \rightarrow V_{flowout} + V_{pcp} \rightarrow V_{evap} \rightarrow V_{seep}$$

where V_{stored} is the initial storage volume; V_{flowin} the daily inflow; $V_{flowout}$ the daily runoff and V_{pcp}, V_{evap}, and V_{seep} refers to the daily rainfall, evaporation, and water loss, respectively, caused by percolation (Kim, 2010).

1.2.7 Aesthetic/Recreation value

Since long, wetlands have been mostly preferred for both mass tourism and ecotourism due to attractive and elegant beauty as well as unique aquatic environments, amphibious landscapes, and distinct biodiversity in these dynamic ecosystems (Ghosh et al., 2024; Ramı´rez & Santana, 2019). Because of their abundant water and diverse vegetation, wetlands not only supply moisture to the surrounding air in cities but also help reduce the intensity of heat waves. Consequently, the local temperature naturally decreases, making wetlands suitable for tourism, fishing, boating, and hunting (Wu et al., 2021). Aesthetic preferences represent an essential connection between people and ecosystems, which could greatly raise public awareness of conservation initiatives (Lee, 2017). Coastal wetlands often attract tourists and recreational activities such as birdwatching, kayaking, and hiking. These activities indirectly support coastal resilience by providing financial incentives for wetland protection, restoration, and management. The livelihoods of neighboring human communities are frequently significantly impacted by coastal wetlands. These communities could rely on wetlands for ecotourism or for resources like fish and shrimp (Secretariat of the Ramsar Convention on Wetlands, 2012).

1.2.8 Wastewater treatment and effluent recycling

Wetlands are increasingly recognized for their effective and sustainable water treatment capability, and are therefore referred as the kidneys of the landscape. Natural wetlands eliminate a variety of pollutants from water resources by natural processes, such as biodegradation, sorption, phytostabilization, phytoextraction, and rhizofiltration (Nyieku et al., 2021; Omondi & Navalia, 2020). The primary factors influencing wetland response and efficacy in wastewater treatment include temperature, microbe availability, surface loading rate, hydrology, substrate, and macrophyte composition (Almuktar et al., 2018). One of the major recent concerns and obstacles to meet the water demand includes the world's water shortage, as well as environmental and public health issues brought on by inadequate sanitation and wastewater disposal systems. Many nations are dealing with this issue, including those in the Middle East, Australia, the Southern United States, North Africa, and Southern Europe (Stikker, 1998). Hence, wetlands are considered one of the most effective, efficient, and cost-effective technologies for pollutant removal, wastewater treatment, and water recycling. As a result, they are widely used in urban wastewater reclamation for agricultural purposes. Wetlands support a diverse range of plant species with varying sizes and growth forms, including floating, emergent, and submerged vegetation, particularly within the littoral zone. (Pandey et al., 2024; Sun et al., 2019). These aquatic plants effectively filter surface water by bioremediation (Vymazal & Březinova, 2015). Additionally, using various aquatic plant species that preserve ecological forms will help in reducing environmental contaminants of surface water as well as purifying function as biodegradation media (Sun et al., 2019). Presently, CWs are one of the best methods for effective removal of organic and suspended solids and meeting the irrigation reuse standards (Almuktar et al., 2018; Cui et al., 2003; Nigam et al., 2023). Use of wetland plants to clean water of rivers or sewage has steadily grown as one of the significant means to control and mitigate water pollution. However, a number of wetland plants have different living conditions, purifying properties, and pollutant removal strategies. Thus, selecting appropriate wetland plants and measuring how various wetland plants alters the pollutant transport pathways seems to be the most important requirements for wastewater management (Xing et al., 2020). The plants used in pollution mitigation in surface water are summarized in Table 1.3.

1.2.9 Cooling service of urban wetlands

Wetlands play a fundamental role in reducing temperature extremes and regulating the local climate through a process referred to as the wetland cooling effect. Wetlands Park constitutes an essential component of the urban blue and green landscape for their cooling benefits to high urban areas (Liu, He et al., 2023). Highly built-up area not only triggers heterogeneity in the regional environment of surrounding areas but also alters its own microclimatic conditions. Exponential urbanization results in a rapid increase in temperature in cities compared to adjacent rural areas. This phenomenon is referred to as the urban heat island (UHI) effect (Bera et al., 2021; Weng, 2009). UHI can increase 4°C temperature of cities, leading to an exacerbation of air quality index as well as a negative impact on human health (Bera et al., 2021; Pataki et al., 2011). Wetlands, including reservoirs, ponds, lakes, and rivers, within urban boundaries, are known as urban cooling island (Bera et al., 2021; Cao et al., 2010) helps in providing cooling services as well reduces human risk from heat stress (Bera et al., 2021; Broadbent et al., 2018). The cooling efficiency of wetlands relies on size, volume, and shape of wetland as well as characteristics of nearby infrastructures and cultural landscape (Cheval & Constantin, 2019). Cooling effect of urban wetlands is higher than that of vegetation and lawn (Bera et al., 2021; Xueru et al., 2018). The wetland cooling efficiency recorded as 0.745°C lower temperature in 300 m buffer of various wetlands

TABLE 1.3 Different aquatic plant species and their purification effect on pollutants.

Vegetation type	Botanical name	Purification effect	References
Emergent plants	*Phragmites australis*	COD, BOD_5, TN	Calheiros et al. (2007)
	Typha angustifolia	TN, TP	Keizer-Vlek et al. (2014)
	Typha orientalis	TN, TP	Xing et al. (2020)
	Polygonum amphibium	TN	Kovacic et al. (2000)
	Lythrum salicaria	TN, TP	Fu and He (2015)
	Nelumbo nucifera	TN, TP, COD, TDS	Thongtha et al. (2014)
Floating plants	*Nymphaea tetragona*	TN, TP	Li et al. (2012)
	Nymphoides peltata	Cd	Gu et al. (2000)
	Lemna minor	TN, TP	Cheng et al. (2002), Xing et al. (2020)
Submerged plants			
	Potamogeton crispus	TN,TP, TSP	Guo et al. (2014)
	Myriophyllum spicatum	TP	Bole and Allan (1978)
	Hydrilla verticillata	TP	Bole and Allan (1978)
	Aldrovanda vesiculosa	–	Cross et al. (2015)
	Ceratophyllum demersum	TN, TP	Xing et al. (2020)

BOD_5, Biochemical oxygen demand; *COD*, chemical oxygen demand; *TN*, total nitrogen; *TP*, total phosphorus.

than the overall urban region in Kolkata (India) (Bera et al., 2021). Moreover, studies on 17 water parks in China also claim park cooling range between 4.52°C in the summer and 3.77°C in the autumn with respect to 14 and 432 m PCD (Park cooling distance) (Liu, He et al., 2023).

1.3 Key drivers for wetlands loss

Wetlands are the most endangered ecosystem . An estimate of 21%–87% wetlands have been lost between 1700 AD and the present (Darrah et al., 2019; Davidson, 2014; Fluet-Chouinard et al., 2023; Wood et al., 2024; Yi et al., 2024). These losses are extensive, pervasive, and prevalent across every continent with wetlands (Davidson, 2014; Hu et al., 2017; Kingsford et al., 2016). Seagrass beds, salt marshes, mangroves, and coral reefs are examples of coastal wetlands that lost more than half of their land area over the 20th century (Wood et al., 2024). The main causes or key drivers of worldwide wetland loss include wetland conversion for agricultural purposes, water abstraction, flow control, eutrophication, pollution, aquaculture, invasive species, and climate change (Davidson, 2014; Reid et al., 2019). These factors have led to the extensive habitat degradation in the remnant wetlands, which has affected the wetland's structure and functionality (Jiang et al., 2015; Nguyen et al., 2016; Nilsson et al., 2005; Wood et al., 2024; Zhang et al., 2017). Intensive anthropogenic activities, such as urbanization, have been considered one of the main reasons for land loss and wetland ecological degradation (Kirwan & Megonigal, 2013; Lin & Yu, 2018). Continental wetlands are more affected than coastal and marine wetlands due to agriculture (25%), urbanization (16.8%), aquaculture (10.7%), and industry (7.6%) (Ballut-Dajud et al., 2022). United States lost 324 Km^2 coastal wetland annually between 2004 and 2009 (Dahl & Stedman, 2013) whereas Australia lost or converted about 30% of its coastal wetlands between 1997 and 2002 (Creighton et al., 2015) as well as due to ecological crises and poor management, Spain lost 45% of its coastal wetlands' ES by 2013 (Gardner & Finlayson, 2018; Gardner et al., 2015). Compared to these countries, developing nations of Africa and Asia are experiencing more exponential urbanization than other regions (Lin & Yu, 2018), resulting in the loss of coastal wetlands loss 1.1% annually of the total area in Asia, greater than Europe (annually 0.99% of the total area) or North America (annually 0.51% of the total area). Additionally, the Mekong River Delta in Vietnam lost 1240 Km^2 of its mangrove wetlands between 1954 and 1974 and the Yellow Sea region in East Asia lost 28% of tidal flats (1.2% annually) of the total area (Lin & Yu, 2018). Thus, wetlands are rapidly disappearing from the ecosphere mainly due to increased anthropogenic activities and some natural factors. The most impacted drivers causing wetland loss are addressed as follows. The key wetland loss drivers are classified into two classes, that is, human associated and natural, which are further divided into 3 subclass and 14 subdivisions listed in Table 1.4.

TABLE 1.4 Classification of wetland impact factors /drivers.

Drivers responsible for wetland loss		
Class	**Subclass**	**Subdivision**
	Land impact area	Human settlements (nonagricultural) encroachment Agriculture encroachment Aquaculture and mangrove deforestation Transportation and service corridors
Human drivers		Microplastics Unwise tree planting Heavy metals
	Environment impact	Demographic explosion Mining (sand and energy)
	Biodiversity impact	Biological resource use Invasive species
Natural factors	-	Climate change Geological events

1.3.1 Human settlements (nonagricultural) encroachment

Wetland loss by nonagricultural human settlements includes two main drivers: industrialization and urbanization. The exponential increase in the human population and the fulfillment of desires to meet all demands through industrialization and urbanization have consequently degraded natural resources and ecosystem services. These drivers also escalated ecological impacts such as destruction and fragmentation of natural habitats, and altered water quality. The gradual ongoing growth in urbanization has worsened the ecological fragility of wetlands and resulted in serious ecological consequences (Dawuda et al., 2024; He et al., 2014; Mao et al., 2018; Wang et al., 2020). Urbanization-driven wetland loss is a global concern, particularly in developing countries. Land used in industrialization in Ghana (a developing country) showed rapid increase in community built-up and industrial land use from 25.19 to 84.21 and 0 to 4.71 ha respectively, resulting a wetland loss of −39.64 ha (−31.10%) between 2015 and 2020. While community-built-up land increased to 78.65 ha and industrial land slightly reduced to 2.54 ha, leading to the wetland degradation of approximately −69.06 ha (−42.53%) during 2015–20 (Dawuda et al., 2024). This rate might be the same for all developing countries and worse in developed countries.

1.3.2 Agriculture encroachment

The wetlands play a key role in agricultural production, food security, and people's livelihood in developing countries with lower per capita incomes or food deficits (Bamutaze et al., 2019). In Asia and Africa, a large amount of economic value of wetland is produced from agriculture (Finlayson et al., 2024; McCartney et al., 2010). Despite high agricultural importance, agriculture has been recognized as one of the main causes of wetland loss in the Conference of the Parties (COP) in 1980. The increased conversion rate of wetland into agricultural land frequently occurred in countries such as North, Central, and South America, Europe, Africa, Russia, China, the Middle East, Southeast Asia, and Australia between 1850 and 2000 (Asselen et al., 2013). The range of wetland conversion area is 1 km^2 to almost 150.000 km^2 (average=4300 km^2, median=97 Km^2). Wetland conversion to agricultural land is mainly determined by two factors, that is, biophysical and socioeconomic factor as follow (Asselen et al., 2013).

1.3.3 Biophysical factors

In this category, independent variables prior to determining the location of wetland conversion are mainly included.

1. *Temperature and precipitation*: Since temperature and precipitation are crucial factors for crop growth, both factors determine wetland conversion for agronomic goals. The majority of wetland conversions to agricultural land occur in wetland areas where plant growth is unrestricted by temperature and precipitation, that is, neither too cold (permafrost land) nor too hot (arid lands) (Asselen et al., 2013).
2. *Slope/terrain*: Priority for wetland conversion to agricultural land and settlements should be higher in flatlands or gentle slopes, which are favorable for agronomic management as well as housing and infrastructure construction.
3. *Histosols*: Rich organic matter content in soil is also considered a favorable factor of wetland conversion for peat harvesting. Highly fertile soil indicates peat extraction for fuel or horticulture. Reclaiming a relatively broad wetland area is cost-effective rather than a much smaller area separately. The small wetlands can easily be lost once human activity starts.

1.3.4 Socioeconomic factor

High vulnerability of wetland conversion is positively related to the distance between cropland and wetland. Since wetlands close to existing cropland can be easily converted and expanded for agriculture, rather than establishing new cropland in more remote and non-agricultural areas. Furthermore, necessary facilities such as travel corridors to the trade center, housing, and manpower are already accessible in existing cropland. The conversion of wetland to cropland requires large-scale inputs, such as drainage and logging. Agricultural yields are expected to be higher in converted wetland than in new, uncultivated land. Additionally, these wetlands are more susceptible to conversion, which has higher accessibility and market influence (Asselen et al., 2013).

1.3.5 Aquaculture and mangrove deforestation

Aquaculture activities negatively impact on local biodiversity of coastal wetlands including unintentional escape of potential invasive species; increased eutrophication due to aquaculture operation's wastewater and aquatic fauna in receiving waters; aquaculture expansion replacing important habitats such as mangroves forest and coastal wetlands;

pathogen transmission of various disease from aquaculture species to wild stocks; increased likelihood of natural stocks' genetic drift (pure line loss) from escaping hatchery-bred species; killing of fish predators near aquaculture farms; excessive use of antibiotic and hormone (Parvez & Mohsin, 2022). Mangrove forest deforestation is one of the major causes for the loss of coastal wetlands. At least 30% of the global mangrove area has been deforested for the aquaculture industry over the past 30 years, most of which has occurred in Southeast Asian countries. In Southeastern Asia, Indonesia and the Philippines have lost 48.6% and 36.7% from 2000 to 2012 due to extensive conversion of Mangrove to aquaculture, respectively (An et al., 2024). Furthermore, rice farming and oil palm plantations are also two other key drivers for mangrove deforestation, which have resulted in to 22% and 16% of mangrove loss respectively in Southeast Asia from 2000 to 2012 (An et al., 2024; Richards & Friess, 2016).

1.3.6 Transportation and service corridors

A canal, as well as a water corridor, is subject to an artificial waterway generally formed by excavation for the purpose of transportation. These corridors play a crucial role in the national economy. In the late 20th century, a total of sea canals were about 63,000 km worldwide (Revenga et al., 2000). In the globe, two of the busiest water transport corridors are the Suez (12% cargo transportation worldwide + 30% annual container trade) and the Panama (facilitating almost 5% of world trade yearly + around 14,000 ships crossing from various nations (Llacer, 2005; Wan et al., 2023). The construction of a water corridor is an intricate process and needs substantial workload. Consequently, these corridors pose critical threats to wetland ecosystems worldwide, affecting biodiversity and its conservation in aquatic ecosystems. Therefore, the effect of canals on the wetland environment is one of the primary concerns for stakeholders and policymakers when they are making strategic decisions (Han et al., 2024). Water canals pose a substantial negative influence on wetland ecology by altering natural habitats and hydrological circumstances, as well as directly or indirectly upsetting plants and animals (Han et al., 2024). Hydrological impacts include water level, inflows, outflows, and interaction with other habitat factors; all of these highly influence wetland functioning, biodiversity, and ecosystem services (William & James, 2000). The hydrological impacts prevail thorough cut off longitudinal hydrological connectivity of the wetland by stepping ship, disrupting the hydrological horizontal connection through the intricate construction and widening of narrow rivers, replacement of lotic habitat with lentic habitat due to dammed water and low hydrological velocity by ship lock operation, as well as water level fluctuation due to sudden lock operation. Due to high and frequent ship crossings, pollution impacts on wetlands have mainly increased through eutrophication from discharges from ships, agriculture, and industry, as well as serious salt intrusion from the sea into connected rivers during ship lock operations. The biodiversity impacts of canals on wetlands generally result in dislocation and biomass loss due to ship-induced waves, intricate canal construction, collision of ships with wild animals, habitat loss due to multiple ships' lock cascades, and change in the original surroundings, biological invasion by cross-basin canals via river connection with ships (Han et al., 2024).

1.3.7 Microplastics in wetland ecosystem soils

Plastics are globally used due to properties such as water resistance, lightweight, and longevity. Plastic manufacturing increased from 2 to 400 million tons between 1950 and 2019 worldwide (Cowger et al., 2024; Dokl et al., 2024; Zhang et al., 2025; Zheng & Suh, 2019). Out of which only 9% of total disposed plastic products are recycled, and the rest of the remaining plastic products enter and gradually accumulate in different ecosystems (Thenkabail, 2024). By 2050, waste plastic deposition is expected to be 3.1 billion tons (Li et al., 2024; Schwarz et al., 2023; Zhang et al., 2024). Plastic particle size <5 mm recognized as Microplastic (MPs) is divided into primary MPs and secondary MPs. Primary MPs are mainly derived from the industrial production process of detergents, care products, and cosmetics, while secondary MPs include MPs <5 mm sized originating by ultraviolet aging, mechanical wear, and large plastics biodegradation, for example, microfibers from textile aging, microparticles from tire wear, and microfilms from agricultural mulch degradation (El Hadri et al., 2020; Xia et al., 2022; Yurtsever, 2019; Zhang et al., 2025). Inland wetlands act as a sink for MPs from water and atmosphere, hence might be significant repositories for MPs. Additionally, MPs also disseminated to inland wetlands from various sources, including industrial sewage, landfills, mulch, sludge compost, tires, asphalt, and road marking coatings (Wu et al., 2024; Zhang et al., 2024; Zhong et al., 2023). MPs deteriorate wetland by entering organisms through skin, breathing, and are carried forward through the food chain, posing critical threats to aquatic organisms, avifauna, and mammals. MPs' detrimental effects include oxidative stress induction, DNA lesions, neurotoxicity, growth inhibition, and immune disorders (Ye et al., 2024). MPs also possess a large surface area and abundant adsorption sites, which increase the adsorption of hazardous substances, including organic pollutants, antibiotics, heavy metals (HMs), and microorganisms in the environment via hydrogen bonding and electrostatic adsorption.

This adsorption potency not only causes movement of MPs within the wetland ecosystems but also increases the serious complex pollutants to wetlands (Wang et al., 2018).

1.3.8 Heavy metals in wetlands

Due to lying between land and ocean, coastal wetlands are highly prone to input of HMs from atmospheric, marine, and terrestrial sources. Coastal human agro-economic, aquaculture, and sewage discharge activities are mainly associated with the use of artificial substances and fuels, and with the production of unacceptable waste and by-products comprising HMs. Cumulative studies infer local coastal human activities as a primary factor for inducing HM pollution in coastal wetlands (Li et al., 2022). Agricultural activities, solar power emissions, and exhaust from vehicles resulted 57.1% of the total HMs in coastal wetlands of a small Province of China (Li et al., 2022; Zhang et al., 2021), 74% of the total HMs in coastal marine sediments in Nigeria contributed by oil exploration and extraction activities and wastes of industries (Li et al., 2022; Ogundele & Ayeku, 2020). Furthermore, the spatial distribution of HMs on the Northern Ionian Coast in southern Italy mainly occurred due to industrial discharge and domestic sewage (Buccolieri et al., 2006; Li et al., 2022).

1.3.9 Population growth

Rapid growth in the human population adversely affects global wetland loss due to increased pollution, habitat loss, demand for food, and other commodities, resulting in the loss of wetland and biodiversity. Excessive demand for resources has put coastal wetlands which serve as natural buffers against natural calamities, in critical danger (Ghosh et al., 2024). Overfishing and habitat destruction gradually affect the sustainability of wetland-based industries. Food and shelter demands of the increased population have rapidly transformed wetlands into dry agricultural fields and human settlements (Padmavathi & Srinu, 2017). Substantial functioning and regulation of wetland ecosystems are mostly affected by the destruction, clogging, and diminishing of river and lowland habitats for the purpose of dams, bridges, resorts, and theme parks construction. Fish diversity and production of certain wetlands in Kolkata (India) have also drastically decreased due to increased population and human activity since 2018 (Sarkar & Borah, 2018; Sarkar et al., 2020b), thus endangering the fishermen's livelihoods, which are entirely dependent on these wetlands (Sarkar et al., 2020a). Additionally, exponential human population growth also accelerates pressure on urban wetlands for household, industrial, and commercial reasons (Ghosh et al., 2024).

1.3.10 Energy and sand mining

The extraction of sand and minerals, crucial for infrastructure and energy production, has accelerated the destruction of wetlands. Sand mining is one of the primary causes of the degradation of freshwater wetland ecosystems and the reduction in biodiversity in the basin by causing changes in the habitat of fish, disrupting benthic animal habitats, and altering the contour of rivers. Due to such ecological difficulties produced by sand mining in rivers, numerous advanced countries have altogether prohibited the mining activity (Deng et al., 2022). Sand mining exacerbates wetland riverine wetlands through inland fragmentation, increasing the lake's outflow capacity (which lowers the water level during the dry season), and shifting terrain distribution (as a result of a drop in water level), as well as grassland spreading toward the lake's center (Deng et al., 2022). Similarly, to fulfill the high demand of Lithium worldwide (used in electric vehicles, rechargeable batteries for mobile phones, and other gadgets), more than 1700 L of lithium (Li) brines are pumped each second from the shallow subsurface in Chile's Salar de Atacama in Chile (South America) (Alam & Sepúlveda, 2022). The largest issue concerning this process is the overuse of water during the Li extraction process. It uses a significant amount of water, roughly 500,000 gallons, or 2000 m^{-3} t of Li, resulting in irreversible harm to the area's groundwater supplies and reliant wetland ecosystems (Alam & Sepúlveda, 2022).

1.3.11 Invasive alien species

The wetland balance is highly sensitive, and invading plant and animal species are a major peril. Indigenous biodiversity (flora and fauna) is subject to being replaced or displaced by invasive species (Ghosh et al., 2024). The main causes of the detrimental effects on native or autochthonous species are invasive floating aquatic plants like *Alternanthera philoxeroides* and *Azolla cristata*, which interfere with fishing gear, reduce oxygen penetration, and change the composition and balance of economically significant fish species by reducing their food sources (Keller et al., 2018). The outcompeting tendency of invasive alien species (IAS) to native taxa negatively impacts coastal wetlands. These species can outcompete native plants and animals for resources like sunshine, space, and nutrients because of their rapid rates of reproduction and exponential development. This results in a loss of biodiversity as native wetland species face displacement or perhaps extinction. Illegal introduction of African catfish has replaced native aquatic species in Vembanad Kol Wetland in Kerala (Krishnakumar et al., 2011); and *Pterygoplichthys,* an invasive fish species, has significantly displaced less aggressive benthic fishes by feeding on fish eggs and also harvesting on periphyton in East Kolkata Wetland (Ghosh et al., 2019). Numerous foreign plant and animal species are intentionally or accidentally introduced onto Indian wetlands. For example, some of the invasive and introduced species in India include water lettuce (*Pistia stratiotes*), water hyacinth (*Eichhornia crassipes*), giant salvinia (*Salvinia molesta*), red tilapia from the Nile (*Oreochromis niloticus*), African catfish (*Clarias gariepinus*), Thai pangus (*Pangasius hypophthalmus*), and common carp (*Cyprinus carpio*) (Ghosh et al., 2024). Moreover, plants (such as *E. crassipes*, *A. philoxeroides*, and *Spartina alterniflora*), mammals (e.g., *Myocastor coypus*, *Ondatra zibethicus*), fishes (e.g., *Oreochromis mossambicus*), insects (e.g., *Lissorhoptrus oryzophilus*), and other invertebrates (e.g., *Procambarus larkia*) and amphibians (e.g., American bullfrog, *Rana catesbeiana*) are invasive species in China (An et al., 2024). The invasive species enters various wetland ecosystems through natural drivers, including flood or currents, wind, accidental introduction through attachment to visitors or boats, or can be intentionally introduced as a target species for conservation, preservation, or management purposes. IAS also exacerbates the climate change impact on coastal wetlands induced by other stressors. IAS species are potentially capable of thriving on altered environmental conditions, possess the ability to capitalize on changed abiotic factors such as temperature, salinity, or sea-level rise, and consequently disrupt the balance of vulnerable ecosystems. To conserve the wetlands from IAS, prevention measures and wise management of the dispersion of invasive species are necessary for the restoration of wetland health and resilience (Ghosh et al., 2024). Effective strategies for IAS prevention include detection and quick response, application and implementation of biosecurity measures, and raising public awareness about the risks and magnitude of impacts associated with IAS. Identifying the link between invasive species and other stressors, such as habitat loss and climate change, is essential for effectively addressing the threats they represent to wetlands (Ghosh et al., 2024).

1.3.12 Tree planting trade-off toward wetland health

Reforestation has been seen as a safe, economical, and efficient natural climate solution to absorb carbon and slow down global warming or climate change. Substantial plantation unprecedentedly increases forest area and reduces flood frequency, soil erosion, dust storms, and desertification. Broad-scale afforestation also enhances evapotranspiration and lowers runoff (Q) and soil moisture (SM), particularly in drylands (Griscom et al., 2017). However, this decrease in available water at the catchment scale invariably leads to a gradual reduction in the water supply to wetlands, which poses a serious threat to the loss of wetlands and a variety of ecosystem services, including food and water security, climate regulation, and their cultural and spiritual significance (Li, Piao, et al., 2018; Xi et al., 2021). The effects of tree planting on wetlands in China show that 1300–1500 km^2 (0.3%–0.4%) of wetland loss occurred as a result of 160,000–190,000 km^2 (25.3%–25.6%) of tree planting over wetland grid cells in the past (Li, Piao, et al., 2018).

1.3.13 Climate change impact

Climate change has been recognized as one of the major threats to wetlands. It impacts wetlands, especially due to elevation in sea level, coral bleaching due to increased sea surface temperatures, and alterations in hydrology in inland waters. These consequences increase the risk to Arctic and montane wetlands, flooding, and drought in many parts of the world (Ramsar Convention, 2018). Compromised hydrology and increased temperature due to climate change not only alter biogeochemistry and function of wetland but also turned wetland ecosystem service into wetland disservice that is, incapable for water filtration and purification, and starting decomposing and release of nutrients to surface water (Salimi et al., 2021). Climate change exacerbates the dynamics of the delicate wetland ecosystem through the direct effects of rising temperatures and changes in rainfall intensity and frequency. Furthermore, extreme and intense conditions such as drought, heat waves, floods, cyclones, and storms also frequently affect the hydrology of wetlands and make them among the most endangered ecosystems (Salimi et al., 2021; Zotos et al., 2024). A high rate of decomposition than production (Photosynthesis) in wetland due to climate change eventually shifted its nature of carbon sink to carbon source (CO_2 and CH_4 emission to the atmosphere) (Flanagan & Syed, 2011).

1.4 Models used in wetland analysis and monitoring

The main objectives of wetland researchers, policymakers, and attorneys are to restore, conserve, and manage the planet's surviving wetlands. Wetland monitoring must be done continuously to accomplish these objectives. However, conventional wetland monitoring techniques have a number of drawbacks, including significant inaccuracies, administrative restrictions, and inconsistent data, and they are unable to collect long-term data on a wide scale (Yu et al., 2024).

Therefore remote sensing (RS) and geographic information systems (GIS) are the essential and mostly preferred modern techniques to obtain accurate long-term large-scale data related to land use/land cover dynamics of the land surface (Ban et al., 2015). For the past five decades, Landsat sensors (such as TM, ETM, and OLI) have been employed to continuously record the Earth's land surface at high spatiotemporal resolution (spatial resolution: 30 m; revisiting period: 16 d) (Sun & Yu, 2024; Zhu, 2017) as well as for generating worldwide and regional maps of typical coastal wetland such as mangrove forests, salt marshes, tidal flats and aquaculture ponds. To comprehend wetland distribution, spatiotemporal dynamics, and degradation area reflecting the present state of damaged wetland ecosystems, researchers can readily extract precise photographs of the earth's surface (Yu et al., 2024; Zhao et al., 2024). There are various RS- and GIS-based models for understanding and analyzing wetland health, dynamics, and the impacts of climate change. These modes are useful and widely used tools for effective environmental planning and prediction of qualitative and quantitative future management activity (Jafarzadeh et al., 2022). Tables 1.5 and 1.6 summarizes some of the commonly used models and enlists the dataset used in the wetland study, respectively.

1.5 Wetland regeneration in the context of climate change

Wetland restoration is essential for building resilient landscapes to combat the severe impact of climate change on wetlands without losing biodiversity or the ecosystem's ability to operate (Bastin et al., 2019; Erwin, 2009). It is critical and vital to comprehend and apply effective and efficient restoration management strategies for wetlands under climate change to mitigate the negative effects of climate change on wetlands (Benayas et al., 2009; Tomscha et al., 2021). Traditionally, two broad-scaled contrasting approaches, such as (1) active restoration and (2) passive (natural) regeneration, are applied in wetland restoration (Holl & Aide, 2011). Active restoration facilitated with significant human efforts, including seedling planting in tube stock or seed regeneration. Whereas natural regeneration merely requires human cooperation to regrow vegetation via available plant propagules once the disturbance ceases. Both the active and passive regeneration need manipulated abiotic factors to commence natural regeneration. The availability of

TABLE 1.5 Different models used for wetland monitoring.

Model Name	Application	References
CA-Markov model	Wetland dynamics, wetland change trends simulation, vegetation restoration, land change analysis, and Wetland conservation priority analysis	Ma et al. (2012), Li et al. (2012), Xu et al. (2024)
IBIS-CH4 (Integrated Biosphere Simulator-Methane) model	Global wetland CH_4 emission	Xiao et al. (2024)
The Energy Exascale Earth System Model (E3SM)	Coastal wetland dynamics by determining the marsh community response to increased CO_2 and temperature	Theresa et al. (2021), Xu et al. (2024)
Topography-based hydrological Model (TOPMODEL)	Hydrological dynamics of a wetland by measuring the depth of the water table in the wetland; hydrological forecasting.	Zhang et al. (2016), Jeziorska and Niedzielski (2018)
Generalized Additive Models (GAMs)	Wetland response (NDVI, Wetland area) and environmental drivers	Doughty et al. (2024)
Storm Water Management Model (SWMM)	Simulation of runoff quantity (rainfall to runoff for each subcatchment and routes this water through conduits and nodes)	Sauer et al. (2024)
HYSITEL/HYDROTEL	In the mitigation simulation of extreme floods and drought, as well as quantifying the hydrological services of wetlands	Wu et al., 2023; Wu, Sun et al., 2023
HEC-RAS model	Simulation for flood control	Kim (2010)
SWAT model	Water balance in a constructed wetland for flood mitigation	Kim (2010)
TERRA/MODIS, NOAA/AVHRR	Land surface temperature for identifying pattern, distribution, dynamics, and heterogeneity of the urban heat island in the wetland cooling effect	Clinton and Peng (2013), Bera et al. (2021)
ORCHIDEE- Hillslope	To analyze tree planting impacts on wetland dynamics	Xi et al. (2022)

TABLE 1.6 Overview of global/regional wetland-related datasets sources.

Name	Spatial resolution	Water body classes	References
GLWD-3	30 arcsec (~1 km)	12 water classes (3 permanent water classes and 9 natural wetland classes)	Lehner and Döll (2004)
GIEMS-1	0.25 degree (~25 km)	Natural and artificial water-saturated area (except larger lakes)	Prigent et al. (2020)
GIEMS-2	0.25 degree (~25 km)	Natural and artificial inundated areas (excluding large lakes)	Prigent et al. (2020)
GIEMS-D15	15 arcsec (~500 m)	Worldwide water saturated	Chouinard et al. (2015)
ESACCI land cover	10 arcsec (~300 m)	Open waterbodies	Li, Macbean, et al. (2018)
JRC	~30 m	Surface water	Pekel et al. (2016)
G2017	~232 m	Wetland and peatland areas in tropical and subtropical (40°N–60°S)	Gumbricht et al. (2017)
RFW	15 arcsec (~500 m)	Permanent flooded wetland (except large permanent lakes)	Tootchi et al. (2019)
SWAMPS	0.25 degree (~25 km)	All terrestrial water fractions	Jensen and Mcdonald (2019)
BAWLD	0.5 degree (~50 km)	Wetlands, including bog, fen, marsh, tundra wetland, and permafrost bog in the northern boreal and tundra biomes (>55°N)	Olefeldt et al. (2021)
WAD2M	0.25 degree (~25 km)	Surface inundated areas, including wetlands under dense canopies, excluding lakes, rivers, ponds, and rice agriculture	Zhang et al. (2021)
NDVI GIMMS NDVI3g		Change in wetland vegetation	Xin et al. (2022a,b)

appropriate plant propagules and establishing locations within the landscape, along with adequate, favorable regeneration conditions such as sufficient time, are prerequisites for natural regeneration. Active restoration is required in areas where ideal conditions have been lost due to prolonged and severe agricultural pressure, substantial erosion, or other constraints. Active restoration may involve a significant trade-off between initial expenditures and continuing upkeep, depending on the degree of involvement needed (Moreno-Mateos et al., 2015). Active restoration initiatives are sometimes limited in scale due to the high costs of acquiring native seeds and tube stock, as well as the labor-intensive process of planting and managing them (Summers et al., 2015).

1.6 Wetland resilience

Anthropogenic activities and climate change caused wanton disruption and rapid decline in wetland characteristics such as floral diversity, abundance, and cover (Paudel et al., 2023; Wu et al., 2024). Gradual deterioration in these attributes results in loss of resilience that keeps wetland stability and triggers ecological succession. Understanding wetland resilience is crucial to preserving the integrity and health of wetland ecosystems. The ability of an ecosystem to adapt and reconstruct itself following a disruption is reflected in its resilience (Holling, 1973). Resilience mechanism consists of (1) the availability of dormant and stored seeds, (2) dispersion of seeds by wind or animals, (3) the growth and development of juvenile plants following seed germination, and (4) the use of resources to compete in a disturbed community (Buma & Wessman, 2011; Falk et al., 2022). Completion of these processes results in state transition or restoration (Falk et al., 2022). Drought and flood make wetland structure and functions vulnerable (Zhang et al., 2023a,b). Prolonged drought reduces water level and increases plant mortality in wetland (Wang et al., 2019), which leads to loss of resilience to the critical point. However, on termination of droughts, wetland resilience gradually increases by hydrological processes such as an increase in SM, seed germination, and plant growth (Sandi et al., 2020; Wu et al., 2024). The wetland vegetation and habitat conditions are controlled by water, air, temperature, and energy. The leave attributes and shedding strategies of different vegetations also impact post draught resilience for example, vegetation with evergreen gymnosperm has more potential resilience in the drought condition than dominated deciduous angiosperm vegetation (Hossain & Li, 2021; Wu et al., 2024). Fluctuations in water level, such as depth, duration, and frequency, are also key influencers of the vegetation growth environment (temperature and light) in which wetland floral diversity performs photosynthesis. Hence, water level fluctuations primarily reflect floral growth, morphology, and community. Prolonged high water levels result in gradual vegetation mortality by inhibiting plant root respiration in deprived soil oxygen levels (Chen et al., 2023; Wu et al., 2024; Zhang et al., 2023a, b). Thus the structure and function of wetland ecosystems are drastically changed by the effect of hydrological processes, which eventually reflects a shift in wetland resilience (Measho et al., 2021). Permanently inundated wetlands have greater resilience potential than intermittently inundated wetlands in the same climate region, suggesting that the impact of wetland resilience is reflected in the hydrological regime. In contrast, intermittently inundated wetlands have intricate hydrological regimes that create a highly variable environment for plant growth (Liu et al., 2020; Wu et al., 2024; Zou et al., 2021). Wetland ecosystems that experience intermittent flooding are especially vulnerable, and vegetation communities may undergo succession due to the significant change in water level caused by prolonged drought and prolonged inundation (Donnelly et al., 2022; Wu et al., 2024; Zhang et al., 2023a, b). Conversely, permanently submerged wetlands are stable due to their sustained water supply, which increases their resilience and resistance to external shocks (Wu et al., 2024). Furthermore, wetlands in wet regions are more resilient than those in arid regions; the water level variations in the former are less pronounced than those in the latter. Additionally, because of the complementarities between species, an ecosystem with more species diversity can recover from external perturbations more quickly (Isbell et al., 2015; Oliver et al., 2015; Wu et al., 2024).

1.7 Wetland conservation and management efforts

To enhance wetland's functions and support nature-based solutions, lost wetlands must be restored and remaining wetlands must be conserved (Thorslund et al., 2017; Wu et al., 2023). The legislative framework, nongovernmental organizations, and international, national, and state policies primarily regulate wetland protection and management.

1.7.1 The Ramsar Convention

With the aim of conservation and sustainable use of wetland, the international treaty or convention on wetlands of international importance, particularly as Peafowl habitat, also known as the Ramsar Convention, was adopted and signed in 1971 in Ramsar city of Iran (An et al., 2024). The Ramsar Convention comprises several measures focusing to support the execution, to achieve Ramsar goals and certain resolutions such as (1) Identify and designate wetland sites as wetlands of international importance; (2) obligating states/parties to submit national reports to COPs (Conference of the Contracting Parties); (3) enforcing international parties to provide regular updates to the Ramsar Information Service; (4) cataloging wetland sites on the Montreux Record (a register maintain changes in ecological character due to anthropogenic activity) (https://www.ramsar.org/sites/default/files/documents/pdf/lib/Matthews-history.pdf); (5) carrying out RAMs (Ramsar Advisory Missions) upon request from a contracting party that reports a change or changes in a site's ecological character (Chapter 3.2) (https://www.ramsar.org/sites/default/files/documents/pdf/lib/Matthews-history.pdf); (6) asking RAMs to provide guidance on how to remove locations from the Montreux Record and evaluating the effectiveness of those steps; (7) establishing objectives at COPs to deal with fresh and developing problems and threats to wetlands (Barry, 2024). However, associated parties abide by strict and serious action to conserve, restore, and protect wetlands through the wise use of wetlands in their countries (An et al., 2024). This international treaty is also subject to all parties with various commitments and an interwoven portfolio, including the formulation and mapping out of planning for better conservation and wise use of wetlands; establishing nature reserves in wetlands to promote training programs related to wetland research and their effective management (Ramsar, 2016). With these portfolios and resolutions, the Ramsar Convention has become the most instrumental and significant international agreement for the preservation of wetlands. The Convention has been signed by 171 contracting countries, and 2418 wetlands totaling more than 2.5 million km^2 have been added to the Ramsar site (An et al., 2024; Singh et al., 2025). Moreover, wetlands become a center for 171 contracting parties to achieving 17 SDGs, mainly focusing poverty, hunger, health, energy, education, and climate change. The agenda set by the Ramsar Convention for the current decade is aimed at associating all wetlands with society and the common public for wise use, conservation, and better management (Ramsar Convention, 2018).

1.7.2 Legal framework and national policies (Indian perspective)

In India, till today, no separate and specific legal provision for wetland conservation has been established. However, the first significant step of MoEFCC that is, Wetlands (Conservation and management) Rule 2017 has become effective to conservation, management and maintenance of the ecological attributes of the wetland. Wetlands (Conservation and Management) Rule, 2017 has enhanced the focus of wetland from central to state bodies (MoEFCC) (Yadav & Goyal, 2022). These rules are subject to providing advisory roles to the national wetland committee to guide and regulate state/ union territory bodies on the integrated management of wetlands. In India wetland conservation and management is directly governed by legal instruments such as Indian Fisheries Act 1857, Indian Forest Act 1927, Wildlife (Protection) Act 1972, Water (Prevention and Control of Pollution) Act 1974, Territorial Water, Continental Shelf, Exclusive Economic Zone and other Marine Zones Act 1976, Water Cess Act 1977, Maritime Zone of India (Regulation and fishing by foreign vessels) Act 1980, Forest (Conservation) Act 1980, Environmental (Protection) Act 1986, Wildlife (Protection) Amendment Act1991, Biodiversity Act 2002, and Scheduled Tribes and Other Traditional Forest Dwellers (Recognition of Forest Rights) Act 2006 (Bassi et al., 2014; MoEF, 2007; Prasad et al., 2002; Prasad et al., 2021). These acts are commuted for wetland biodiversity preservation through provisions including water quality protection and management, and strict monitoring of ecologically sensitive landscapes for preservation, upkeep, and enhancement of the nation's aquatic ecosystem's biodiversity in terms of flowers, animals, and birds (Bassi et al., 2014; Chowdhury & Behera, 2022). In respect to national policies including National Conservation Strategy and Policy Statement on Environment and Development, 1992 (https://elibrary.sansad.in/items/acdc882a-8a3b-4f7f-9247-a9624840b23b); Coastal Zone Regulation Notification, 1991 (https://environmentclearance.nic.in/writereaddata/SCZMADocument/CRZ%20Notification,%201991.pdf); National Policy and Macro Level Action Strategy on Biodiversity, 1999 (http://www.nbaindia.org/nbsap/uploaded/nbsap/1999.pdf); and National Water Policy, 2002 (https://prod-qt-images.s3.amazonaws.com/indiawaterportal/import/sites/default/files/iwp2/National_Water_Policy__MoWR__2002.pdf) etc. having biggest impact on conservation and management of wetlands due to international obligations made under the Ramsar Convention (Bassi et al., 2014; MoEF, 2007; Prasad et al., 2002; Prasad et al., 2021). To extend conservation and management of Indian wetland, the Government of India has recently launched the Amrit Dharohar implementation scheme 2023–24 to promote unique conservation and cultural heritage of Ramsar sites; creation of nature tourism and income generation for local people; safeguard and enhancement of biodiversity, etc. Thus, progressive steps toward wetland conservation and maintenance are regularly being taken at the national level.

Declaration of generative AI and AI-assisted technologies in the writing process

During the preparation of this work, the author(s) have used the Large Language Model tool ChatGPT to improve the readability and language of the work. After using this tool/service, the author(s) reviewed and edited the content as needed and take full responsibility for the content of the publication.

Acknowledgment(s)

Authors, S.S., A.S., I.T., V.K., and K.T. are thankful to the funding agency for the grant NMHS 2023-24/SC-XI/LG12/02 from the National Mission on Himalayan Studies (NMHS) for the project titled "Multidimensional Assessment of Ecological Dynamics and Ecosystem Health of Selected High-Altitude Wetlands of Indian Himalayan Region (IHR) for Effective Conservation and Management planning."

References

Alam, M. A., & Sepúlveda, R. (2022). Environmental degradation through mining for energy resources: The case of the shrinking Laguna Santa Rosa wetland in the Atacama Region of Chile. *Energy Geoscience, 3*(2), 182–190.

Almuktar, S. A., Abed, S. N., & Scholz, M. (2018). Wetlands for wastewater treatment and sub-sequent recycling of treated effluent: A review. *Environmental Science and Pollution Research, 25*, 23595–23623.

An, S., Yin, S., Verhoeven, Jos, T. A., & Jeelani, N. (2024). Chapter 29 - Wetlands. In (Fourth Edition). Ian D. Jones, & John P. Smol (Vol. Eds.), *Wetzel's limnology: 2024*, (pp. 999–1013). Academic Press ISBN 9780128227015.

Asselen, S. V., Verburg, P. H., Vermaat, J. E., & Janse, J. H. (2013). Drivers of wetland conversion: A global meta-analysis. *PLoS One, 8*(11), e81292.

Azubuike, C. C., Chikere, C. B., & Okpokwasili, G. C. (2016). Bioremediation techniques–classification based on site of application: Principles, advantages, limitations and prospects. *World Journal of Microbiology Biotechnology, 32*, 180. https://doi.org/10.1007/s11274-016-2137-x.

Ballut-Dajud, G. A., Sandoval Herazo, L. C., Fernández-Lambert, G., Marín-Muñiz, J. L., López Méndez, M. C., & Betanzo-Torres, E. A. (2022). Factors affecting wetland loss: A review. *Land, 11*(3), 434.

Bamutaze, Y., Kyamanywa, S., Singh, B. R., Nabanoga, G., & Lal, R. (2019). *Agriculture and Ecosystem Resilience in Sub Saharan Africa,* 1–399.

Ban, Y., Gong, P., & Giri, C. (2015). Global land cover mapping using Earth observation satellite data: Recent progresses and challenges. *ISPRS Journal of Photogrammetry and Remote Sensing, 103,* 1–6. https://doi.org/10.1016/j.isprsjprs.2015.01.001.

Barry, T. (2024). Arctic wetlands, an evaluation of progress towards implementation of the Ramsar convention on wetlands: 1978–2022. *International Environmental Agreements: Politics, Law and Economics, 24*(2), 423–448.

Bassi, N., Kumar, M. D., Sharma, A., & Pardha-Saradhi, P. (2014). Status of wetlands in India: A review of extent, ecosystem benefits, threats and management strategies. *Journal of Hydrology: Regional Studies, 2,* 1–19.

Bastin, J., Finegold, Y., Garcia, C., Mollicone, D., Rezende, M., Routh, D., et al. (2019). The global tree restoration potential. *Science (New York, N.Y.), 365*(6448), 76–79. https://doi.org/10.1126/science aax0848.

Benayas, J. R., Newton, A. C., Diaz, A., & Bullock, J. M. (2009). Enhancement of biodiversity and ecosystem services by ecological restoration: A meta-analysis. *Science (New York, N.Y.), 325*(5944), 1121–1124. https://doi.org/10.1126/science.1172460.

Bera, B., Shit, P. V., Saha, S., & Bhattacharjee, S. (2021). Exploratory analysis of cooling effect of urban wetlands on Kolkata metropolitan city region, eastern India. *Current Research in Environmental Sustainability, 3,* 100066 ISSN 2666-0490.

Bezzalla, A., Silabdi, B., et al. (2025). Species diversity within ecosystem diversity of wetlands within drylands: Waterbird ecological attributes in an ecological complex of wetland habitats. *Estuarine, Coastal and Shelf Science, 313,* 109103. https://doi.org/10.1016/j.ecss.2024.109103 ISSN 0272-7714.

Binilkumar, A.S. (2010). Economic Valuation of a Wetland in West Bengal, India (dissertation/ thesis Indian Institute of Technology, Bombay).

Bole, J. B., & Allan, J. R. (1978). Uptake of phosphorus from sediment by aquatic plants, Myriophyllum spicatum and Hydrilla verticillata. *Water Research, 12*(5), 353–358.

Broadbent, A. M., Coutts, A. M., Tapper, N. J., Demuzere, M., & Beringer, J. (2018). The microscale cooling effects of water sensitive urban design and irrigation in a suburban environment. *Theoretical Applied Climatology, 134,* 1–23. https://doi.org/10.1007/s00704-017-2241-3.

Buccolieri, A., Buccolieri, G., Cardellicchio, N., Dell'Atti, A., Di Leo, A., & Maci, A. (2006). Heavy metals in marine sediments of Taranto Gulf (Ionian Sea, Southern Italy). *Marine Chemistry, 99*(1–4), 227–235 ISSN 0304-4203.

Buma, B., & Wessman, C. A. (2011). Disturbance interactions can impact resilience mechanisms of forests. *Ecosphere, 2*(5), 1–13.

Burks, R. L., Mulderij, G., Gross, E., Jones, I., Jacobsen, L., Jeppesen, E., et al. (2006). *Wetlands: Functioning, biodiversity conservation, and restoration*. Berlin: Springer-Verlag.

Calheiros, C. S., Rangel, A. O., & Castro, P. M. (2007). Constructed wetland systems vegetated with different plants applied to the treatment of tannery wastewater. *Water Research, 41*(8), 1790–1798.

Cao, X., Onishi, A., Chen, J., & Imura, H. (2010). Quantifying the cool island intensity of urban parks using ASTER and IKONOS data. *Landscape and Urban Planning, 96,* 224–231. https://doi.org/10.1016/j.landurbplan.2010.03.008.

Chen, M., Zeng, S., Jiang, B., Wen, Z., Wu, J., & Xia, J. (2023). The comprehensive evaluation of how water level fluctuation and temperature change affect vegetation cover variations at a lake of ecological importance (Poyang Lake), China. *Ecological Indicators, 148,* 110041. https://doi.org/10.1016/j.ecolind.2023.110041.

Cheng, J., Landesman, L., et al. (2002). Nutrient removal from swine lagoon liquid by lemna minor 8627. *Transactions of the American Society of Agricultural Engineers, 45,* 1003–1010.

Cherry, J. A. (2011). Ecology of Wetland Ecosystems: Water, Substrate, and Life. *Nature Education Knowledge, 3*(10). 16.

Cheval, S., & Constantin, S. (2019). Black Sea impact on its west-coast land surface temperature. *Theoretical and Applied Climatology, 135*(3–4), 1583–1593.

Chislock, M. F., Doster, E., Zitomer, R. A., & Wilson, A. E. (2013). Eutrophication: Causes, consequences, and controls in aquatic ecosystems. *Nature Education Knowledge, 4*(4), 10.

Chouinard, E. F., Lehner, B., Rebelo, L. M., Papa, F., & Hamilton, S. K. (2015). Development of a global inundation map at high spatial resolution from topographic downscaling of coarse-scale remote sensing data. *Remote Sensing of Environment, 158,* 348–361. https://doi.org/10.1016/j.rse.2014.10.015.

Chowdhury, K., & Behera, B. (2022). Institutional dynamics and water resource management: The case of traditional water bodies in West Bengal, India. *International Journal of Water Resources Development, 38*(5), 836–860.

Christofilopoulos, S., Kaliakatsos, A., Triantafyllou, K., Gounaki, I., Venieri, D., & Kalogerakis, N. (2019). Evaluation of a constructed wetland for wastewater treatment: Addressing emerging organic contaminants and antibiotic resistant bacteria. *New Biotechnology, 52,* 94–103. https://doi.org/10.1016/j.nbt.2019.05.006.

Clinton, N., & Peng, G. (2013). MODIS detected surface urban heat islands and sinks: Global locations and controls. *Remote Sensing Environment, 134*(5), 294–304. https://doi.org/10.1016/j.rse.2013.03.008.

Collen, B., Whitton, F., Dyer, E. E., Baillie, et al. (2014). Global patterns of freshwater species diversity, threat and endemism. *Global Ecology Biogeography. 23,* 40–51.

Costanza, R., de Groot, R., Sutton, P., van der Ploeg, S., Anderson, S. J., et al. (2014). Changes in the global value of ecosystem services. *Global Environmental Change, 26,* 152–158.

Cowardin, L. M., Gilmer, D. S., & Shaiffer, C. W. (1985). The agricultural environment of North Dakota. *Mallard Recruitment in Wildl. Monogr. 92.*

Cowger, W., et al. (2024). Global producer responsibility for plastic pollution. *Science in Advances, 10,* eadj8275.

Creighton, C., Boon, P. I., Brookes, J. D., & Sheaves, M. (2015). Repairing Australia's estuaries for improved fisheries production – What benefits, at what cost? *Marine and Freshwater Research, 66*, 493–507. https://doi.org/10.1071/MF14041.

Cross, A. T., et al. (2015). Population ecology of the endangered aquatic carnivorous macrophyte Aldrovanda vesiculosa at a naturalised site in North America. *Freshwater Biology, 60*(9), 1772–1783.

Cui, B. S., & Yang, Z. F. (2006). *Wetland Science*. Beijing, China: Beijing Normal University Press.

Cui, L., Li, G., Zhao, M., & Xu, X. (2024). Natural and anthropogenic contributions to the landscape evolution in the Yellow River Delta wetland, China. *Global Ecology and Conservation, 56*, e03347. https://doi.org/10.1016/j.gecco.2024.e03347.

Cui, L. H., Luo, S. M., Zhu, X. Z., & Liu, Y. H. (2003). Treatment and utilization of septic tank effluent using vertical-flow constructed wetlands and vegetable hydroponics. *Journal of Environmental Sciences, 15*(1), 75–82 PMID: 12602607.

Culhane, F., Teixeira, H., Nogueira, A. J. A., Borgwardt, F., et al. (2019). Risk to the supply of ecosystem services across aquatic ecosystems. *The Science of the Total Environment, 660*, 611–621.

Dahdouh-Guebas, F., et al. (2005). How effective were mangroves as a defense against the recent tsunami? *Current Biology CB, 15*(12), 1337–1338. https://doi.org/10.1016/j.cub.2005.07.025 PMID: 15964259.

Dahl, T.E., & Stedman, S.M. (2013). Status and trends of wetlands in the coastal watersheds of the conterminous United States 2004 to 2009 (US Department of the Interior, US Fish and Wildlife Service and National Oceanic and Atmospheric Administration, National Marine Fisheries Service.

Dan, T. H., Quang, L. N., Chiem, N. H., & Brix, H. (2011). Treatment of high-strength wastewater in tropical constructed wetlands planted with Sesbania sesban: Horizontal subsurface flow versus vertical down flow. *Ecological Engineering, 37*(5), 711–720. https://doi.org/10.1016/j.ecoleng.2010.07.030.

Darrah, S. E., Shennan-Farpón, Y., Loh, J., Davidson, N. C., et al. (2019). Improvements to the Wetland extent trends (WET) index as a tool for monitoring natural and human-made wetlands. *Ecological Indicators, 99*, 294–298. https://doi.org/10.1016/j.ecolind.2018.12.032.

Das, S., & Vincent, J. R. (2009). Mangroves protected villages and reduced death toll during Indian super cyclone. *Proceedings of the National Academy of Sciences of the United States of America, 106*(18), 7357–7360.

Davidson, N. C. (2014). How much wetland has the world lost? Long-term and recent trends in global wetland area. *Marine and Freshwater Research, 65*, 934–941.

Davidson, N. C., Fluet-Chouinard, E., & Finlayson, C. M. (2018). Global extent and distribution of wetlands: Trends and issues. *Marine and Freshwater Research, 69*, 620–627. https://doi.org/10.1071/MF17019.

Dawuda, U. K., Yonghong, C., et al. (2024). Assessment of wetland loss through industrialization and community expansion: A case study of the greater amanzule wetland in the Ellembelle district of Ghana. *Frontiers Environmental Sciences, 12*, 1451394.

Deng, M., Li, Q., Li, W., Lai, G., & Pan, Y. (2022). Impacts of sand mining activities on the wetland ecosystem of Poyang Lake (China). *Land, 11*(8), 1364.

Díaz-Delgado, R., Aragon´es, D., Af´an, I., & Bustamante, J. (2016). Long-term monitoring of the flooding regime and hydroperiod of Do˜nana marshes with landsat time series (1974-2014). *Remote Sens, 8*, 775.

Dise, N. B. (2009). Peatland response to global change. *Science (New York, N.Y.), 326*(5954), 810–811.

Dokl, M., Copot, A., et al. (2024). Global projections of plastic use, end-of-life fate and potential changes in consumption, reduction, recycling and replacement with bioplastics to 2050. *Sustainable Production and Consumption, 51*, 498–518.

Donato, D. C., Kauffman, J. B., Murdiyarso, D., Kurnianto, S., Stidham, M., & Kanninen, M. (2011). Mangroves among the most carbon-rich forests in the tropics. *Nature Geosciences, 4*, 293–297.

Donnelly, J. P., Moore, J. N., Casazza, M. L., & Coons, S. P. (2022). Functional wetland loss drives emerging risks to waterbird migration networks. *Frontiers Ecology Evolution, 10*, 844278. https://doi.org/10.3389/fevo.2022.844278.

Doughty, C. L., Gu, H., Ambrose, R. F., et al. (2024). Climate drivers and human impacts shape 35-year trends of coastal wetland health and composition in an Urban Region. *Ecosphere, 15*(4), e4832. https://doi.org/10.1002/ecs2.4832.

El Hadri, H., Gigault, J., Maxit, B., Grassl, B., & Reynaud, S. (2020). Nanoplastic from mechanically degraded primary and secondary microplastics for environmental assessments. *NanoImpact, 17*, 100206. https://doi.org/10.1016/j.impact.2019.100206.

Erwin, K. L. (2009). Wetlands and global climate change: The role of wetland restoration in a changing world. *Wetlands Ecology and management, 17*(1), 71–84.

Falfán, I., & Zambrano, L., (2024). Urban Wetlands for Sustainable Cities in Mexico. 10.1007/978-3-031-69590-2_3.

Falk, D. A., van Mantgem, P. J., Keeley, J. E., Gregg, R. M., Guiterman, C. H., Tepley, A. J., et al. (2022). Mechanisms of forest resilience. *Forest Ecology and Management, 512*, 120129. https://doi.org/10.1016/j.foreco.2022.120129.

Finlayson, C., Everard, M., Irvine, K., McInnes, R. J., Middleton, B. A., van Dam, A., et al. (2018). *The wetland book I: Structure and function, management and methods*. Springer.

Finlayson, C. M. (2012). Forty years of wetland conservation and wise use. *Aquatic Conservation: Marine. Freshwater Ecosystems, 22*, 139–143.

Finlayson, C. M., Fennessy, M. S., Gardner, R. C., Kumar, R., McCartney, M. P., & Van Dam, A. A. (2024). Closing the driver–response loop for halting and reversing wetland degradation and loss from agriculture. *Marine and Freshwater Research, 75*(15), MF24050.

Flanagan, L. B., & Syed, K. H. (2011). Stimulation of both photosynthesis and respiration in response to warmer and drier conditions in a boreal peatland ecosystem. *Global Change Biology, 17*, 2271–2287.

Fluet-Chouinard, E., Stocker, B. D., et al. (2023). Extensive global wetland loss over the past three centuries. *Nature, 614*, 281–286.

Fu, X., & He, X. (2015). Nitrogen and phosphorus removal from contaminated water by five aquatic plants. In *2015 International Conference on Mechatronics, Electronic, Industrial and Control Engineering (MEIC-15)*. Atlantis Press, 1274–1277.

Gao, J., Li, X. L., et al. (2013). Geomorphic-centered classification of wetlands on the Qinghai-Tibet Plateau, Western China. *Journal of Mountain Science, 10*. https://doi.org/10.1007/s11629-013-2561-4.

Gardner, R.C. et al. (2015). State of the world's wetlands and their services to people: A compilation of recent analyses. Ramsar Briefng Note no. 7 (Ramsar Convention Secretariat 2015).

Gardner, R.C. & Finlayson, C. (2018). Global wetland outlook: State of the world's wetlands and their services to people. Ramsar Convention Secretariat, 2018, Stetson University College of Law Research Paper No. 2020-5, Available at SSRN:

Ghermandi, A., Van Den Bergh, J. C., Brander, L. M., De Groot, H. L., & Nunes, P. A. (2010). Values of natural and human-made wetlands: A metaAanalysis. *Water Resources Research, 46*(12). https://doi.org/10.1029/2010WR009071.

Ghosh, S., Anju, P., Pattanayak, R., & Sahu, N. C. (2024). Fisheries and aquaculture in wetland ecosystems: A review of benefits, risks, and future prospects in India. *Journal of Coastal Research, 40*.

Ghosh, S., Sahu, N. C., Rahaman, F. H., & Das, K. S. (2019). Periphyton based climate smart aquaculture for the farmers of Indianrural Sunderban areas. *Indian Research Journal of Extension Education, 19*(1), 60–72.

Golden, H. E., Lane, C. R., Rajib, A., & Wu, Q. (2021). Improving global flood and drought predictions: Integrating non-floodplain wetlands into watershed hydrologic models. *Environmental Research Letters, 16*. https://doi.org/10.1088/1748-9326/ac1fbc.

Green, A. J., Guardiola-Albert, C., Bravo-Utrera, M. A., Bustamante, J., et al. (2024). Groundwater abstraction has caused extensive ecological damage to the Do˜nana world heritage site, Spain. *Wetlands, 44*, 1–13.

Griscom, B. W., Adams, J., Ellis, P. W., Houghton, R. A., Lomax, G., Miteva, D. A., et al. (2017). Natural climate solutions. *Proceedings of the National Academy of Sciences, 114*(44), 11645–11650.

Gu, G. P., Wu, G. R., Lu, C. M., Zhou, C. F., Xu, Q. S., & Shi, G. X. (2000). Purifying effect of nymphoides peltatum (Gmel) O. Kuntze on Cd^{2+} containing sewage and its mechanism. *Journal of Ecology and Rural Environment, 16*(3), 9–14.

Gumbricht, T., Roman-Cuesta, R. M., Verchot, L., et al. (2017). An expert system model for mapping tropical wetlands and peatlands reveals South America as the largest contributor. *Glob Change Biol, 23*, 3581–3599. https://doi.org/10.1111/gcb.13689.

Guo, Y., Liu, Y., Zeng, G., Hu, X., Li, X., Huang, D., et al. (2014). A restoration-promoting integrated floating bed and its experimental performance in eutrophication remediation. *Journal of Environmental Sciences, 26*(5), 1090–1098.

Guo, Z., Cui, G., Zhang, M., & Li, X. (2019). Analysis of the contribution to conservation and effectiveness of the wetland reserve network in China based on wildlife diversity. *Global Ecology and Conservation, 20*, e00684.

Gupta, G., Khan, J., Upadhyay, A. K., & Singh, N. K. (2020). *Wetland as a sustainable reservoir of ecosystem services: Prospects of threat and conservation. restoration of wetland ecosystem: A trajectory towards a sustainable environment*, 31–43.

Güsewell, S., Koerselman, W., & Verhoeven, J. T. A. (2003). Biomass N:P ratios as indicators of nutrient limitation for plant populations in wetlands. *Ecological Applications, 13*(2), 372–384. http://www.jstor.org/stable/3099905.

Han, Z., Yin, D., Zhao, R., Liu, Z., Cheng, R., Lin, G., et al. (2024). The ecological effect of ship canals on wetlands. *Water, 16*(22), 3324. https://doi.org/10.3390/w16223324.

Hassan, I., Chowdhury, S. R., Prihartato, P. K., & Razzak, S. A. (2021). Wastewater treatment using constructed wetland: *Current trends and future potential. Processes, 9*(11), 1917.

He, C., Liu, Z., Tian, J., & Ma, Q. (2014). Urban expansion dynamics and natural habitat loss in China: A multiscale landscape perspective. *Global Change Biology, 20*(9), 2886–2902. https://doi.org/10.1111/gcb.12553.

Holl, K. D., & Aide, T. M. (2011). When and where to actively restore ecosystems? *Forntiers Ecology Management, 261*(10), 1558–1563. https://doi.org/10.1016/j.foreco.2010.07.004.

Holling, C. S. (1973). Resilience and stability of ecological systems. *Annual Review of Ecology and Systematics*, 1–23.

Hossain, M. L., & Li, J. (2021). NDVI-based vegetation dynamics and its resistance and resilience to different intensities of climatic events. *Global Ecology and Conservation, 30*, e01768 ISSN 2351-9894,.

Howard, J., Sutton-Grier, A., Herr, D., Kleypas, J., Landis, E., Mcleod, E., et al. (2017). Clarifying the role of coastal and marine systems in climate mitigation. *Frontiers in Ecology and the Environment, 15*(1), 42–50.

Hu, S., Niu, Z., Chen, Y., Li, L., & Zhang, H. (2017). Global wetlands: Potential distribution, wetland loss, and status. *The Science of the Total Environment, 15*(586), 319–327.

Isbell, F., Craven, D., Connolly, J., et al. (2015). Biodiversity increases the resistance of ecosystem productivity to climate extremes. *Nature, 526*, 574–577.

Jafarzadeh, H., Mahdianpari, M., Gill, E. W., Brisco, B., & Mohammadimanesh, F. (2022). Remote sensing and machine learning tools to support wetland monitoring: A meta-analysis of three decades of research. *Remote Sensing, 14*(23), 6104.

Jain, C. K., Singhal, D. C., & Sharma, M. K. (2007). Estimating nutrient loadings using chemical mass balance approach. *Environmental Monitoring and Assessment, 134*(1–3), 385–396. https://doi.org/10.1007/s10661-007-9630-5 Epub 2007 Jul 7. PMID: 17616829.

Jarvie, H.P., Pallett, D.W., Schäfer, S.M., Macrae, M.L., Bowes, M.J., Farrand, P., et al. (2020). Biogeochemical and climate drivers of wetland phosphorus and nitrogen release: Implications for nutrient legacies and eutrophication risk. 49(6), 1703–1716).

Jensen, K., & Mcdonald, K. (2019). Surface water microwave product series version 3: A near-real time and 25-year historical global inundated area fraction time series from active and passive microwave remote sensing. *IEEE Geoscience and Remote Sensing Letters, 16*(9), 1402–1406.

Jeziorska, J., & Niedzielski, T. (2018). Applicability of TOPMODEL in the mountainous catchments in the upper Nysa Kłodzka river basin (SW Poland). *Acta Geophysics, 66*, 203–222. https://doi.org/10.1007/s11600-018-0121 6.

Jiang, B., Ouyang, Z. Y., Miao, H., Zheng, H., Bai, Y., Zhuang, C., et al. (2011). Ecosystem services valuation of the Haihe River basin wetlands. *Shengtai Xuebao/Acta Ecologica Sinica, 31*(8), 2236–2244.

Jiang, T. T., Pan, J. F., Pu, X. M., Wang, B., & Pan, J. J. (2015). Current status of coastal wetlands in China: Degradation, restoration, and future management. *Estuarine, Coastal and Shelf Science, 164*, 265–275 ISSN 0272-7714.

Junk, W. J. (2024). World wetlands classification: A new hierarchic hydro-ecological approach. *Wetlands Ecology Management, 32*, 975–1001. https://doi.org/10.1007/s11273-024-10010-7.

Junk, W. J., An, S., Finlayson, C. M., et al. (2013). Current state of knowledge regarding the world's wetlands and their future under global climate change: *A synthesis. Aquat Sci, 75*, 151–167. https://doi.org/10.1007/s00027-012-0278-z.

Keizer-Vlek, H. E., Verdonschot, P. F. M., Verdonschot, R. C. M., & Dekkers, T. B. M. (2014). The contribution of plant uptake to nutrient removal by floating treatment wetlands. *Ecological Engineering, 73*, 684–690.

Keller, R. P., Masoodi, A., & Shackleton, R. T. (2018). The impact of invasive aquatic plants on ecosystem services and human well-being in Wular Lake, India. *Regional Environmental Change, 18*. https://doi.org/10.1007/s10113-017-1232-3.

Kim, S. G. (2010). The evolution of coastal wetland policy in developed countries and Korea. *Ocean & Coastal Management, 53*(9), 562–569.

Kingsford, R. T., Basset, A., & Jackson, L. (2016). Wetlands: Conservation's poor cousins. *Aquatic Conservation: Marine Freshwater. Ecosystemes, 26*, 892–916. https://doi.org/10.1002/aqc.2709.

Kirwan, M., & Megonigal, J. (2013). Tidal wetland stability in the face of human impacts and sea-level rise. *Nature, 504*, 53–60. https://doi.org/10.1038/nature12856.

Kovacic, D. A., David, M. B., Gentry, L. E., Starks, K. M., & Cooke, R. A. (2000). Effectiveness of constructed wetlands in reducing nitrogen and phosphorus export from agricultural tile drainage. *Journal of Environmental Quality, 29*(4), 1262–1274.

Krishnakumar, K., Ali, A., Pereira, B., & Raghavan, R. (2011). Unregulated aquaculture and invasive alien species: A case study of the African Catfish *Clarias gariepinus* in Vembanad Lake (Ramsar wetland), Kerala, India. *Journal of Threatened Taxa, 3*(5), 1737–1744.

Kulik, M., Urban, D., Grzywaczewski, G., Bochniak, A., Grzywna, A., & Sender, J. (2024). Half a century of wetland degradation: The present state and trends of changes in Western Polesie – Long-term wetland degradation. *Global Ecology and Conservation, 56*, e03324.

Kumar, M. D., Bassi, N., Sivamohan, M., & Venkatachalam, L. (2013). Agriculture in West Bengal: Can the new policies trigger a second green revolution? *Review of Development and Change, 18*(1), 37–51.

Lavery, P. S., Mateo, M.-Á., Serrano, O., & Rozaimi, M. (2013). Variability in the carbon storage of seagrass habitats and its implications for global estimates of blue carbon ecosystem service. *PLoS One, 8*(9), e73748.

Lee, L. (2017). Ecological aesthetics perspective for coastal wetland conservation. *Coastal Wetlands: Alteration and Remediation*, 455–478.

Lehner, B., & Döll, P. (2004). Development and validation of a global database of lakes, reservoirs and wetlands. *Journal of Hydrology, 296*(1–4), 1–22.

Leopold, K., Foulkes, M., & Worsfold, P. (2010). Methods for the determination and speciation of mercury in natural waters–a review. *Analytica Chimica Acta, 663*(2), 127–138. https://doi.org/10.1016/j.aca.2010.01.048 Epub 2010 Feb 1. PMID: 20206001.

Le Landes, A. A., Aquilina, L., De Ridder, J., Longuevergne, L., Pag´e, C., & Goderniaux, P. (2014). Investigating the respective impacts of groundwater exploitation and climate change on wetland extension over 150 years. *J. Hydrol. 509*, 367–378.

Li, C., Wang, H., Liao, X., Xiao, R., Liu, K., Bai, J., et al. (2022). Heavy metal pollution in coastal wetlands: A systematic review of studies globally over the past three decades. *Journal of Hazardous Materials, 424*, 127312.

Li, N. Y., Zhong, B., Guo, Y., Li, X. X., Yang, Z., & He, Y. X. (2024). Non-negligible impact of microplastics on wetland ecosystems. *Science of the Total Environment*, 171252.

Li, S., Cui, L., Song, H., Zhang, Y., Gao, C., Guo, J., et al. (2012). Comparison on purification capacity of soil nitrogen and phosphorus in different wetland plants. *Ecology and Environmental Science, 21*(11), 1870–1874.

Li, W., MacBean, N., Ciais, P., Defourny, P., Lamarche, C., Bontemps, S., et al. (2018). Gross and net land cover changes in the main plant functional types derived from the annual ESA CCI land cover maps (1992–2015). *Earth System Science Data, 10*(1), 219–234.

Li, X., Bellerby, R., Craft, C., et al. (2018). Coastal wetland loss, consequences, and challenges for restoration. *Anthropocene Coasts, 1*, 1–15. https://doi.org/10.1139/anc-2017-0001.

Li, Y., Piao, S., Li, L. Z., Chen, A., Wang, X., Ciais, P., et al. (2018). Divergent hydrological response to large-scale afforestation and vegetation greening in China. *Science Advances, 4*(5), eaar4182.

Lin, Q., & Yu, S. (2018). Losses of natural coastal wetlands by land conversion and ecological degradation in the urbanizing Chinese coast. *Scientific Reports, 8*(1), 15046. https://doi.org/10.1038/s41598-018-33406-x 30301927; PMCID: PMC6177474.

Liu, L., He, H., Cai, Y., Hang, J., Liu, J., Liu, L., et al. (2023). Cooling effects of wetland parks in hot and humid areas based on remote sensing images and local climate zone scheme. *Building and Environment, 243*, 110660.

Liu, Q., Liu, J., Liu, H., Liang, L., Cai, Y., Wang, X., et al. (2020). Vegetation dynamics under water-level fluctuations: Implications for wetland restoration. *Journal of Hydrology, 581*, 124418 ISSN 0022–1694.

Liu, S., Zhao, Y., Li, T., Hu, T., Zheng, K., Shen, M., et al. (2023). Removal of micro/nanoplastics in constructed wetland: Efficiency, limitations and perspectives. *Chemical Engineering Journal, 475*, 146033. https://doi.org/10.1016/j.cej.2023.146033.

Liu, X., Qiu, Y., Zheng, Z., Hong, Q., Zhang, Y., Qian, Q., et al. (2022). Spatiotemporal changes in waterfowl habitat suitability in the Caohai Lake wetland and responses to human activities. *Sustainability, 14*(21), 14409.

Llacer, F. J. M. (2005). Panama canal management. *Marine policy, 29*(1), 25–37.

Ma, C., Zhang, G. Y., Zhang, X. C., Zhao, Y. J., & Li, H. Y. (2012). Application of Markov model in wetland change dynamics in Tianjin Coastal Area, China. *Procedia Environmental Sciences, 13*, 252–262.

Maltby, E., & Barker, T. (Eds.). (2009). *The Wetlands Handbook, 2 Volume Set*. John Wiley & Sons.

Mandishona, E. (2021). Rural wetlands use and sustainability: A case study of zindi area in Honde Valley, Zimbabwe (Doctoral dissertation, University of the Witwatersrand, Johannesburg (South Africa)).

Mao, D., Wang, Z., Wu, J., Wu, B., Zeng, Y., Song, K., et al. (2018). China's wetlands loss to urban expansion. *Land Degradation & Development, 29*(8), 2644–2657. https://doi.org/10.1002/ldr.2939.

Massel, S. R., Furukawa, K., & Brinkman, R. M. (1999). Surface wave propagation in mangrove forests. *Fluid Dynamics Reserach, 24*, 219–249.

McCartney, M., Rebelo, L. M., Sellamuttu, S. S., & Silva, S. D. (2010). Wetlands, agriculture and poverty reduction. In *IWMI Research Report 137*. Colombo, Sri Lanka: International Water Management Institute. http://doi.org/10.5337/2010.230.

Mcleod, E., Chmura, G. L., et al. (2011). A blueprint for blue carbon: Toward an improved understanding of the role of vegetated coastal habitats in sequestering CO_2. *Frontiers in Ecology and the Environment, 9*, 552–560. https://doi.org/10.1890/110004.

Measho, S., Chen, B., Pellikka, P., Guo, L., Zhang, H., Cai, D., et al. (2021). Assessment of vegetation dynamics and ecosystem resilience in the context of climate change and drought in the horn of Africa. *Remote Sensing, 13*(9), 1668. https://doi.org/10.3390/rs13091668.

Mitsch, W. J., & Gosselink, J. G. (2015). *Wetlands*. John Wiley & Sons.

Mitsch, W. J. (1994). *Global wetlands: Old world and new*. Amsterdam, The Netherlands: Elsevier Science.

MoEF. (2007). Conservation of wetlands in India: A profile (approach and guidelines). MoEF, Government of India, New Delhi.

Möller, I., Kudella, M., Rupprecht, F., Spencer, T., Paul, M., van Wesenbeeck, B. K., et al. (2014). Wave attenuation over coastal salt marshes under storm surge conditions. *Nature Geoscience, 7*, 727–731. https://doi.org/10.1038/ngeo2251.

Moreno-Mateos, D., Meli, P., Vara-Rodríguez, M. I., & Aronson, J. (2015). Ecosystem response to interventions: Lessons from restored and created wetland ecosystems. *The Journal of Applied Ecology, 52*(6), 1528–1537. https://doi.org/10.1111/1365-2664.12518.

Mudd, D'Alpaos, S. M., & Morris, J. T. (2010). How does vegetation affect sedimentation on tidal marshes? Investigating particle capture and hydrodynamic controls on biologically mediated sedimentation. *Journal of Geophysical Research, 115*, F03029.

Nabuyanda, M.M. (2024). Investigating the fate of Cobalt, Copper, and Lead in Sediments, Phragmites mauritianus and Typha spp. from three Zambian Wetlands (Doctoral dissertation, Wageningen University and Research).

Navarro, N., & Rodríguez-Santalla, I. (2023). Coastal Wetlands. *Journal of Marine Science and Engineering, 11*(4), 767.

Nguyen, H. H., Dargusch, P., & Moss, P. (2016). A review of the drivers of 200 years of wetland degradation in the Mekong Delta of Vietnam. *Regional Environmental Change, 16*. https://doi.org/10.1007/s10113-016-0941-3.

Nigam, M., Mishra, P., Kumar, P., Rajoriya, S., Pathak, P., Singh, S. R., et al. (2023). Comprehensive technological assessment for different treatment methods of leather tannery wastewater. *Environmental Science and Pollution Research, 30*(60), 124686–124703.

Nilsson, C., Reidy, C. A., Dynesius, M., & Revenga, C. (2005). Fragmentation and flow regulation of the world's large river systems. *Science (New York, N.Y.), 308*(5720), 405–408. https://doi.org/10.1126/science.1107887 PMID: 15831757.

Ninan, K. N. (Ed.). (2014). *Valuing ecosystem services: Methodological issues and case studies* (pp. 427). Edward Elgar Publishing Limited.

Nyieku, F. E., Essandoh, H. M. K., Armah, F. A., et al. (2021). Environmental conditions and the performance of free water surface flow constructed wetland: A multivariate statistical approach. *Wetlands Ecology and Management, 29*, 381–395.

Ogundele, L. T., & Ayeku, P. O. (2020). Source apportionment and associated potential ecological risk assessment of heavy metals in coastal marine sediments samples in Ondo, Southwest, Nigeria. Stoch. *Stochastic Environmental Research and Risk Assessment, 34*, 2013–2022.

Olefeldt, D., Hovemyr, M., et al. (2021). The boreal–arctic wetland and lake dataset (BAWLD). *Earth System Science Data, 13*, 5127–5149. https://doi.org/10.5194/essd-13-5127-2021.

Oliver, T., Isaac, N., August, T., et al. (2015). Declining resilience of ecosystem functions under biodiversity loss. *Nature Communication, 6*, 10122. https://doi.org/10.1038/ncomms10122.

Omondi, D. O., & Navalia, A. C. (2020). *Constructed wetlands in wastewater treatment and challenges of emerging resistant genes filtration and reloading*. IntechOpen.

Orihuela-Torres, A., Pérez-García, J. M., Arrondo, E., Pessano-Serrat, T., Green, A. J., Naves-Alegre, L., et al. (2025). Vertebrate scavenger assemblages and their functioning differ between artificial and natural wetlands: Implications for ecosystem management. *Biological Conservation, 302*, 110929.

Padmavathi, P., & Srinu, G. (2017). Wetlands of India: Biodiversity, ecological services and strategies for conservation. In Sagar A. Vhanalakar, & Sharadrao A. Vhanalakar (Eds.). *Biodiversity assessment: Tool for conservation* (pp. 189–204). Bhumi Publishing 416207.

Pandey, P. C., Srivastava, P. K., & Srivastava, S. K. (Eds.). (2024). *Aquatic ecosystems monitoring: Conventional assessment to advanced remote sensing*. CRC Press.

Pant, N., & Verma, R.K. (2010). Tanks in Eastern India: A study in exploration. Hyderabad, India: International Water Management Institute, IWMI-TATA Water Policy Research Program; Lucknow, Uttar Pradesh, India: Centre for Development Studies. 215p.

Parvez, M. T., & Mohsin, A. B. M. (2022). Expansion of aquaculture threatens the existence of wetlands in Bangladesh. *Journal of Fisheries, 10*(2), 102208. https://doi.org/10.17017/j.fish.449.

Pataki, D. E., Carreiro, M. M., et al. (2011). Coupling biogeochemical cycles in urban environments: Ecosystem services, green solutions, and misconceptions. *Frontiers in Ecology and the Environment, 9*, 27–36. https://doi.org/10.1890/090220.

Paudel, A., Richardson, M., & King, D. (2023). Vegetation structure and composition in small forested wetlands and their associations with water table dynamics. *Wetlands, 43*, 82. https://doi.org/10.1007/s13157-023-01729-9.

Pekel, J. F., Cottam, A., Gorelick, N., et al. (2016). High-resolution mapping of global surface water and its long term changes. *Nature, 540*, 418–422.

Pendleton, L., Donato, D. C., Murray, B. C., et al. (2012). Estimating global "blue carbon" emissions from conversion and degradation of vegetated coastal ecosystems. *PLoS One, 7*, e43542.

Poffenbarger, H. J., Needelman, B. A., & Megonigal, J. P. (2011). Salinity influence on methane emissions from tidal marshes. *Wetlands, 31*(5), 831–842. https://doi.org/10.1007/s13157-011-0197-0.

Prasad, R., Jhariya, M. K., & Banerjee, A. (Eds.). (2021). *Advances in sustainable development and management of environmental and natural resources: Economic outlook and opinions, 2-volume set*. CRC Press.

Prasad, S. N., Ramachandra, T. V., Ahalya, N., Sengupta, T., Kumar, A., Tiwari, A. K., et al. (2002). Conservation of wetlands of India-a review. *Tropical Ecology, 43*(1), 173–186.

Prigent, C., Jimenez, C., & Bousquet, P. (2020). Satellite-derived global surface water extent and dynamics over the last 25 years (GIEMS-2). *Journal of Geophysical Research: Atmospheres, 125*, e2019JD030711. https://doi.org/10.1029/2019JD030711.

Ragavan, P., Kathiresan, K., Mohan, P. M., Ravichandran, K., Jayaraj, R. S. C., & Rana, T. S. (2021). Ensuring the adaptive potential of Coastal wetlands of India-the need of the hour for sustainable management. *Wetlands Ecology and Management, 29*, 641–652.

Ramı´rez, F., & Santana, J. (2019). *Environmental education and ecotourism. Springer*, 59. http://doi.org/10.1007/978-3- 030-01968-6.

Ramsar (2012). Destination wetlands: Supporting sustainable tourism. In *Secretariat of the Ramsar Convention on Wetlands*. Gland, Switzerland, & World Tourism Organization (UNWTO).

Ramsar. (2016). Ramsar handbook: The fourth ramsar strategic plan 2016–2024 (Fifth ed., Vols. Sub-series I: Handbook 2, International Cooperation on Wetlands). Gland: RamsarSecretariat. https://www.ramsar.org/sites/default/files/documents/library/hb2_5ed_strategic_plan_2016_24_e.pdf.

Ramsar Convention. (2008). Resolution X.3 The Changwon Declaration on human well-being and wetlands. In '10th Meeting of the Conference of the Parties to the Convention on Wetlands', 28October–4 November 2008, Changwon, Republic of Korea. (Ramsar Convention on Wetlands: Gland, Switzerland) Available at https://www.ramsar.org/sites/default/files/documents/pdf/res/key_res_x_03_e.pd.

Ramsar Convention. (2015). Wetlands for our future: Act now to prevent, stop, and reservewetland loss; ramsar convention bureau: Gland, Switzerland, 2015; Available from:https://www.ramsar.org/news/press-releasewetlands-for-our-future-act-now-to-prevent-stop-and-reverse-wetland-loss.

Ramsar Convention. (2018). Global wetland outlook: State of the world's wetlands and their services to people. ramsar convention secretariat, Gland, Switzerland.

Reddy, K. R., Newman, S., Osborne, T. Z., White, J. R., & Fitz, H. C. (2011). Phosphorous cycling in the greater everglades ecosystem: legacy phosphorous implications for management and restoration. *Critical Reviews in Environmental Science and Technology, 41*(sup1), 149–186. https://doi.org/10.1080/10643389.2010.530932.

Regnier, P., Resplandy, L., Najjar, R. G., et al. (2022). The land-to-ocean loops of the global carbon cycle. *Nature, 603*, 401–410. https://doi.org/10.1038/s41586-021-04339-9.

Reid, A. J., Carlson, A. K., et al. (2019). Emerging threats and persistent conservation challenges for freshwater biodiversity. *Biological Reviews of the Cambridge Philosophical Society, 94*(3), 849–873. https://doi.org/10.1111/brv.12480.

Reithmaier, G., et al. (2024). Small wetland-fringed estuaries deliver disproportionately large carbon loads to the ocean. *Limnology and Oceanography, 69*, 2229–2242.

Reithmaier, G. M. S., et al. (2023). Carbonate chemistry and carbon sequestration driven by inorganic carbon outwelling from mangroves and saltmarshes. *Nature Communication, 14*, 8196. https://doi.org/10.1038/s41467-023-44037-w.

Revenga, C. B. J., Henninger, N., Kassem, K., & Payne, R. (Eds.). (2000). *Pilot analysis of global ecosystems: Freshwater systems* (pp. 1–78). World Resources Institute.

Richards, D. R., & Friess, D. A. (2016). Rates and drivers of mangrove deforestation in Southeast Asia, 2000–2012. *Proceedings of the National Academy of Sciences of the United States of America, 113*(2), 344–349.

Rupprecht, F., et al. (2017). Vegetation-wave interactions in salt marshes under storm surge conditions. *Ecological Engineering, 100*, 301–315. https://doi.org/10.1016/j.ecoleng.2016.12.030.

Russi, D., ten, Brink, P., Farmer, A., Badura, T., Coates, D., et al. (2013). *The economics of ecosystems and biodiversity for water and wetlands.* , UK: IEEP.

Salimi, S., Almuktar, S., & Scholz, M. (2021). Impact of climate change on wetland ecosystems: A critical review of experimental wetlands. *Journal of Environmental Management, 286*, 112160. https://doi.org/10.1016/j.jenvman.2021.112160 ISSN 0301-4797.

Sandi, S. G., et al. (2020). Resilience to drought of dryland wetlands threatened by climate change. *Scientific Reports, 10*(1), 13232. https://doi.org/10.1038/s41598-020-70087-x 32764646; PMCID: PMC7414121.

Sarkar, U. K., et al. (2020b). Fisheries management of a sewage fed Gangetic floodplain wetland in the context of changing climate: A case study using multiple approaches. *Journal of the Inland Fisheries Society of India, 52*(1), 28–41.

Sarkar, U. K., Mishal, P., et al. (2020a). Status, potential, prospects, and issues of floodplain wetland fisheries in India: Synthesis and review for sustainable management. *Reviews in Fisheries Science & Aquaculture, 29*(1), 1–32.

Sarkar, U. K., & Borah, B. C. (2018). Flood plain wetland fisheries of India: With special reference to impact of climate change. *Wetlands Ecology and Management, 26*. https://doi.org/10.1007/s11273-017-9559-6.

Sauer, J., et al. (2024). Estimating combined effects of climate change and land cover change on water regulation services of urban wetlands in Valdivia, Chile. *Earth's Future, 12*, e2023EF003801. https://doi.org/10.1029/2023EF003801.

Schwarz, A. E., Lensen, S. M. C., Langeveld, E., Parker, L. A., & Urbanus, J. H. (2023). Plastics in the global environment assessed through material flow analysis, degradation and environmental transportation. *Science of The Total Environment, 875*, 162644. https://doi.org/10.1016/j.scitotenv.2023.162644 ISSN 0048-9697.

Scott, D., Frail-Gauthier, J., & Mudie, P. (2014). *Coastal wetlands of the world: Geology, ecology, distribution and applications*. https://doi.org/10.1017/CBO9781107296916.

Sial, T. A., Teewno, A. M., Memon, S. A., Mahar, R. B., & Korai, M. S. (2023). Municipal solid waste landfill leachate treatment by *Phragmites australis*, *Typha latifolia* and *Scirpus validus* through constructed wetlands. *Journal of Ecological Engineering, 24*(6), 303–314. https://doi.org/10.12911/22998993/162653.

Singh, R., Yusuf, M., & Shahid, M. (2025). *Wetland chemistry: Limnological, bio-geochemical, and economic perspectives*. Jenny Stanford Publishing.

Stefanakis, A., Akratos, C. S., & Tsihrintzis, V. A. (2014). Vertical flow constructed wetlands: Eco-engineering systems for wastewater and sludge treatment. In *Newnes. Elsevier*.

Stikker, A. (1998). Water today and tomorrow: Prospects for overcoming scarcity. *Futures, 30*, 43–62.

Strassburg, B. B. N., Iribarrem, A., Beyer, H. L., et al. (2020). Global priority areas for ecosystem restoration. *Nature, 586*, 724–729. https://doi.org/10.1038/s41586-020-2784-9.

Subcommittee, W. (2013). Classification of wetlands and deepwater habitats of the United States. Federal Geographic Data Committee: Reston, VA, USA, 1–91.

Summers, D. M., Bryan, B. A., Nolan, M., & Hobbs, T. J. (2015). The costs of reforestation: A spatial model of the costs of establishing environmental and carbon plantings. *Land Use Policy, 44*, 110–121. https://doi.org/10.1016/j.landusepol.2014.12.002.

Sun, J., et al. (2019). Microplastics in wastewater treatment plants: Detection, occurrence and removal. *Water Research, 152*, 21–37.

Sun, K., & Yu, W. (2024). A satellite view of the wetland transformation path and associated drivers in the greater bay area of china during the past four decades. *Remote Sensing, 16*(6), 1047. https://doi.org/10.3390/rs16061047.

Tanaka, N., Sasaki, Y., Mowjood, M., Jinadasa, K., & Homchuen, S. (2007). Coastal vegetation structures and their function in tsunami protection: experience of the recent Indian Ocean tsunami. *Landsc. Ecol. Eng. 3*, 33–45.

Thenkabail, P. S. (Ed.). (2024). *Remote sensing handbook, volume V: Water, hydrology, floods, snow and ice, wetlands, and water productivity*. CRC Press.

Theresa, A., et al. (2021). Considering coasts: Adapting terrestrial models to characterize coastal wetland ecosystems. *Ecological Modelling, 450*, 109561.

Thongtha, S., Teamkao, P., et al. (2014). Phosphorus removal from domestic wastewater by *Nelumbo nucifera* Gaertn. and *Cyperus alternifolius* L. *Journal of Environmental Management, 137*, 54–60. https://doi.org/10.1016/j.jenvman.2014.02.003 Epub 2014 Mar 4. PMID: 24603027.

Thorslund, J., Jarsjo, J., Jaramillo, F., Jawitz, J. W., Manzoni, S., Basu, N. B., et al. (2017). Wetlands as large-scale nature-based solutions: Status and challenges for research, engineering and management. *Ecological Engineering, 108*, 489–497.

Tiner, R. W. (2016). *Wetland indicators: A guide to wetland formation, identification, delineation, classification, and mapping*. CRC press.

Tolhurst, T. J., Gust, G., & Paterson, D. M. (2002). The influence of an extracellular polymeric substance (EPS) on cohesive sediment stability. *Proceedings in Marine Science, 5*, 409–425.

Tomscha, S., et al. (2021). Multiple methods confirm wetland restoration improves ecosystem services. *Ecosystems. People, 17*(1), 25–40. https://doi.org/10.1080/26395916.2020.1863266.

Tootchi, A., Jost, A., & Ducharne, A. (2019). Multi-source global wetland maps combining surface water imagery and groundwater constraints. *Earth Systems Sciences Data, 11*, 189–220. https://doi.org/10.5194/essd-11-189-2019.

Trakala, G., Tsiroukis, A., & Martinis, A. (2023). Eco-Cultural development of a restored lake environment: The case study of lake Karla (Thessaly, Greece). *Land, 12*(6), 1227.

Turetsky, M., Benscoter, B., Page, S., et al. (2015). Global vulnerability of peatlands to fire and carbon loss. *Nature Geosciences, 8*. https://doi.org/10.1038/ngeo2325.

Turner, R. E., & Ohimain, E. I. (2024). Dredged canals, wetland loss, and legacy. *Estuaries and Coasts, 47*(8), 2147–2159.

Turner, R. E., Kearney, M. S., & Parkinson, R. W. (2018). Sea level rise tipping point of delta survival. *Journal of Coastal Research, 34*, 470–474.

Turner, R. K., Georgiou, S., & Fisher, B. (2008). *Valuing Ecosystem Services. The Case of Multifunctional Wetlands*. Earthscan ISBN 978-1-84407-615-4.

Verhoeven, J.T.A. (2009). Wetland Biogeochemical Cycles and their interactions, 266 pp.

Verschuuren, B. (2016). Religious and spiritual aspects of wetland management. 10.1007/978-94-007-6172-8_242-2.

Vymazal, J., & Březinová, T. (2015). The use of constructed wetlands for removal of pesticides.

Vymazal, J., Zhao, Y., & Mander, U. (2021). Recent research challenges in constructed wetlands for wastewater treatment: A review. *Ecological Engineering, 169*, 106318. https://doi.org/10.1016/j.ecoleng.2021.106318 ISSN 0925-8574.

Wan, Z., Su, Y., Li, Z., Zhang, X., Zhang, Q., & Chen, J. (2023). Analysis of the impact of Suez Canal blockage on the global shipping network. *Ocean & Coastal Management, 245*, 106868. https://doi.org/10.1016/j.ocecoaman.2023.106868.

Wang, F., et al. (2018). Interaction of toxic chemicals with microplastics: A critical review. *Water Research, 1*(139), 208–219. https://doi.org/10.1016/j.watres.2018.04.003 Epub 2018 Apr 6. PMID: 29653356.

Wang, X., Yan, F., & Su, F. (2020). Impacts of urbanization on the ecosystem services in the Guangdong-Hong Kong-Macao greater bay area, China. *Remote Sens, 12*(19), 3269. https://doi.org/10.3390/rs12193269.

Wang, X., Zhang, D., Qi, Q., Tong, S., An, Y., Lu, X., et al. (2019). The restoration feasibility of degraded *Carex* Tussock in soda-salinization area in arid region. *Ecological Indicators, 98*, 131–136.

Wang, Y., Fukuda, H., et al. (2022). Urban wetlands as a potential habitat for an endangered aquatic plant, Isoetes sinensis. *Global Ecology and Conservation, 34*, e02012. https://doi.org/10.1016/j.gecco.2022.e02012.

Weng, Q. (2009). Thermal infrared remote sensing for urban climate and environmental studies: Methods, applications, and trends. *ISPRS Journal of Photogrammetry and Remote Sensing, 64*(4), 335–344. https://doi.org/10.1016/j.isprsjprs.2009.03.007.

White, M. P., Elliott, L. R., Gascon, M., Roberts, B., & Fleming, L. E. (2020). Blue space, health and well-being: A narrative overview and synthesis of potential benefits. *Environmental Research, 191*, 110169. https://doi.org/10.1016/j.envres.2020.110169.

William, J. M., & James, G. G. (2000). *Wetlands*. WILEY Press, 107–155.

Wood, K. A., et al. (2024). A global systematic review of the cultural ecosystem services provided by wetlands. *Ecosystem Services, 70*, 101673. https://doi.org/10.1016/j.ecoser.2024.101673.

Wu, J., Liu, Q., Liang, L., Li, S., Gan, L., & Wu, H. (2024). Alteration of wetland resilience for the intermittently and permanently inundated wetland. *Environmental Research Letters, 19*(12), 124077.

Wu, S., et al. (2021). The effects of the cooling efficiency of urban wetlands in an inland megacity: A case study of Chengdu, Southwest China. *Building and Environment, 204*, 108128.

Wu, W., & Wang, S. S. (2006). Formulas for sediment porosity and settling velocity. *Journal of Hydraulic Engineering, 132*(8), 858–862.

Wu, Y., et al. (2023a). Wetland mitigation functions on hydrological droughts: From drought characteristics to propagation of meteorological droughts to hydrological droughts. *Journal of Hydrology, 617*, 128971.

Wu, Y., Sun, J., Hu, B., Zhang, G., & Rousseau, A. N. (2023b). Wetland-based solutions against extreme flood and severe drought: Efficiency evaluation of risk mitigation. *Climate Risk Management, 40*, 100505.

Wu, Y., Zhang, G., Rousseau, A. N., & Xu, Y. J. (2020a). Quantifying streamflow regulation services of wetlands with an emphasis on quickflow and baseflow responses in the Upper Nenjiang River Basin, Northeast China. *Journal of Hydrology, 583*, 124565. https://doi.org/10.1016/j.jhydrol.2020.124565.

Wu, Y., Zhang, G., Rousseau, A. N., Xu, Y. J., & Foulon, E. (2020b). On how wetlands can provide flood resilience in a large river basin: A case study in Nenjiang River Basin, China. *Journal of Hydrology, 587*, 125012. https://doi.org/10.1016/j.jhydrol.2020.125012.

Xi, Y., Peng, S., Ciais, P., & Chen, Y. (2021). Future impacts of climate change on inland Ramsar wetlands. *Nature Climate Change. 11*, 45–51.

Xi, Y., Peng, S., Liu, G., et al. (2022). Trade-off between tree planting and wetland conservation in China. *Nature Communication, 13*, 1967. https://doi.org/10.1038/s41467-022-29616-7.

Xia, B., et al. (2022). Secondary PVC microplastics are more toxic than primary PVC microplastics to Oryzias melastigma embryos. *Journal of Hazardous Materials, 424*(Pt B), 127421. https://doi.org/10.1016/j.jhazmat.2021.127421.

Xiao, H., et al. (2024). Global wetland methane emissions from 2001 to 2020: Magnitude, dynamics and controls. *Earth's Future, 12*(9), e2024EF004794.

Xin, P., et al. (2022). Surface water and groundwater interactions in salt marshes and their impact on plant ecology and coastal biogeochemistry. *Reviews of Geophysics, 60*, e2021RG000740. https://doi.org/10.1029/2021RG000740.

Xin, Y., Rui, W., & Zhenguo, N. (2022). Response of China's wetland NDVI to climate changes. *Wetlands, 42*, 55. https://doi.org/10.1007/s13157-022-01568-0.

Xing, W., Han, Y., Guo, Z., & Zhou, Y. (2020). Quantitative study on redistribution of nitrogen and phosphorus by wetland plants under different water quality conditions. *Environmental Pollution, 261*, 114086.

Xu, J., Morris, P. J., Liu, J., & Holden, J. (2018). PEATMAP: Refining estimates of global peatland distribution based on a meta-analysis. *Catena, 160*, 134–140.

Xu, N., et al. (2024). Wetland distribution prediction based on CA–Markov model under current land use and protection policy in sanjiang plain. *Sustainability, 16*(13), 5750. https://doi.org/10.3390/su16135750.

Xu, T., Weng, B., Yan, D., Wang, K., Li, X., Bi, W., et al. (2019). Wetlands of international importance: Status, threats, and future protection. *International Journal of Environmental Research and Public Health, 16*(10), 1818.

Xu, X., Chen, M., Yang, G., Jiang, B., & Zhang, J. (2020). Wetland ecosystem services research: A critical review. *Global Ecology and Conservation, 22*, e01027.

Xueru, Z., Xiao, J., & Yue, Q. (2018). Ecological land cold island effect evaluation based on land surface temperature retrieval. *In: 2018 7th International Conference on Agro- Geoinformatics. Agro-Geoinformatics*, 1–4.

Yadav, S., & Goyal, V. C. (2022). Current status of ponds in India: A framework for restoration, policies and circular economy. *Wetlands, 42*, 107. https://doi.org/10.1007/s13157-022-01624-9.

Yang, M., Guo, B., Gao, N., Yu, Y., Song, X., & Gu, Y. (2025). A prediction of Estuary wetland vegetation with satellite images. *Journal of Marine Science and Engineering, 13*(2), 287. https://doi.org/10.3390/jmse13020287.

Yang, S. L., Shi, B. W., et al. (2012). Wave attenuation at a salt marsh margin: A case study of an exposed coast on the Yangtze Estuary. *Estuaries and Coasts, 35*, 169–182.

Ye, C., Lin, J., Li, Z., Wang, G., & Li, Z. (2024). Characteristics of microplastic pollution in agricultural soils in xiangtan, China. *Sustainability, 16*(17), 7254.

Yi, Q., Huixin, G., Yaomin, Z., et al. (2024). Global conservation priorities for wetlands and setting post-2025 targets. *Communication of Earth Environmental, 5*. https://doi.org/10.1038/s43247-023-01195-5.

Yu, B., Zang, Y., Wu, C., & Zhao, Z. (2024). Spatiotemporal dynamics of wetlands and their future multi-scenario simulation in the Yellow River Delta, China. *Journal of Environmental Management, 353*, 120193. https://doi.org/10.1016/j.jenvman.2024.120193 27.

Yurtsever, M. (2019). Glitters as a source of primary microplastics: An approach to environmental responsibility and ethics. *Journal of Agricultural and Environmental Ethics, 32*. https://doi.org/10.1007/s10806-019-09785-0.

Zedler, J. B., & Kercher, S. (2005). Wetland resources: Status, trends, ecosystem services, and restorability. Annual Review of Environment and Resources, *30*, 39–74.

Zhang, D., et al. (2023a). Ecological and photosynthetic characteristics of Carex schmidtii with respect to hydrological fluctuations. *Journal of Freshwater Ecology, 38*(1), 2271502.

Zhang, H., Yin, A., Yang, X., Fan, M., Shao, S., Wu, J., et al. (2021). Use of machine-learning and receptor models for prediction and source apportionment of heavy metals in coastal reclaimed soils. *Ecological Indicators, 122*, 107233. https://doi.org/10.1016/j.ecolind.2020.107233 ISSN 1470-160X.

Zhang, K. Q., Liu, H. Q., & Li, Y. P. (2012). The role of mangroves in attenuating storm surges. *Estuarine, Coastal and Shelf Science. 102–103*, 11–23. https://doi.org/10.1016/j.ecss.2012.02.021.

Zhang, S., Li, T., Xie, H., Song, M., Huang, S., Guo, Z., et al. (2024). The crucial factor for microplastics removal in large-scale subsurface-flow constructed wetlands. *Journal of Hazardous Materials, 480*, 136023.

Zhang, X., & Song, Y. (2014). Optimization of wetland restoration siting and zoning in flood retention areas of river basins in China: A case study in Mengwa, Huaihe River Basin. *Journal of Hydrology, 519*, 80–93. https://doi.org/10.1016/j.jhydrol.2014.06.043.

Zhang, X., Li, C., Liu, X., Zhao, Y., Wang, Y., Zhang, Y., et al. (2025). Distribution characteristics and sources of microplastics in inland wetland ecosystem soils. *Water, 17*, 231. https://doi.org/10.3390/w17020231.

Zhang, Y., et al. (2017). Global loss of aquatic vegetation in lakes. *Earth-Science Reviews, 173*, 259–265. https://doi.org/10.1016/j.earscirev.2017.08.013.

Zhang, Z., et al. (2023b). Development of the global dataset of wetland area and dynamics for methane modeling (WAD2M). *Earth System Science Data, 13*, 2001–2023.

Zhang, Z., Zimmermann, N. E., Kaplan, J. O., & Poulter, B. (2016). Modeling spatiotemporal dynamics of global wetlands: Comprehensive evaluation of a new sub-grid TOPMODEL parameterization and uncertainties. *Biogeosciences, 13*(5), 1387–1408.

Zhao, L., Gao, H., Liu, J., Wang, F., & Fu, T. (2024). Modeling, assessment, and prediction of carbon storage in hebei–tianjin coastal wetlands. *Remote Sensing, 16*(23), 4428.

Zheng, J., & Suh, S. (2019). Strategies to reduce the global carbon footprint of plastics. *Nature Climate Change, 9*, 374–378. https://doi.org/10.1038/s41558-019-0459-z.

Zhong, L., et al. (2023). Recent advances towards micro (nano) plastics research in wetland ecosystems: A systematic review on *sources, removal, and ecological impacts. Journal of Hazardous Materials, 452*, 131341. 15 https://doi.org/10.1016/j.jhazmat.2023.131341, 37023576.

Zhu, Z. (2017). Change detection using landsat time series: A review of frequencies, preprocessing, algorithms, and applications. *ISPRS Journal of Photogrammetry and Remote Sensing, 130*, 370–384.

Zotos, A., et al. (2024). Wetlands in crisis: The silent desertification threat on the greek wetlands. *Land, 13*(10), 1567. https://doi.org/10.3390/land13101567.

Zou, Z., et al. (2021). Characterizing wetland inundation and vegetation dynamics in the arctic coastal plain using recent satellite data and field photos. *Remote Sensing, 13*(8), 1492. https://doi.org/10.3390/rs13081492.

Chapter 2

Wetlands and water quality: ecological functions, threats, and conservation strategies

Gaurav Rawat[1], Rashi Miglani[2], Ankit Kumar[3], Nagma Parveen[2], Uzma Siddiqui[1], Seeta Dewali[1], Satpal Singh Bisht[4], Harish Chandra Singh Bisht[1], and Inderjeet Tyagi[5]

[1]*Department of Zoology, D.S.B Campus, Kumaun University, Nainital, Uttarakhand, India,* [2]*Department of Forensic Science, D.S.B. Campus, Kumaun University, Nainital, Uttarakhand, India,* [3]*Department of Pharmaceutical Sciences, Sir J.C. Bose Technical Campus, Bhimtal, Uttarakhand, India,* [4]*Vice Chancellor, Soban Singh Jeena University, Almora, Uttarakhand, India,* [5]*Centre for DNA Taxonomy, Molecular Systematics Division, Zoological Survey of India, Kolkata, West Bengal, India*

2.1 Introduction

Water resources are vital yet fragile components of the environment, playing a crucial role in the growth and development of human societies (Mishra et al., 2021; Molnar et al., 2010). Wetlands, in particular, are recognized as highly productive and valuable ecosystems that are integral to effective water resource management (Bassi et al., 2014; Sharma & Naik, 2024). A wetland or transition zone is generally a land area that is either fully submerged in water or entirely flooded, either year-round or during particular seasons (Balwan & Kour, 2021). As defined by the Ramsar Convention Secretariat (2016), wetlands are "areas of marsh, fen, peatland, or water, whether natural or artificial, permanent or temporary, containing static or flowing water, which may be fresh, brackish, or saline, including marine regions with a depth not exceeding six meters at low tide." More than 95% of the estimated value of wetland ecosystem services is attributed to functions such as nutrient cycling, food production (including storage and carbon fixation), flood mitigation, water retention, enhancement of water quality through purification and contaminant retention, recreation, provision of habitat and shelter for diverse flora and fauna, conservation of migratory birds, educational and cultural significance, and carbon sequestration (Ferreira et al., 2023; Kingsford et al., 2016).

The global population has reached 8.01 billion in 2025 (World Population Review, 2025). Yet, freshwater, limited to just 2.5% of Earth's total water, has declined further due to the growing impacts of climate change and pollution (Zhang et al., 2023). Water pollution results from a combination of anthropogenic and natural processes. Human-induced factors include the excessive application of fertilizers, accelerated urbanization, industrial expansion, and rapid population growth, whereas natural processes primarily involve atmospheric deposition, geogenic leaching, mining activities, and rock weathering (Patil et al., 2012; Verma et al., 2025). Inorganic contaminants, such as heavy metals (Pb, Cd, Cu) and ions (NO_3^-, PO_4^{3-}, SO_4^{2-}), account for a significant portion of water pollution, surpassing that of organic pollutants (Rahmanian et al., 2015).

Wetlands enhance water quality through pollutant removal, nutrient cycling, and sediment retention. By slowing the flow of water, they promote sedimentation and reduce turbidity. Nitrogen is primarily removed through microbial nitrification–denitrification processes, while phosphorus is retained via plant assimilation and adsorption to substrate surfaces (Deslippe & Bentley, 2025; Rani et al., 2024; Wang et al., 2022). Additionally, wetlands mitigate contaminants, including heavy metals and organic micropollutants, through mechanisms such as substrate adsorption, biosorption, and microbial degradation (Kushwaha et al., 2024). Moreover, they regulate surface runoff and promote groundwater recharge, thereby enhancing natural subsurface filtration and providing an effective buffer against flooding (Ferreira et al., 2023; Nepal et al., 2024).

Wetlands: Biodiversity, Health and their Role in Ecosystem Services. DOI: https://doi.org/10.1016/B978-0-443-34222-6.00011-5

Keeping in view the ecological importance and role of wetlands in maintaining the water quality, in this chapter, an attempt has been made to elucidate the importance of water quality indicators (physical, chemical, and biological) for long-term monitoring thereby playing a significant role in the conservation and wise-use of wetlands. Additionally, this chapter will provide insights about different ecological roles such as nutrient cycling, retention, groundwater recharge, and water and wastewater treatment.

2.2 Ecological functions of wetlands

Wetlands are ecologically essential ecosystems that perform a variety of functions that are crucial to the preservation of hydrological and biogeochemical equilibrium. The essence of nature-based climate resilience strategies is their capacity to regulate water, filter pollutants, recharge groundwater, and mitigate flooding. Acting as biogeochemical hot spots, wetlands contribute to both ecosystem health and human well-being (Fig. 2.1).

2.2.1 Water filtration and purification

Wetlands play a vital role in water and wastewater treatment and are often referred to as "Kidneys of the Earth." Through a series of different synergistic processes, namely, sedimentation, filtration, substrate adsorption, biosorption, and interaction between plant and microbe, the water quality of wetlands shows significant improvement (Haberl et al., 2003; Hatvani et al., 2022; Kujala et al., 2019; Kumar, 2022; Préau et al., 2022). Additionally, the microbial degradation of organic debris and metabolic transformation of toxic compounds into nontoxic end products or other beneficial metabolites further contribute to the improved water quality in the wetland ecosystem (Weber, 2016). Ferreira et al. (2023) concluded that in wetlands with dense vegetation, the water flow gets slowed down, thereby promoting sedimentation and the settling of particulates (Ferreira et al., 2023). This also facilitates the uptake of the nutrients and their retention in agricultural crops and plant species commonly found near wetlands (Moreno-Mateos & Comin, 2010).

The efficiency of pollutant removal in wetlands is influenced by multiple factors such as vegetation type, substrate, nature of pollutants, and water flow regime (surface or subsurface), as they govern the key physical, chemical, and biological processes. Additionally, climatic variations alter the aerobic and anaerobic conditions, thereby driving different pollutant removal mechanisms. (Ferreira et al., 2023).

2.2.2 Biogeochemical cycling

The complex interaction of different processes, namely, hydrological, mineral transformations, microbial community structure, vegetation, and soil in wetlands, makes it an ideal ecosystem for maintaining the biogeochemical cycles

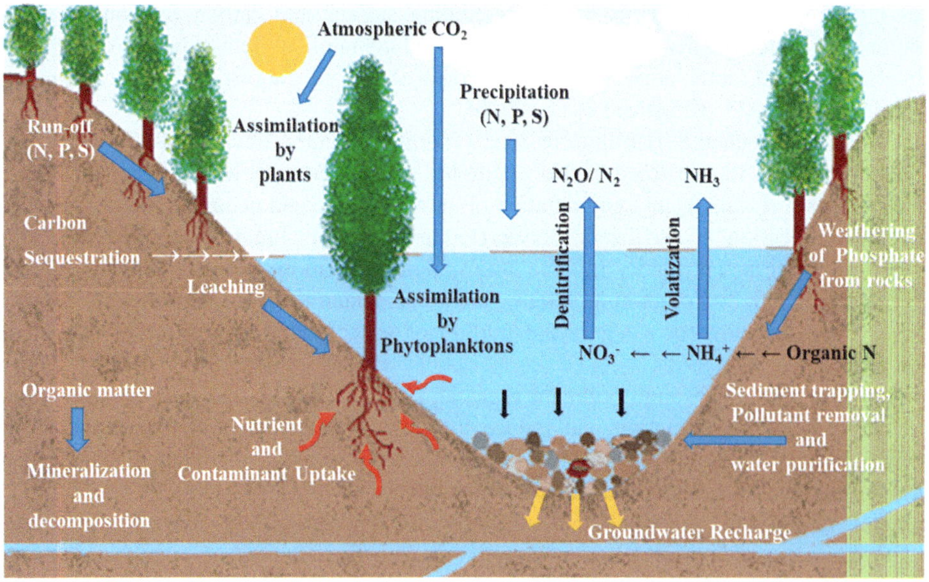

FIGURE 2.1 Biogeochemical cycling in wetlands and associated ecological functions.

(Rezanezhad et al., 2020). Wetlands act as sources and sinks for carbon sequestration, greenhouse gas (GHG) emissions, and nitrogen assimilation via significant metabolic pathways (Everard, 2016; Ma et al., 2025). The excess agricultural nitrogen within these ecosystems is managed through a series of processes, including nitrogen fixation, nitrification, denitrification, and ammonification (Everard, 2016). Additionally, processes such as redox gradients (interactions of ions such as iron, sulfur, carbon, nitrogen, and heavy metals), soil adsorption, and precipitation (influencing nutrient bioavailability) play a significant role in deciphering the fate of contaminants and nutrient transformations, thereby affecting wetland productivity (Santoro et al., 2024; Wu et al., 2019).

2.2.3 Groundwater recharge and flow regulation

Wetlands are crucial for groundwater recharge, particularly in regions where permeable substrates facilitate downward infiltration. Subsurface lateral flow and SW-GW (surface water-groundwater) interactions facilitate percolation, which replenishes aquifers and preserves the hydrology of wetlands (Wu et al., 2020). Globally, groundwater comprises nearly 70% of the total freshwater supply (Gleeson et al., 2016) and it is rapidly deteriorating, with hundreds of cubic kilometers extracted annually (Sutanudjaja et al., 2018). The recharge rates can be increased by 10%–30% in wetland-rich catchments in comparison to areas with less or no wetland extent (Trásy-Havril et al., 2022).

2.2.4 Flood mitigation

Wetlands act as natural sponges, intercepting excess precipitation and delaying surface runoff through their position in floodplains and topographic depressions. By temporarily storing stormwater and releasing it gradually, they flatten hydrographs and lower peak flood heights (Flotemersch et al., 2016). As nature-based solutions (NBS), wetlands are most effective at reducing flood risk from short-return-period events (≤5 years). Their capacity declines for intermediate floods (10–50 years) and becomes negligible during extreme events (>100 years) (Ferreira et al., 2023).

Sustaining 3%–7% of a temperate catchment as wetlands is recommended to ensure effective flood regulation and water purification. This regulatory function is contingent on wetland size, vegetation, topography, and antecedent soil saturation (Jisha & Puthur, 2021). A single acre can retain an estimated 330,000 gal (~1.25 million liters) of floodwater (Sargent & Carter, 1999). In the Prairie Pothole Region of North America, wetland complexes substantially attenuate downstream flooding and support baseflow in river systems (Acreman & Holden, 2013). Likewise, the East Kolkata Wetlands in India, extending over 12,500 ha, process more than 600 million liters of sewage each day while reducing flood risk in adjoining urban areas (Chakraborty et al., 2023).

2.2.5 Constructed wetlands as nature-based solutions

Constructed wetlands (CWs) have emerged as ecofriendly and sustainable systems for treating wastewater from domestic, industrial, agricultural, and urban drainage sources (Adrados et al., 2014). While their overall global contribution remains uncertain (Turcios et al., 2021), CWs are increasingly recognized as cost-effective alternatives, particularly in developing countries and across diverse climatic settings (Shukla et al., 2021). By mimicking the functions of natural wetlands, they facilitate microbial degradation, nutrient assimilation, and organic pollutant breakdown through anaerobic zones, vegetated beds, and sedimentation units.

Reported removal efficiencies are high, with biochemical oxygen demand (BOD) reductions of 85%–95%, total suspended solids (TSS) removal of 70%–90%, and total nitrogen removal of 50%–80% (Saeed et al., 2022; Tang et al., 2021b). Integration of CWs with microbial fuel cells further adds an energy-generating function, while simultaneously treating wastewater (Doherty et al., 2015). When appropriately designed and managed, CWs represent a scalable and economical approach to enhancing water quality.

2.2.6 Wetlands as biodiversity hot spots

Wetlands are biodiversity hot spots, accounting for nearly 40% of global biodiversity, and are recognized as among the most biologically productive ecosystems (Convention on Wetlands Secretariat, 2025). They support more than 100,000 documented freshwater species, with approximately 200 new species described each year. Wetlands provide essential habitats for threatened amphibians and reptiles, sustain both migratory and resident waterbirds, and host a remarkable diversity of plant species. Coastal wetlands, such as mangroves and coral reefs, represent some of the most biologically diverse ecosystems on Earth, while many endemic species are confined to specific wetland habitats (https://www.ramsar.org/sites/default/files/ramsar_50_factsheet_biodiversity_english_as_v7.pdf). Wetlands are at significant risk due to overexploitation of resources in terms of rapid urbanization, industrialization, and anthropogenic activities. These multiple drivers play a crucial role in the degradation/deterioration of wetlands, which in turn results in biodiversity loss. In the current context, species were declining at the exponential rates and over 1 million animal and plant species are at the verge of extinction that is, considered as "threatened with extinction" (https://www.ramsar.org/sites/default/files/ramsar_50_factsheet_biodiversity_english_as_v7.pdf).

2.2.7 Wetlands for climate resilience

Wetlands play a pivotal role in regulating climate through their capacity to influence atmospheric concentrations of key GHGs, including methane, carbon dioxide, and nitrous oxide, which contribute approximately 60%, 20%, and 6% of global warming potential, respectively (Pachauri & Reisinger, 2007). Although they cover only 5%–8% of the Earth's land surface, wetlands act as major carbon sinks, storing an estimated 20%–30% of global terrestrial organic carbon and thereby making a substantial contribution to climate change mitigation (Guan et al., 2025). With the highest carbon density among ecosystems, wetlands are central to global biogeochemical and carbon cycles, underscoring their importance in climate regulation (Kayranli et al., 2010; Salimi et al., 2021). However, climate change poses a significant threat to these ecosystems, as altered hydrology and rising temperatures can disrupt their biogeochemistry and ecological functions. Such shifts may transform critical ecosystem services into disservices; for example, wetlands may lose their water purification capacity and instead release stored nutrients into surface waters. Furthermore, when decomposition outpaces primary production, wetlands risk transitioning from carbon sinks to net carbon sources (Salimi et al., 2021).

2.3 Key water quality factors affecting wetland health

Water quality underpins biodiversity conservation, regulates biogeochemical cycles, and governs the ecosystem services derived from wetlands. Numerous floral and faunal taxa are highly sensitive to pollution, and elevated contaminant levels can disrupt community structure and ecological balance. To evaluate water quality, parameters are classified into physical, chemical, and biological categories, as outlined in the following subsections (Fig. 2.2).

2.3.1 Physical parameters

Physical parameters such as temperature, color, odor, turbidity, total dissolved solids (TDS), TSS, and electrical conductivity (EC) (Fig. 2.2) play a crucial role in assessing the health and ecological dynamics of wetlands. Variations in these parameters strongly influence processes such as thermal regulation, light penetration, photosynthesis, gas solubility, and nutrient transfer (Alam et al., 2015; Singh et al., 2022; Son et al., 2025; Verma & Singh, 2013). Slight variation in temperature can act as a catalyst (positive/negative) and impacts the wetland dynamics to a greater extent in terms of community structure variation (Singh et al., 2022). Moreover, high turbidity results in enhanced thermal stress, thereby

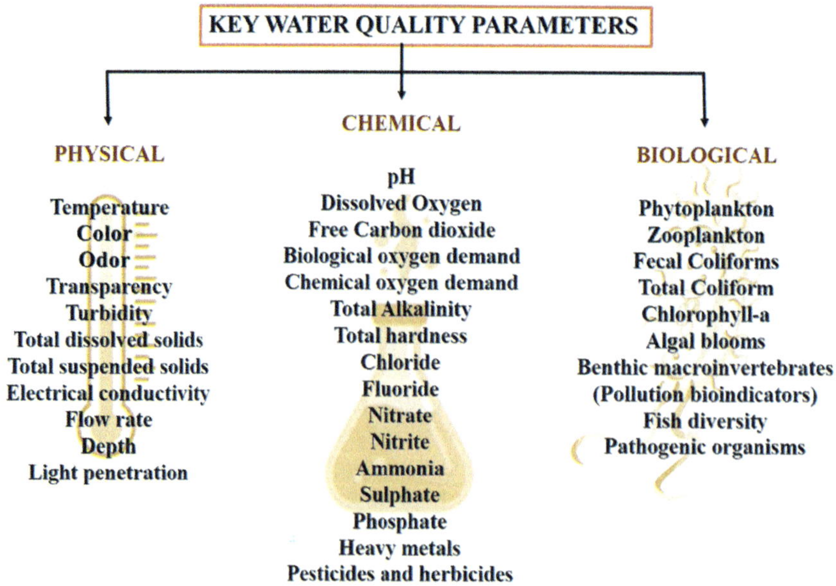

FIGURE 2.2 Physical, chemical, and biological water quality parameters for the estimation of wetland water health.

impeding productivity and dissolved oxygen (DO) (Jain et al., 2022; Léziart et al., 2019; Panda et al., 2018). In addition, parameters such as EC, TDS also play a vital role in wetland dynamics, the higher concentration of these parameter results in to polluted ecosystem and they are considered indicators to anthropogenic pollution and natural weathering (Adjovu et al., 2023; Jaffar et al., 2020; Krishna Kumar et al., 2015; Wang et al., 2021). Furthermore, these factors strongly regulate the bioavailability of nutrients and trace elements, rendering them robust indicators of environmental change.

2.3.2 Chemical parameters

Chemical parameters (Fig. 2.2) play a pivotal role in wetland dynamics, as they determine the ionic concentration, reactivity, and buffering capacity of wetland water. The pH of the water, governed by the equilibrium among carbonic acid (H_2CO_3), bicarbonate ions (HCO_3^-), and carbonate ions (CO_3^{2-}), influences several factors such as metal solubility and microbial activity (Saalidong et al., 2022). Furthermore, the concentration of DO reflects the overall health of wetland water and serves as one of the most critical indicators of pollution within wetland ecosystems, as it represents the balance between photosynthetic oxygen production and oxygen consumption during organic matter decomposition (Jain et al., 2022; Verma & Singh, 2013). Conversely, a high biological oxygen demand (BOD) indicates a high concentration of organic matter, meaning more oxygen is required for microbial degradation. This process often leads to hypoxic conditions and imposes considerable stress on the ecosystem (Mishra et al., 2023; Tanalgo et al., 2024). Similarly, chemical oxygen demand (COD) is frequently employed in conjunction with BOD to evaluate the biodegradability and pollution load (Alotaibi et al., 2023; Lv et al., 2024). It is a metric that evaluates both biodegradable and persistent organic pollutants. Alkalinity, which is regulated by bicarbonates and hydroxides, serves as a buffer against abrupt pH fluctuations and promotes aquatic stability (Sharma & Rajan, & Nayak, 2024).

The composition and survival of species are influenced by total hardness (predominantly caused by calcium and magnesium ions), which influences corrosion and scaling in aquatic infrastructure (Chanu et al., 2024; Rajora & Sarma, 2024). Chlorides, fluorides, nitrates, sulfates and phosphates are among the other ions that have distinct roles in the functioning of ecosystems and the health of humans. Eutrophication and degradation of aquatic habitats are facilitated by a variety of chloride and nitrate sources, including natural mineral leaching, sewage, and fertilizer discharge (Pang et al., 2013; Prasanth et al., 2012). Sulfates promote sulfate-reducing microorganisms, which in turn reduce methane emissions, whereas excess fluoride causes health disorders such as fluorosis (Kothari et al., 2021; Shen et al., 2025). Moreover, elevated phosphate concentrations resulting from effluents such as detergents and agricultural runoff contribute to algal blooms and subsequent oxygen depletion (CPCB, 2012; Singh, 2013). In addition, phosphate plays a significant role in nutrient cycling within wetland ecosystems (Singh, 2013). The importance of these chemical parameters in comprehensive wetland assessments is further emphasized by their interdependence with temperature, DO, and microbial activity.

2.3.3 Biological parameters and water quality indicators

Biological indicators such as microbes, algae, macroinvertebrates, zooplankton, phytoplankton, and fecal coliforms are essential markers for assessing wetland ecosystem health and dynamics, providing valuable insights into long-term ecological changes. Among these, algal and cyanophyte communities serve as effective indicators of trophic status and nutrient enrichment in wetlands due to their high sensitivity to nutrient fluctuations and light conditions (Goshtasbi et al., 2022). Algal groups such as Chlorophyta and Bacillariophyta are particularly responsive to environmental variables, including water transparency, DO, and nutrient concentrations (phosphorus and nitrogen), making them significant indicators for long-term monitoring of wetland ecosystems (Delgado-Fernández et al., 2025).

Furthermore, microbial communities play a crucial role in the degradation of organic matter and toxic pollutants, as well as in maintaining nutrient cycling and ecosystem dynamics. Bacterial genera such as Bacillus, Pseudomonas, and Acinetobacter exhibit strong bioremediation potential and are capable of degrading hazardous contaminants, including heavy metals and complex organic compounds (Tyagi et al., 2021). Microbial consortia also facilitate bioaugmentation and biosorption processes, leading to reductions in biochemical oxygen demand (BOD) and nutrient loads, thereby improving overall water quality (Krishnaswamy & Da Silva, 2025).

In addition, fecal coliform bacteria are widely used as microbiological indicators to detect contamination from human and animal waste. Their presence signifies potential pathogenic risks and indicates that the water is unsuitable for recreational or consumption purposes (Davis & McCuen, 2005; Kothari et al., 2021).

Macroinvertebrates and fish complement microbial and algal indicators by reflecting broader ecological responses within aquatic systems. Aquatic macroinvertebrates such as Ephemeroptera, Trichoptera, and Diptera larvae are widely used as bioindicators for assessing wetland health due to their varying tolerance levels to contaminants, limited mobility, and strong association with specific habitat conditions (Sharma & Rawat, 2009). In polluted waters, sensitive taxa

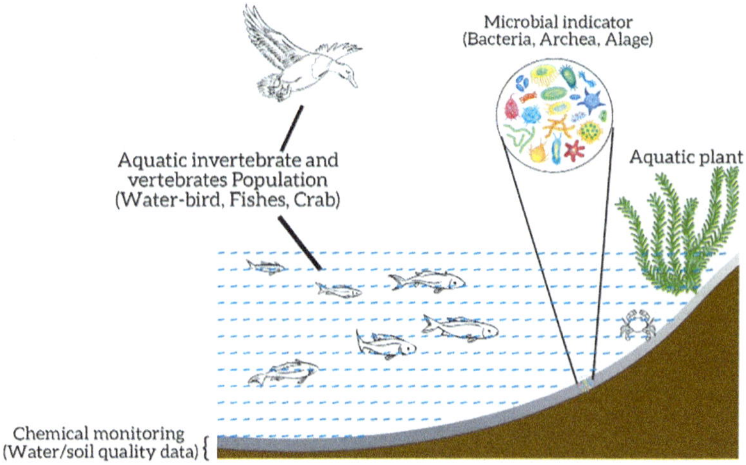

FIGURE 2.3 Conceptual framework of wetland water quality assessment using biological indicators.

including Ephemeroptera, Plecoptera, and Trichoptera are typically absent, whereas tolerant groups such as Chironomidae tend to thrive under stressful conditions (Ambelu et al., 2013; Mereta et al., 2012). These community shifts provide valuable insights into processes such as organic enrichment, habitat degradation, and pollution intensity.

The histological, behavioral, and biochemical responses of fish serve as apex bioindicators, capable of reflecting sublethal stress. Agricultural discharge and industrial effluents have been associated with the bioaccumulation of heavy metals in the tissues of *Cyprinus carpio* and *Oreochromis mossambicus* from wetlands such as Vembanad and Loktak (Sarkar & Maji, 2022). Additionally, multimetric indices that incorporate fish diversity, abundance, and health offer a comprehensive assessment of ecosystem integrity. The reliability of wetland assessments is improved by the combined application of these biological groups, which enables the early detection of degradation and informs conservation policies under frameworks such as the Ramsar Convention and India's wetlands (Conservation and Management) Rules, 2017 (MoEF&CC, 2017). Fig. 2.3 illustrates the contribution of microbial, algal, invertebrate, and fish-based indicators to a comprehensive wetland monitoring system.

2.4 Threats to wetland water quality

Wetlands are essential for preserving biodiversity, mitigating floods, recharging groundwater, and purifying water. Nevertheless, the capacity of these ecosystems to execute these functions is being progressively compromised by anthropogenic pressures. The factors behind wetland degradation are more complex and necessitate focused analysis, the integrated framework for the water quality of wetland, pollution sources, ecological impacts and management strategies is provided in Fig. 2.4.

2.4.1 Urbanization and sewage/wastewater discharge

Surface water bodies have become significantly polluted by untreated sewage and urban runoff, rendering them unsuitable for use (Agrawal et al., 2021). Wetlands located downstream of expanding urban areas often act as sinks for stormwater carrying oil waste, pathogens, heavy metals, pesticides, nutrients, pharmaceuticals, and other organic contaminants (Levin et al., 2020; Vijayaraghavan et al., 2021). In Shangri-La County, China, urban encroachment led to increased concentrations of nutrients and metals in the Napahai Wetland, thereby compromising both water quality and biodiversity (Fang et al., 2015). Similarly, in Mengxi Town, inadequate wastewater infrastructure was identified as the primary cause of wetland eutrophication, resulting from urban discharges (Wang et al., 2024). These disturbances collectively reduce the filtration capacity of wetlands and disrupt natural hydrological cycles.

Estrada-Rivera et al. (2022) reported that water quality parameters such as TSS, BOD, COD, and fecal coliforms were positively correlated with urbanization levels in the Alto Atoyac watershed. Likewise, Khan et al. (2022) found that unplanned urbanization severely affects the water quality of wetland ecosystems in India. These findings are consistent with those of Cerqueira et al. (2020), who identified urbanization as a major threat to wetland water quality.

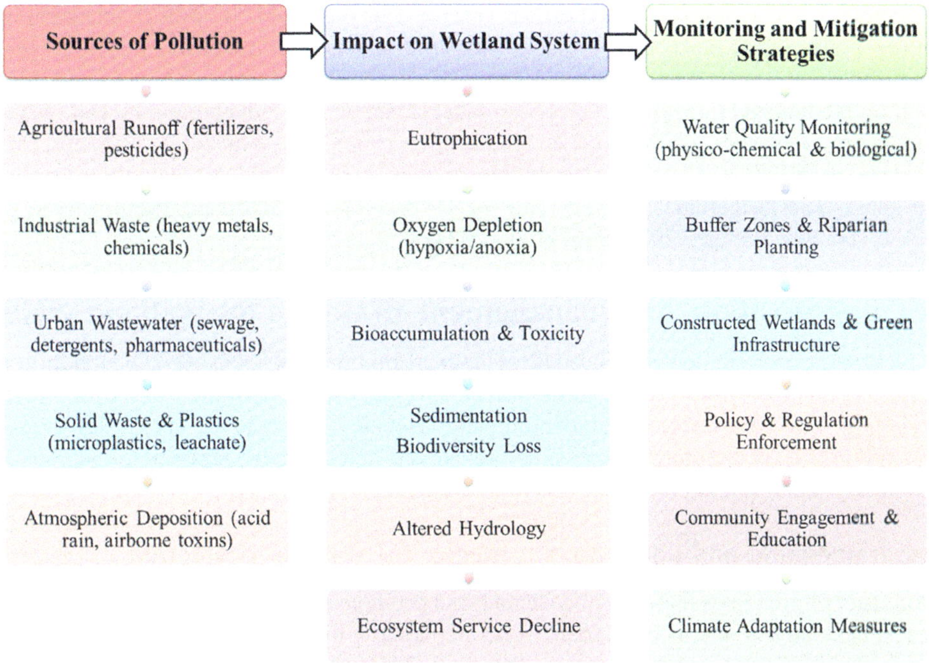

FIGURE 2.4 Integrated framework for wetland water quality: Pollution sources, ecological impacts, and management strategies.

2.4.2 Agricultural runoff and chemical inputs

Agricultural runoff that is, discharges of agrochemicals (primarily pesticides and fertilizers) into surface waters is one of the primary contributor of nonpoint source (NPS) pollution (Ioannidou & Stefanakis, 2020). One of the case studies of the Mount Abu region, India, highlighted the reduction in the wetland ecosystem services and biodiversity as a result of pollution due to agricultural runoff (Imdad et al., 2023). Heavy metals, such as cadmium and chromium, are also present in numerous fertilizers, which accumulate in biota and sediment (Rashmi et al., 2020). Moreover, nitrogen and phosphorus fertilizers contribute to water pollution through leaching and runoff, with nitrate posing health risks, including methemoglobinemia. Phosphorus accumulation, on the other hand, promotes algal blooms and leads to ecosystem degradation. The environmental persistence of these agrochemicals underscores the urgent need for environmentally friendly agricultural practices to mitigate water contamination (Akhtar et al., 2021; Rashmi et al., 2020).

CWs have emerged as one of the most sustainable and effective management strategies for treating agricultural runoff and wastewater sedimentation. They integrate the characteristics of basins, grass filter strips, created wetlands, and retention ponds (Higgins et al., 2020). The removal efficiency of CWs depends on several factors, including contaminant properties, aeration, CW type and design, hydraulic parameters, substrate media, and vegetation (Tang et al., 2021a).

2.4.3 Industrial pollutants

Rapid industrialization results in the discharge of toxic contaminants such as heavy metals, dyes, persistent organic pollutants, emerging contaminants such as pharmaceutical personal care products (PPCPs) and microplastics into the freshwater ecosystem. These pollutants disrupt the wetland dynamics and alter the microbial community composition, thereby resulting in the increased COD (Hu et al., 2021; Krishnaswamy & Da Silva, 2025; Picó et al., 2020; Shi et al., 2023; Tyagi et al., 2021).

PPCPs being persistent to biodegradation gets bioaccumulated in the aquatic species and disrupt the microbial and nutrient cycle in wetlands (Picó et al., 2020; Shi et al., 2023). CWs play a significant role in the reduction of PPCP loads; however, their effectiveness is restricted to the limited microbial activity and cold climates (Hu et al., 2021). On the other hand, microplastics, derived from synthetic textiles, packaging, and personal care products, accumulate in aquatic organisms and sediments, thereby obstructing metabolism and transferring toxic compounds like heavy metals and hydrocarbons (Dalvand & Hamidian, 2023; Li et al., 2024). Their prevalence indicates regulatory deficiencies in the management of plastic and solid waste. Microfibers, a form of microplastic that is shed from synthetic textiles such as polyester and nylon during washing and plastic breakdown, are significant emerging pollutants. They contain bioaccumulative, hazardous compounds that endanger marine life and humans through consumption or inhalation (Priyadarshini et al., 2024; Samal et al., 2024). Rapid industrial and urban growth has increased synthetic microfibre pollution, causing oxidative stress, DNA damage, and cell death. The European Union and the United Nations Environment Program recommend integrated microfibre management and remediation policy and research (Yadav et al., 2024). Another category of plastics, microbeads, are found in personal care items and cosmetics, do not disintegrate, escape wastewater treatment, often infiltrate water bodies, and are known to release and absorb pollutants that spread through the food chain. The Netherlands, Europe, and the US have prohibited these manufactured plastics but many countries still lack regulations (Ghosh, 2021; Patil et al., 2024).

2.5 Policies, conservation, and management strategies for wetland water quality

The persistence of wetland degradation is indicative of both institutional deficiencies and pollution pressures. Illegal waste, unregulated construction, and fertilizer misuse frequently persist uninterrupted due to ineffective enforcement, fragmented jurisdiction, and underresourced environmental agencies (Terrisse et al., 2025). This emphasizes the immediate necessity of effective policy frameworks, integrated management strategies, and sustained monitoring mechanisms to regularly monitor and protect the health of wetlands.

2.5.1 Policies, framework, and integration strategies for wetland protection

Policy is a structured set of principles that inform decisions and promote coordinated actions across governance levels. In the case of wetlands, national and international frameworks particularly the Ramsar Convention on Wetlands (1971) form the backbone for legislative and conservation strategies (Ramsar, 2016). As of 2025, over 2500 Ramsar sites spanning 172 Convention Contracting parties across the world have been designated, covering approximately 2.5 million km^2 globally (Convention on Wetlands Secretariat, 2025).

Wetland protection across the globe is supported by various legal and community-based frameworks. The Environmental Protection Agency (EPA) and the Army Corps are responsible for the regulation of wetland discharges in the United States under the Clean Water Act (Section 404), with Public input ensured through the National Environmental Policy Act (NEPA) (USEPA, 2022). The European Union implements integrated and participatory management through the implementation of the Water Framework Directive and Natura 2000 (European Commission, 2020), while the Environment Protection and Biodiversity Conservation Act (1999) in Australia is supported by land care programs and Indigenous participation, which in turn support wetland conservation (Finlayson & van Dam, 2023).

In India, the Wetlands (Conservation and Management) Rules, 2017, under the Environment (Protection) Act, 1986, empower State Wetland Authorities to notify and manage wetlands sustainably (MoEF&CC, 2017). The National Plan for Conservation of Aquatic Ecosystem (NPCA) and policies like the National Water Policy (2012) support central and state-level actions. However, only 0.4% of the country's natural wetlands (219 out of 55,862) are currently under conservation (Vasumathi et al., 2023).

2.5.2 Community-based conservation, management, and nature-inspired solutions for water quality

Community-based conservation (CBC) amalgamates local stewardship with ecological management, fostering environmental sustainability and socioeconomic advantages (Philip & Mukundi, 2024). Effective models like mangrove cooperatives in Southeast Asia demonstrate that empowering local populations enhances environmental outcomes (Selvam & Thamizoli, 2021). In India, the incorporation of CBC into initiatives such as the Mahatma Gandhi National Rural Employment Guarantee Act (MGNREGA) has improved wetland management via labor-intensive restoration and decentralized governance (MRD, 2019). Many NGOs, self-help groups, and local communities are actively involved in the conservation of these wetlands, for example., Conservation Alliance International (2015–20) collaborated with local shoreline communities, traditional authorities, and government agencies to restore the Songor Ramsar Site in Ghana through mangrove cultivation, agroforestry, and education (Owusu-Achiaw & Osei-Owusu, 2023). Tsomgo Pokhri Sanrakshan Samiti (TPSS) is the inaugural lake conservation committee in Sikkim, established in 2008 in accordance with the state's lake conservation guidelines. TPSS promotes community-led wetland management with the support of NGOs and the government, preserving local culture and traditions (FED Sikkim, 2025).

NBS enhance CBC by utilizing natural processes to augment water quality. Examples encompass riparian buffers, which eliminate up to 85% of nitrates (Brears, 2023); CWs, which remediate 80%–95% of BOD, COD (65%–80%), TSS (90%–95%), TP (35%–50%), TN (65%–80%) and pathogens from urban runoff (de Campos & Soto, 2024; Shukla et al., 2022); and floating treatment wetland in eutrophic wetlands, which substantially diminish sediment and nitrogen loads

(Grosshans et al., 2019). Research on riparian buffer design has demonstrated that structured, multispecies buffers, which consist of approximately 10 m of Zone 3 (perennial herbs), 20 m of Zone 2 (harvested deciduous forest), and 10 m of Zone 1 (nonharvested deciduous forest), can decrease sediment by approximately 86%–91% and dissolved nitrogen by approximately 85%–92% in catchments (Liu et al., 2023). Buffer widths of 10–30 m are prescribed for erosion control and 30–100 m for effective nutrient retention, with modifications for vegetation structure, soil, slope, and land use intensity, according to a global assessment (Majumdar & Avishek, 2025). Similarly, dredged-system wetlands constructed from dredged sediments provide a natural solution for restoring degraded wetlands and mitigating land subsidence. These systems increase the resilience of wetlands to sea-level rise by improving hydrology, supporting native vegetation, and enhancing wetland elevation (Harris et al., 2025). These integrated strategies not only restore hydrological equilibrium but also provide local livelihoods through ecotourism, aquaculture, and artisanal businesses, bolstered by corporate social responsibility and adaptation financing.

2.5.3 Climate change and the future of wetlands

Climate change can affect the spatiotemporal distribution of inland wetlands (Grill et al., 2015; Pekel et al., 2016; Pickens et al., 2020; Prigent et al., 2012); however, the direction and severity of these changes remain unknown (Salimi et al., 2021). The wetland regime, periodically to permanently inundated by water, is significantly governed by surface water dynamics, which are influenced by fluctuations across small spatial scales. Nonetheless, certain wetlands may not be completely submerged in water, as saturated soils are adequate to establish wetland ecosystems. Prior research utilizing satellite measurements at the subkilometer scale has determined that annual precipitation is the primary element influencing large-scale surface water dynamics (Tao et al., 2015; Zou et al., 2018). Temperature can impact wetlands by influencing many processes such as evapotranspiration, snowmelt, infiltration, soil melting and freezing, and precipitation (Alexander et al., 2013; Xu et al., 2024).

Climate change poses a significant threat to mountain ecosystems (Huggel et al., 2019), and within a decade or more, it will be considered one of the primary drivers of biodiversity loss (Malhi et al., 2020). Due to temperature variation, the hydrological cycles of wetlands get changed as a result of glacier retreat, and precipitation changes are observed in Andean wetlands (Bradley et al., 2006; Rivera et al., 2021). High-altitude wetlands exhibit significant sensitivity to global warming due to their year-round low temperatures. As a result of global warming, deterioration of these delicate ecosystems has been observed globally (Durigan et al., 2022). Thus, they face significant global threats, prompting a demand to prevent the diminishing availability of freshwater (Finlayson et al., 2019). Future decades are expected to witness alterations and reductions in these regions, resulting in detrimental effects on natural habitats and a significant risk of extinction for numerous endemic species (Cuesta et al., 2023). Research indicates that the average temperature in the Central and Coastal Area will rise markedly under projected climate conditions, especially under high-emissions scenarios (Rivera et al., 2020). Consequently, climate change is anticipated to induce significant structural and functional alterations in wetland dynamics, thereby constraining the provision of ecosystem services (Cuesta et al., 2023; Salimi et al., 2021).

2.6 Conclusion

Wetlands are essential ecosystems that serve a variety of ecological functions, including carbon storage, water flow regulation, pollutant purification, and habitat and hot spots for biodiversity. Nevertheless, their capacity to provide critical ecosystem services is being compromised by the increasing degradation of these systems due to climate-induced stresses, urbanization, hydrological disruptions, pollution, land use change, and emerging contaminants. The protection and restoration of wetlands must be a strategic priority as climate change continues to escalate. Wetlands contribute to resilience by sustaining baseflows during droughts, mitigating floods, and sequestering substantial quantities of carbon. Their degradation not only worsens the loss of biodiversity but also undermines efforts to improve water security and climate adaptation.

A comprehensive strategy that integrates stringent regulatory enforcement, active community participation, and the implementation of science-based and NBS is essentially required to safeguard the purity of wetland water in the face of climate change. The practical efficacy of these interventions can be significantly improved by customizing them to the characteristics of the watershed, including its area, pollutant type, and land-use intensity. Additionally, the maintenance of long-term ecological viability and water security in wetland ecosystems necessitates adaptive governance, intersectoral collaboration, and climate-resilient planning. In addition to being an ecological necessity, the preservation of wetlands is a socio-economic necessity for the development of sustainable and resilient futures.

Declaration of AI and AI-based tools

During the preparation of this work, the author(s) used the Large Language Model (LLM) tool ChatGPT and Grammarly to assist with grammatical corrections and to improve the readability of the text. After using these tools, the author(s) reviewed and edited the content as needed, and take full responsibility for the final content of the publication.

References

Acreman, M., & Holden, J. (2013). How wetlands affect floods. *Wetlands, 33*, 773–786.

Adjovu, G. E., Stephen, H., James, D., & Ahmad, S. (2023). Overview of the application of remote sensing in effective monitoring of water quality parameters. *Remote Sensing, 15*(7), 1938.

Adrados, B., Sánchez, O. L. G. A., Arias, C. A., Becares, E. L. O. Y., Garrido, L. A. U. R. A., Mas, J. O. R. D. I., Brix, H., & Morató, J. O. R. D. I. (2014). Microbial communities from different types of natural wastewater treatment systems: Vertical and horizontal flow constructed wetlands and biofilters. *Water Research, 55*, 304–312.

Agrawal, K. K., Panda, C., & Bhuyan, M. K. (2021). Impact of urbanization on water quality. In *Current advances in mechanical engineering: Select proceedings of ICRAMERD 2020*. Springer Singapore, 665–673.

Akhtar, N., Syakir Ishak, M. I., Bhawani, S. A., & Umar, K. (2021). Various natural and anthropogenic factors responsible for water quality degradation: A review. *Water, 13*(19), 2660.

Alam, M. T., Hussain, M. A., Sultana, S., Hasan, M. T., Haque, M. Z., Simon, K. D., & Mazumder, S. K. (2015). Water quality parameters and their correlation matrix: A case study in two important wetland beels of Bangladesh. *Ciencia e Tecnica Vitivinicola, 30*, 463–489.

Alexander, L., Allen, S., Bindoff, N. L., Breon, F. M., et al. (2013). IPCC. Climate change 2013: The physical science basis, in contribution of Working Group I (WGI) to the Fifth. In T. F. Stocker, (Ed.). *Assessment report (AR5) of the intergovernmental panel on climate change*.

Alotaibi, M., Munshi, F., Refaat, A., Said, M. E., & Nasir, M. (2023). Correlation between biochemical oxygen demand and chemical oxygen demand, at high salinity bioreactor-based, wastewater treatment plant in Al-Hasa Saudi Arabia. *Journal of Water, 1*(2), 35–47.

Ambelu, A., Mekonen, S., G/Silassie, A., Malu, A., & Karunamoorthi, K. (2013). Physicochemical and biological characteristics of two Ethiopian wetlands. *Wetlands, 33*, 691–698.

Balwan, W. K., & Kour, S. (2021). Wetland-an ecological boon for the environment. *East African Scholars Journal of Agriculture and Life Sciences, 4*(3), 38–48.

Bassi, N., Kumar, M. D., Sharma, A., & Pardha-Saradhi, P. (2014). Status of wetlands in India: A review of extent, ecosystem benefits, threats and management strategies. *Journal of Hydrology: Regional Studies, 2*, 1–19.

Bradley, R. S., Vuille, M., Díaz, H. F., & Vergara, W. (2006). Threats to water supplies in the tropical Andes. *Science, 312*, 1755–1756. https://doi.org/10.1126/science.1128087.

Brears, R. C. (2023). Nature-based solutions: Harnessing ecosystems for enhanced water quality and flood mitigation. *Medium. Mark and Focus*. https://medium.com/mark-and-focus/nature-based-solutions-harnessing-ecosystems-for-enhanced-water-quality-and-flood-mitigation-60d4924ecbba.

de Campos, S. X., & Soto, M. (2024). The use of constructed wetlands to treat effluents for water reuse. *Environments, 11*(2), 35. https://doi.org/10.3390/environments11020035.

Cerqueira, T. C., Mendonça, R. L., Gomes, R. L., De Jesus, R. M., & Da Silva, D. M. L. (2020). Effects of urbanization on water quality in a watershed in northeastern Brazil. *Environmental monitoring and assessment, 192*(1), 65.

Chakraborty, S. K., Sanyal, P., & Ray, R. (2023). Ecosystem services and values of wetlands with special reference with east Kolkata wetlands. In *Wetlands ecology: Eco-biological uniqueness of a ramsar Site (East Kolkata Wetlands, India*. Springer International Publishing, 227–255.

Convention on Wetlands Secretariat. (2025). *Wetlands: Key enablers for a resilient and sustainable recovery*. Switzerland: The Convention on Wetlands. https://www.ramsar.org/wetlands-key-enablers-resilient-sustainable-recovery.

CPCB. (2012). *Central pollution control board. guide manual: Water and wastewater analysis*1–183. https://cpcb.nic.in/openpdffile.php?id=UmVwb3J0RmlsZXMvMjA0XzE1MjQ2NTA4OTNfbWVkaWFwaG90bzEyODI3LnBkZg.

Chanu, T. N., Gogoi, P., Koushlesh, S. K., Jana, C., Nair, S. M., Das, S. K., Sinha, A., Devi, H. M., & Das, B. K. (2024). Revealing the temporal dynamics of fish assemblage pattern, feeding guilds in relation to habitat quality parameters in a coastal wetland of the Indian Sundarban eco-region. *Environmental monitoring and assessment, 196*(11), 1094.

Cuesta, F., Carilla, J., LLambí, L. D., Muriel, P., Lencinas, M. V., Meneses, R. I., Feeley, K. J., Pauli, H., Aguirre, N., Beck, S., & Bernardi, A. (2023). Compositional shifts of alpine plant communities across the high Andes. *Global Ecology and Biogeography, 32*(9), 1591–1606.

Dalvand, M., & Hamidian, A. H. (2023). Occurrence and distribution of microplastics in wetlands. *Science of the Total Environment, 862*, 160740.

Davis, A. P., & McCuen, R. H. (2005). Water quality parameters. *Stormwater Management for Smart Growth*, 12–36.

Delgado-Fernández, E., Cruz, D., Ayavaca, R., Benítez, Á., & Hernández, B. (2025). Microalgal diversity as bioindicators for assessing and sustaining water quality in the high mountain lakes of Quimsacocha, Azuay, Ecuador. *Sustainability, 17*(4).

Deslippe, J. R., & Bentley, S. B. (2025). The role of wetland restoration in mediating phosphorus ecosystem services in agricultural landscapes. *Current Opinion in Biotechnology, 91*, 103227.

Doherty, L., Zhao, Y., Zhao, X., Hu, Y., Hao, X., Xu, L., & Liu, R. (2015). A review of a recently emerged technology: Constructed wetland–microbial fuel cells. *Water Research, 85*, 38–45.

Durigan, G., Munhoz, C. B., Zakia, M. J. B., Oliveira, R. S., et al. (2022). Cerrado wetlands: Multiple ecosystems deserving legal protection as a unique and irreplaceable treasure. *Perspectives in Ecology and Conservation. 20*(3), 185–196.

Estrada-Rivera, A., Díaz Fonseca, A., Treviño Mora, S., García Suastegui, W. A., Chávez Bravo, E., Castelán Vega, R., ... Handal-Silva, A. (2022). The impact of urbanization on water quality: Case study on the alto atoyac basin in puebla, Mexico. *Sustainability, 14*(2), 667.

European Commission. (2020). Water Framework Directive. https://environment.ec.europa.eu/topics/water/water-frameworkdirectiveen#:~:text=The%20Water%20Framework%20Directive%20(WFD,same%20time%20as%20human%20needs.

Everard, M. (2016). Nutrient cycling in wetlands: Supporting services. In *The Wetland Book*. Dordrecht: Springer, 1–4.

Fang, H., Xu, Y., Ye, Z., Zhang, Z., Pan, S., Deng, L., Luo, Z., & Chen, S. (2015). Impact of urbanization on nutrients and heavy metal pollution of Napahai Wetland, Shangri-La County, China. *International Journal of Sustainable Development & World Ecology, 22*(2), 117–126.

FED Sikkim. (2025). Forest and Environment Department, Government of Sikkim. https://forest.sikkim.gov.in/psshome/index?name=pss-aboutus.

Ferreira, C. S., Kašanin-Grubin, M., Solomun, M. K., Sushkova, S., Minkina, T., Zhao, W., & Kalantari, Z. (2023). Wetlands as nature-based solutions for water management in different environments. *Current Opinion in Environmental Science & Health, 33*, 100476.

Finlayson, C. M., & van Dam, A. A. (2023). Climate change and wetlands: Vulnerability, adaptation, mitigation, resolutions, and scientific societies. In *Ramsar Wetlands*. Elsevier, 495–524.

Finlayson, C. M., Davies, G. T., Moomaw, W. R., Chmura, G. L., Natali, S. M., Perry, J. E., Roulet, N., & Sutton-Grier, A. E. (2019). The second warning to humanity–providing a context for wetland management and policy. *Wetlands, 39*(1), 1–5.

Flotemersch, J. E., Leibowitz, S. G., Hill, R. A., Stoddard, J. L., Thoms, M. C., & Tharme, R. E. (2016). A watershed integrity definition and assessment approach to support strategic management of watersheds. *River Research and Applications, 32*(7), 1654–1671.

Ghosh, S. (2021). Tiny microbeads-threat to human life, threat to environment. *IInternational Journal of Engineering Research & Technology, 10*(1).

Gleeson, T., Befus, K. M., Jasechko, S., Luijendijk, E., & Cardenas, M. B. (2016). The global volume and distribution of modern groundwater. *Nature Geoscience, 9*(2), 161–167.

Goshtasbi, H., Atazadeh, E., Fathi, M., & Movafeghi, A. (2022). Using physicochemical and biological parameters for the evaluation of water quality and environmental conditions in international wetlands on the southern part of Lake Urmia, Iran. *Environmental Science and Pollution Research*, 1–15.

Grill, G., Lehner, B., Lumsdon, A. E., MacDonald, G. K., Zarfl, C., & Liermann, C. R. (2015). An index-based framework for assessing patterns and trends in river fragmentation and flow regulation by global dams at multiple scales. *Environmental Research Letters, 10*(1), 015001.

Grosshans, R., Lewtas, K., Gunn, G., & Stanley, M. (2019). Floating treatment wetlands and plant bioremediation: nutrient treatment in eutrophic freshwater lakes. *International Institute of Sustainable Development*, 1–30.

Guan, Y., Jin, S., Zhang, Z., Chen, S., Chen, Y., Liu, S., Fan, M., Ding, Y., & Yuan, X. (2025). A comprehensive review of climate warming and carbon dynamics in wetland ecosystems. *Discover Sustainability, 6*(1), 1–13.

Haberl, R., Grego, S., Langergraber, G., Kadlec, R. H., Cicalini, A. R., Dias, S. M., Novais, J. M., Aubert, S., Gerth, A., Thomas, H., & Hebner, A. (2003). Constructed wetlands for the treatment of organic pollutants. *Journal of Soils and Sediments, 3*(2), 109–124.

Harris, B. D., Ostojic, A., Tedesco, L. P., VanDerSys, K., Bailey, S., Shawler, J. L., Jafari, N. H., & Chasten, M. (2025). Wetland elevation change following beneficial use of dredged material nourishment. *Frontiers in Ecology and Evolution, 13*, 1518759.

Hatvani, I. G., Dokulil, M. T., & Clement, A. (2022). The role of wetlands in mitigating impacts from diffuse agricultural loads. *book: Encyclopedia of Inland Waters (Second Edition)*, 285–299.

Higgins, M. J., Rock, C. A., Bouchard, R., & Wengrezynek, B. (2020). Controlling agricultural runoff by use of constructed wetlands. In *Constructed wetlands for water quality improvement*. CRC Press, 359–367.

Hu, X., Xie, H., Zhuang, L., Zhang, J., Hu, Z., Liang, S., & Feng, K. (2021). A review on the role of plant in pharmaceuticals and personal care products (PPCPs) removal in constructed wetlands. *Science of the Total Environment, 780*, 146637.

Huggel, C., Muccione, V., Carey, M., James, R., Jurt, C., & Mechle, R. (2019). Loss and damage in the mountain cryosphere. *Regional Environmental Change, 19*, 1387–1399. https://doi.org/10.1007/s10113-018-1385.

Imdad, K., Rihan, M., Sahana, M., Parween, S., Ahmed, R., Costache, R., Chaudhary, A., & Tripathi, R. (2023). Wetland health, water quality, and resident perceptions of declining ecosystem services: A case study of Mount Abu, Rajasthan, India. *Environmental Science and Pollution Research, 30*(55), 116617–116643.

Ioannidou, V., & Stefanakis, A. I. (2020). The use of constructed wetlands to mitigate pollution from agricultural runoff. In *Contaminants in agriculture: Sources, impacts and management*. Springer International Publishing, 233–246.

Jaffar, A., Thamrin, N. M., Ali, M. S. A. M., Misnan, M. F., & Yassin, A. I. M. (2020). The influence of physico-chemical parameters to determine water quality: A review. *Journal of Electrical And Electronic Systems Research (JEESR, 17*, 116–121.

Jain, N., Yevatikar, R., & Raxamwar, T. S. (2022). Comparative study of physico-chemical parameters and water quality index of river. *Materials Today: Proceedings, 60*, 859–867.

Jisha, K. C., & Puthur, J. T. (2021). Ecological importance of wetland systems. *Wetlands Conservation: Current Challenges and Future Strategies*, 40–54.

Kayranli, B., Scholz, M., Mustafa, A., & Hedmark, Å. (2010). Carbon storage and fluxes within freshwater wetlands: A critical review. *Wetlands, 30*(1), 111–124.

Khan, I., Zakwan, M., Pulikkal, A. K., & Lalthazula, R. (2022). Impact of unplanned urbanization on surface water quality of the twin cities of Telangana state, India. *Marine Pollution Bulletin, 185*, 114324.

Kingsford, R. T., Basset, A., & Jackson, L. (2016). Wetlands: Conservation's poor cousins. *Aquatic Conservation: Marine and Freshwater Ecosystems, 26*(5), 892–916.

Kothari, V., Vij, S., Sharma, S., & Gupta, N. (2021). Correlation of various water quality parameters and water quality index of districts of Uttarakhand. *Environmental and Sustainability Indicators, 9*, 100093.

Krishna Kumar, S., Logeshkumaran, A., Magesh, N. S., Godson, P. S., & Chandrasekar, N. (2015). Hydro-geochemistry and application of water quality index (WQI) for groundwater quality assessment, Anna Nagar, part of Chennai City, Tamil Nadu, India. *Applied Water Science, 5*, 335–343.

Krishnaswamy, U. R., & Da Silva, G. H. R. (2025). Floating treatment wetland with microbial fuel cell for removing pollutants, energy utilization, recovery of nutrients-review on sustainable technology. *Desalination and Water Treatment,* 101264.

Kujala, K., Karlsson, T., Nieminen, S., & Ronkanen, A. K. (2019). Design parameters for nitrogen removal by constructed wetlands treating mine waters and municipal wastewater under Nordic conditions. *Science of the Total Environment, 662*, 559–570.

Kumar, V. (2022). Wetlands: Kidneys of the earth is at risk. *Agriculture & Environment, 3*(7), 19–25.

Kushwaha, A., Goswami, L., Kim, B. S., Lee, S. S., Pandey, S. K., & Kim, K. H. (2024). Constructed wetlands for the removal of organic micropollutants from wastewater: Current status, progress, and challenges. *Chemosphere, 360*, 142364.

Levin, P. S., Howe, E. R., & Robertson, J. C. (2020). Impacts of stormwater on coastal ecosystems: The need to match the scales of management objectives and solutions. *Philosophical Transactions of the Royal Society B, 375*(1814) 20190460.

Li, N. Y., Zhong, B., Guo, Y., Li, X. X., Yang, Z., & He, Y. X. (2024). Non-negligible impact of microplastics on wetland ecosystems. *Science of the Total Environment,* 171252.

Liu, C., Qu, L., Clausen, J., Lei, T., & Yang, X. (2023). Impact of riparian buffer zone design on surface water quality at the watershed scale, a case study in the Jinghe Watershed in China. *Water, 15*(15), 2696.

Lv, Z., Ran, X., Liu, J., Feng, Y., Zhong, X., & Jiao, N. (2024). Effectiveness of chemical oxygen demand as an indicator of organic pollution in aquatic environments. *Ocean-Land-Atmosphere Research, 3*, 0050.

Léziart, T., Dutheil de la Rochere, P. M., Cheswick, R., Jarvis, P., & Nocker, A. (2019). Effect of turbidity on water disinfection by chlorination with the emphasis on humic acids and chalk. *Environmental Technology, 40*(13), 1734–1743.

Ma, S., Mistry, P., Badiou, P., Bansal, S., & Creed, I. F. (2025). Factors regulating the potential for freshwater mineral soil wetlands to function as natural climate solutions. *Wetlands, 45*(1), 11.

Majumdar, A., & Avishek, K. (2025). Mitigating riparian buffer zone degradation through policy interventions and learnings from best practices. *Discover Environment, 3*(1), 1–29.

Malhi, Y., Franklin, J., Seddon, N., Solan, M., et al. (2020). Climate change and ecosystems: Threats, opportunities and solutions. *Philosophical Transactions of the Royal Society B, 375*, 20190104. https://doi.org/10.1098/rstb.2019.0104.

Mereta, S. T., Boets, P., Ambelu, B. A., Malu, A., Ephrem, Z., Sisay, A., Endale, H., Yitbarek, M., Jermal, A., De Meester, L., & Goethals, P. L. M. (2012). Analysis of environmental factors determining the abundance and diversity of macroinvertebrate taxa in natural wetlands of Southwest Ethiopia. *Ecological Informatics, 7*, 52–61.

Mishra, A. P., Kumar, S., Patra, R., Kumar, A., Sahu, H., Chandra, N., Pande, C. B., & Alshehri, F. (2023). Physicochemical parameters of water and its implications on avifauna and habitat quality. *Sustainability, 15*(12), 9494.

Mishra, B. K., Kumar, P., Saraswat, C., Chakraborty, S., & Gautam, A. (2021). Water security in a changing environment: Concept, challenges and solutions. *Water, 13*(4), 490.

MoEF&CC. Wetlands (Conservation and Management) Rules. (2017). Ministry of Environment, Forest and Climate Change, Government of India. https://moef.gov.in/uploads/2020/01/final-version-and-printed-wetland-guidelines-rules-2017-03.01.20.pdf.

Molnar, J. J. (2010). Climate change and societal response: Livelihoods, communities, and the environment. *Rural Sociology, 75*, 1–16.

Moreno-Mateos, D., & Comin, F. A. (2010). Integrating objectives, and scales for planning and implementing wetland restoration and creation in agricultural landscapes. *Journal of Environmental Management.91,2087–2095.*

MRD. (2019). *Water conservation under MGNREGS, Ministry of rural development.* https://www.pib.gov.in/PressReleasePage.aspx?PRID=1576612#:~:text=The%20MGNREGA%20has%20been%20working,in%20planning%20and%20implementation%20efforts.

Nepal, D., Parajuli, P., Ouyang, Y., To, F., Wijewardane, N., & Venishetty, V. (2024). Evaluation of wetland area effects on hydrology and water quality at watershed scale. *Resources, 13*(8), 114.

Owusu-Achiaw, R., & Osei-Owusu, Y. (2023). Community-based approach to wetland restoration: Case study of the Songor Wetland, Ghana. In M. Nishi (Ed.). *Governing Sustainability in the Global South* (pp. 157). *Routledge*.

Pachauri, R.K. & Reisinger, A. (2007). Climate change 2007: Synthesis report. Contribution of working groups I, II and III to the fourth assessment report of the Intergovernmental Panel on Climate Change. Ipcc.

Panda, P. K., Panda, R. B., & Dash, P. K. (2018). The river water pollution in India and abroad: A critical review to study the relationship among different physico-chemical parameters. *American Journal of Water Resources, 6*(1), 25–38.

Pang, Z., Yuan, L., Huang, T., Kong, Y., Liu, J., & Li, Y. (2013). Impacts of human activities on the occurrence of groundwater nitrate in an alluvial plain: A multiple isotopic tracers approach. *Journal of Earth Science, 24*(1), 111–124.

Patil, P. M., Mahamuni-Badiger, P., Ingavale, R. R., Patel, P. R., & Dhanavade, M. J. (2024). Usage of microplastic beads in pharmaceuticals and cosmetics industry: A review. *Microplastic Pollution,* 51–72.

Patil, P. N., Sawant, D. V., & Deshmukh, R. N. (2012). Physico-chemical parameters for testing of water–A review. *International journal of Environmental Sciences, 3*(3), 1194–1207.

Pekel, J. F., Cottam, A., Gorelick, N., & Belward, A. S. (2016). High-resolution mapping of global surface water and its long-term changes. *Nature, 540*, 418–422. https://doi.org/10.1038/nature20584.

Philip, J. S., & Mukundi, M. B. (2024). Community-based approaches to environmental conservation: Empowering local initiatives. *Greener Journal of Social Sciences, 14*(2), 289–299. https://doi.org/10.15580/gjss.2024.2.122024211.

Pickens, A. H., Hansen, M. C., Hancher, M., Stehman, S. V., Tyukavina, A., Potapov, P., Marroquin, B., & Sherani, Z. (2020). Mapping and sampling to characterize global inland water dynamics from 1999 to 2018 with full landsat time-series. *Remote Sensing of Environment, 243*, 111792.

Picó, Y., Alvarez-Ruiz, R., Alfarhan, A. H., El-Sheikh, M. A., Alshahrani, H. O., & Barceló, D. (2020). Pharmaceuticals, pesticides, personal care products and microplastics contamination assessment of Al-Hassa irrigation network (Saudi Arabia) and its shallow lakes. *Science of the Total Environment, 701*, 135021.

Prasanth, S. V. S., Magesh, N. S., Jitheshlal, K. V., & Chandrasekar, N. (2012). Evaluation of groundwater quality and its suitability for drinking and agricultural use in the coastal stretch of Alappuzha District, Kerala, India. *Applied Water Science, 2*(3), 165–175.

Prigent, C., Papa, F., Aires, F., Jimenez, C., Rossow, W. B., & Matthews, E. (2012). Changes in land surface water dynamics since the 1990s and relation to population pressure. *Geophysical Research Letters, 39*(8).

Priyadarshini, S., Jagatee, S., & Das, A. P. (2024). Synthetic fabrics and microfiber pollution–an assessment of their global impact. In *Renewable energy generation and value addition from environmental microfiber pollution through advanced greener solution*. Springer Nature Switzerland, 137–157.

Préau, C., Tournebize, J., Lenormand, M., Alleaume, S., Boussada, V. G., & &Luque, S. (2022). Habitat connectivity in agricultural landscapes improving multi-functionality of constructed wetlands as nature-based solutions. *Ecological Engineering, 182*, 106725.

Rahmanian, N., Ali, S. H. B., Homayoonfard, M., Ali, N. J., Rehan, M., Sadef, Y., & Nizami, A. S. (2015). Analysis of physiochemical parameters to evaluate the drinking water quality in the State of Perak, Malaysia. *Journal of chemistry, 2015*(1), 716125.

Rajora, K., & Sarma, K. (2024). Assessment of Pollution Status in Surajpur Wetland using multivariate statistical analysis, Uttar Pradesh, India. *Environmental Earth Sciences, 83*(12), 381.

Ramsar Convention Secretariat. (2016). *An introduction to the ramsar convention on wetlands*, 1–110 7th ed.

Rani, A., Chauhan, M., Sharma, P. K., Kumari, M., Mitra, D., & Joshi, S. (2024). Microbiological dimensions and functions in constructed wetlands: A review. *Current Research in Microbial Sciences, 7*, 100311.

Rashmi, I., Roy, T., Kartika, K. S., Pal, R., Coumar, V., Kala, S., & Shinoji, K. C. (2020). Organic and inorganic fertilizer contaminants in agriculture: Impact on soil and water resources. *Contaminants in Agriculture: Sources, Impacts and Management*, 3–41.

Rezanezhad, F., McCarter, C. P., & Lennartz, B. (2020). Wetland biogeochemistry: Response to environmental change. *Frontiers in Environmental Science, 8*, 55.

Rivera, J. A., Naranjo-Tamayo, E., & Viale, M. (2020). Water resources change in Central-Western Argentina under the Paris agreement warming targets. *Frontiers in Climate, 2*, 587126. https://doi.org/10.3389/fclim.2020.587126.

Rivera, J. A., Otta, S., Lauro, C., & Zazulie, N. (2021). A decade of hydrological drought in Central-Western Argentina. *Frontiers Water, 3*, 640544. https://doi.org/10.3389/frwa.2021.640544.

Saalidong, B. M., Aram, S. A., Otu, S., & Lartey, P. O. (2022). Examining the dynamics of the relationship between water pH and other water quality parameters in ground and surface water systems. *PLoS ONE, 17*(1), e0262117.

Saeed, T., Miah, M. J., & Yadav, A. K. (2022). Free-draining two-stage microbial fuel cell integrated constructed wetlands development using biomass, construction, and industrial wastes as filter materials: Performance assessment. *Chemical Engineering Journal, 437*, 135433.

Salimi, S., Almuktar, S. A., & Scholz, M. (2021). Impact of climate change on wetland ecosystems: A critical review of experimental wetlands. *Journal of Environmental Management, 286*, 112160.

Samal, K., Samal, S. R., Mishra, S., & Nayak, J. K. (2024). Sources, transport, and accumulation of synthetic microfiber wastes in aquatic and terrestrial environments. *Water, 16*(16), 2238.

Santoro, V., Schiavon, M., & Celi, L. (2024). Role of soil abiotic processes on phosphorus availability and plant responses with a focus on strigolactones in tomato plants. *Plant and Soil, 494*(1), 1–49.

Sargent, M. S., & Carter, K. S. (1999). *Managing Michigan wildlife: A landowners guide*.

Sarkar, D., & Maji, N. (2022). Status and threats of wetland change in land use pattern and planning: Impact of land use patterns and urbanization. In *Handbook of research on monitoring and evaluating the ecological health of wetlands*. IGI Global Scientific Publishing, 106–127.

Selvam, V., & Thamizoli, P. (2021). Science-based and community-centred approach to restore and sustain mangrove wetlands of India. *Current Science, 121*(10), 1288–1295. https://doi.org/10.18520/cs/v121/i10/1288-1296.

Sharma, K., Rajan, S., & Nayak, S. K. (2024). Water pollution: Primary sources and associated human health hazards with special emphasis on rural areas. In *Water resources management for rural development*. Elsevier, 3–14.

Sharma, L. K., & Naik, R. (2024). Wetland Ecosystems. In *Conservation of saline wetland ecosystems: An initiative towards un decade of ecological restoration*. Springer Nature Singapore, 3–32.

Sharma, R. C., & Rawat, J. S. (2009). Monitoring of aquatic macroinvertebrates as bioindicator for assessing the health of wetlands: A case study in the Central Himalayas, India. *Ecological Indicators, 9*(1), 118–128.

Shen, L., Peng, S., Zhang, Z., Tong, C., Lin, J., Li, Y., Zhong, H., Ma, S., Zhuang, M., & Gauci, V. (2025). The large role of declining atmospheric sulfate deposition and rising CO_2 concentrations in stimulating future wetland CH_4 emissions. *Science Advances, 11*(6), eadn1056.

Shi, Q., Yuan, Y., Zhou, Y., Yuan, Y., Liu, L., Liu, X., ... Wang, H. (2023). Pharmaceutical and personal care products (PPCPs) degradation and microbial characteristics of low-temperature operation combined with constructed wetlands. *Chemosphere, 341*, 140039.

Shukla, A., Parde, D., Gupta, V., Vijay, R., & Kumar, R. (2022). A review on effective design processes of constructed wetlands. *International Journal of Environmental Science and Technology, 19*, 12749–12774. https://doi.org/10.1007/s13762-021-03549-y.

Shukla, R., Gupta, D., Singh, G., & Mishra, V. K. (2021). Performance of horizontal flow constructed wetland for secondary treatment of domestic wastewater in a remote tribal area of Central India. *Sustainable Environment Research, 31*, 1–10.

Singh, A. L. (2013). Nitrate and phosphate contamination in water and possible remedial measures. In N. Dwivedi (Ed.). *book: Environmental problems and plant edition: Ist chapter: Chapter publisher* (pp. 44–56). Springer Verlag.

Singh, Y., Singh, G., Khattar, J. S., Barinova, S., Kaur, J., Kumar, S., & Singh, D. P. (2022). Assessment of water quality condition and spatiotemporal patterns in selected wetlands of Punjab, India. *Environmental Science and Pollution Research, 29*(2), 2493–2509.

Son, Y. B., Kim, E., Cho, J. H., Choi, S. K., & Kang, D. (2025). Analysis of regional variation of water transparency in the Yellow Sea and East China Sea based on MODIS data. *Regional Studies in Marine Science, 84*, 104093.

Sutanudjaja, E. H., Van Beek, R., Wanders, N., Wada, Y., Bosmans, J. H., Drost, N., Van Der Ent, R. J., De Graaf, I. E., Hoch, J. M., De Jong, K., & Karssenberg, D. (2018). PCR-GLOBWB 2: A 5 arcmin global hydrological and water resources model. *Geoscientific Model Development, 11*(6), 2429–2453.

Tanalgo, K. C., Manampan-Rubio, M., Alvaro-Ele, R. J., Hilario-Husain, B. A., Murray, S. A., Delos Reyes, J. L., Pangato, N. M., Magkidong, N. S., Angcaco, K. L. D., Catulos, A. J., & Dimacaling, A. D. (2024). Ecological indicators of water quality and marshland impact area (MARia) index of Ligawasan Marsh: A critically important wetland in the Southern Mindanao, Philippines. *Discover Environment, 2*(1), 113.

Tang, X., Li, J., Liu, W., Yu, H., & Wang, F. (2021a). A method to increase the number of positive samples for machine learning-based urban waterlogging susceptibility assessments. *Stochastic Environmental Research and Risk Assessment*, 1–18.

Tang, Z., Wood, J., Smith, D., Thapa, A., & Aryal, N. (2021b). A review on constructed treatment wetlands for removal of pollutants in the agricultural runoff. *Sustainability, 13*(24), 13578.

Tao, S., Fang, J., Zhao, X., & Guo, Q. (2015). Rapid loss of lakes on the mongolian plateau. In: *Proceedings of the National Academy of Sciences of the United States of America112:* 2281–2286. https://doi.org/10.1073/pnas.1411748112.

Terrisse, A., Karner, M., Kaufmann, J., & Ernoul, L. (2025). Characterizing governance models for upscaling wetland restoration. *Environmental Management, 75*(5), 1155–1167.

Trásy-Havril, T., Szkolnikovics-Simon, S., & Mádl-Szőnyi, J. (2022). How complex groundwater flow systems respond to climate change induced recharge reduction? *Water, 14*(19), 3026.

Turcios, A. E., Miglio, R., Vela, R., Sánchez, G., Bergier, T., Włodyka-Bergier, A., & &Papenbrock, J. (2021). From natural habitats to successful application-Role of halophytes in the treatment of saline wastewater in constructed wetlands with a focus on Latin America. *Environmental and Experimental Botany, 190*, 104583.

Tyagi, I., Tyagi, K., Bhutiani, R., Chandra, K., & Kumar, V. (2021). Bacterial diversity assessment of world's largest sewage-fed fish farms with special reference to water quality: A Ramsar site. *Environmental Science and Pollution Research, 28*(31), 42372–42386.

USEPA. (2022). *Wetlands protection and restoration*. https://www.epa.gov/wetlands/basicinformation-about-wetland-restoration-and-protection.

Vasumathi, V., Kalpana, R., Pazhanivelan, S., Kumaraperumal, R., Priya, M. V., Varshini, P. A., & Pon Arasan, A. (2023). A Review on wetlands – Threats, conservation, strategies and policies. *Journal of Current Crop Science & Technology, 2023, 110*(1-3), p58 1-3, p58.

Verma, A. K., & Singh, T. N. (2013). Prediction of water quality from simple field parameters. *Environmental Earth Sciences, 69*, 821–829.

Verma, S., Verma, S., Ramakant, R., Pandey, V., & Verma, A. (2025). Comprehensive assessment of physico-chemical and biological parameters in water quality monitoring: A review of contaminants, indicators, and health impacts. *International Journal of Horticulture, Agriculture and Food science, 9*(2), 611019.

Vijayaraghavan, K., Biswal, B. K., Adam, M. G., Soh, S. H., Tsen-Tieng, D. L., Davis, A. P., Chew, S. H., Tan, P. Y., Babovic, V., & Balasubramanian, R. (2021). Bioretention systems for stormwater management: Recent advances and future prospects. *Journal of Environmental Management, 292*, 112766.

Wang, H., Sun, J., Xu, J., & Sheng, L. (2021). Study on clogging mechanisms of constructed wetlands from the perspective of wastewater electrical conductivity change under different substrate conditions. *Journal of Environmental Management, 292*, 112813.

Wang, J., Long, Y., Yu, G., Wang, G., Zhou, Z., Li, P., Zhang, Y., Yang, K., & Wang, S. (2022). A review on microorganisms in constructed wetlands for typical pollutant removal: Species, function, and diversity. *Frontiers in Microbiology, 13*, 845725.

Wang, S., Shentu, H., Yu, H., Wang, L., Wang, J., Ma, J., Zheng, H., Huang, S., Dong, L., & Wei, J. (2024). Impact of urbanization and land use on wetland water quality: A case study in Mengxi town. *Urban Climate, 55*, 101855.

Weber, K. P. (2016). Microbial community assessment in wetlands for water pollution control: Past, present, and future outlook. *Water, 8*(11), 503.

World Population Review. (2025). Current World Population: 8,005,176,000. (2025-06-01). https://worldpopulationreview.com/.

Wu, S., Vymazal, J., & Brix, H. (2019). Critical review: Biogeochemical networking of iron in constructed wetlands for wastewater treatment. *Environmental Science & Technology, 53*(14), 7930–7944.

Wu, X., Ma, T., & Wang, Y. (2020). Surface water and groundwater interactions in wetlands. *Journal of Earth Science, 31*(5), 1016–1028.

Xu, D., Gautam, B., Zeli, T., Eva, S., Alan, V. D. V., Tian, Z., Ivanov, V. Y., & Leung, L. R. (2024). Climate change will reduce North American inland wetland areas and disrupt their seasonal regimes. *Nature Communications, 15*(1), 2438.

Yadav, M., Negi, V., & Verma, H. (2024). Impact of synthetic microfibers and their toxicity on the environment. In *Sustainable microbial technology for synthetic and cellulosic microfiber bioremediation*. Springer Nature Switzerland, 123–156.

Zhang, W., ElDidi, H., Masuda, Y. J., Meinzen-Dick, R. S., Swallow, K. A., Ringler, C., DeMello, N., & Aldous, A. (2023). Community-based conservation of freshwater resources: Learning from a critical review of the literature and case studies. *Society & Natural Resources, 36*(6), 733–754.

Zou, Z., Xiao, X., Dong, J., & Wang, J. (2018). Divergent trends of open-surface water body area in the contiguous United States from 1984 to 2016. In: *Proceedings of the National Academy of Sciences of the United States of America.115,* (15), 3810–3815. https://doi.org/10.1073/pnas.1719275115.

Chapter 3

Strengthening coastal ecosystem resilience: integrating biodiversity, water, soil health, ecosystem services, and stressors in coastal wetlands

Sushree Swati Rout[1] and Bhaskar Behera[1,2]
[1]*P.G. Department of Zoology, Fakir Mohan University, Balasore, Odisha, India,* [2]*P.G. Department of Bioscience and Biotechnology, Fakir Mohan University, Balasore, Odisha, India*

3.1 Introduction

Services provided by ecosystems are critical to human health and welfare. Ecosystems under stress provide deteriorated services that may have detrimental consequences on mental and physical health, while healthy ecosystems offer high-quality ecosystem services (Zhou et al., 2025). For convenience, biodiversity-supported ecosystem services that promote human well-being can be grouped into several categories, including better mental and physical health; reduced rates of inflammatory and other noninfectious diseases; reduced spread of some infectious diseases; aesthetic, cultural, recreational, socioeconomic, and spiritual benefits; and material goods like food, fiber, and medications (Buffière et al., 2025). To provide ecosystem services that humans want from coastal ecosystems, such as fish nursery habitat, water purification, flood risk reduction, climate modulation, nutrient cycling, and others, the biodiversity in coastal and marine ecosystems—including marshes, mangrove forests, and seagrass meadows—is crucial (DSouza et al., 2025). The operation of marine ecosystems and the services they provide are supported by marine biodiversity; without biodiversity, ecosystem services would not exist. In terms of the ecosystem services they provide, wetlands—particularly mangroves and tidal marshes—are regarded as the most important, or at least among the most, of our ecosystems (Chen et al., 2025). Their biodiversity serves as a crucial basis for the ongoing generation and provision of ecosystem services. Yet, the loss of natural biodiversity along our coastlines has already reduced the ecosystem's ability to provide ecosystem services and increased the risks to human health from dangerous microorganisms, harsh weather, and chemical pollutants (Rossi et al., 2025). Moreover, biodiversity could have intricate connections to the development and spread of illness. Numerous studies have examined how local biodiversity affects the spread of infectious diseases in humans, typically by examining how species and habitat diversity influence the abundance, behavior, or condition of hosts, vectors, or parasites (Xiong & Li, 2024). Numerous studies have shown that biodiversity may help prevent the spread of illness. When compared to corals at locations without nearby seagrasses, corals with seagrasses had twice as many disease indicators (Heysham & Elkadi, 2025). The researchers created this study after doing fieldwork on Indonesian coral reefs, which resulted in amebic dysentery for all participants.

With a vast array of marine (seamounts, frontal zones, and pelagic habitats) and coastal (mangrove forests, seagrass beds, and salt marshes) ecosystems and species, the seas connect all coastal zones and provide the biggest livable region on Earth (Larsen et al., 2025). Spatial management tools, restoration projects, and marine protected areas may all be built on a wealth of expertise in managing marine and coastal ecosystems using tried-and-true methods. In the first place, little is known about how marine ecosystems and biodiversity provide ecosystem services. As a result, parameterizing models of marine and coastal ecosystems to evaluate change and quantify advantages in monetary terms to support cost–benefit analysis is more difficult (Halekotte et al., 2025). Additionally, compared with terrestrial systems, this leads to greater

Wetlands: Biodiversity, Health and their Role in Ecosystem Services. DOI: https://doi.org/10.1016/B978-0-443-34222-6.00016-4

ambiguity regarding the optimal design and implementation of successful interventions. While this has long been understood and implemented via the integrated coastal zone management concept, its effectiveness has varied (Issah et al., 2025). Lastly, increased public and policy knowledge of the importance of marine and coastal ecosystems is necessary for the successful implementation of the marine and coastal environment. This will involve ocean literacy, participation, cooperation, and conceptual understanding.

3.2 Integrating biodiversity

It is widely acknowledged that nature has great value and provides ecosystem services that promote both human and planetary well-being. Therefore using nature's power has become a key strategy for addressing social issues (Gunawan & Nakajima, 2025). By preserving the supply of ecosystem services, restoring or maintaining nature in a comprehensive and sustainable manner will contribute to biodiversity and human well-being. To accomplish more for cross-cutting challenges, it combines various methods and calls for a more comprehensive integrated viewpoint (Bai & Li, 2023). For some stakeholder groups, restoring a coastal environment to improve biodiversity, slow down climate change, and strengthen coastal protection may have detrimental economic effects. Interventions that satisfy this criterion must be planned to improve biodiversity while tackling other social issues at the same time (Tran et al., 2024).

3.3 Biodiversity and coastal resilience

Biodiversity is the cornerstone of ecosystem resilience, enabling coastal wetlands to recover from disturbances and adapt to changing conditions. Wetlands host an extensive range of vegetation, including mangroves, salt marshes, and seagrasses, which play crucial roles in preserving environmental functions (El-Sharkawy et al., 2025). Mangroves stabilize shorelines with their dense root systems, while salt marshes provide breeding grounds for fish and migratory birds (Aburto-oropeza et al., 2025). The loss of biodiversity due to habitat destruction and invasive species disrupts these critical functions, reducing the adaptive capacity of wetlands. Protecting native species, restoring degraded habitats, and fostering ecological connectivity are essential strategies for maintaining biodiversity and enhancing ecosystem resilience (Samper-villarreal et al., 2025).

3.4 Water health in coastal wetlands

Water quality and hydrological processes are vital for the functionality of coastal wetlands. Nutrient pollution, often from agricultural runoff and urban wastewater, leads to eutrophication, algal blooms, and hypoxic conditions that degrade aquatic habitats (Leng et al., 2025). Altered hydrology from damming, dredging, and water diversion further impacts wetland ecosystems by disrupting natural flow regimes and salinity gradients. Restoring natural hydrological patterns through sustainable water management practices. Implementing buffer zones, reducing nutrient inputs, and enhancing freshwater inflows can improve water health and maintain the ecological integrity of coastal wetlands (Aprijanto et al., 2025).

3.5 Sediment contamination and carbon sequestration

Coastal wetland soils, rich in organic matter, act as significant carbon sinks, mitigating environmental change by sequestering large quantities of carbon under anaerobic conditions. However, sediment health is threatened by land reclamation, pollution, and rising salinity due to sea-level rise (Fuady et al., 2025). Soil degradation reduces the wetland's ability to sequester carbon and support plant growth, ultimately impacting biodiversity and ecosystem functions (Fig. 3.1). Strategies to preserve soil health include preventing soil erosion, reducing chemical inputs, and promoting the restoration of wetland vegetation, which stabilizes sediments and enhances organic matter accumulation (Toukabri & Chaabi, 2025).

3.6 Sediment pollution due to mining

Mining, quarrying, and manufacturing processes significantly disturb sediments and release pollutants into surrounding ecosystems, posing severe environmental threats (Wang et al., 2025). These activities often involve the extraction and processing of raw materials, which can lead to soil erosion, sediment displacement, and the contamination of water bodies with heavy metals, chemicals, and other toxic substances (He et al., 2025). For example, mining operations generate tailings that frequently contain hazardous elements like mercury, arsenic, and lead, which leach into nearby rivers and wetlands, impacting aquatic habitats (Table 3.1) and biodiversity (Visave & Aldrich, 2025). Quarrying disturbs large areas of land, increasing sediment runoff into waterways and altering water quality by introducing suspended solids that reduce

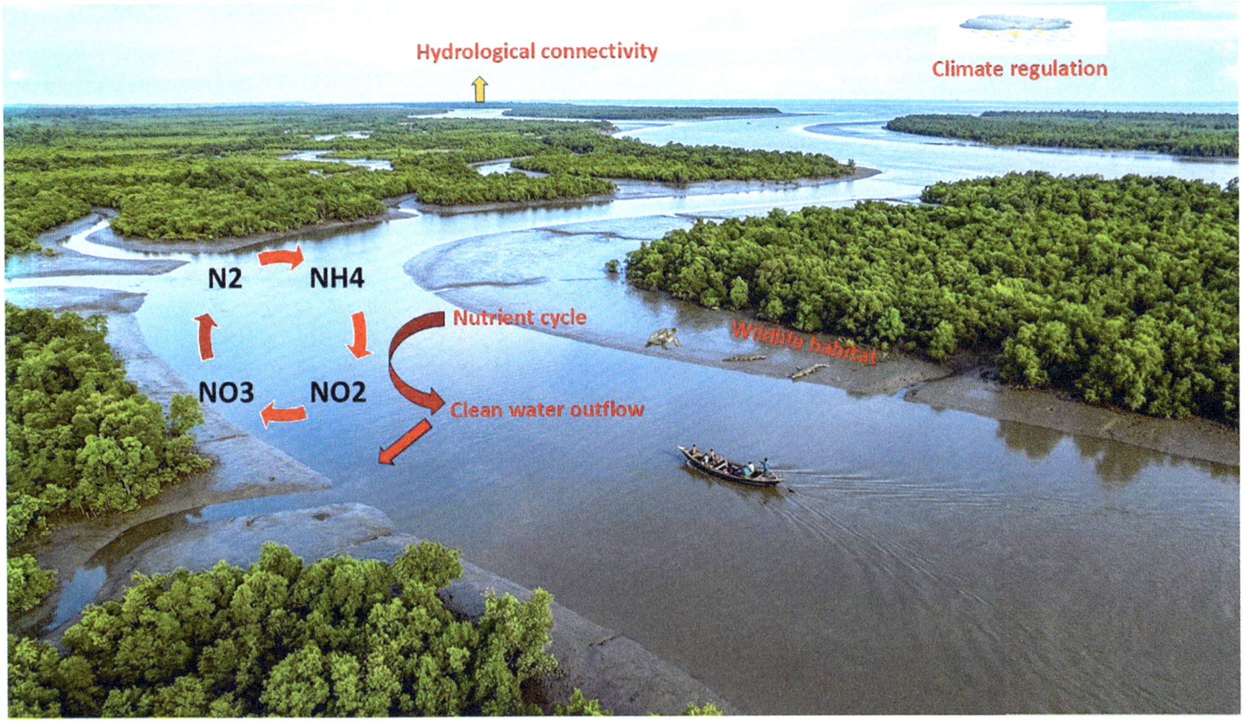

FIGURE 3.1 Restoring natural hydrological regimes and implementing practices.

TABLE 3.1 Different aspects of sediment pollution and their effect (Mirzaei & Ajdari, 2024).

Different aspects	Activities
Source of pollution	Mining activities (e.g., excavation, waste disposal, erosion from disturbed land)
Types of sediment	Fine particles (clay, silt), coarse particles (sand, gravel), tailings, and debris
Processes of pollution	Erosion, runoff from stockpiles, waste rock, and tailing ponds, transportation by water and wind
Impacts on ecosystems	Sediment smothering aquatic habitats, decreased water clarity, altered water temperature, and pH levels
Impacts on water quality	Increased turbidity, reduced oxygen levels, contamination with metals and chemicals from mining waste
Impacts on human health	Contaminated water sources, accumulation of heavy metals in food chains
Mitigation measures	Erosion control (e.g., silt fences, revegetation), proper waste management (e.g., tailings containment), buffer zones around water bodies, water treatment, and sediment filtration systems

light penetration and oxygen levels. Similarly, manufacturing processes release industrial effluents, including harmful chemicals, dyes, and microplastics, into aquatic systems, affecting both the chemical composition and ecological balance of these environments (Mirzaei & Ajdari, 2024). The cumulative effects of these pollutants can disrupt aquatic food chains, harm fish and other wildlife, and render water unsafe for human consumption. Preventing these impacts requires stringent regulatory frameworks, proper waste management practices, and the adoption of environmentally sustainable technologies in mining, quarrying, and manufacturing operations (Jahan & Islam, 2025).

3.7 Runoff pollution

Runoff-induced coastal marine pollution is a significant environmental issue resulting from the flow of water, often from agricultural, industrial, and urban areas, into coastal ecosystems. This runoff carries contaminants like excess micronutrients, pesticides, heavy metal compounds, and untreated sewage, which have severe ecological consequences (Russo et al., 2025). Nutrient overloads from agricultural fertilizers cause injurious blooms and depletion of the ozone layer in marine waters, a phenomenon known as hypoxia or "dead zones." Industrial and urban runoff introduces toxic chemicals and microplastics, which harm aquatic organisms and disrupt food chains (Gijón Mancheño et al., 2025). Coastal wetlands and estuaries, vital for filtering pollutants and maintaining ecosystem health, are often overwhelmed, losing their natural capacity to mitigate pollution. Addressing runoff pollution requires sustainable land use practices, stricter regulation of pollutants, and the restoration of natural buffers like mangroves, salt marshes, and riparian zones to protect coastal marine ecosystems (Liu et al., 2021).

3.8 Ecosystem services

Coastal wetland soils, rich in organic matter, act as significant carbon sinks, mitigating environmental transformation by loading large quantities of carbon in anaerobic conditions (Martin et al., 2025). However, soil health is threatened by land reclamation, pollution, and rising salinity due to sea-level rise. Soil degradation reduces the wetland's ability to sequester carbon and support plant growth, ultimately impacting biodiversity and ecosystem functions (Feng et al., 2025). Strategies to preserve soil health include preventing soil erosion, reducing chemical inputs, and promoting the restoration of wetland vegetation, which stabilizes sediments and enhances organic matter accumulation.

3.9 Stressors in coastal wetlands

Human activities and climate change are the two primary stressors that threaten coastal wetland resilience. Rising sea levels, increased storm intensity, and temperature fluctuations alter wetland dynamics, leading to habitat loss, salinization, and species migration (Smith et al., 2025). Additionally, urbanization and industrial development encroach upon wetlands, causing fragmentation and pollution (Table 3.2). Addressing these stressors requires a multidisciplinary approach that includes adaptive management, policy interventions, and community engagement. For instance, adopting nature-based solutions like living shorelines, which combine vegetation and natural structures to enhance coastal resilience, can mitigate the effects of climate change while preserving wetland ecosystems (Kemarau et al., 2025).

3.10 Integration for resilience building

Strengthening the resilience of coastal wetlands demands a comprehensive, integrated approach that connects biodiversity, water, and soil health and the mitigation of stressors (Guo et al., 2025). Efforts to restore and conserve wetlands should

TABLE 3.2 Stressors in coastal wetlands and their impact on biodiversity (Omoregie et al., 2025).

Stressor	Source	Impacts
Habitat loss	Urban development, agriculture, and infrastructure projects	Loss of biodiversity, decreased water filtration capacity, and altered hydrology
Pollution	Agricultural runoff (pesticides, fertilizers), industrial discharges, oil spills	Contamination of water and sediment, toxic effects on aquatic organisms, and eutrophication
Invasive species	Human introduction of nonnative species (e.g., plants, animals)	Displacement of native species, alteration of food webs, and changes in habitat structure
Sea-level rise	Climate change	Loss of land due to inundation, increased salinity in freshwater areas, and erosion of coastal lines.
Coastal erosion	Natural processes, human activities (e.g., construction, dredging)	Loss of wetlands and coastal ecosystems, increased vulnerability to storm surges and flooding
Nutrient loading	Agricultural and urban runoff (excess nitrogen and phosphorus)	Eutrophication, algal blooms, and oxygen depletion lead to dead zones
Hydrological alteration	Dams, water diversion, and drainage projects	Altered water flow patterns, decreased sediment deposition, and reduced freshwater input

prioritize biodiversity hotspots, enhance hydrological and soil management, and account for climate projections. Policies and frameworks should integrate ecosystem services into decision-making processes, ensuring that the social, economic, and ecological benefits of coastal wetlands are sustained (Albrecht et al., 2025). Collaborative actions involving governments, researchers, local communities, and NGOs are critical to achieving these goals. By focusing on the interlinkages between these components, we can ensure the long-term sustainability and resilience of coastal wetlands, securing their invaluable contributions to both human well-being and environmental stability (Aprizal et al., 2025).

3.11 Effect on fish diversity

Marine fish diversity is profoundly affected by various environmental and anthropogenic factors, which have led to significant declines in populations and altered the structure and functioning of marine ecosystems (Takal et al., 2025). Pollution from industrial, agricultural, and urban runoff contributes to habitat degradation by introducing toxic substances, such as heavy metals, pesticides, and microplastics, which accumulate in the tissues of marine fish, leading to physiological stress, reduced reproductive success, and mortality (Heysham & Elkadi, 2025). Nutrient pollution leads to eutrophication, causing harmful algal blooms and hypoxic dead zones that reduce oxygen availability, forcing fish to migrate, adapt, or perish. Overfishing, a major driver of biodiversity loss, disrupts food chains and depletes key species, while bycatch practices threaten nontarget species, including endangered fish and other marine organisms (Xiong & Li, 2024). Climate change exacerbates these challenges by altering ocean temperatures, leading to shifts in species distributions, coral bleaching, and the damage of sensitive territories corresponding to coral reefs, seagrass meadows, and mangroves. Ocean acidification, driven by increased CO_2 absorption, affects calcifying organisms, destabilizing ecosystems that marine fish rely on for shelter and food (Chen et al., 2025). Additionally, habitat destruction caused by coastal development, bottom trawling, and dredging results in the loss of essential breeding and feeding grounds. Invasive species introduced through ballast water or aquaculture activities outcompete native fish, further reducing diversity. The cumulative effects of these stressors not only diminish marine fish populations but also threaten the maintenance of many homeless poor people who hinge on fisheries for food security and economic stability. Protecting marine fish diversity requires an integrated approach, including the establishment of marine protected areas, sustainable fishing practices, regulation of contamination, restoration of degraded habitats, and global efforts to combat climate change (Camacho-Muñoz et al., 2020). Only through coordinated actions that address the multifaceted threats to marine ecosystems can we ensure the preservation of fish diversity, which is critical for maintaining healthy oceans and the services they provide to humanity.

Wetland fish diversity is increasingly threatened by a combination of natural and anthropogenic factors, resulting in significant ecological imbalances and biodiversity loss. Wetlands, which serve as critical habitats for breeding, feeding, and sheltering diverse fish species, are being degraded by anthropogenic activities such as urbanization, agricultural expansion, and industrial development (DSouza et al., 2025). Habitat loss due to wetland drainage, land reclamation, and hydrological alterations, including damming and water diversion, severely impacts fish populations by disrupting their life cycles and fragmenting habitats. Pollution from agricultural runoff, laden with pesticides, fertilizers, and sediments, causes eutrophication, leading to oxygen depletion and harmful algal blooms that make wetland waters uninhabitable for many fish species (Buffière et al., 2025). Additionally, the accumulation of noxious substances, such as toxic metals and hydrocarbons, contaminates aquatic environments, affecting fish health, reproduction, and survival. Climate change further exacerbates the pressures on wetland fish diversity by altering temperature regimes, hydrological cycles, and precipitation patterns (Zhu et al., 2025). Increased temperatures and prolonged droughts can dry up wetlands, while rising sea levels lead to salinization of freshwater systems, forcing freshwater fish to either adapt or face local extinction. Invasive species, introduced through human activities, also pose a significant threat by competing with native fish for resources, preying on them, or introducing diseases (Zhou et al., 2025). Overfishing and unsustainable harvesting practices in wetlands reduce fish stocks and disproportionately impact larger, slow-reproducing species, thereby altering community composition and ecosystem dynamics. The decline in wetland fish diversity has cascading effects on the broader ecosystem, as fish play crucial roles in nutrient cycling, algal control, and the maintenance of ecological balance (DSouza et al., 2025). It also undermines the livelihoods and food security of communities that depend on wetland fisheries for sustenance and income. Conservation efforts, including the restoration of degraded coastal areas, the maintenance of protected areas, and sustainable water management practices, are essential to preserving fish diversity (Rossi et al., 2025). Addressing pollution, mitigating the impacts of climate change, and controlling invasive species are also critical steps. By adopting an integrated approach that values wetlands for their ecological, economic, and cultural importance, we can safeguard wetland fish diversity and the vital services these ecosystems provide (Heysham & Elkadi, 2025).

3.12 Discussion

Biodiversity is central to the resilience of coastal wetlands. A diverse array of plant and animal species allows these ecosystems to maintain their functionality, adapt to environmental changes, and recover from disturbances. Species like mangroves and seagrasses provide structural integrity, while microbial communities drive nutrient cycling and organic matter decomposition (Larsen et al., 2025). The presence of high biodiversity also enhances ecosystem stability by distributing ecological functions across species, creating redundancy that buffers against species loss. However, habitat degradation, pollution, and invasive species threaten biodiversity, leading to weakened ecosystem resilience (Bai & Li, 2023). Protecting and restoring biodiversity is therefore essential for ensuring the long-term health and adaptability of marine wetlands. Effective strategies include conserving native species, restoring degraded habitats, and creating ecological corridors to promote connectivity and genetic diversity (Tran et al., 2024). Water and soil health are the foundations upon which coastal wetland ecosystems thrive. Hydrological processes regulate salinity, nutrient flow, and sediment deposition, all of which influence plant and animal communities. Alterations to these processes, such as damming, dredging, or excessive water extraction, disrupt the delicate balance required for wetland functionality (Issah et al., 2025). Similarly, soil health is critical because wetland soils are significant carbon reservoirs that store vast amounts of organic carbon under anaerobic conditions. Degradation of wetland soils through erosion, contamination, or salinization reduces their capacity to sequester carbon and support vegetation, further compromising ecosystem resilience (Larsen et al., 2025). Restoring natural hydrological regimes and implementing practices that enhance soil stability and organic matter content are necessary steps to sustain wetland health and resilience. The interconnectedness of biodiversity, water and soil health, and ecosystem services necessitates integrated approaches to coastal wetland conservation and management. Single-focus interventions are often insufficient to address the complexity of wetland dynamics and the multiple stressors they face (Heysham & Elkadi, 2025). Integrated approaches combine ecological, social, and economic considerations, ensuring that conservation efforts are both effective and sustainable. For example, integrating wetland restoration into coastal development plans can balance the needs of urban growth with ecosystem protection. Similarly, fostering partnerships among governments, local communities, researchers, and NGOs can enhance the effectiveness of conservation initiatives by pooling resources, knowledge, and expertise (Chen et al., 2025). Additionally, climate change adaptation and mitigation strategies should prioritize coastal wetland conservation. Wetlands serve as natural carbon sinks, playing a significant role in reducing greenhouse gas concentrations (DSouza et al., 2025). Protecting these ecosystems not only supports global climate goals but also enhances local resilience to climate impacts. Incorporating wetlands into national and international climate strategies. Strengthening coastal wetland resilience requires a forward-looking approach that anticipates future challenges and builds adaptive capacity. Research into the ecological, social, and economic aspects of wetland systems can provide valuable insights for improving management practices (Díaz de León-Martínez et al., 2020). Developing and implementing green infrastructure solutions, such as constructed wetlands for wastewater treatment, can expand the benefits of wetland conservation. Additionally, community awareness and enlightening activities can foster a deeper appreciation for the importance of wetlands, encouraging greater community involvement in conservation efforts.

3.13 Conclusion

Coastal wetlands are vital ecosystems that serve as ecological, economic, and social lifelines for communities and biodiversity worldwide. Their unique ability to deliver serious environmental facilities, with carbon sequestration, marine sanitization, coastal protection, and habitats for diverse species, underscores their importance in addressing and demanding international encounters like microclimate transformation and ecological deprivation (Veríssimo & Roseta-Palma, 2023). However, these ecosystems are increasingly threatened by human-induced stressors such as urbanization, pollution, and unsustainable land use, compounded by the impacts of temperature variation, rising sea levels, and extreme weather events. Strengthening the resilience of coastal wetlands requires an integrated, multidimensional approach that combines biodiversity conservation, water and soil health management, and the safeguarding of ecosystem services. Protecting biodiversity ensures ecological stability and enhances the capacity of wetlands to adapt to changing conditions. Maintaining water and soil health not only supports wetland functionality but also strengthens their role as natural carbon sinks, mitigating the activities of microclimate revolution (Zhao et al., 2024). Additionally, recognizing the commercial and biological value of environmental facilities fosters sustainable management practices and incentivizes wetland conservation. Addressing the stressors that threaten coastal wetlands involves both proactive and adaptive strategies, such as restoring degraded habitats, adopting nature-based solutions like living shorelines, and reducing nutrient and pollutant loads (Hintze et al., 2020). Collaborative efforts among governments, researchers, local communities, and stakeholders are essential to implement these strategies effectively. Policy frameworks that integrate wetland conservation with coastal development planning, alongside community-driven initiatives, can ensure long-term resilience and sustainability (Jones et al., 2004). Coastal wetlands are indispensable assets for both human well-being and environmental stability. By prioritizing their protection and restoration through holistic and integrated approaches, we can not only safeguard these ecosystems but also enhance their capacity to support thriving biodiversity, mitigate climate risks, and sustain the livelihoods of millions (Liu et al., 2022). The resilience of coastal wetlands is ultimately a shared responsibility and a critical step toward a sustainable, climate-resilient future.

References

Aburto-Oropeza, O., Platzgummer, V., Favoretto, F. & Giron-Nava, A., 2025. Marine prosperity areas: A framework for aligning ecological restoration and human well-being using area-based protections, 1–16. https://doi.org/10.3389/fmars.2024.1491483.

Albrecht, E., Belinskij, A., & Heikkilä, E. (2025). Policy coherence for ecosystem-based management: Implementing EU water and marine policies in the Archipelago Sea. *Marine Policy, 171*. https://doi.org/10.1016/j.marpol.2024.106427.

Aprijanto, Prijambodo, T., Wibawa, B., et al. (2025). Innovative strategies for tidal flood protection: A systematic literature review on spatial management in coastal city (case study: Semarang City, Central Java, Indonesia). *Marine Systems & Ocean Technology, 20*. https://doi.org/10.1007/s40868-024-00158-5.

Aburto-Oropeza, O., Platzgummer, V., Favoretto, F. & Giron-Nava, A., 2025. Marine prosperity areas: A framework for aligning ecological restoration and human well-being using area-based protections, 1–16. https://doi.org/10.3389/fmars.2024.1491483.

Bai, J., & Li, X. (2023). The evolution of China's policies on marine and coastal ecosystems in climate change adaptation. *Frontiers Marine Sciences, 10*, 1–16. https://doi.org/10.3389/fmars.2023.1190132.

Buffière, P., Ramirez, D. A., Franco, R. T., et al. (2025). Oxygen traces impact on biological methanation from hydrogen and CO_2. *Bioresource Technology, 419*. https://doi.org/10.1016/j.biortech.2025.132080.

Camacho-Muñoz, D., Lawton, L. A., & Edwards, C. (2020). Degradation of okadaic acid in seawater by UV/TiO_2 photocatalysis – Proof of concept. *The Science of the Total Environment, 733*. https://doi.org/10.1016/j.scitotenv.2020.139346.

Chen, M., Jin, L., Liu, X., et al. (2025). Immobilization of ammonia-oxidizing bacteria using mycelial pellets: Preparation, characteristics, and application for nitritation. *Bioresource Technology, 419*, 132083. https://doi.org/10.1016/j.biortech.2025.132083.

Díaz de León-Martínez, L., de la Sierra-de la Vega, L., Palacios-Ramírez, A., et al. (2020). Critical review of social, environmental and health risk factors in the Mexican indigenous population and their capacity to respond to the COVID-19. *The Science of the Total Environment, 733*, 139357. https://doi.org/10.1016/j.scitotenv.2020.139357.

DSouza, G. C., Chio, C., Venkatesh, A., et al. (2025). Microbially degradable phenolic foams based on depolymerized Kraft lignin for hydrophilic applications. *Bioresource Technology, 419*, 132082. https://doi.org/10.1016/j.biortech.2025.132082.

El-Sharkawy, M., Alotaibi, M. O., Li, J., et al. (2025). Heavy metal pollution in coastal environments: Ecological implications and management strategies: A review. *Sustainability, 17*, 1–29. https://doi.org/10.3390/su17020701.

Feng, J., Yu, H., Wu, L., et al. (2025). Evaluating ecosystem characteristics and ecological carrying capacity for marine fauna stock enhancement within a marine ranching system. *Animals, 15*, 1–48. https://doi.org/10.3390/ani15020165.

Fuady, M., Kevin, M.A., Farrel, M.R. & Triaputri, A., 2025. Enhancing urban resilience: opportunities and challenges in adapting to natural disasters in Indonesian cities, pp.1–15.

Gijón Mancheño, A., Jafino, B. A., Hofland, B., et al. (2025). Nature meets infrastructure: The role of mangroves in strengthening bangladesh's coastal flood defenses. *Sustain, 17*, 1–14. https://doi.org/10.3390/su17041567.

Gunawan, R., & Nakajima, K. (2025). Asset pricing in a country embracing religious beliefs and social norms: Evidence from the Indonesian stock market. *Borsa Istanbul Review, 25*, 227–239. https://doi.org/10.1016/j.bir.2024.12.005.

Guo, X., Liu, Y., Xie, T. *et al.*, 2025. Impact of ecological restoration on carbon sink function in coastal wetlands: A review.

Halekotte, L., Mentges, A., & Lichte, D. (2025). Do we practice what we preach? The dissonance between resilience understanding and measurement. International Journal of Disaster Risk Reduction, *118*. https://doi.org/10.1016/j.ijdrr.2025.105265.

He, Q., Li, Z., Daleo, P., et al. (2025). Coastal wetland resilience through local, regional and global conservation. *Nature Reviews Biodiversity, 1*. https://doi.org/10.1038/s44358-024-00004-x.

Heysham, N., & Elkadi, H. (2025). Social capital geospatial network mapping for furniture-making: Impacts of long-term social bonds on city fabric and industry competitiveness. *Cities (London, England), 159*, 105802. https://doi.org/10.1016/j.cities.2025.105802.

Hintze, S., Glauser, G., & Hunkeler, D. (2020). Influence of surface water – Groundwater interactions on the spatial distribution of pesticide metabolites in groundwater. *The Science of the Total Environment, 733*, 139109. https://doi.org/10.1016/j.scitotenv.2020.139109.

Issah, M., Sulaiman, L. A. R., Aliu, F., et al. (2025). Deconstructing a single-actor resource ownership model: A study of the proposed uranium mining in the karoo region of South Africa. *The Extractive Industries and Society, 22*. https://doi.org/10.1016/j.exis.2024.101609 101609.

Jahan, F. & Islam, R., 2025. Carbon sequestration in coastal ecosystems: a review of modeling techniques and applications. 1, 41–70. https://doi.org/10.63125/4z73rb29

Jones, G. P., McCormick, M. I., Srinivasan, M., & Eagle, J. V. (2004). Coral decline threatens fish biodiversity in marine reserves. *Proceedings of the National Academy of Sciences of the United States of America, 101*, 8251–8253. https://doi.org/10.1073/pnas.0401277101.

Kemarau R.A., Suab S.A., Eboy O.V., et al. (2025). Integrative approaches in remote sensing and GIS for assessing climate change impacts across Malaysian ecosystems and societies.

Larsen, R. B., Ravn, S. H., & Santoro, E. (2025). House prices, endogenous productivity, and the effects of government spending shocks. *European Economic Review, 172*, 104937. https://doi.org/10.1016/j.euroecorev.2024.104937.

Leng, Z., Liu, J., He, C., et al. (2025). Deposition of sulfur by Spartina alterniflora promoted its ecological adaptability in cadmium-polluted coastal wetlands. *Bioresource Technology, 419*, 132069. https://doi.org/10.1016/j.biortech.2025.132069.

Liu, H. Y., Jay, M., & Chen, X. (2021). The role of nature-based solutions for improving environmental quality, health and well-being. *Sustainability, 13*, 1–56. https://doi.org/10.3390/su131910950.

Liu, Y., Wu, G., Fan, X., et al. (2022). Hydrological impacts of land use/cover changes in the Lake Victoria basin. *Ecological Indicaters, 145*, 109580. https://doi.org/10.1016/j.ecolind.2022.109580.

Martin, T., Henri, D. A., Martinez, L. M., et al. (2025). A common future for coastal peoples and seabirds facing extreme climatic events. *ICES J Marine Sciences, 82*, 1–7. https://doi.org/10.1093/icesjms/fsae198.

Mirzaei M.R., & Ajdari A. (2024). Marine ecosystem resilience: Integrative insights on climate change impacts and adaptive strategies, 2, 1–9.

Omoregie, A. I., Ouahbi, T., Kan, F. K., et al. (2025). Advancing slope stability and hydrological solutions through biocementation: A bibliometric review. *Hydrology, 12*. https://doi.org/10.3390/hydrology12010014.

Rossi, T. S., Francescato, L., Gupte, A. P., et al. (2025). Harnessing the potential of Cupriavidus necator for CO_2 capture from alcoholic fermentation and its bioconversion into poly(3-hydroxybutyrate). *Bioresource Technology, 419*, 132060. https://doi.org/10.1016/j.biortech.2025.132060.

Russo, L., Bellardini, D., Casotti, R., et al. (2025). The spatiotemporal variability of marine plankton ecosystem services at the regional scale: A combined approach using a systematic review and network analysis. *Sustaintabilty, 17*, 1–22. https://doi.org/10.3390/su17031182.

Samper-Villarreal, J., Adame, M.F. & Crooks, S. *et al.*, 2025. Review recommendations for strengthening blue carbon science. https://doi.org/10.1016/j.oneear.2025.101175.

Smith, G., Atkins, J., Gregory, A., & Elliott, M. (2025). The minimum complexity necessary: The value of a simple Social-Ecological systems analysis in holistic marine environmental management. *Sustaintabilty Future, 9*. https://doi.org/10.1016/j.sftr.2025.100476.

Takal, S. U., Tahiru, A. W., Fattah, I. R., et al. (2025). Enhancing resilience to climate change: A comprehensive PRISMA review of agricultural and non-agricultural adaptation strategies for Ghana. *Cogent Social Science, 11*. https://doi.org/10.1080/23311886.2025.2453905.

Toukabri M., & Chaabi A. (2025). Strengthening ecosystems and biodiversity conservation in response to climate change in Saudi Arabia, 357–372. https://doi.org/10.14207/ejsd.2025.v14n1p357.

Tran, T. M. A., Gagnon, V. S., & Schelly, C. (2024). A review of traditional ecological knowledge in resilient livelihoods and forest ecosystems: Lessons for restoration sciences and practices. *For Trees Livelihoods, 34*, 27–53. https://doi.org/10.1080/14728028.2024.2408725.

Veríssimo, D., & Roseta-Palma, C. (2023). Rewilding with the beaver in the iberian peninsula - Economic potential for river restoration. *Nature-Based Solution, 3*, 100055. https://doi.org/10.1016/j.nbsj.2023.100055.

Visave, J., & Aldrich, D. P. (2025). The role of social capital in strengthening community resilience against floods: A case study of Mumbai, India. *Climate Risk Management, 47*, 100685. https://doi.org/10.1016/j.crm.2024.100685.

Wang, R., Zhang, J., Zhang, D., et al. (2025). Problem-based techniques and practices of coastal ecological engineering construction in China. Environment, Development and Sustainability. https://doi.org/10.1007/s10668-025-06006-9.

Xiong, A., & Li, Y. (2024). The role of social capital in building community disaster resilience –empirical evidences from rural China. International Journal of Disaster Risk Reduction, *110*, 104623. https://doi.org/10.1016/j.ijdrr.2024.104623.

Zhao, Y., Han, Z., Zhang, C., et al. (2024). Coastal cultural ecosystem services: A bridge between the natural ecosystem and social ecosystem for sustainable development. *Land, 13*, 1352. https://doi.org/10.3390/land13091352.

Zhou, Z., Liu, S., Saleem, M., et al. (2025). Unraveling phase-dependent variations of viral community, virus-host linkage, and functional potential during manure composting process. *Bioresource Technology, 419*, 132081. https://doi.org/10.1016/j.biortech.2025.132081.

Zhu, L., Liu, L., Tan, C., et al. (2025). Sustainable decentralized food waste composting using a pulse alternating ventilation pilot-scale device: Case study based on LCA and LCC analysis. *Bioresource Technology, 419*, 132078. https://doi.org/10.1016/j.biortech.2025.132078.

Chapter 4

Monitoring of wetlands with integrated remote sensing and geographic information system applications: a case study of Kirat Sagar wetland during 1994–2024 in Mahoba, India

Surendra Kumar Chandniha[1], Diwakar Naidu[2], Arun Mondal[3], Harithalekshmi V.[4], Anupam Pandey[5], and Sananda Kundu[6]

[1]*BRSM College of Agricultural Engineering and Technology & Research Station, IGKV, Mungeli, Chhattisgarh, India,* [2]*SV College of Agricultural Engineering and Technology & Research Station, IGKV, Raipur, Chhattisgarh, India,* [3]*Department of Geography, University of Allahabad (A Central University), Prayagraj, Uttar Pradesh, India,* [4]*Project Scientist, Institute of Climate Change Studies, Kottayam, Kerala, India,* [5]*Department of Geography, University of Allahabad, Prayagraj, Uttar Pradesh, India,* [6]*Department of Geography, Manipur University, Imphal, Manipur, India*

4.1 Introduction

Wetlands are among the most critical ecosystems on Earth, offering a wide range of ecological, hydrological, and socioeconomic benefits. These dynamic landscapes serve as natural water reservoirs, carbon sinks, and habitats for diverse flora and fauna, making them indispensable for environmental stability. They play a vital role in flood control, groundwater recharge, water purification, and climate regulation. However, wetlands across the globe are facing severe threats from anthropogenic pressures and climate change, leading to their degradation and loss. Monitoring wetlands is, therefore, essential for their conservation and sustainable management. The advent of remote sensing (RS) and geographic information systems (GIS) technologies has revolutionized wetland monitoring, providing a cost-effective and efficient means of assessing spatial and temporal changes. Traditional methods relied on field surveys, which were often time-consuming, labor-intensive, and geographically limited. In contrast, RS and GIS applications enable continuous, large-scale monitoring with high accuracy. These technologies facilitate the detection of wetland changes over time, allowing researchers and policymakers to understand wetland dynamics, assess threats, and formulate effective conservation strategies. The present study focuses on monitoring the Kirat Sagar wetland in the Mahoba district using integrated RS and GIS approaches. The primary objective is to analyze wetland dynamics over the past four decades (1994–2024) across different seasons (summer and winter) using multitemporal satellite imagery. The study also assesses the impact of climate variability on the wetland through the analysis of minimum and maximum temperature and rainfall data from 1990 to 2024. The Modified Normalized Difference Water Index (MNDWI) was employed to delineate and evaluate fluctuations in wetland area. Landsat-5, Landsat-7, Landsat-8, and Landsat-9 imagery were processed using GIS-based techniques to analyze spatial and temporal variations in wetland extent. Meteorological data from official records were incorporated to validate significant changes in wetland area with respect to climatic parameters.

4.1.1 Significance of wetland monitoring

Wetlands contribute significantly to global biodiversity, water resource management, and climate resilience. They serve as breeding grounds for numerous aquatic species and migratory birds, supporting both local and global biodiversity (Demarquet

Wetlands: Biodiversity, Health and their Role in Ecosystem Services. DOI: https://doi.org/10.1016/B978-0-443-34222-6.00014-0

et al., 2023; Xu et al., 2020). Additionally, wetlands regulate hydrological cycles, acting as natural sponges that store excess rainfall and reduce the intensity of floods (Ferreira et al., 2023). Their ability to sequester carbon helps mitigate climate change by reducing greenhouse gas concentrations in the atmosphere (Pant, 2023; Were et al., 2019). However, the rapid expansion of urban settlements, agricultural encroachment, and industrial activities has led to widespread wetland degradation. Rising temperatures, altered precipitation patterns, and extreme weather events have further increased the vulnerability of wetlands (Ballut-Dajud et al., 2022; Kundu et al., 2024; Xiong et al., 2023). RS and GIS applications provide a scientific basis for wetland conservation by enabling continuous monitoring and timely assessment. These technologies facilitate the mapping and classification of wetlands based on spectral signatures, allowing for accurate detection of changes in wetland size, vegetation cover, and water quality (Adam et al., 2010; Guo et al., 2017). The integration of meteorological data further enhances monitoring efforts by linking observed spatial variations with climatic trends.

4.1.2 Remote sensing and geographic information system in wetland monitoring

RS provides a synoptic view of wetlands, enabling the assessment of wetland extent, vegetation health, and water quality over time. Satellite-based observations enable classification of wetlands using spectral reflectance, distinguishing open water, emergent vegetation, and terrestrial areas. Since water bodies exhibit high absorption and low reflectance in the near-infrared (NIR) band, whereas bare ground and vegetation have higher reflectance in this band, the NIR band is frequently used for studying water bodies (Avtar et al., 2014; Ma et al., 2019). The MNDWI was employed in this study to delineate water bodies and evaluate their seasonal variations. MNDWI is an improved index for water body extraction, enhancing the discrimination of water features from nonwater features and minimizing misclassification errors (Ali et al., 2019; Haibo et al., 2011; Jiang et al., 2021; Yue et al., 2020). Landsat satellite imagery has been instrumental in tracking wetland changes in numerous studies (Baker et al., 2007; Demarquet et al., 2023; Jafarzadeh et al., 2022; Kayastha et al., 2012; Ozesmi & Bauer, 2002). Landsat-5, Landsat-7, Landsat-8, and Landsat-9 have provided both historical and recent datasets, enabling comprehensive analysis of wetland transformations over the past four decades (Demarquet et al., 2023; Hemati et al., 2021; Wulder et al., 2022). Image processing techniques, including supervised classification, change detection, and time-series analysis, were applied to identify significant shifts in wetland boundaries. GIS-based spatial analysis further facilitated the integration of climatic data with wetland observations, allowing for a multidimensional understanding of wetland changes.

4.1.3 Climate variability and its impact on wetlands

Climate change poses a significant threat to wetland ecosystems by altering precipitation patterns, increasing temperature extremes, and intensifying drought events (Day et al., 2008; Moomaw et al., 2018; Wang & Gu, 2021; Xiong et al., 2023).

Validating wetland changes using meteorological data provides a scientific basis for understanding the impact of climate variability on wetland ecosystems. Such an approach is critical for formulating adaptive management strategies that can mitigate the adverse effects of climate change on wetlands.

4.1.4 Anthropogenic interventions and wetland degradation

Apart from climate change, human activities have significantly contributed to wetland degradation. Urbanization, deforestation, industrial pollution, and excessive water extraction have led to habitat fragmentation, biodiversity loss, and reduced wetland functionality (Ballut-Dajud et al., 2022; Butt et al., 2021; Chakraborty et al., 2023; Faulkner, 2004). In the case of Kirat Sagar, unregulated land use changes and agricultural encroachments have encroached upon wetland boundaries, leading to a decline in its spatial extent. Poor waste management and pollution from nearby settlements have further deteriorated water quality, threatening the health of aquatic ecosystems.

Integrating RS and GIS techniques allows for the identification and mapping of anthropogenic stressors affecting wetlands. Land use classification and change detection analyses highlight areas where human activities have encroached upon wetland zones (Kassaye, 2014; Rebelo et al., 2009; Wu et al., 2017). This information is vital for policymakers and conservationists to implement effective restoration measures and sustainable land management practices.

4.1.5 The need for a sustainable conservation framework

To ensure the long-term sustainability of Kirat Sagar wetland, a holistic conservation framework is required. This framework should incorporate RS-based monitoring, climate-resilience strategies, community participation, and policy interventions. The study underscores the importance of:

1. Strengthening conservation policies: Implementing strict regulations to prevent encroachments and pollution.

2. Sustainable water management practices: Promoting rainwater harvesting, controlled irrigation, and groundwater recharge initiatives.
3. Community engagement: Involving local stakeholders in wetland conservation programs through awareness campaigns and participatory decision-making.
4. Long-term monitoring programs: Establishing a dedicated wetland monitoring system using advanced RS and AI-driven predictive modeling.

4.2 Study area: Kirat Sagar wetland, Mahoba district

Kirat Sagar is a significant freshwater wetland located in the Mahoba district of Uttar Pradesh, India. This wetland plays a vital role in regional water management, agriculture, and biodiversity conservation. It supports diverse aquatic life and provides livelihood opportunities for local communities engaged in fishing, farming, and tourism. However, recent observations indicate substantial changes in the wetland's hydrology, primarily attributed to climatic fluctuations and human interventions. Encroachments for agriculture, urban expansion, and excessive water extraction have significantly altered the wetland's natural hydrological cycle. Understanding the long-term changes in Kirat Sagar is crucial for designing effective conservation strategies. Location map of the study area is shown in Fig. 4.1.

4.2.1 Data acquisition

The study relies on a combination of satellite imagery and meteorological data spanning four decades (1994–2024). The primary datasets used include:

- Satellite data:
  - o Landsat-5 TM (1994–2013)
  - o Landsat-7 ETM+ (1999–2024)
  - o Landsat-8 OLI (2013–24)
  - o Landsat-9 OLI-2 (2021–24)
- Meteorological data:
  - o Minimum and maximum temperature records (1990–2024)
  - o Annual and seasonal rainfall data (1990–2024)
  - o Data sourced from the India Meteorological Department (IMD) and other official repositories
- Ancillary data:
  - o Topographic maps and land use/land cover data from the Survey of India
  - o Existing wetland inventories and Study Area Selection

4.3 Methodology

The methodology for monitoring the Kirat Sagar wetland involves the integration of RS and GIS techniques along with meteorological data analysis. The step-by-step procedure is as follows, shown in Fig. 4.2:

4.3.1 Satellite imagery acquisition

Multitemporal satellite imagery was acquired for wetland analysis over four decades (1994–2024). The selected datasets include:

- 1994: Summer (25-04-1994), Winter (04-02-1994)
- 2003: Summer (26-04-2003), Winter (05-02-2003)
- 2014: Summer (02-05-2014), Winter (25-12-2013)
- 2024: Summer (05-05-2024), Winter (30-01-2024)

Landsat-5, Landsat-7, Landsat-8, and Landsat-9 satellite images were used to analyze wetland variations over time. Further details of the datasets are illustrated in Table 4.1.

4.3.1.1 Preprocessing of satellite data

To ensure accuracy and consistency, all satellite images were preprocessed using standardized techniques:

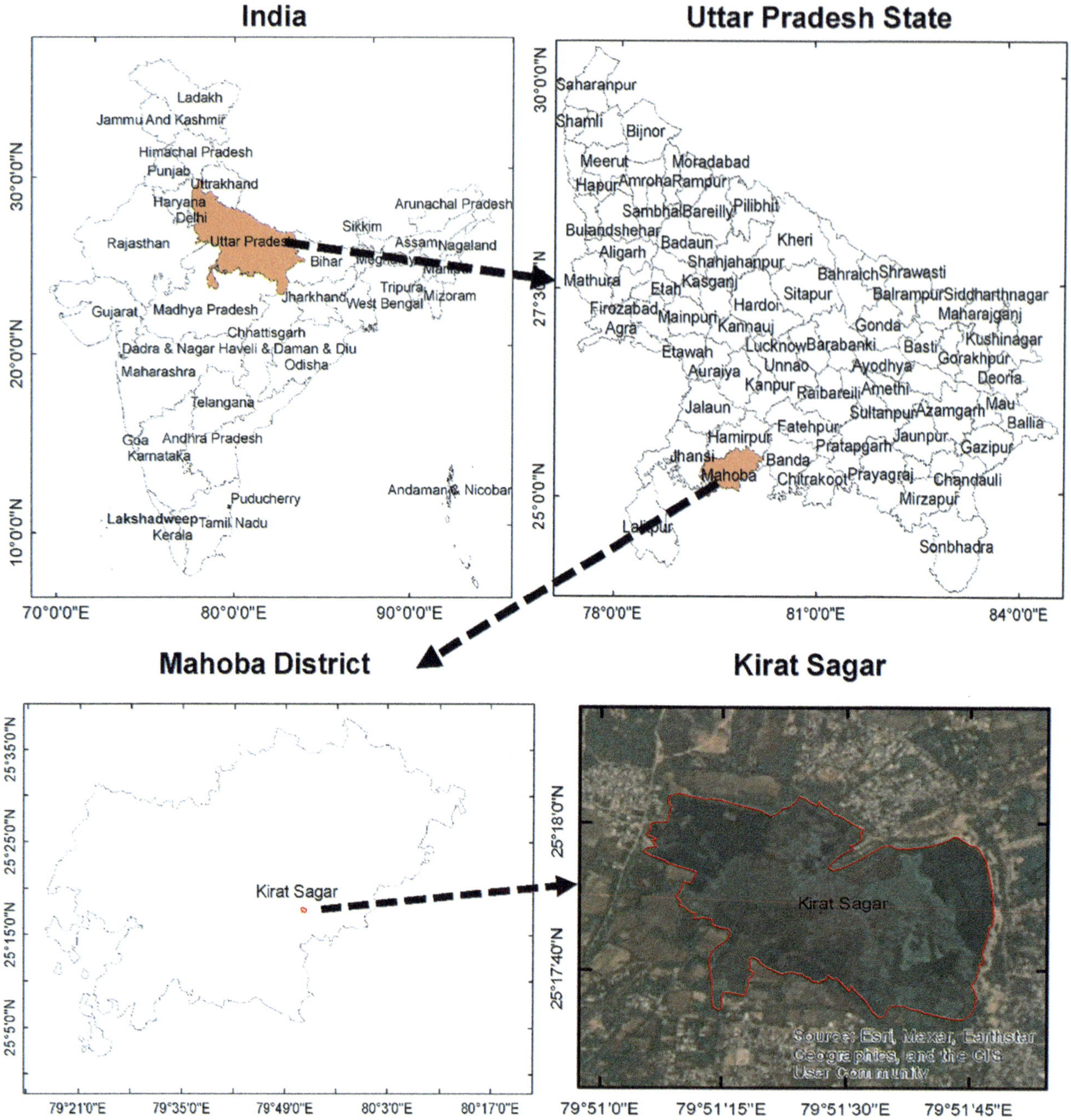

FIGURE 4.1 Kirat Sagar wetland in Google image. Map lines delineate study areas and do not necessarily depict accepted national boundaries.

1. Radiometric and geometric correction:
   - Atmospheric corrections were applied to remove haze and radiometric distortions.
   - Georeferencing was performed to align images with standard coordinate systems (WGS 84, UTM Zone 44 N).
2. Image enhancement:
   - Contrast stretching and histogram equalization were applied to improve image interpretability.
   - Cloud cover masking was performed to eliminate interference.

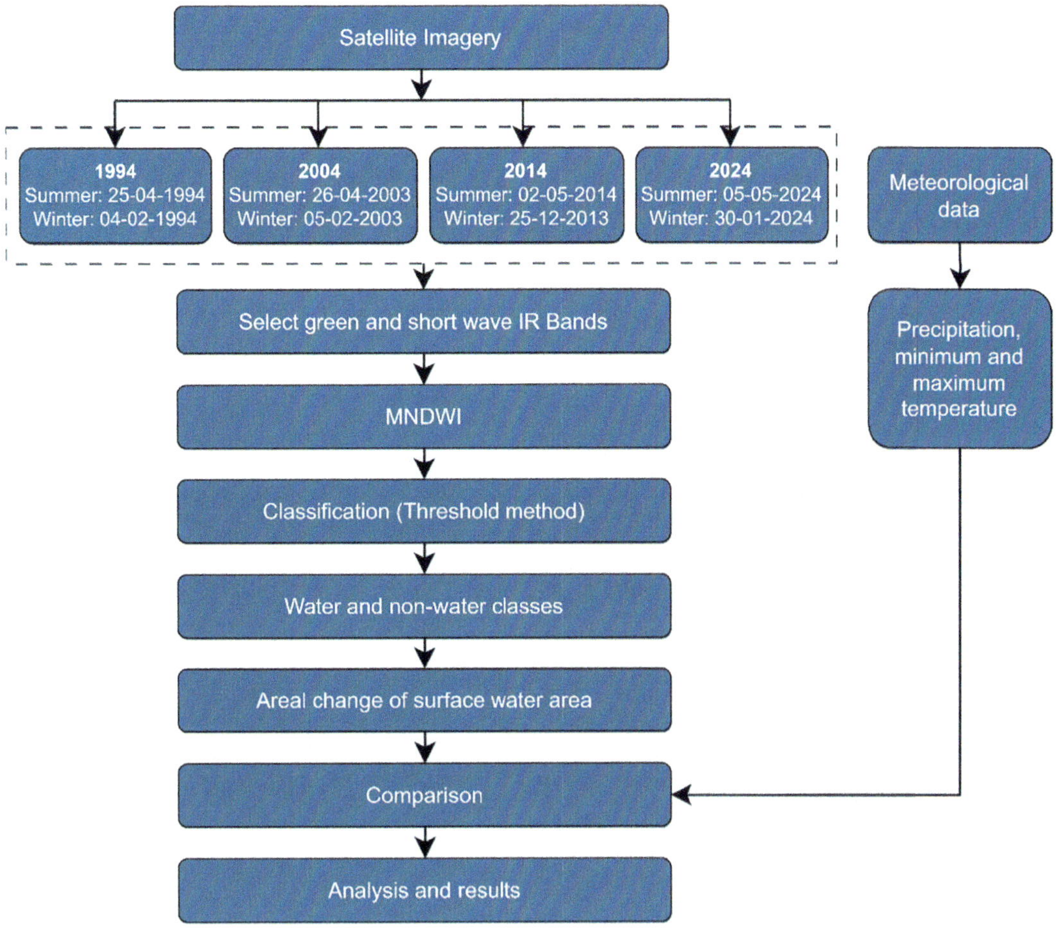

FIGURE 4.2 Methodology of the study. *MNDWI*, Modified Normalized Difference Water Index.

4.3.1.2 Wetland delineation using Modified Normalized Difference Water Index

The MNDWI was used to delineate the Kirat Sagar wetland and analyze its spatial-temporal variations. In the MNDWI, the green band enhances the identification of water features, while the shortwave infrared (SWIR) band improves discrimination between water and nonwater areas. Threshold values for classification were derived from historical wetland boundary data and further validated through field observations.

4.3.1.3 Change detection analysis

To quantify wetland changes over time, a multitemporal change detection approach was adopted:

1. Supervised classification:
   - Maximum likelihood classification was applied to categorize wetland areas.
   - Accuracy assessment was performed using ground truth data.
2. Wetland extent mapping:
   - Seasonal variations were analyzed by comparing summer and winter imagery.
   - Historical trends were identified by assessing wetland shrinkage and expansion patterns.
3. Statistical analysis:
   - Correlation analysis was conducted between wetland changes and meteorological parameters.
   - Regression models were used to evaluate the impact of temperature and rainfall variability on wetland sustainability.

TABLE 4.1 Specification of satellite images.

Landsat scene ID	Date of acquisition	Coordinated universal time (UTC)	Path/row	Sun elevation (degree)	Sun Azimuth (degree)	Cloud cover (%)	Earth-sun distance (astronomical unit)	Resolution of VNIR bands (m)	Resolution of TIR bands (m)
Summer season									
LT51440421994115ISP00	25-Apr-94	04:27:10	144/042	56.89	106.64	8.00	1.0059886	30	120
LE71440432003116ASN00	26-Apr-03	04:56:26	144/043	63.48	110.18	0.00	1.0061637	30	120
LC81440422014122LG-N01	02-May-14	05:07:00	144/042	66.68	112.77	5.48	1.0077811	30	120
LC91440432024126LG-N00	05-May-24	05:07:17	144/043	67.49	106.97	10.26	1.0086205	30	120
Winter season									
LT51440421994035ISP01	04-Feb-94	04:28:27	144/042	35.01	136.30	56.00	0.9857910	30	120
LE71440422003036SGS00	05-Feb-03	04:55:53	144/042	39.21	142.63	2.00	0.9858970	30	120
LC81440422013359LG-N01	25-Dec-13	05:08:44	144/042	36.00	154.20	0.59	0.9835339	30	120
LC91440422024030LG-N00	30-Jan-24	05:07:34	144/042	39.37	147.14	44.09	0.9849821	30	120

4.3.1.4 Geographic information system-based spatial analysis

GIS techniques were used to integrate and analyze multisource data:

- Buffer analysis: Examined anthropogenic pressures around the wetland boundary.
- Zonal statistics: Assessed wetland area fluctuations in relation to rainfall and temperature.
- Overlay analysis: Mapped land use changes and identified wetland encroachments over time.

4.3.1.5 Validation and accuracy assessment

To ensure reliability, the results were validated using:

- Ground truthing surveys conducted at selected locations.
- Historical records and topographic maps for cross-verification.
- Kappa coefficient and overall accuracy assessment for classification validation.

4.3.2 Meteorological data collection

Meteorological data, including precipitation, minimum and maximum temperatures, were obtained from official records such as the IMD. These data were used to assess climate variability and its impact on the wetland.

4.3.3 Selection of spectral bands

For wetland delineation, the Green Band and SWIR Band were selected from the satellite imagery. These bands enhance the identification of water features and separate them from nonwater areas.

4.3.4 Calculation of Modified Normalized Difference Water Index

The MNWDI formula was applied to distinguish water bodies from surrounding land features (Xu, 2006):

$$\text{MNWDI} = \frac{(\text{Green} - \text{SWIR})}{(\text{Green} + \text{SWIR})}$$

Threshold values were determined based on historical wetland boundary data.

4.3.5 Classification using the threshold method

A classification technique based on thresholding was used to categorize land features into water and nonwater classes (Bijeesh & Narasimhamurthy, 2020; Dronova, 2015). The classification ensured precise delineation of the wetland.

4.3.6 Determination of areal changes in surface water

The classified images were analyzed to estimate the changes in the surface water area of the wetland. The variations were assessed separately for summer and winter to identify seasonal trends.

4.3.7 Comparison of wetland dynamics over time

Wetland changes were compared across different years (1994, 2003, 2014, 2024) using spatial overlays in GIS. This helped to identify trends in wetland shrinkage or expansion.

4.3.8 Integration of climate data with wetland changes

Meteorological data (temperature and precipitation records) were correlated with wetland changes to assess the impact of climate variability. Regression models were applied to analyze the relationship between temperature, rainfall, and wetland area fluctuations.

4.3.9 Analysis and interpretation of results

The final step involved interpreting the results to determine the key factors affecting wetland sustainability. The findings were used to provide insights into wetland conservation and management strategies.

4.4 Results and discussion

4.4.1 False color composite image analysis of Kirat Sagar wetland

The analysis of false color composite (FCC) images for the summer and winter seasons of 1994, 2003, 2014, and 2024 reveals significant temporal changes in the spatial extent of the Kirat Sagar wetland. The variations in the wetland area are influenced by seasonal differences, climate variability, and anthropogenic pressures. The interpretation of FCC images provides crucial insights into wetland dynamics over the past three decades, as shown in Figs. 4.3 and 4.4. In this visualization (Fig. 4.3), vegetation appeared in shades of red due to its high reflectance in the NIR band. Water bodies appeared dark because they strongly absorbed light in both the NIR and Red bands. Built-up and barren land areas were visible in shades of blue, gray, or white, indicating surfaces with lower vegetation cover and different reflectance properties. The FCC images of the Kirat Sagar wetland from 1994, 2003, 2014, and 2024 revealed significant changes in vegetation, water bodies, and land use over the past 30 years. These images helped in understanding how the wetland changed due to natural and human activities. In 1994 and 2003, the FCC images showed bright red areas around the wetland, indicating dense, healthy vegetation. This suggested that the ecosystem was stable, with good coverage of aquatic plants, marshes, and greenery. However, in 2014 and 2024, the red color appeared less intense in some areas, indicating a loss of vegetation. This decline was likely caused by land use changes, deforestation, agricultural expansion, or changes in water availability that affected plant growth. The analysis of water bodies in the FCC images was also crucial for understanding wetland health. In 1994 and 2003, the dark blue or black areas, which represented water, appeared larger and more continuous. However, in 2014 and 2024, the water extent seemed to have decreased, especially during the summer season. This suggested that the wetland's water storage had decreased over time, possibly due to increased evaporation, reduced rainfall, or human activities such as water diversion for agriculture or urban use. Seasonal variations were also evident, with more water in winter than in summer, as expected given natural seasonal changes. However, the wetland's water area had been shrinking over the years, indicating long-term drying trends caused by climate change or excessive groundwater use.

Another key observation from the FCC images was the increase in built-up areas and changes in land use. In earlier years, there were fewer cyan, gray, or white patches, which typically indicated urban infrastructure, barren land, or degraded areas. However, by 2014 and 2024, these features had become more noticeable, suggesting increased urban expansion, land encroachment, or agricultural activity near the wetland. The growth of settlements and roads likely affected the wetland through pollution, changes in drainage patterns, or reduced groundwater recharge. The loss of natural vegetation buffers due to human activities might also have reduced the wetland's ability to retain water and support biodiversity.

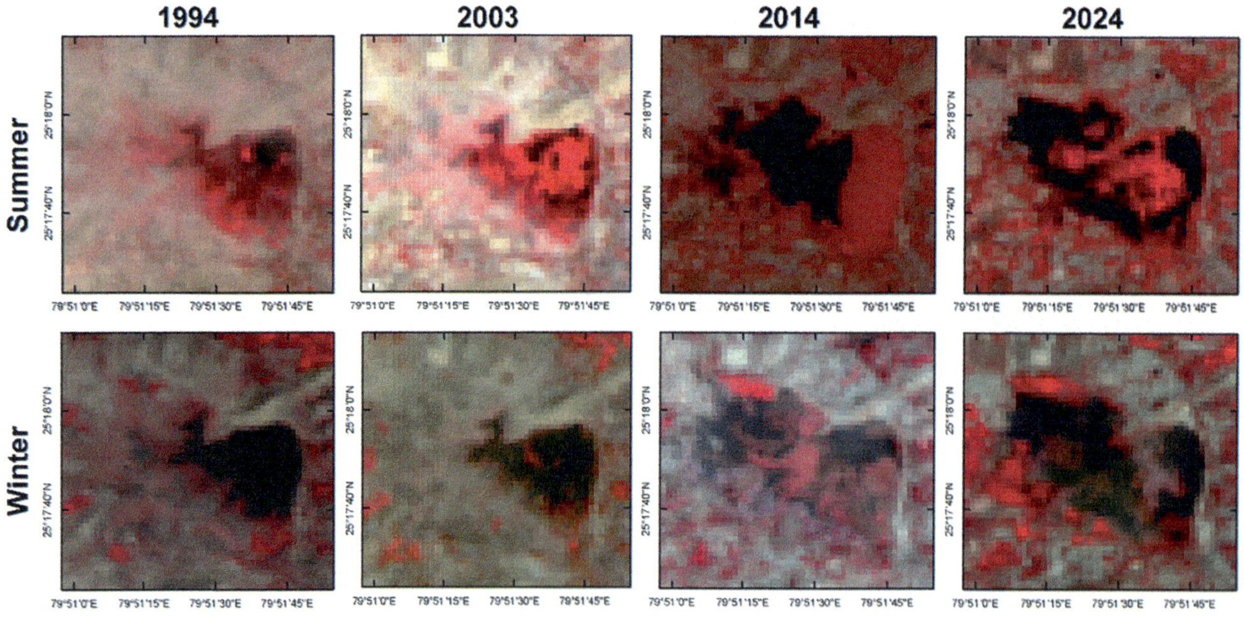

FIGURE 4.3 False color composite images of the summer and winter seasons of 1994, 2003, 2014, and 2024 in the Kirat Sagar wetland.

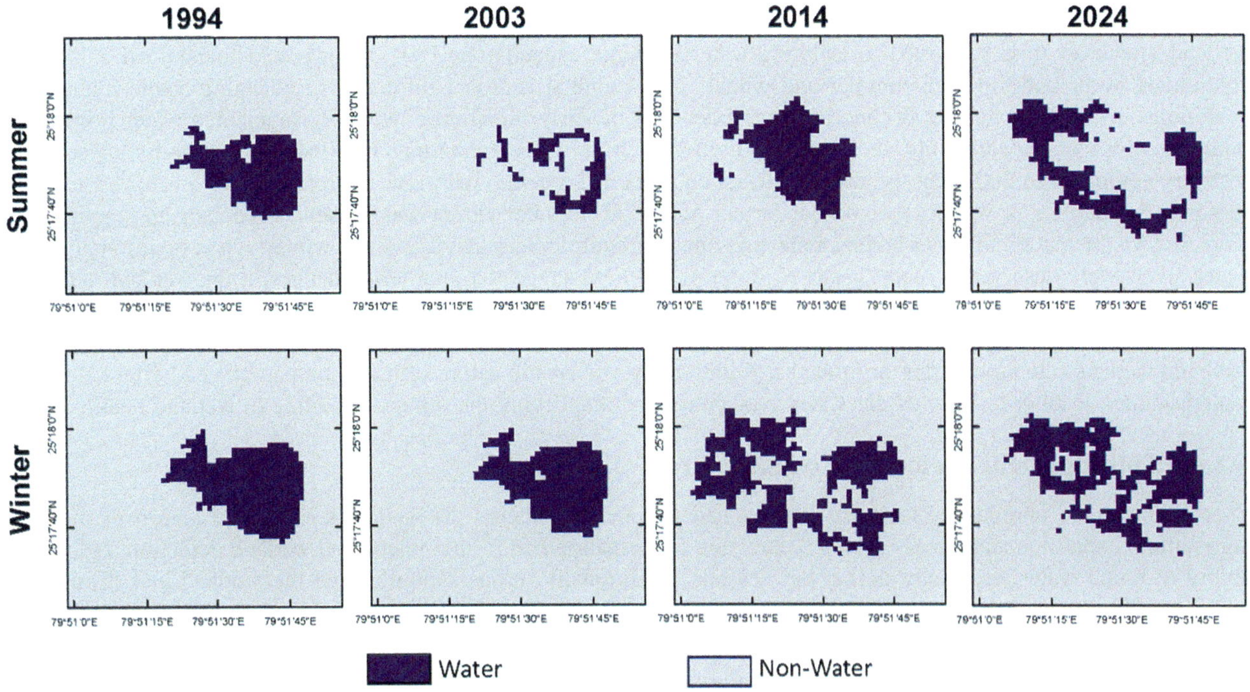

FIGURE 4.4 Spatial variation of Kirat Sagar wetland in two seasons (summer and winter) from 1994 to 2024.

Overall, the FCC images from 1994 to 2024 provided clear evidence of significant environmental changes in the Kirat Sagar wetland. The decline in vegetation, reduction in water bodies, and increase in built-up areas indicated a combination of natural and human-driven impacts. These findings emphasized the need for conservation efforts, better water management, and sustainable land use planning to protect the wetland.

Fig. 4.4 showed the spatial variation of the Kirat Sagar wetland in summer and winter from 1994 to 2024. The *blue areas* represented water bodies, while the *gray areas* indicated nonwater regions. By analyzing the changes over time, several important trends in wetland dynamics became clear. The seasonal and temporal changes in the water extent of the Kirat Sagar wetland from 1994 to 2024 revealed significant variations.

4.4.2 Temporal changes in wetland extent

The temporal analysis of the Kirat Sagar wetland from 1994 to 2024 highlights significant changes in its extent and surrounding land use. In 1994, the wetland exhibited a well-defined and stable water body, with a larger extent during winter, indicating effective water retention. The surrounding vegetation cover was prominent, suggesting minimal human encroachment. However, by 2003, the first signs of degradation emerged, with a noticeable reduction in water extent, particularly in summer. Increased patches of nonwater areas hinted at higher evaporation rates and possibly reduced inflow, while agricultural expansion and urbanization became more evident. The degradation accelerated by 2014, as the wetland's surface area continued to decline in both summer and winter, indicating growing stress. Encroachment along the periphery became more pronounced, reducing the wetland's natural ability to retain water. Increased human activity further transformed buffer zones into nonwetland areas. By 2024, the wetland showed an alarming reduction in extent, with summer images depicting only small, isolated water patches. Even in winter, when water levels were traditionally higher, a significant decline was observed compared to previous years. Rapid land use changes, including built-up areas and agricultural expansion, further restricted the wetland's natural boundary, emphasizing the urgent need for conservation efforts to prevent further degradation.

4.4.3 Seasonal variations and climatic influence

The seasonal variations in the Kirat Sagar wetland over the years highlighted the influence of climatic factors on its extent and water retention capacity. The FCC images revealed distinct seasonal trends, with winter consistently showing a larger wetland extent due to higher precipitation and lower evaporation rates. However, the gradual reduction in the winter wetland area over time suggested a declining water recharge capacity. In 1994, the wetland maintained a large and continuous water body in both summer and winter, with minimal seasonal differences, indicating stable hydrological conditions. By 2003, a notable decline in water extent, particularly in summer, led to a fragmented water body, likely influenced by climate variability and human activities such as water extraction. In 2014, the wetland showed partial recovery compared to 2003, but the water distribution remained uneven, with lower summer water levels and increased seasonal fluctuations in winter. By 2024, a further reduction in water extent was evident, especially in summer, with increased fragmentation of water bodies, reflecting ongoing hydrological stress. Even in winter, when traditionally higher water levels were observed, a more dispersed water distribution suggested long-term changes in the wetland's ability to retain water. These trends aligned with meteorological data, which indicated rising temperatures and fluctuating rainfall patterns over the years. Higher maximum temperatures had contributed to elevated evapotranspiration rates, further reducing wetland retention during dry months. Additionally, the erratic nature of monsoon patterns had affected wetland replenishment, leading to inconsistent water availability and amplifying the observed decline in wetland health.

4.4.4 Anthropogenic factors and wetland stress

Encroachment and unregulated land use changes had significantly altered the wetland's catchment area over the years. Agricultural expansion and increased water extraction for irrigation had further intensified wetland depletion, reducing its ability to retain water, especially during dry seasons. Additionally, urbanization around the wetland had disrupted its hydrological balance, restricting natural water flow and contributing to habitat degradation. These human-induced changes had accelerated the wetland's shrinking trend, affecting both its ecological functions and water availability.

4.4.5 Percentage of area of water and nonwater in two seasons

The table illustrates the percentage of water and nonwater areas in the Kirat Sagar wetland during summer and winter across four different years: 1994, 2003, 2014, and 2024. In 1994 the wetland maintained 8.6% water area in summer and 11.42% in winter, indicating a stable hydrological condition with minimal seasonal fluctuations. By 2003, a significant decline in water extent was observed, with the summer water area reducing to 3.56%, while winter water coverage decreased to 10.06%. This decline suggested increased water loss, likely due to climate variability, rising temperatures, increased evaporation, and human activities such as water extraction and urbanization. However, in 2014, the wetland showed partial recovery, with the water area increasing to 9.69% in summer and 14.92% in winter, marking the highest recorded water extent in the dataset shown in Table 4.2. This recovery may have resulted from improved rainfall conditions, conservation efforts, or temporary shifts in land use patterns that allowed for better water retention. By 2024, a slight reduction in wetland extent was again observed, with 10.06% water area in summer and 13.09% in winter. Although this was still higher than the 2003 levels, it indicated continued stress on the wetland, with ongoing fragmentation and a declining ability to retain water.

The overall shrinking trend, particularly in summer, indicated drying effects likely driven by increased evaporation, reduced rainfall, or excessive groundwater use, as suggested by Apurv and Cai (2020). Additionally, the continuous water body became more fragmented over the years, suggesting disruptions in hydrological connectivity, possibly due to sedimentation, land use changes, or human encroachment (Ballut-Dajud et al., 2022; Chakraborty et al., 2023). Seasonal

TABLE 4.2 Percentage of area under water and nonwater during summer and winter seasons from 1994 to 2024 in the Kirat sagar wetland.

Summer			Winter		
Year	Water(%)	Nonwater(%)	Year	Water(%)	Nonwater(%)
1994	8.60	91.40	1994	11.42	88.58
2003	3.56	96.44	2003	10.06	89.94
2014	9.69	90.31	2014	14.92	85.08
2024	10.06	89.94	2024	13.09	86.91

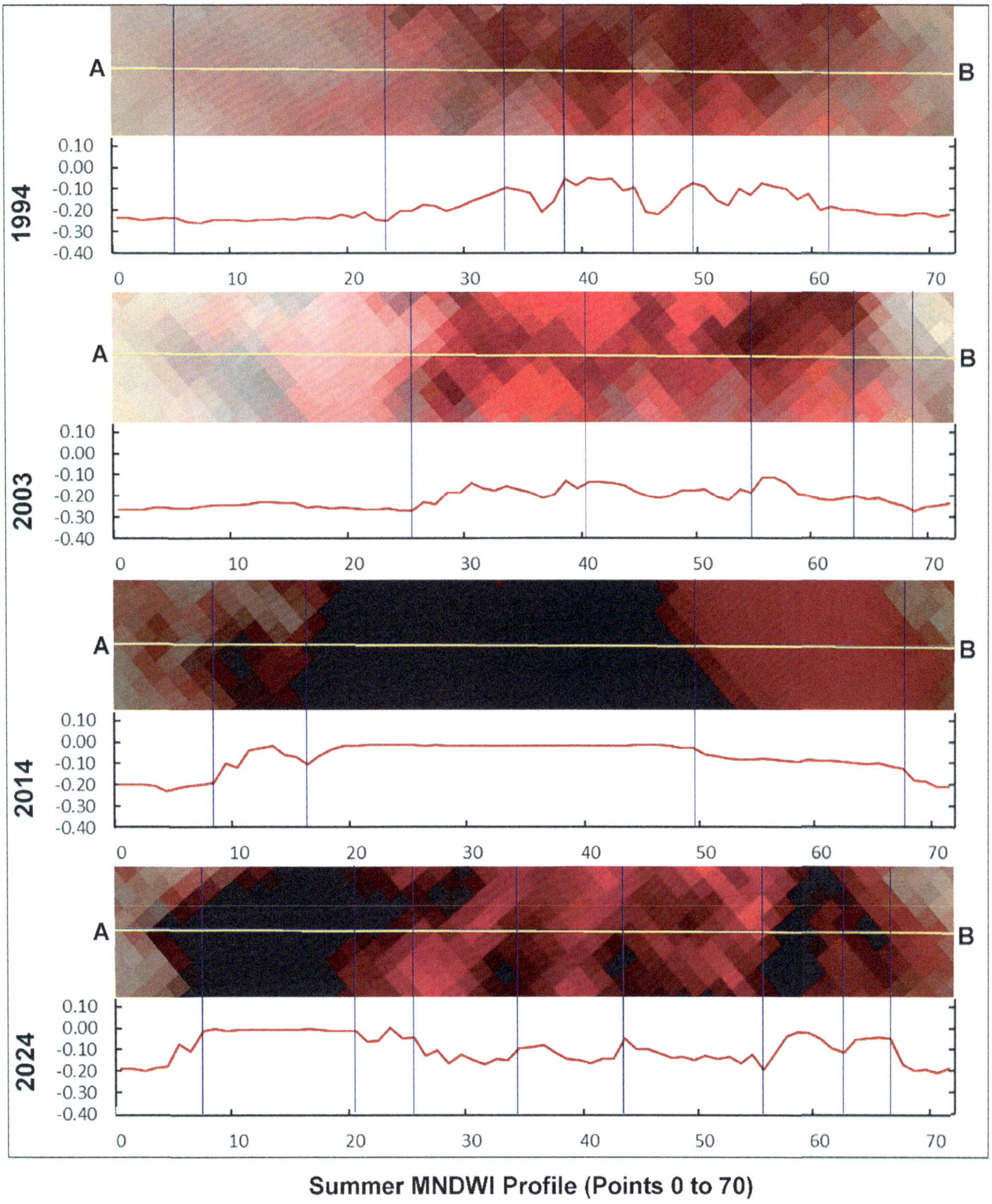

FIGURE 4.5 Variation of Modified Normalized Difference Water Index (*MNDWI*) along A to B cross-section line in Kirat Sagar in summer season. The *blue vertical line* indicates the significant fluctuation of values.

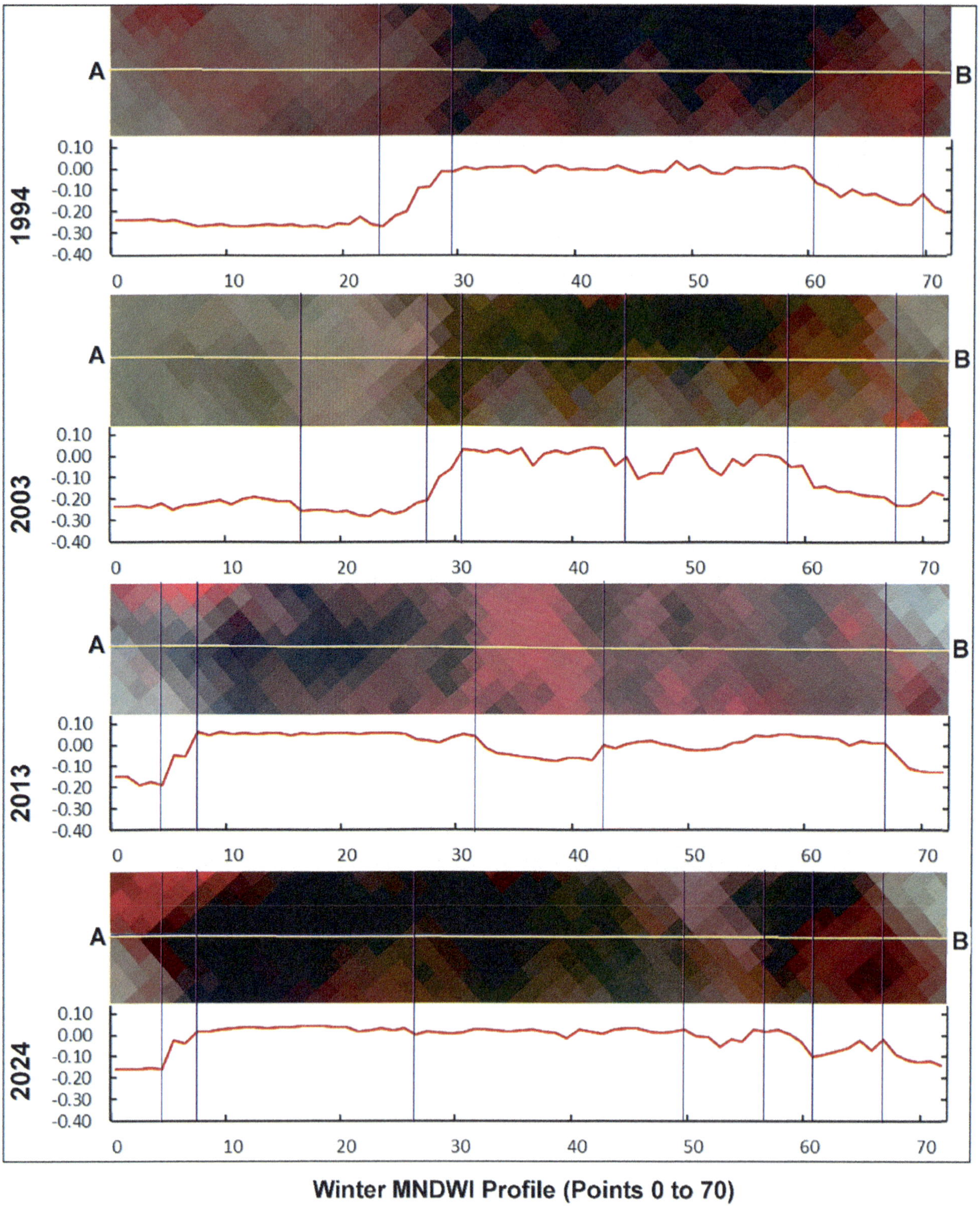

FIGURE 4.6 Variation of Modified Normalized Difference Water Index (*MNDWI*) along A to B cross-section line in Kirat Sagar in winter season. The *blue vertical line* indicates the significant fluctuation of values.

variability also became more pronounced, with a widening difference between summer and winter water extents, highlighting a weakened ability of the wetland to retain water during dry seasons. These changes suggested increasing ecological stress, which, if left unchecked, could lead to severe wetland degradation, impacting biodiversity and water availability. To mitigate these effects, conservation measures such as sustainable water management, restoration of natural inflows, and careful monitoring of land use changes were necessary. Further studies using RS techniques like modified NDWI (Normalized Difference Water Index) and change detection analysis could provide deeper insights into the wetland's long-term transformation and support conservation planning.

4.4.6 Monitoring the water body using Modified Normalized Difference Water Index

Figs. 4.5 and 4.6 show the variation of MNDWI along a cross-section (A to B) in the Kirat Sagar wetland during the summer and winter seasons, respectively, from 1994 to 2024. During the summer season (Fig. 4.5), the MNDWI values, ranging from −0.40 to 0.10, indicated the presence of water (higher values) and nonwater areas such as land or built-up regions (lower values). The blue vertical lines marked points of significant fluctuations, representing abrupt changes in water presence along the cross-section. In 1994 the MNDWI values remained stable with fewer fluctuations, suggesting that the wetland had a continuous water body with minimal fragmentation. The image showed a uniform color pattern, indicating a balanced hydrological condition. By 2003, the MNDWI values declined slightly, particularly in some sections where the index became more negative. This indicated a partial reduction in water extent, leading to the beginning of fragmentation in the wetland. The changes in red shades suggested the start of drying trends and seasonal water loss. In 2014, a significant drop in MNDWI values was observed, with large portions turning black, representing dry or nonwater areas. The red line in the graph confirmed a sharp decline in water coverage, and the fragmentation of water bodies became more pronounced, with major portions of the wetland drying up. By 2024, although some water patches remained, the MNDWI profile exhibited pronounced fluctuations, indicating severe hydrological stress. The increase in dark areas suggested continued shrinkage, with only small, scattered bodies of water remaining. Compared to 1994, the wetland's ability to retain water had drastically declined, making it more vulnerable to climate and seasonal variations. Over time, the wetland became smaller and more fragmented, with 2014 and 2024 showing the most severe reductions. The blue vertical lines highlighted points of major changes in MNDWI values, indicating disruptions in water connectivity and wetland stability. The increasing black areas in 2014 and 2024 suggested a progressive drying trend, likely driven by climate variability, reduced rainfall, and human activities such as land use changes and water extraction.

The figure shows the variation of MNDWI along a cross-section (A to B) in the Kirat Sagar wetland during the winter season from 1994 to 2024. The MNDWI values, ranging from −0.40 to 0.10, indicated water presence (higher values) and nonwater areas such as land and vegetation (lower values). The blue vertical lines represented points of significant fluctuation in water presence along the cross-section. In 1994 the MNDWI values showed stable and higher readings, indicating a continuous water body with minimal fragmentation. The presence of black areas in the upper section confirmed higher water retention during winter. By 2003, the MNDWI profile showed slight variations, with some sections experiencing a decline in water presence. The greenish and brownish shades indicated increasing vegetation cover, suggesting partial drying of the wetland in certain areas. In 2013 the MNDWI values further declined, with more fragmentation in water distribution. The appearance of more red shades suggested reduced water retention and increased exposure of land surfaces. By 2024, the MNDWI profile showed more pronounced fluctuations, with large sections turning

TABLE 4.3 Descriptive statistics of Modified Normalized Difference Water Index (*MNDWI*) in two seasons (summer and winter) from 1994 to 2024 in the Kirat Sagar wetland.

Summer					Winter				
Year	Min	Max	Mean	Std	Year	Min	Max	Mean	Std
1994	−0.270	−0.013	−0.305	0.043	1994	−0.350	0.052	−0.183	0.079
2003	−0.312	−0.017	−0.233	0.039	2003	−0.351	0.068	−0.190	0.077
2014	−0.284	−0.008	−0.172	0.069	2014	−0.311	0.108	−0.106	0.089
2024	−0.294	0.001	−0.172	0.069	2024	−0.276	0.073	−0.108	0.075

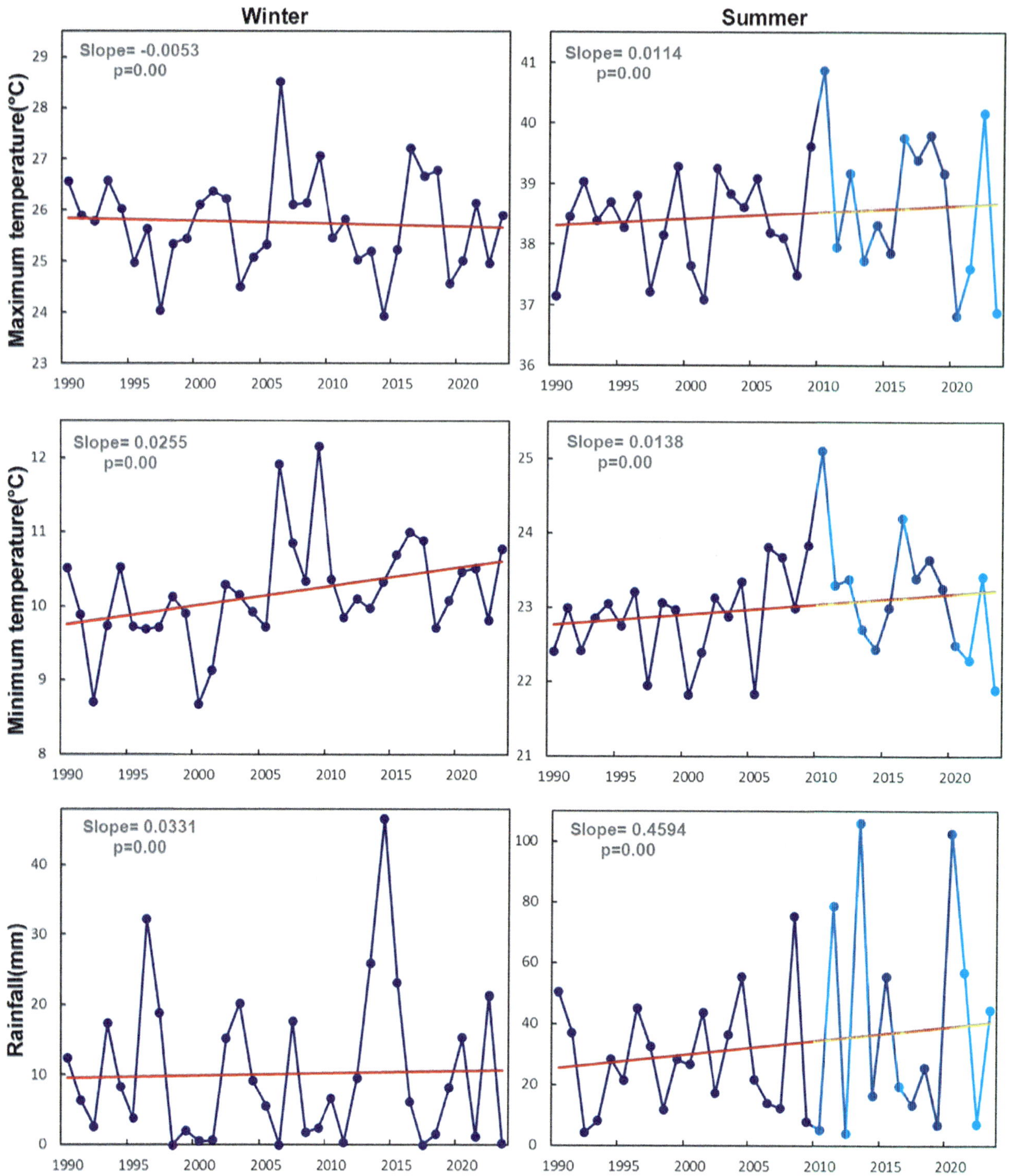

FIGURE 4.7 Trend of minimum temperature, maximum temperature, and rainfall data of two seasons (summer and winter) from 1990 to 2023.

black and brown, indicating further drying and fragmentation of water bodies. The water distribution became highly uneven, reflecting severe hydrological stress and declining wetland health. Over time, the wetland's ability to retain water in winter had weakened, with the most significant reductions observed in 2013 and 2024.

Table 4.3 presents the descriptive statistics of MNDWI for the summer and winter seasons from 1994 to 2024 in the Kirat Sagar wetland. The minimum MNDWI values remained negative in both seasons, indicating persistent seasonal drying trends. In summer, the maximum values slightly increased over time, reaching 0.001 in 2024, which suggests the presence of small water bodies. The mean MNDWI values became less negative, shifting from −0.305 in 1994 to −0.172 in 2024, indicating relatively higher wetness or surface water presence compared to earlier years. The standard deviation increased in 2014 and 2024, reflecting greater variation in water distribution. In winter, the minimum values slightly increased from −0.350 in 1994 to −0.276 in 2024, suggesting reduced drying intensity compared to earlier years. The maximum values also increased, indicating that certain areas retained surface water during the season. The mean MNDWI values became less negative, shifting from −0.183 in 1994 to −0.108 in 2024, again reflecting relatively higher wetness. The standard deviation fluctuated, with the highest value recorded in 2014, indicating greater spatial variation in water presence. Overall, the wetland showed seasonal fluctuations and localized increases in wetness over time, yet the general pattern indicates fragmentation of water bodies and ongoing seasonal drying. These findings highlight the importance of conservation measures to sustain and restore the wetland's hydrological balance.

4.4.7 Correlation with climate trends

Fig. 4.7 shows the trends in maximum temperature, minimum temperature, and rainfall in summer and winter from 1990 to 2023. The slope values indicated the rate of change per year, and the P values showed that these trends were statistically significant (P=.00). In winter, the maximum temperature showed a slight decreasing trend (−0.0053°C year^{-1}), while the minimum temperature increased (0.0255°C year^{-1}). This suggested that winter nights became warmer, which could have affected water loss through evaporation and reduced wetland water retention. Rainfall in winter increased slightly (0.0331 mm year^{-1}), but large variations indicated irregular rainfall patterns, which may have impacted wetland recharge. In summer, both maximum and minimum temperatures increased (0.0114 and 0.0138°C year^{-1}, respectively), showing an overall warming trend. Higher summer temperatures may have increased evaporation, leading to faster wetland shrinkage. Summer rainfall also increased (0.4594 mm year^{-1}), but the fluctuations suggested inconsistent water availability, which may have contributed to hydrological imbalances in the wetland. The increase in temperature appeared closely linked to wetland shrinkage, as higher temperatures, especially in summer, likely led to greater water loss. Although rainfall increased, its irregular pattern may not have helped in wetland recovery. One unusual trend was a slight decrease in winter maximum temperatures, which may have been due to regional climate changes or local environmental factors.

If these climate and human activities continued, the wetland might shrink further, with reduced water retention and increased fragmentation. Rising temperatures and unpredictable rainfall could worsen seasonal drying, leading to loss of wetland habitat. Conservation efforts, such as improved water management and controlled land use, would be necessary to protect the wetland and maintain its ecological balance.

4.5 Conclusion

Wetlands are indispensable ecosystems that require continuous monitoring and conservation efforts. The spatio-temporal analysis of the Kirat Sagar watershed from 1994 to 2024 revealed significant changes in water extent, land use, and hydrological patterns. RS and GIS techniques played a crucial role in detecting and quantifying these changes over time. The chapter showed a consistent decline in water extent, fragmentation of water bodies, and increasing human encroachment, with the most significant reduction observed during summer months. Seasonal variations indicated that while winter retained higher water levels, the overall trend reflected a decrease in the watershed's capacity to store water. The FCC images highlighted urban expansion, vegetation loss, and land use changes as major drivers of watershed degradation. The MNDWI further confirmed increasing hydrological stress, with pronounced fluctuations in water presence over the decades. The integration of RS and GIS techniques enabled precise mapping and temporal assessment of these changes, facilitating the identification of critical areas affected by climate variability and anthropogenic activities. The percentage analysis of water and nonwater areas indicated a significant decline in wetland extent in 2003, a slight recovery in 2014, and further reduction by 2024. This pattern underscored the impact of urbanization, agricultural expansion, and unregulated land use changes on watershed dynamics. The shrinking water extent, particularly in summer, reflected the influence of rising temperatures, altered rainfall patterns, and increased evaporation rates. These findings emphasized the importance of RS and GIS in monitoring watershed changes and informing conservation strategies. Sustainable water management, restoration of natural inflows, and strict land use monitoring are necessary to mitigate further degradation. Advanced RS techniques and long-term hydrological assessments will be essential for understanding future watershed dynamics and ensuring effective conservation planning. Protecting the Kirat Sagar watershed is crucial for maintaining ecological balance, biodiversity, and regional water security.

References

Adam, E., Mutanga, O., & Rugege, D. (2010). Multispectral and hyperspectral remote sensing for identification and mapping of wetland vegetation: A review. *Wetlands Ecology and Management, 18*, 281–296.

Ali, M. I., Dirawan, G. D., Hasim, A. H., & Abidin, M. R. (2019). Detection of changes in surface water bodies urban area with NDWI and MNDWI methods. *International Journal on Advanced Science Engineering Information Technology, 9*(3), 946–951.

Apurv, T., & Cai, X. (2020). Impact of droughts on water supply in US watersheds: The role of renewable surface and groundwater resources. *Earth's Future, 8*(10) e2020EF001648.

Avtar, R., Herath, S., Saito, O., Gera, W., Singh, G., Mishra, B., & Takeuchi, K. (2014). Application of remote sensing techniques toward the role of traditional water bodies with respect to vegetation conditions. *Environment, Development and Sustainability, 16*, 995–1011.

Baker, C., Lawrence, R. L., Montagne, C., & Patten, D. (2007). Change detection of wetland ecosystems using landsat imagery and change vector analysis. *Wetlands, 27*(3), 610–619.

Ballut-Dajud, G. A., Sandoval Herazo, L. C., Fernández-Lambert, G., Marín-Muñiz, J. L., López Méndez, M. C., & Betanzo-Torres, E. A. (2022). Factors affecting wetland loss: A review. *Land, 11*(3), 434.

Bijeesh, T. V., & Narasimhamurthy, K. N. (2020). Surface water detection and delineation using remote sensing images: A review of methods and algorithms. *Sustainable Water Resources Management, 6*(4), 68.

Butt, M. A., Zafar, M., Ahmed, M., Shaheen, S., Sultana, S., Butt, M. A., & Sultana, S. (2021). Threats to biodiversity of wetland. *Wetland Plants: A Source of Nutrition and Ethno-medicines*, 117–133.

Chakraborty, S. K., Sanyal, P., & Ray, R. (2023). Pollution, environmental perturbation and consequent loss of wetlands. In *Wetlands Ecology: Eco-biological uniqueness of a Ramsar site (East Kolkata* Wetlands, India). Cham: Springer International Publishing, 521–582.

Day, J. W., Christian, R. R., Boesch, D. M., Yáñez-Arancibia, A., Morris, J., Twilley, R. R., & Stevenson, C. (2008). Consequences of climate change on the ecogeomorphology of coastal wetlands. *Estuaries and Coasts, 31*, 477–491.

Demarquet, Q., Rapinel, S., Dufour, S., & Hubert-Moy, L. (2023). Long-term wetland monitoring using the landsat archive: A review. *Remote Sensing, 15*(3), 820.

Dronova, I. (2015). Object-based image analysis in wetland research: A review. *Remote Sensing, 7*(5), 6380–6413.

Faulkner, S. (2004). Urbanization impacts on the structure and function of forested wetlands. *Urban Ecosystems, 7*(2), 89–106.

Ferreira, C. S., Kašanin-Grubin, M., Solomun, M. K., Sushkova, S., Minkina, T., Zhao, W., & Kalantari, Z. (2023). Wetlands as nature-based solutions for water management in different environments. *Current Opinion in Environmental Science & Health, 33*, 100476.

Guo, M., Li, J., Sheng, C., Xu, J., & Wu, L. (2017). A review of wetland remote sensing. *Sensors (Basel, Switzerland), 17*(4), 777.

Haibo, Y., Zongmin, W., Hongling, Z., & Yu, G. (2011). Water body extraction methods study based on RS and GIS. *Procedia Environmental Sciences, 10*, 2619–2624.

Hemati, M., Hasanlou, M., Mahdianpari, M., & Mohammadimanesh, F. (2021). A systematic review of landsat data for change detection applications: 50 years of monitoring the earth. *Remote Sensing, 13*(15), 2869.

Jafarzadeh, H., Mahdianpari, M., Gill, E. W., Brisco, B., & Mohammadimanesh, F. (2022). Remote sensing and machine learning tools to support wetland monitoring: A meta-analysis of three decades of research. *Remote Sensing, 14*(23), 6104.

Jiang, W., Ni, Y., Pang, Z., Li, X., Ju, H., He, G., & Qin, X. (2021). An effective water body extraction method with new water index for sentinel-2 imagery. *Water, 13*(12), 1647.

Kassaye, H.A. (2014). Assessment of spatio-temporal changes of wetlands using GIS and RS techniques in Dawa Chefa Area in the Northern Central Ethiopian Highlands (Doctoral dissertation, The Institute of Development Studies and Partner Organisations).

Kayastha, N., Thomas, V., Galbraith, J., & Banskota, A. (2012). Monitoring wetland change using inter-annual landsat time-series data. *Wetlands, 32*, 1149–1162.

Kundu, S., Barnali, K., Narendra, K. R., & Susanta, M. (2024). Wetland degradation and its impacts on livelihoods and sustainable development goals: An overview. *Sustainable Production and Consumption.*

Ma, S., Zhou, Y., Gowda, P. H., Dong, J., Zhang, G., Kakani, V. G., & Jiang, W. (2019). Application of the water-related spectral reflectance indices: A review. *Ecological Indicators, 98*, 68–79.

Moomaw, W. R., Chmura, G. L., Davies, G. T., Finlayson, C. M., Middleton, B. A., Natali, S. M., & Sutton-Grier, A. E. (2018). Wetlands in a changing climate: Science, policy and management. *Wetlands, 38*(2), 183–205.

Ozesmi, S. L., & Bauer, M. E. (2002). Satellite remote sensing of wetlands. *Wetlands Ecology and Management, 10*, 381–402.

Pant, H. K. (2023). Wetland preservations: Solutions to tackling greenhouse gas emissions. In *Handbook of Nature-Based Solutions to Mitigation and Adaptation to Climate Change*. Cham: Springer International Publishing, 1–15.

Rebelo, L. M., Finlayson, C. M., & Nagabhatla, N. (2009). Remote sensing and GIS for wetland inventory, mapping and change analysis. *Journal of Environmental Management, 90*(7), 2144–2153.

Wang, Y. S., & Gu, J. D. (2021). Ecological responses, adaptation and mechanisms of mangrove wetland ecosystem to global climate change and anthropogenic activities. *International Biodeterioration & Biodegradation, 162*, 105248.

Were, D., Kansiime, F., Fetahi, T., Cooper, A., & Jjuuko, C. (2019). Carbon sequestration by wetlands: A critical review of enhancement measures for climate change mitigation. *Earth Systems and Environment, 3*, 327–340.

Wu, W. T., Zhou, Y. X., & Tian, B. (2017). Coastal wetlands facing climate change and anthropogenic activities: A remote sensing analysis and modelling application. *Ocean & Coastal Management, 138*, 1–10.

Wulder, M. A., Roy, D. P., Radeloff, V. C., Loveland, T. R., Anderson, M. C., Johnson, D. M., & Cook, B. D. (2022). Fifty years of landsat science and impacts. *Remote Sensing of Environment, 280*, 113195.

Xiong, Y., Mo, S., Wu, H., Qu, X., Liu, Y., & Zhou, L. (2023). Influence of human activities and climate change on wetland landscape pattern—A review. *Science of the Total Environment, 879*, 163112.

Xu, H. (2006). Modification of normalised difference water index (NDWI) to enhance open water features in remotely sensed imagery. *International Journal of Remote Sensing, 27*(14), 3025–3033.

Xu, X., Chen, M., Yang, G., Jiang, B., & Zhang, J. (2020). Wetland ecosystem services research: A critical review. *Global Ecology and Conservation, 22*, e01027.

Yue, H., Li, Y., Qian, J., & Liu, Y. (2020). A new accuracy evaluation method for water body extraction. *International Journal of Remote Sensing, 41*(19), 7311–7342.

Chapter 5

Wetlands and Sustainable Development Goals with special reference to roles and conservation strategies

Rishabh Sen[1], Raghib Hussain[1], Amit Kumar Ghosh[1], Kaomud Tyagi[1], Vikas Kumar[1], Shantabala Devi Gurumayum[2], and Inderjeet Tyagi[1]

[1]*Centre for DNA Taxonomy, Molecular Systematics Division, Zoological Survey of India, Kolkata, West Bengal, India,* [2]*Arunachal Pradesh Regional Centre (APRC), Zoological Survey of India, Itanagar, Arunachal Pradesh, India*

5.1 Introduction

Wetlands support a wide variety of terrestrial as well as aquatic life forms and are considered a vital global freshwater-related ecosystem. Although wetlands cover a small percentage of the earth surface, they contribute significantly as biodiversity hotspots across the globe (Gibbs, 2000; Thorslund et al., 2017). Wetlands are known for providing multiple ecosystem services, holding the promise to be utilized as natural remedies to a wide range of environmental, social, and economic problems. The major ecosystem services provided by wetlands include carbon sequestration (Kreplin et al., 2021; Mitsch et al., 2013), water purification (Hatvani et al., 2022), coastal protection (Gedan et al., 2011; Temmerman et al., 2013), maintenance of groundwater reservoirs and soil moisture (Rojas et al., 2022), flood regulation (Acreman & Holden, 2013; De Groot et al., 2002), regulation of air quality (Fang et al., 2023) and biodiversity support (Gibbs, 2000; Hambäck et al., 2023) along with many social and economic benefits, to develop the sustainable future.

To mitigate the global challenges related to poverty, climate change, quality education, and energy as well as to support the sustainable growth, the United Nations has proposed Sustainable Development Goals (SDGs) on September 25th, 2015, targeting a total of 17 areas to achieve a sustainable future till 2030. The main agenda behind these goals is ending poverty and inequality, protecting the environment, and encouraging peace and prosperity (Ghosh & Chakravarty, 2024). Thus, SDGs primarily focused on decreasing or eradicating the major challenges the world is currently facing (Confraria et al., 2024). Wetland services align with most of these goals, providing multiple benefits and services to achieve different SDG targets. Wetland areas, with their diverse nature and rich resources, provide multiple solutions to global environmental and social problems and support many agricultural and farming activities, thereby providing livelihoods for local communities. These wide range of services provided by wetlands are primarily important to deal with food and water safety, long-term human settlements and livelihood, poverty eradication, climate change mitigation, and the establishment of suitable infrastructure, as well as ensuring the safety and availability of water (Sarabhai, 2014) along with multiple indirectly associated targets such as gender equality and cultural issues and their remedies in accordance with these global goals. Additionally, different SDGs have significant interactions with one another, many of which result in trade-offs and disputes. It is crucial to comprehend and express these connections to assist investors, policymakers, and other stakeholders in recognizing and controlling the risks and rewards of completing the various goals (International Council for Science ICSU, 2017). In this chapter, different wetland services based on their contributions to achieve SDGs (Table 5.1) have been discussed in detail.

5.2 Materials and methods

A comprehensive literature review was carried out to elucidate the role of wetlands in achieving the SDGs in both aquatic and terrestrial habitats. Peer-reviewed research articles and scientific reports from renowned journals and sources were

Wetlands: Biodiversity, Health and their Role in Ecosystem Services. DOI: https://doi.org/10.1016/B978-0-443-34222-6.00015-2

TABLE 5.1 Summary of wetland services in maintaining different Sustainable Development Goals (*SDGs*).

SDGs	Objectives	Services provided by wetlands
SDG 1	*No Poverty*	Wetlands play a key role in providing livelihoods to the local communities
SDG 2	*Zero Hunger*	Availability of water and moist soil helps in agricultural practices and fishery practices
SDG 3	*Good Health and Well Being*	Wastewater treatment characteristics of wetland and recreational attributes reduce the risk of psychological disorders and pathological disease outbreaks
SDG 4	*Quality Education*	Educational excursions and scientific research on wetlands provide practical knowledge on wetland functioning and other associated nature-based activities.
SDG 5	*Gender Equality*	Various aspects of wetlands have been observed playing a pivotal role in empowering women through the course of water management and drinking water availability, as well as wetland-based occupations mainly handled by women.
SDG 6	*Clean Water and Sanitation*	Wetlands are the key source in maintaining groundwater reservoirs and drinking water availability
SDG 7	*Affordable and Clean Energy*	Wetlands make it feasible to generate hydropower, or renewable electricity, for use in residencial, commercial buildings, and industries, establishing the foundation for the advancement of human society in the future.
SDG 8	*Decent Work and Economic Growth*	This goal highlights the policies that support sustainable ecotourism, which generates jobs and helps in the augmentation of local culture and local goods, directly increasing the economic growth of the nation.
SDG 9	*Industry, Innovation, and Infrastructure*	Wetlands work as a natural buffer that aids in achieving this Sustainable Development Goal and reducing the impact of these natural disasters
SDG 10	*Reduced Inequality*	Wetlands provide fresh, clean drinking water and food for each and everyone. Moreover, wetlands also help in economic upliftment by engaging them in ecotourism, agriculture, and other practices.
SDG 11	*Sustainable Cities and Communities*	Coral reefs, mangroves, and salt marshes are examples of nature-based solutions that contribute to establishing sustainable cities and communities. They also offer coastal protection at no cost by mitigating the effects of storm surges, wave heights, and capacity, and retention of water.
SDG 12	*Responsible Consumption and Production*	Wetland areas fulfill the global water demands in all sectors if they are managed efficiently
SDG 13	*Climate Action*	By reducing greenhouse gas emissions, creating blue carbon habitats, and sequestering carbon, wetlands help to offset the effects of climate change
SDG 14	*Life Below Water*	Mangrove forests, coral reefs, estuaries, river deltas, lagoons, and coastal wetlands are the homes of various flora and fauna, rich in nutrients, and help to remove the leaching of toxic organic and inorganic materials by bioremediation.
SDG 15	*Life on Land*	Freshwater ecosystems help significantly in agricultural practices, logging, food, and habitats for the various species, as well as human beings. Apart from this, the wetland also helps in reducing the risk of severe floods as well as droughts.
SDG 16	*Peace, Justice, and Strong Institutions*	Effective management of transboundary or interconnected wetlands improves safety and peace and helps avoid conflicts between countries. As a result, several nations simultaneously profit from the same wetland and improve their economic circumstances and cooperations
SDG 17	*Partnerships for the Goals*	It should be important to establish partnerships that facilitate easy access to and acquisition of products to successfully implement the other SDGs targets for the sustainable and effective use of the wetland

searched using different keywords such as SDG and wetlands in the title or abstract section. For more comprehensive search results, different types of the aforementioned SDG have been considered (SDG 1–17) combined with the keyword "Wetland" was used for better insights regarding the role of wetlands in achieving the sustainable growth, further irrelevant subject areas were removed from search criteria. With these selections, a total of approximately 65 publications, not more than a decade old, with an additional 32 references to validate the work, were considered.

5.3 Roles of wetlands in achieving Sustainable Development Goals

5.3.1 Livelihoods and economic growth

5.3.1.1 SDG 1 (no poverty)

Goal 1 of the SDGs deals with eradication of poverty from the world in a sustainable manner. Wetland activities provide crucial outcomes for the sustainable achievement of this goal. Kumar et al. (2011) investigated the potential roles of a wetland ecosystem in eradicating poverty through sustainable development of livelihood in Lake Chilika, India, as a case study. Chilika Lake holds a dynamic and diverse community of different vertebrate (fish) and invertebrate (crustacean) assemblages due its unique ecological characteristics. These biological resources sustain nearly 200,000 local fisher folks, generating ~6% of the state's international revenue exports. The lake shorelines have been inhabited by seven major subcast communities with nearly 141 villages (Mitra & Mohapatra, 1957), majorly belonging to poorest socio-economic groups. Traditionally, these families have catered to distinct fishing niches based on various spatial or temporal limits such as seasonality, gears, and fish size. In addition, the conjunctions near the lake (Mahanadi Delta) facilitated the introduction of large-scale shrimp farming to the households with the government's intervention. Based on the surveys (2002–07), the authors reported that fish, prawns, and crabs comprised nearly 67.7%, 31.1%, and 1.2% of annual fishery landings by weight, which was calculated to be monetarily above the international poverty line per day for a family of 6. Overall, the study highlights how the dynamic nature of the lake systematically uplifts the economic status of communities below the poverty line, with the evaluation of socio-ecological interrelations that reflect the ideology of how human societies are linked to the wetland ecosystems for the economic well-being of the community (Kumar et al., 2011).

In another study, the role of the Volta River Basin was investigated to ease the poverty (Welling et al., 2012). The river basin, spanning nearly 400,000 km^2 across six countries, have experienced extensive exploitation of the natural resources in the past decades due to the increasing population and poverty. For greater societal impact, a governance plan was launched to monitor the wetland and develop socioeconomic benefits. The outcomes of this integrated water resource management demonstrated a significant positive effect in eradicating poverty and raised awareness among the local communities regarding the improvement of farming processes (Welling et al., 2012).

Taken together, these studies demonstrate the critical roles played by wetlands in sustaining local livelihoods, reducing poverty, and delivering economically viable ecosystem services through sustainable use of natural resources.

5.3.1.2 SDG 2 (zero hunger)

Humans rely on nature and its products to produce food. On the other hand, food production is largely dependent on different types of water sources. Wetlands, both natural and constructed, are major sources of water for agriculture in many parts of the world. (https://www.ramsar.org/sites/default/files/documents/library/wetlands_sdgs_e.pdf). Rice is a good example: it is commonly cultivated in wetland-based systems and is a staple food for a large share of the world's population (McCartney and Rebelo, 2010).

Wetlands also serve as a nutrient reservoir for majorityof the world's population. For example, in Cambodia, 60%–80% animal protein is derived from the Tonle Sap Lake and the surrounding floodplains (McCartney and Rebelo, 2010; Millennium Ecosystem Assessment MEA, 2005). Furthermore, wetlands help maintain groundwater levels by regulating the water cycle and keeping the surrounding soil moist (Hefting et al., 2004; Seifollahi-Aghmiuni et al., 2019; Xiong et al., 2003), they also help reduce water stress caused by increasing irrigation demands and recurring droughts by storing rainwater and snowmelt runoff, allowing it to recharge groundwater reserves.(Seifollahi-Aghmiuni et al., 2019). Natural flood regulations provide good conditions for agricultural activities along the shorelines by depositing nutrients and freshwater for irrigation (De Groot et al., 2002; Russi et al., 2013). Jaramillo et al. (2019) conducted a survey-based study on 45 various types of wetlandscapes spanning over 21 countries and five continents, reporting significant positive effects of wetlands on sustainable food production (2.4), which directly influences the reinforcement of other goals. Thus, it can be concluded wetlands play crucial role in the achievement of SDG 2 (zero hunger).

5.3.1.3 SDG 8 (decent work and economic growth)

In recent years, nature-based "green economy," "green growth," and "natural capital" have been emphasized for economic growth strategies that prioritize the sustainable management and conservation of natural resources. The key target of these strategic plans is based on natural resources and nature-based developments. Wetlands, for instance, are a rich natural resource, playing critical roles by supplying water essential for agricultural production and supporting cooling processes in manufacturing and energy generation. A key objective under this goal is to separate economic growth from the environmental harm. Wetlands support this goal by delivering valuable services to agriculture and industrial activities, such as mitigation of climate change, nutrient cycling, flood protection, water purification, and prevention of environmental degradation (Kreinin & Aigner, 2022). These services would otherwise require costly human-engineered solutions, making wetlands a cost-effective and sustainable alternative. Furthermore, wetlands also act as a tourist attraction for a large portion of the global population, therefore providing a significant part of revenue to the economic sector of society through ecotourism (Kreinin & Aigner, 2022). According to a report by the World Travel and Tourism (2014), nearly 266 million livelihoods are supported by wetland tourism and travel. The report also estimated an increase in tourism's contribution of ~4.2% each year in the GDP, resulting in a total of USD 3379.3 bn by 2024 end (World Travel & Tourism; Economic Impact 2014). Overall, these studies outline the contribution of various wetland services and ecotourism-based activities to sustainable economic growth, thus achieving the objectives of SDG 8.

5.3.1.4 SDG 10 (reduce inequalities)

SDG 10 focuses on reducing inequality within and among countries the in every sector. With this goal, the foremost benefits gained by wetlands are clean drinking water and water for other activities across the world. Wetlands' ability to manage the water cycle and recharge water reservoirs plays a key role in ensuring clean water availability for poor or lower-middle-class communities in a developing country where proper purification systems are unavailable (Millennium Ecosystem Assessment, MEA, 2005). SDG 10 emphasizes that healthy wetlands mitigate the risk to an estimated 5 billion people living with poor access to water by 2050 (https://allianceforscience.org/blog/2024/02/wetlands-crucial-for-the-achievement-ofsustainable-development-goals/). Moreover, wetland sources are extensively used for irrigation and agricultural purposes, therefore reducing inequality in food sectors and making food accessible to everyone. Additionally, by mitigating the risks of floods, droughts, and soil erosion (https://www.ramsar.org/sites/default/files/documents/library/wetlands_sdgs_e.pdf), wetlands reduce the vulnerability of marginalized groups, particularly those in flood-prone or coastal areas, safeguarding the livelihoods and reducing socioeconomic disparities. Therefore various natural services provided by wetlands—water supply, disaster risk reduction, agricultural support, and livelihood security- help in the mitigation of inequality in society in diverse ways.

5.3.2 Environmental and ecological health

5.3.2.1 SDG 6 (clean water and sanitation)

Goal 6 is focused towards ensuring the availability of drinking water and sanitation for everyone across the globe (United Nations, 2015). Wetlands play a pivotal role in water availability, and a significant part of the globe relies upon wetlands for drinking water. Furthermore, wetlands with their dense and rich vegetation and sediments act as nature-based bio-filters for water. For example, the Musi River in India was found to play a key role in improving the water quality (CBD, 2010). The observations made revealed significant improvement in water quality of samples obtained from downstream in comparison to the water samples obtained from upstream, located near urban pollution (CBD, 2010). However, based on the extent of hazardous component limits and microbial diversities, heavily polluted wetlands possess a potential risks for human health. Therefore, safeguarding the wetlands can improve the the water quality and maintain a steady drinking water supply across the world (Ramsar Convention Secretariat, 2016).

Turpie et al. (2010) estimated the economic values of wetland-oriented water treatment services in the Fynbos Biome of the Western Cape. The remediation potential of wetlands in terms of removing nitrogen, phosphorus, and suspended solids from the effluents was explored. Based on the findings, the removal of N (in the form of ammonia) content was observed to be nearly 33 mg/L, whereas for phosphorus, the removal rate was observed to be around 67%. Conventional water treatment cost was estimated to range between ZAR 7 million (USD 0.9 million) to ZAR 10 million (USD 1.37 million) per megaliter. In that context, the wetlands ability to remove nitrogen and phosphorus was found to be equivalent to a cost of ZAR 26 (USD 3.47) per kg of N (ammonium nitrogen) and ZAR 71 (USD 9.47) per kg of P. With the conventional methods, the cost of removal of ammonium nitrogen and phosphorus was found to be equivalent to ZAR 328 million (USD 43.7 million) for the total catchment.

Additionally, wetlands ability to maintain groundwater levels through water-cycle processes provides a rich source of drinking water (Hefting et al., 2004; Xiong et al., 2003). For example, in drought-affected areas, wetlands help in refilling the groundwater reservoirs through percolating the excess rainwater from the ground (Seifollahi-Aghmiuni et al., 2019).

Thus, based on these findings, it can be suggested that the services provided by the wetlands are valuable as a cost-efficient water treatment facility and they play a crucial role in remediation of toxic pollutants from water and wastewater, thereby facilitating the availability of clean and safe water across the globe.

5.3.2.2 SDG 12 (responsible consumption and production)

Goal 12 aims to ensure sustainable consumption and production of resources, including those derived from wetland-associated activities. Achieving this goal relies significantly on the wetland services in the agricultural and clean water sectors. Wetlands provide a range of natural resources, which are crucial for local livelihoods, and their sustainable use is therefore essential to limit overexploitation of resources and reduce environmental degradation. Properly managed wetland ecosystems can efficiently help in maintaining water availability (The Ramsar Convention, 2016; Seifollahi-Aghmiuni et al., 2019), contributing to the sustainable use of water sources. In addition, other wetland-associated services, including climate change mitigation, habitat for aquatic communities, and numerous medicinal plant species, also contribute to the sustainable use of these resources, thereby maintaining a steady production rate, thus contributing to the agenda of Goal 12.

5.3.2.3 SDG 13 (climate action)

SDG 13 emphasizes the urgent need for climate action to mitigate the impacts of climate change. Climate change typically leads to intensification of the hydrological cycle, resulting in high-intensity short-duration storms and high temperatures, which directly affect the groundwater resources. To mitigate the climatic action, wetlands act as one of the most significant natural resources of methane (Seifollahi-Aghmiuni et al., 2019) on the earth that has the capacity to absorb methane and carbon from the atmosphere, but improper management can turn them into a source of greenhouse gases (Adhikari et al., 2009). Peatlands, one of the major wetland types, naturally regulate the climate by slowing down atmospheric warming. It has been estimated that peatlands store up to 88.6 billion metric tons of carbon worldwide, of which the majority (~60 billion metric tonnes) comes from Indonesian peat (Merriman et al., 2017; UN Environment Annual Report, 2016), therefore, acting as nature-based solutions (NbS) for the mitigation of climate change. To stop the global annual temperature increase rate (~1.5°C annually), the CO_2 emissions must decrease by 45% within 2030 and 0% by 2050 (Asif, 2022). For this, carbon projects could provide a wide range of co-benefits for carbon mitigation, including fish and wildlife habitat protection in the wetland (IUCN Commission on Ecosystem Management CEM, 2023). The most effective natural carbon sinks on Earth per area are coastal blue carbon ecosystems, which include tidal marshes, seagrasses, and mangroves. Blue carbon ecosystem restoration and conservation provide effective means of preventing and reducing the release of greenhouse gases, especially for countries with major coastal vegetation regions and rapid rates of erosion (Blue Carbon Initiative, Conservation International, & IPBC, 2023). The preservation of biodiversity, nutrients, economy, and human well-being are just a few of the advantages that coastal blue carbon ecosystems offer to people and coasts, in addition to their capacity to mitigate climate change. Additionally, these ecosystems also help with climate adaptation and resilience, shielding millions of people worldwide from the effects of storms, erosion, and flooding along the coast (Blue Carbon Initiative, Conservation International, & IPBC, 2023). In addition, it has also been noted that wetland-dwelling waterbirds act as one of the most important indicators of climatic actions. The availability of the waterbirds (Family: Ardeidae) was significantly affected as a result of severe climatic deterioration, therefore, acting as important natural indicators of climate actions (De la Cruz & Numa, 2024).

5.3.2.4 SDG 14 (life below water)

Out of all the goals, SDG 14 focuses on the sustainable use of the sea, oceans, and marine resources. Water is a natural home for many plants and animals. It also plays an important role in the water and chemical cycles. (Mukherjee & Dash, 2024). Wetlands are important because they support nutrition, breathing, water movement, waste removal, and reproduction in living organisms (Kılıç, 2020). The South Asian Sea (SAS) region has rich biodiversity, including mangrove forests, coral reefs, estuaries, river deltas, lagoons, and coastal wetlands. These ecosystems provide habitats for many species and are protected under Sustainable Development Goal (SDG) 14. The region is also home to about 6.8% of the world's coral reefs and mangrove forests. (Farooqui, 2014).

A report from Alaska's Bristol Bay shows that salmon fishing generates about **$2.2 billion each year** for local communities. This supports the local economy (SDG 1) while also helping to protect marine life under **SDG 14**.However, in the recent times, salmon habitats have been compromised causing both economic and underwater deterioration due to the mining developments near the Bristol Bay wetland (https://preserve.nature.org/page/81370/action/1?chain).

Therefore, wetlands and waterbodies are important for the maintenance of life below water, and any changes in these natural habitats causes a significant effect on the aquatic life. For example, pollution from land-based sources (e.g., plastic debris) disturbs the pH of water and increases the methane amount, which automatically hinders the aquatic communities (NITI Aayog, 2018). Moreover, the role of wetlands in mitigating the elevated level of anthropogenic CO_2 is commendable; however, the emission of carbon affects the life of flora and fauna of the wetland and increases the marine pollution, therefore, hindering the goals of SDG 14. These findings show that wetlands provide a home for many aquatic plants and animals. Although,the advancement of human occupation and poor management of water in these natural ecosystems (Moretti et al., 2024) create a hindrance to the achievement of this global goal.

5.3.2.5 SDG 15 (life on land)

In a similar approach to SDG 14, SDG 15 targets the various sources of life on land and promotes sustainable use of terrestrial ecosystems (Ezbakhe, 2018). Both land and water are essential for the existence of life, such as for the mitigation of climate change, floods, and environmental pollution (Pandey & Asif, 2022). Sustainable use of terrestrial ecosystems such as forests and prevention of land degradation is the primary target of this goal. Terrestrial ecosystems sustains nearly 126,000 inland freshwater species, representing nearly 9.5% of all known animal species across the globe (Meli et al., 2014). Wetlands are essential for the maintenance of the natural equilibrim, and have been estimated to provide nearly 40% of global renewable ecosystem activities (Costanza et al., 2014). Moreover, it has been estimated that 76% of the threatened species communities rely on wetlands for their habitat (https://www.ramsar.org/sites/default/files/documents/library/wetlands_sdgs_e.pdf).

Wetlands are important breeding and resting places for many migratory birds, helping them complete their life cycles and support future generations. For example, after the polluted El Cascajo wetland in Peru was restored in 2010, it became a migration stop for hundreds of bird species and several fish species, supporting both land and aquatic life. (Government of Peru, 2015).

Overall, wetlands provide important habitats for many land animals and help maintain natural resources, supporting the goals of **SDG 15 (Life on Land)**.

5.3.3 Human and infrastructure

5.3.3.1 SDG 3 (good health and well-being)

The objective behind SDG 3 is to ensure healthy lives and promote well-being for all , targeting the reduction in maternal and noncommunicable mortality, putting a stop to preventable deaths under 5 years of age, fighting and ending the epidemics of communicable diseases, reducing health problems and deaths from hazardous chemicals and pollution, etc. (United Nations, 2015). This section highlights the recent studies based on the effects of wetlands on maintaining the SDG 3. Human and wetlands have been interrelated for a long time for resources, clean water, agricultural developments etc (Gopal et al., 2008; Junk, 2002). Based on the United Nations World Water Development report, healthy and unpolluted natural environment is essential for human well-being and sustainable development with further emphasis on the wetlands as valuable and irreplaceable resource for the same (World Water Assessment Programme WWAP, 2003). In the context of wetlands and the well-being of humans, Horwitz et al. (2012) explored the importance of wetlands for humans with special reference to the health system. The authors reported that wetlands supported human well-being in a number of ways, such as safe water and food, livelihoods, including areas where pollution levels and infectious disease rates are high. However, they also suggested that the activities around the wetlands can also influence the functioning of the waterbodies, resulting in both positive and negative impacts on human health and support for livelihoods (Horwitz et al., 2012).

Anthonj et al. (2017) demonstrated that wetlands associated health risks are largely linked to agricultural practices (irrigation), exposure to vector mosquitoes, diarrheal diseases, and schistosomiasis, and domestic use of wetlands for drinking can spike the cases of typhoid fever and diarrhea. Further, Anthonj et al. (2018) emphasized that safe water supply and storage, adequate sanitation, and good hygiene (WASH) are a pivotal step to prevent disease outbreaks from wetland waters with heavy loads of infectious agents and polluted particles.

de Jesús Crespo et al. (2019) investigated the relationship between the dengue outbreak and wetlands in surrounding areas of San Juan Bay Estuary, Puerto Rico. The results revealed lower dengue occurrence in neighboring areas with higher wetland covers as compared to regions with lesser wetland covers (de Jesús Crespo et al., 2019). In a similar approach, Skaff and Cheruvelil (2016) compared the West Nile Virus outbreak in US with wetland sizes, which revealed that the infection of West Nile Virus in human was significantly lower in regions with higher average wetland size than the regions with smaller wetland size, however the effect of wetlands were also related to those of the mosquito strain (dominant) or the drought conditions. Few lines of studies have also deciphered effect of wetlands on the mental and physical health of the humans, which revealed that exposure to different water resources (marine, coastal, lake, river, or other freshwater sources) also called as "blue space" or "green space" alleviates mental distastefulness and has many psychological health benefits (Nichols, 2014; Peng et al., 2016; Völker et al., 2016; Wheeler et al., 2012). Thus, it can be concluded that wetlands play a significant role in decreasing disease outbreaks and elevating the mental or physical well-being.

However, it should also be noted that different activities, either natural or anthropogenic, may cause the wetlands to behave antagonistically, with an elevated rate of human health-related issues associated with the wetlands' dynamic nature.

5.3.3.2 SDG 7 (affordable and clean energy)

SDG 7 aims to ensure affordable and sustainable energy resources for everyone across the globe by 2030. Wetlands play a major role in controlling the flood, which makes it possible to produce renewable electricity (hydropower) for commercial and household use, laying the foundation for future human society development (goal SDGs 1 and 7) without emitting harmful chemicals into the atmosphere (target SDG 14). However, it's critical to evaluate the surrounding ecology of hydropower plants and balance the biodiversity benefits with the downsides. Hydropower production puts additional strain on water reservoirs (river, lakes, dams etc); which is efficiently countered by the wetland's water management and refilling nature, charging these reservoirs back to their normal states (Acreman & Holden, 2013; De Groot et al., 2002). Based on a report, the US Energy Information Administration demonstrated that the hydropower industry in the United States currently generates 38% of the country's electricity from renewable sources, mostly from hydroelectric reservoirs that are connected to dams. (https://allianceforscience.org/blog/2024/02/wetlands-crucial-for-the-achievement-of-sustainable-development-goals/). Furthermore, wetlands naturally help in purification and filtration of water, therefore reducing the need of energy-intensive processes to perform such activities (United Nations, 2023). Thus, significantly reduces the energy requirements for water management. Additionally, wetlands resilience to floods, droughts, and other natural calamities also adds up to minimize the damage of energy-based infrastructures. Collectively, the services provided by wetlands in various sectors are directly or indirectly associated with the sustainable development of energy on a global scale and the mitigation of the problems associated with the sector. The services are well aligned with the aspects of SDG 7.

5.3.3.3 SDG 9 (industry, innovation, and infrastructure)

Infrastructure and industries heavily rely upon the natural conditions, and any extremities in the natural conditions, such as floods or climate change, can cause serious damage to these sectors. Wetlands specifically serve as natural buffers against the physical effects of hazards and offer life-sustaining ecosystem services, including freshwater, habitat, forage, and climate change mitigation through carbon sequestration (Millennium Ecosystem Assessment MEA, 2005). Their conservation and sustainable management are considered NbS for disaster risk reduction (Cohen-Shacham et al., 2016). The Convention on Wetlands, through the Fourth Ramsar Strategic Plan (2016–24) and Ramsar Resolution XII.13 on "Wetlands and disaster risk reduction," underscores that effectively managed and restored wetlands hold relevance for biodiversity conservation, disaster risk reduction, livelihoods, and climate change mitigation and adaptation, which in turn helps in boosting the infrastructural and innovation sectors (Ramsar Convention Secretariat, 2016). International agreements such as the Convention on Biological Diversity (CBD), the Paris Agreement, and the Sendai Framework for Disaster Risk Reduction 2015–30 also recognize the role of healthy ecosystems, including wetlands, in reducing disaster risk (Pörtner et al., 2021; Sebesvari et al., 2019; United Nations Office for Disaster Risk Reduction UNISDR, 2015). Similar to natural wetlands, constructed wetlands are ecological engineering systems designed to enhance water quality with minimal external energy requirements, low construction costs, easy operation, and substantial environmental benefits (Corbella & Puigagut, 2018; Ji et al., 2022). To mitigate the problems of degraded or affected natural waterbodies, constructed wetlands can efficiently alter the roles of these degraded natural reservoirs and promote sustainable growth of infrastructures in both urban and semiurban localities, preferably in the developing countries where water treatment facilities are unavailable, hence acting as natural water treatment infrastructure.

5.3.3.4 SDG 11 (sustainable cities and communities)

SDG 11 aims to promote safe, resilient, and sustainable urban ecosystems by 2030. Urban wetlands play significant roles in making the urban communities resilient and safe from natural disasters. Ninety percent (90%) of all disasters throughout the world over the past two decades have been connected to water. Coral reefs, mangroves, and salt marshes are examples of NbS that provide affordable coastline protection by reducing wave height and power, reducing storm surges, and absorbing part of the extra water (Narayan et al., 2017). During a hurricane, wetlands prevented a significant amount of damage, including a loss of nearly USD 625 million due to the floods, by buffering storm surge and lowering the flood heights in the United States (Narayan et al., 2017). Moreover, wetlandsability to retain excess rainwater also helps in reducing flood risks and disaster management. Mitsch and Gosselink (2015) delineated the role of wetlands in the Mississippi River watershed in lowering flood risks by retaining extra water during heavy rainfall.

In the coastal areas, cities with a high rate of urbanization are particularly vulnerable to risks associated with water. For instance, Egypt's second-largest city and important port, Alexandria, is extremely vulnerable to sea-level rise due to the elevated rate of climate change. The surrounding wetlands are also impacted by the intrusion of seawater, sewage water, and pollution, posing significant health and environmental challenges. Efforts are currently being focused on integrated coastal zone management and wetland restoration to support fisheries, agriculture, and wildlife while also strengthening the city's resilience to climate change and increased environmental challenges (United Nations Development Programme UNDP Egypt, no, date). Thus, wetlands role in supporting the livelihoods and a healthy environment in an urban setup is crucial for the achievement of sustainable growth in line with the objectives of Goal 11.

5.3.4 Social equity and governance

5.3.4.1 SDG 4 (quality education)

The utility of wetlands is not only confined to its directly related goals, but also associated with some of the distant targets. SDG 4 targets the enrichment of educational qualities for each and every child across the world. Wetlands provide rich natural resources to mankind, which are essential for the colonization of human communities in those regions, subsequently developing better education for the local communities and their future generations. The unique nature of the wetland ecosystem provides valuable insights to young minds. Hails (2018) highlighted different types of formal and informal education offered by the wetland ecosystems. Formal education is primarily associated with wetland dynamics and management-based studies in academic institutions (schools, colleges, etc.). Informal education, on the other hand, occurs through ecotourism and education based practical opportunities in wetlands; such as bird watching, fishing, nature study, data collection and interpretation from both social and scientific approaches; performing practical activities (fishing, farming etc.) with local communities in wetlands and NGOs or local government authorities; covering the anthropogenic or natural reasons in wetlands degradation (climate change, pollution, freshwater availability etc. (Hails, 2018).

Furthermore, globally, multiple study-oriented excursions are being conducted on wetlands by educational institutes for students for their development. In Sydney Olympic Park Australia, multiple wetlands are present, the authorities performed multiple excursions with around 30,000 students to enrich the students educational experience in multi-disciplinary manner (geography, science, and mathematics). In India, educational excursions are being conducted annually by educational institutes for the enrichment of ideas of nature and its tributaries in the growing generation (https://www.tribuneindia.com/news/govt-college-pupils-visit-harike-wetland-490289/; https://mhsforgirls.edu.in/archive/mhs-archive/east-kolkata-wetlands-an-ecological-wonder).

Gitau et al. (2019) conducted a survey in Nairobi (Kenya) on 34 adult visitors based on the importance of Lily wetland, Karura forest, to evaluate the visitors' knowledge about the environment, sightings, aesthetic values, and recreational jobs. Out of the 34 participants, 24 responded with recreation as the main reason behind the visit, whereas the remaining 10 participants selected education as their purpose of visit, considering the wetland as an ideal site for learning about nature and the environment (Gitau et al., 2019). Overall, the services of the wetland ecosystem enhance the educational qualities, provide significant knowledge about nature, its functioning, and natural resources among the growing generation.

5.3.4.2 SDG 5 (gender equality)

SDG 5 focuses on gender equality and women's empowerment everywhere around the world. Women play a critical role in food collection, agriculture, and water management in many regions around the world; however, their contribution in wetlands management is often overlooked, with the social and economic barriers reinforcing unequal involvement and decision-making (https://www.ramsar.org/sites/default/files/documents/library/wetlands_sdgs_e.pdf). To improve women's participation in wetland and water management, the foremost principles include recognizing their central role in the management and safeguarding of these vital resources, and actively supporting their full participation in governance of the wetlands (Ali & Grobicki, 2016).

In many parts of the world, women play critical roles in water and wetland-associated activities. For instance, in Burkina Faso, West Africa, the Local Water Committees established in 2013 for the betterment of women have efficiently helped in resolving the water collection and management-related issues. The dynamic behavior of this committee has contributed to the management of the Lake Dem, a Ramsar site; and the establishment of three new Ramsar sites, namely, Tougouri Dam Lake, Yalgo Dam Lake, and Nakanbé-Mané basin; which play a vital role in the local community's water accessibility (Ali & Grobicki, 2016).

Similarly, in Northern Greece, women have been playing pivotal roles for the protection of the threatened (in Europe) Lesser White Fronted Goose in the Ramsar sites Kerkini Lake and Evros Delta, with the majority of the wardens patrolling the region being women (Ali & Grobicki, 2016). In India, an initiative taken by "Help Us Green" for the prevention of Ganga river pollution and maintenance of groundwater reservoirs employed women for the collection of floral wastes from different worshiping places (e.g., Temples) to produce usable products (handcrafted, fertilizers, bioincense, etc.). This act was developed to create a healthy aquatic ecosystem in Ganga and its associated regions; along with equality, respect, and social upliftment to the women (Aguilar, 2021). In conclusion, gender sensitivity needs to be implemented in wetland management-related activities, recognizing the ideologies from both ends for a better outcome in governance and decision-making, generating effective and efficient conservation, sustainable use, and equal access to these wetland ecosystems.

5.3.4.3 SDG 16 (peace, justice, and strong institutions)

Goal 16 seeks to promote peaceful and inclusive societies for sustainable development by lowering violence against children, guaranteeing equal access to justice, decreasing corruption in natural resources, and providing legal identification for everyone (Ezbakhe, 2018). Wetlands are regions where enormous amounts of resources are available, making these areas more prone to conflict between the political parties and countries. For this UN Environment Program (United Nations Environment Programme UNEP, no, date) organized an effective program for the resource management through strong governance structures to help develop the nations and improve their living standards without compromising their future (United Nations Environment Programme UNEP, no, date). Conflicts over the wetland resources were common, particularly in areas for competing land uses like agriculture, fishing, etc., for which SDG 16 ensures the establishment of strong institutions that are capable of wise use and managing the natural resources and cooperation between governments, local communities, and non-governmental organizations (https://www.ramsar.org/). Wetland conservation is easier and more successful when local people, broader civil society, and the scientific communities are communicating with each other and are involved in decision-making steps (Bridgewater et al., 2015).

5.3.4.4 SDG 17 (partnerships for the goals)

In Goal 17, nineteen different targets have been set, related to finance (17.1–17.5), technology (17.6–17.8), capacity building (17.9), trade (17.10–17.12), and policy coherence (17.13–17.19) (Ezbakhe, 2018). The main focus of this goal is to advance the progress of all the other 16 SDGs by fostering partnerships and collaboration that support the conservation and management of wetland ecosystems. However, the achievements of these SDGs are often hampered by challenges such as habitat loss, unsustainable resource use, and pollution. Therefore, to conserve and manage wetlands, certain partnerships have been established, including national and international collaborations, government initiatives, community participation, and public–private partnerships (https://unstats.un.org/sdgs/report/2016/Goal-17/).

To ensure the sustenance of such initiatives, involvement of stakeholders at all levels in the process of integrating wetland conservation, prudent use, and restoration into livelihood strategies must be implemented. The involvement and expertise of indigenous and local communities, in particular, are increasingly recognized as essential for long-term success in conservation and management of wetlands (www.ramsar.org).

Shams and Muhammad (2023) investigated the transboundary water issue between Afghanistan and Pakistan based on geographic, hydrographic, hydrologic, historic, institutional, and political aspects. The instability of the then war-stricken nation (Afghanistan) caused disruption in data collection, maintenance, and preservation, resulting in a lack of technical, engineering, and hydrologic data for the Kabul River Basin (KRB) on the Afghan side. To counteract these issues, international agencies and institutional finance (The World Bank) came forward for support to both countries through financial, technical, political, and other aids to resolve the issues associated with the river basin (Shams & Muhammad, 2023).

Similarly, Vinca et al. (2021) had investigated the condition of cross-boundary cooperation among riparian countries in the Indus basin and its potential to support integrated development policies across water, energy, and food sectors. The study used an integrated water–energy–land modeling approach to mitigate multi-sectoral and cross-boundary issues, thereby alleviating developmental costs, water stress, and soil and climatic pollution through a cooperative approach. According to the analysis, the authors reported that investments to mitigate water scarcity and improve access to resources in the basin countries by 2050 can be significantly reduced through strong cross-border collaboration, yielding economic and ecological benefits for all the countries involved (Vinca et al., 2021).

In summary, effective monitoring and conservation of the wetlands with the intervention of different international agencies are essential for the maintenance of their ecological services and achievement of SDGs.

5.4 Wetlands and one health

One health is an approach to study the infectious diseases, acknowledging the close interconnection of humans, animals, plants, and environment (Centers for Disease Control & Prevention CDC, 2023). The concept signifies the close relationship of human and animals in context to their health, behavioral patterns, and environment (Evans & Leighton, 2014). The One Health approach is closely aligned with the UN SDGs as they encompass targets related to health and well-being (SDG 3), clean water and sanitation (SDG 6), climate actions (SDG 13), and management of different ecosystems (terrestrial or aquatic) (SDG 14, 15) (Pitt & Gunn, 2024; https://sdgs.un.org/goals). This implies that the parameters, such as biodiversity loss, ecosystem disruptions, and disease outbreaks, which indicate the health of an ecosystem, are also determining factors of one's health. As discussed in the previous section, wetlands play pivotal roles in the achievement of SDGs through various services, either in direct or indirect connections. Therefore it may be stated that wetland services are also associated with the one health approach in a multidisciplinary way as with different SDGs.

Changes in the wetland ecosystem characteristics can have a significant impact on human health. For instance, changes in climate or water flow structure and physicochemical parameters may enhance the risk of increase in disease vectors populations and pathogen exposure (such as malaria, West Nile virus and schistosomiasis) to humans (Keesing et al., 2010; Lafferty, 2009), whereas any disruptions in the trophic structures from over predation or commercial activities can further generate health risks for humans (Stauffer et al., 2006). Additionally, wetland degradation and decline in productivity can undermine livelihoods, with health consequences for dependent communities (Béné & Friend, 2009).

Viral infections such as Dengue spread between humans via the vector mosquito *Aedes* spp. are induced by the changes in the surrounding environmental parameters. Environmental factors such as climate change and elevated temperature have been found to influence reproductive activities in mosquitoes, causing higher dengue incidence and human exposure to mosquitoes. Moreover, *Aedes* mosquito's preference of organic content rich pools formed of flood water to lay eggs also implies toward the higher risks of dengue incidences across the globe (Damtew et al., 2023; Kulkarni et al., 2022; Pitt & Gunn, 2024). Various wetland services are associated with the mitigation of climate change and global warming (SDG 13) related issues; however, proper monitoring of wetlands is crucial to make steady contributions of these natural (or artificial) aquatic ecosystems to control the dengue outbreaks and vector abundance.

Other microbial diseases, such as Trachoma, caused by the bacteria *Chlamydia trachomatis,* often spread through clothes and beddings, or even hands. One of the major causes for this disease outbreak is the lack of access to safe water (SDG 6), where washing activities of the household belongings can be done (Pitt & Gunn, 2024; Solomon et al., 2022). Most of these activities are performed in small waterbodies (ponds) or lakes and river basins, where disease accumulation can be a problem if unchecked and untreated waters are used for such activities, thus elevating the risk of disease spread and hindering the concept of one health and SDG programs (SDG 3 and 6).

Furthermore, awareness against water related diseases such as human schistosomiasis is typically targeted toward mass drug administration drives (promoting SDG 3), safe water (for drinking) provisions (SDG 6), transmission ecology, multihost pathogenicity and zoonotic nature demonstrations (Keenan et al., 2013; Lu et al., 2011; Queenan et al., 2017; Standley et al., 2012; Webster et al., 2016), which takes a wider view in implementing the One health and SDG (goal 3, 6, and 14) related approaches associated with wetlands.

In addition to microbial diseases, antimicrobial drugs can further induce antibiotic resistance and other complexities. High dosage of antibiotics results in the release of metabolic products into the sewage system, which enhances the emergence rate of resistance genes in the sewage microbiome. These resistant bacteria can enter the food chain through microplastics or nanoparticles, and may cause severe pathogenic activities if exposed to human health (Nath et al., 2023; Perveen et al., 2023; Pitt & Gunn, 2024; Pitt, 2018). Moreover, increased use of antibiotics as growth promoters in aquaculture programs also negatively affects the ecosystem, contaminating the waterbodies and entering into humans through the food chains (Shen et al., 2020). Therefore, despite having numerous activities promoting the one health approach, wetlands under heavy doses of antibiotics or metabolic remnants of antibiotics can pose a setback to one health and a number of SDGs (SDG 3, 6, and 14).

Therefore, it is understandable that wetlands play a multidimensional role in advancing the One Health framework by supporting disease regulation, water purification, and ecosystem resilience. However, without proper management and monitoring, these services can be compromised, declining the progress toward SDGs and increasing public health risks. Thus integrating wetland conservation with One Health strategies is essential for achieving long-term global health and environmental sustainability (SDGs).

5.5 Conclusion and future directions

Wetlands are crucial for the world, playing a significant role in keeping the environment habitable and providing crucial services to mankind. Based on the above discussions, the impact of wetland-based services on the achievement of different SDG targets is significant and aligned with the contexts of the goals. Moreover, different wetland services are observed to benefit multiple goals at once, thereby creating cross-links between different SDGs for smooth and effortless implementation. Although, the elevated anthropological activities hinder the proper functioning of wetlands in multiple regions of the world, proper care and rebuilding the degraded wetlands, along with the construction of artificial wetlands, are crucial for the speedy recovery of the environment and keeping up with the sustainable growth of different sectors. Therefore, the findings emphasize that wetlands are not just passive natural resources but active, dynamic contributors to achieving the global sustainability agenda. Further, the multipolar services of wetlands make significant contributions to the goals (SDGs, CBD) set by international agencies such as the UN, thereby supporting their smooth implementation within the desired timeline.

References

Acreman, M., & Holden, J. (2013). How wetlands affect floods. *Wetlands, 33*(5), 773–786.

Adhikari, S., Bajracharaya, R. M., & Sitaula, B. K. (2009). A review of carbon dynamics and sequestration in wetlands. *Journal of Wetlands Ecology, 2*(1), 42–46.

Aguilar, L., (2021) *Guidance on mainstreaming gender under the Ramsar Convention on Wetlands.* Gland, Switzerland: Secretariat of the Convention on Wetlands.

Ali, M., & Grobicki, A. (2016). Women's roles in managing wetlands. *Solutions, 7*(6), 58–63.

Anthonj, C., Githinji, S., & Kistemann, T. (2018). The impact of water on health and ill-health in a sub-Saharan African wetland: Exploring both sides of the coin. *Science of the Total Environment, 624*, 1411–1420.

Anthonj, C., Rechenburg, A., Höser, C., & Kistemann, T. (2017). Contracting infectious diseases in Sub-Saharan African wetlands: A question of use? A review. *International Journal of Hygiene and Environmental Health, 220*(7), 1110–1123.

Asif, M. (Ed.). (2022). *Handbook of Energy and Environmental Security*. Academic Press.

Béné, C., & Friend, R. M. (2009). Water, poverty and inland fisheries: Lessons from Africa and Asia. *Water International, 34*(1), 47–61.

Blue Carbon Initiative, Conservation International, & Intergovernmental Panel on Biodiversity and Ecosystem Services. (2023). Actionable blue carbon ecosystems for climate mitigation and adaptation. https://bluecarbonpartnership.org/document/actionable-blue-carbon-ecosystems-for-climate-mitigation-and-adaptation/.

Bridgewater, P., Régnier, M., & García, R. C. (2015). August. Implementing SDG 15: Can large-scale public programs help deliver biodiversity conservation, restoration and management, while assisting human development? *In Natural Resources Forum, Vol. 39*(No. 3-4), 214–223.

CBD (Secretariat of the Convention on Biological Diversity). (2010). *Drinking water, biodiversity and development: A good practice guide*. Montreal: CBD.

Centers for Disease Control and Prevention (CDC). (2023). *One health*. https://www.cdc.gov/onehealth/index.html.

Cohen-Shacham, E., Walters, G., Janzen, C., & Maginnis, S. (2016). Nature-based solutions to address global societal challenges. *IUCN: Gland, Switzerland, 97*(2016), 2036.

Confraria, H., Ciarli, T., & Noyons, E. (2024). Countries' research priorities in relation to the sustainable development goals. *Research Policy, 53*(3), 104950.

Corbella, C., & Puigagut, J. (2018). Improving domestic wastewater treatment efficiency with constructed wetland microbial fuel cells: Influence of anode material and external resistance. *Science of the Total Environment, 631*, 1406–1414.

Costanza, R., De Groot, R., Sutton, P., Van der Ploeg, S., Anderson, S. J., Kubiszewski, I., Farber, S., & Turner, R. K. (2014). Changes in the global value of ecosystem services. *Global Environmental Change, 26*, 152–158.

Damtew, Y. T., Tong, M., Varghese, B. M., Anikeeva, O., Hansen, A., Dear, K., Zhang, Y., Morgan, G., Driscoll, T., Capon, T., & Bi, P. (2023). Effects of high temperatures and heatwaves on dengue fever: A systematic review and meta-analysis. *EBioMedicine, 91.*

De Groot, R. S., Wilson, M. A., & Boumans, R. M. (2002). A typology for the classification, description and valuation of ecosystem functions, goods and services. *Ecological Economics, 41*(3), 393–408.

de Jesús Crespo, R., Méndez Lázaro, P., & Yee, S. H. (2019). Linking wetland ecosystem services to vector-borne disease: Dengue fever in the San Juan Bay estuary, Puerto Rico. *Wetlands, 39*(6), 1281–1293.

De la Cruz, A., & Numa, C. (2024). Habitat availability decline for waterbirds in a sensitive wetland: Climate change impact on the Ebro Delta. *Ecological Modelling, 498*, 110896.

Evans, B. R., & Leighton, F. A. (2014). A history of One Health. *Revue Scientifique et Technique (International Office of Epizootics), 33*(2), 413–420.

Ezbakhe, F. (2018). Addressing water pollution as a means to achieving the sustainable development goals. *Journal of Water Pollution Controll, 1*(1), 6.

Fang, J., Tao, Y., Liu, J., Lyu, T., Yang, X., Ma, S., Dong, J., Dou, H., & Zhang, H. (2023). Effects of emergent plants on soil carbon-fixation and denitrification processes in freshwater and brackish wetlands in a watershed in northern China. *Geoderma, 430*, 116311.

Farooqui, Z. (2014). Mangroves of keti-bunder, indus delta: A present status. *European Scientific Journal.*

Gedan, K. B., Kirwan, M. L., Wolanski, E., Barbier, E. B., & Silliman, B. R. (2011). The present and future role of coastal wetland vegetation in protecting shorelines: Answering recent challenges to the paradigm. *Climatic Change, 106*(1), 7–29.

Ghosh, P., & Chakravarty, T. (2024). Sustainable development goals: Challenges, opportunities, and the way forward. *Sustainability: Science, Policy, and Practice in India: Challenges and Opportunities,* 235–241.

Gibbs, J. P. (2000). Wetland loss and biodiversity conservation. *Conservation Biology, 14*(1), 314–317.

Gitau, P. N., Ndiritu, G. G., & Gichuki, N. N. (2019). Ecological, recreational and educational potential of a small artificial wetland in an urban environment. *African Journal of Aquatic Science, 44*(4), 329–338.

Gopal, B., Junk, W. J., Finlayson, C. M., & Breen, C. M. (2008). Present state and future of tropical wetlands. In N. V. C. Polunin (Ed.). Aquatic ecosystems: trends and global prospects. *Cambridge* (pp. 142–154). Cambridge University Press.

Government of Peru. (2015). National report to the 12th COP of the Ramsar Convention. Lima: Government of Peru.

Hails, S. (2018). Educational benefits of wetlands. In C. M. Finlayson, G. R. Milton, R. C. Prentice, & N. C. Davidson (Eds.). *The Wetland Book*. Dordrecht: Springer.

Hambäck, P. A., Dawson, L., Geranmayeh, P., Jarsjö, J., Peacock, M., Collentine, D., Destouni, G., Futter, M., Hedman, S., Jonsson, S., & Kačergytė, I. (2023). Trade-offs and synergies in the design of multifunctional wetlands: A scaling issue. *Science of the Total Environment, 862*.

Hatvani, I.G., Dokulil, M.T. and Clement, A. (2022). The role of wetlands in mitigating impacts from diffuse agricultural loads.

Hefting, M., Clement, J. C., Dowrick, D., Cosandey, A. C., Bernal, S., Cimpian, C., Tatur, A., Burt, T. P., & Pinay, G. (2004). Water table elevation controls on soil nitrogen cycling in riparian wetlands along a European climatic gradient. *Biogeochemistry, 67*(1), 113–134.

Horwitz, P., Finlayson, C. M., & Weinstein, P. (2012). Healthy wetlands, healthy people: A review of wetlands and human health interactions. In *Ramsar Technical Report No. 6. Gland, Switzerland: Secretariat of the Ramsar Convention on Wetlands and Geneva*. Switzerland: World Health Organization.

International Council for Science (ICSU). (2017). *A guide to SDG interactions: From science to implementation*.

IUCN Commission on Ecosystem Management (CEM). (2023). *Climate mitigation and biodiversity conservation: a review of progress and key issues in global carbon markets and potential impacts on ecosystems*. Gland, Switzerland: IUCN. https://iucn.org/resources/information-brief/climate-mitigation-and-biodiversity-conservation.

Jaramillo, F., Desormeaux, A., Hedlund, J., Jawitz, J. W., Clerici, N., Piemontese, L., Rodríguez-Rodriguez, J. A., Anaya, J. A., Blanco-Libreros, J. F., Borja, S., & Celi, J. (2019). Priorities and interactions of sustainable development goals (SDGs) with focus on wetlands. *Water, 11*(3), 619.

Ji, Z., Tang, W., & Pei, Y. (2022). Constructed wetland substrates: A review on development, function mechanisms, and application in contaminants removal. *Chemosphere, 286*, 131564.

Junk, W. J. (2002). Long-term environmental trends and the future of tropical wetlands. *Environmental Conservation, 29*(4), 414–435.

Keenan, J. D., Hotez, P. J., Amza, A., Stoller, N. E., Gaynor, B. D., Porco, T. C., & Lietman, T. M. (2013). Elimination and eradication of neglected tropical diseases with mass drug administrations: A survey of experts. *PLoS Neglected Tropical Diseases, 7*(12), e2562.

Keesing, F., Belden, L. K., Daszak, P., Dobson, A., Harvell, C. D., Holt, R. D., Hudson, P., Jolles, A., Jones, K. E., Mitchell, C. E., & Myers, S. S. (2010). Impacts of biodiversity on the emergence and transmission of infectious diseases. *Nature, 468*(7324), 647–652.

Kılıç, Z. (2020). The importance of water and conscious use of water. *International Journal of Hydrology, 4*(5), 239–241.

Kreinin, H., & Aigner, E. (2022). From "Decent work and economic growth" to "Sustainable work and economic degrowth": A new framework for SDG 8. *Empirica, 49*(2), 281–311.

Kreplin, H. N., Santos Ferreira, C. S., Destouni, G., Keesstra, S. D., Salvati, L., & Kalantari, Z. (2021). Arctic wetland system dynamics under climate warming. *Wiley Interdisciplinary Reviews: Water, 8*(4), e1526.

Kulkarni, M. A., Duguay, C., & Ost, K. (2022). Charting the evidence for climate change impacts on the global spread of malaria and dengue and adaptive responses: A scoping review of reviews. *Globalization and Health, 18*(1), 1.

Kumar, R., Horwitz, P., Milton, G. R., Sellamuttu, S. S., Buckton, S. T., Davidson, N. C., Pattnaik, A. K., Zavagli, M., & Baker, C. (2011). Assessing wetland ecosystem services and poverty interlinkages: A general framework and case study. *Hydrological Sciences Journal, 56*(8), 1602–1621.

Lafferty, K. D. (2009). The ecology of climate change and infectious diseases. *Ecology, 90*(4), 888–900.

Lu, D. B., Wang, T. P., Rudge, J. W., Donnelly, C. A., Fang, G. R., & Webster, J. P. (2011). Genetic diversity of Schistosoma japonicum miracidia from individual rodent hosts. *International Journal for Parasitology, 41*(13-14), 1371–1376.

McCartney, M., Rebelo, L. M., Senaratna Sellamuttu, S., & De Silva, S. (2010). *Wetlands, agriculture and poverty reduction. Vol. 137*. Iwmi.

Meli, P., Rey Benayas, J. M., Balvanera, P., & Martínez Ramos, M. (2014). Restoration enhances wetland biodiversity and ecosystem service supply, but results are context-dependent: A meta-analysis. *PLoS One, 9*(4), e93507.

Merriman, L. S., Moore, T. L. C., Wang, J. W., Osmond, D. L., Al-Rubaei, A. M., Smolek, A. P., Blecken, G. T., Viklander, M., & Hunt, W. F. (2017). Evaluation of factors affecting soil carbon sequestration services of stormwater wet retention ponds in varying climate zones. *Science of the Total Environment, 583*, 133–141.

Millennium Ecosystem Assessment (MEA). (2005). *Ecosystems and human well-being: wetlands and water synthesis*. World Resources Institute.

Millennium ecosystem assessment (MEA). (2005). *Ecosystems and human well-being* (Vol. 5, p. 563). Island press.

Mitra, G. N., & Mohapatra, P. (1957). *Bulletin on the development of Chilika Lake: Survey report on the fishing industry*. Cuttack, India: Orissa Government Press.

Mitsch, W. J., & Gosselink, J. G. (2015). *Wetlands*. Hoboken, NJ: John Wiley & Sons.

Mitsch, W. J., Bernal, B., Nahlik, A. M., Mander, Ü., Zhang, L., Anderson, C. J., Jørgensen, S. E., & Brix, H. (2013). Wetlands, carbon, and climate change. *Landscape Ecology, 28*(4), 583–597.

Moretti, V., Corraini, N. R., Melo, E. L., Scherer, M. E. G., & Colmenero, J. C. (2024). Progress towards the sustainable development goal 14 (Life below water) in the context of Brazil: A multicriteria approach. *Sustainable Futures, 8*, 100410.

Mukherjee, A., & Dash, A. (2024). Water matters: The thirst, demand, and the society. In A. Mukherjee (Ed.). Water matters: Achieving the sustainable development goals. *Netherlands* (pp. 1–8). Elsevier B.V.

Narayan, S., Beck, M. W., Wilson, P., Thomas, C. J., Guerrero, A., Shepard, C. C., Reguero, B. G., Franco, G., Ingram, J. C., & Trespalacios, D. (2017). The value of coastal wetlands for flood damage reduction in the northeastern USA. *Scientific Reports, 7*(1), 9463.

Nath, J., De, J., Sur, S., & Banerjee, P. (2023). Interaction of microbes with microplastics and nanoplastics in the agroecosystems—impact on antimicrobial resistance. *Pathogens, 12*(7), 888.

Nichols, W. J. (2014). *Blue mind: How water makes you happier, more connected and better at what you do.* Hachette UK.

NITI Aayog. (2018). *SDG India Index: Baseline Report* 2018. New Delhi: NITI Aayog.

Pandey, A., & Asif, M. (2022). Assessment of energy and environmental sustainability in South Asia in the perspective of the sustainable development goals. *Renewable and Sustainable Energy Reviews, 165*, 112492.

Peng, C., Yamashita, K., & Kobayashi, E. I. I. C. H. I. (2016). Effects of the coastal environment on well-being. *Journal of Coastal Zone Management, 19*(2), 1–7.

Perveen, S., Pablos, C., Reynolds, K., Stanley, S., & Marugán, J. (2023). Growth and prevalence of antibiotic-resistant bacteria in microplastic biofilm from wastewater treatment plant effluents. *Science of the Total Environment, 856*, 159024.

Pitt, S. J. (2018). *Clinical Microbiology for Diagnostic Laboratory Scientists*. John Wiley & Sons.

Pitt, S. J., & Gunn, A. (2024). The one health concept. *British Journal of Biomedical Science, 81*, 12366.

Pörtner, H.-O., Scholes, R. J., Agard, J., Archer, E., Arneth, A., Bai, X., Barnes, D., Burrows, M., Chan, L., Cheung, W. L. W., & Diamond, S. (2021). Scientific outcome of the IPBES–IPCC co-sponsored workshop on biodiversity and climate change.

Queenan, K., Garnier, J., Nielsen, L. R., Buttigieg, S., Meneghi, D. D., Holmberg, M., Zinsstag, J., Rüegg, S., Häsler, B., & Kock, R. (2017). Roadmap to a one health agenda 2030. *CABI Reviews, 2017*, 1–17.

Ramsar Convention Secretariat. (2016). The fourth Ramsar strategic plan 2016–2024 (Ramsar handbooks for the wise use of wetlands, 5th ed., Vol. 2). Ramsar Convention Secretariat.

Rojas, O., Soto, E., Rojas, C., & López, J. J. (2022). Assessment of the flood mitigation ecosystem service in a coastal wetland and potential impact of future urban development in Chile. *Habitat International, 123*, 102554.

Russi, D., ten, Brink, P., Farmer, A., Badura, T., Coates, D., Förster, J., Kumar, R., & Davidson, N. (2013). The economics of ecosystems and biodiversity for water and wetlands. In *London and Brussels: Institute for European Environmental Policy; Gland*. Switzerland: Ramsar Convention Secretariat.

Sarabhai, K. V. (2014). ESD and sustainable development goals. *Journal of Education for Sustainable Development, 8*, 1–2.

Sebesvari, Z., Woelki, J., Walz, Y., Sudmeier-Rieux, K., Sandholz, S., Tol, S., García, V. R., Blackwood, K., & Renaud, F. G. (2019). Opportunities for considering green infrastructure and ecosystems in the sendai framework monitor. *Progress in Disaster Science, 2*, 100021.

Seifollahi-Aghmiuni, S., Nockrach, M., & Kalantari, Z. (2019). The potential of wetlands in achieving the sustainable development goals of the 2030 Agenda. *Water, 11*(3), 609.

Shams, A. K., & Muhammad, N. S. (2023). Towards sustainable transboundary water cooperation between Afghanistan and Pakistan: A case study of Kabul River. *Ain Shams Engineering Journal, 14*(2), 101842.

Shen, Y., Zhang, R., Schwarz, S., Wu, C., Shen, J., Walsh, T. R., & Wang, Y. (2020). Farm animals and aquaculture: Significant reservoirs of mobile colistin resistance genes. *Environmental Microbiology, 22*(7), 2469–2484.

Skaff, N. K., & Cheruvelil, K. S. (2016). Fine-scale wetland features mediate vector and climate-dependent macroscale patterns in human West Nile virus incidence. *Landscape Ecology, 31*(7), 1615–1628.

Solomon, A. W., Burton, M. J., Gower, E. W., Harding-Esch, E. M., Oldenburg, C. E., Taylor, H. R., & Traoré, L. (2022). Trachoma. *Nature Reviews Disease Primers, 8*(1), 32.

Standley, C. J., Dobson, A. P., & Stothard, J. R. (2012). Out of animals and back again: Schistosomiasis as a zoonosis in Africa. In *Schistosomiasis. InTech*, 209–230.

Stauffer, J. R., Jr, Madsen, H., McKaye, K., Konings, A., Bloch, P., Ferreri, C. P., Likongwe, J., & Makaula, P. (2006). Schistosomiasis in Lake Malawi: Relationship of fish and intermediate host density to prevalence of human infection. *EcoHealth, 3*(1), 22–27.

Temmerman, S., Meire, P., Bouma, T. J., Herman, P. M., Ysebaert, T., & De Vriend, H. J. (2013). Ecosystem-based coastal defence in the face of global change. *Nature, 504*(7478), 79–83.

Thorslund, J., Jarsjo, J., Jaramillo, F., Jawitz, J. W., Manzoni, S., Basu, N. B., Chalov, S. R., Cohen, M. J., Creed, I. F., Goldenberg, R., & Hylin, A. (2017). Wetlands as large-scale nature-based solutions: Status and challenges for research, engineering and management. *Ecological Engineering, 108*, 489–497.

Turpie, J., Day, E., Ross-Gillespie, V., Louw, A., & Environmental, A. (2010). 4. Estimation Of The Water Quality Amelioration Function And Value Of Wetlands: A Case Study Of The Western Cape, South Africa. *Wetland Valuation Volume II Wetland Valuation Vase Studies*, 67.

UN Environment Annual Report. (2016). Environmental Cooperation for Peacebuilding Programme FINAL REPORT 2016.

United Nations. (2015). *Transforming our world: the 2030 Agenda for Sustainable Development*. A/RES/70/1. United Nations. https://www.refworld.org/docid/57b6e3e44.html.

United Nations Development Programme (UNDP) Egypt. (no date). http://www.eg.undp.org/content/egypt/en/home.html.

United Nations Environment Programme (UNEP). (no date). *Frequently Asked Questions | About UNEP*. A https://www.unep.org/who-we-are/frequently-asked-questions.

United Nations Office for Disaster Risk Reduction (UNISDR). (2015). *The human cost of natural disasters: A global perspective*. Geneva: UNISDR.

Vinca, A., Parkinson, S., Riahi, K., Byers, E., Siddiqi, A., Muhammad, A., Ilyas, A., Yogeswaran, N., Willaarts, B., Magnuszewski, P., & Awais, M. (2021). Transboundary cooperation a potential route to sustainable development in the Indus basin. *Nature Sustainability, 4*(4), 331–339.

Völker, S., Matros, J., & Claßen, T. (2016). Determining urban open spaces for health-related appropriations: A qualitative analysis on the significance of blue space. *Environmental Earth Sciences, 75*(13), 1067.

Webster, J. P., Gower, C. M., Knowles, S. C., Molyneux, D. H., & Fenton, A. (2016). One health–an ecological and evolutionary framework for tackling Neglected Zoonotic Diseases. *Evolutionary Applications, 9*(2), 313–333.

Welling, R., Cartin, M., Baykono, D., & Diallo, O. (2012). Volta River Basin. In *Ghana & Burkina Faso: Transboundary water management through multi-level participatory governance and community projects*. Gland, Switzerland: IUCN Water and Nature Initiative.

Wheeler, B. W., White, M., Stahl-Timmins, W., & Depledge, M. H. (2012). Does living by the coast improve health and well-being? *Health & Place, 18*(5), 1198–1201.

World Water Assessment Programme (WWAP). (2003). *Water for people, water for life: the United Nations World Water Development Report 1*. Paris: UNESCO. https://unesdoc.unesco.org/ark:/48223/pf0000129556.

Xiong, S., Johansson, M. E., Hughes, F. M., Hayes, A., Richards, K. S., & Nilsson, C. (2003). Interactive effects of soil moisture, vegetation canopy, plant litter and seed addition on plant diversity in a wetland community. *Journal of Ecology, 91*(6), 976–986.

Chapter 6

Conventional and advanced biodiversity assessment techniques with special emphasis on environmental DNA

Sk Saruk Islam[1], Sumana Mahato[1], Sk Md Abu Imam Saadi[2], and Sujoy Midya[3]

[1]*Department of Natural & Applied Sciences, Raja Narendra Lal Khan Women's College (Autonomous), Midnapore, West Bengal, India,* [2]*Department of Biological Sciences, Aliah University, Kolkata, West Bengal, India,* [3]*Department of Zoology, Raja Narendra Lal Khan Women's College (Autonomous), Midnapore, West Bengal, India*

6.1 Introduction

Species extinction rates across the globe are increasing, humans are altering vast areas of natural habitat, and habitat degradation is continuing at an accelerated rate (Brondizio et al., 2019). Industrialization and land use practices contribute to the degradation of biodiversity and increase global temperatures. These global changes have caused a 100–1000-fold increase in the extinction rate, leading to a significant and rapid loss of biodiversity (Halder & Jha, 2023). Consequently, if biodiversity is lost, the genetic diversity and extinction rate gradually increase, and ecosystems tend to destabilize and risk collapse. Many natural resource monitoring programs aim to detect biodiversity loss and monitor the distribution of species and the associated flora and fauna of ecological communities within specific habitats (Hata et al., 2022). These programs are crucial to understanding the inhabited population variations, and classifying the stressors contributing to taxa declines. Examples of existing systems designed to monitor riverine macroinvertebrates as indicators of water quality include the Riverfly Monitoring Initiative and the Environment Agency's BIOSYS program in the United Kingdom. Additional examples are the Australian River Assessment System and Europe's National Rivers Monitoring Program under the Water Framework Directive (WFD). These programs generally involve sampling the riverbed and water column using kick-nets and identifying the collected organisms through macroscopic examination. However, this approach is time-consuming and costly, and it requires expert knowledge for species identification (Fediajevaite et al., 2021). To overcome the limitations of traditional methods, biologists have developed a molecular technique to monitor ecosystems, communities, and populations with minimal environmental impact. Environmental DNA (eDNA) technology, one of the innovative methods in extracting and analyzing the genomic substances from the open environment (Chen et al., 2020).

eDNA refers to all the collective DNA present in an environment that comes from the native organisms that inhabit that area. It can be extracted from virtually any habitat containing biological life, including water, soil, sediments, air, feces, or gut contents, as well as ancient sources like cores from ice, sediments, or permafrost (Taberlet et al., 2012; Thomsen & Willerslev, 2015; Balint et al., 2018; Deiner et al., 2021; Beng and Corlett, 2020; Yang et al., 2021). The origins of eDNA and the mechanisms by which it enters the environment can vary between species and can be cryptic, but researchers have identified plant and animal eDNA derived from both living and dead tissue as well as various bodily fluids (Ruppert et al., 2019). Despite the cryptic nature of eDNA, recent technological advancements—particularly the innovative technique known as metabarcoding—have facilitated its application in large-scale biodiversity assessments. Built on the same molecular techniques and principles as genetic barcoding, metabarcoding of eDNA has been transformative for biodiversity research (Ruppert et al., 2019; Taberlet et al., 2012). Metabarcoding is especially valuable in biodiversity studies because it allows for both accurate and noninvasive data collection. Traditional biodiversity studies require taxonomists with the expertise to identify every species found in an ecosystem through morphological means (Zhang et al., 2024). However, there has been a consistent trend in the reduction of expert taxonomists—scientists who study the

Wetlands: Biodiversity, Health and their Role in Ecosystem Services. DOI: https://doi.org/10.1016/B978-0-443-34222-6.00001-2

relationships and organizations of organisms—which may either result from or contribute to the shift towards molecular-based phylogenetic research (Wheeler & Wilson, 2004). Metabarcoding mitigates reliance on expert taxonomists for species-level identification by using DNA and existing barcode libraries. Additionally, traditional studies often depend on visual surveys to detect species presence in the environment (Tibone et al., 2024). While robust statistical methodologies exist for determining species presence this way, they frequently fail to detect rare species that can be visually elusive (Pfleger et al., 2016). Because DNA barcoding can identify minute quantities of unique DNA, eDNA metabarcoding is an optimal approach for detecting rare species.

Furthermore, unlike traditional biodiversity studies, which can often be highly invasive despite contrary intentions, metabarcoding approaches tend to have little to no impact on organisms and ecosystems. The application of eDNA relies on various marker genes available for identifying and monitoring environmental organisms; for instance, riverine invertebrates can be monitored using CO1 (cytochrome c oxidase subunit I), 18 S (18 S rRNA), or 16 S (16 S rRNA), among others (Neice & McRae, 2021; Takahashi et al., 2020; Ahn, 2020). For plant systems, standardized barcode regions, including the widely recognized COI (Cytochrome c Oxidase Subunit I), ITS (internal transcribed spacer), matK (Maturase K), and rbcL (Ribulose-1,5-bisphosphate carboxylase/oxygenase large subunit), sequences are used as eDNA markers (Ogata et al., 2021). These barcode regions have become the cornerstone of DNA barcoding, enabling a uniform and standardized approach to species identification across diverse taxonomic groups (Ratnasingham and Hebert, 2007). The current study highlights the innovation of eDNA technology compared to traditional methods for biodiversity assessment in the environment. It also provides a concise overview of the species identification process using eDNA technology, along with its applications and limitations.

6.2 Literature search strategy and selection criteria

The aim of our present investigation was to detect the best biodiversity assessment tools with special emphasis on the e-DNA method across the varied environments. To assess these, firstly, we search the available published literature from online databases like Google Scholar (Karim et al., 2020). Literature was searched in the database with the article title, abstract, and Keywords by the search engine as follows:

Conventional biodiversity assessment techniques with emphasis on eDNA:

TS1=("biodiversity" OR "conventional" OR "assessment" OR "technique") AND TS2=("conventional" OR "technique" OR *environmental DNA* OR "eDNA") AND TS3=("biodiversity" OR "conventional" OR "eDNA") AND TS4=("biodiversity" OR "techniques" OR "eDNA") AND TS5=("ecosystems" OR "assessment" OR "eDNA")

Advanced biodiversity assessment techniques with emphasis on eDNA:

TS1=("biodiversity" OR "advanced" OR "assessment" OR "technique") AND TS2=("advanced" OR "technique" OR *environmental DNA* OR "eDNA") AND TS3=("biodiversity" OR "advanced" OR "eDNA") AND TS4=("biodiversity" OR "techniques" OR "environmental DNA") AND TS5=("ecosystems" OR "assessment" OR "molecular" OR "eDNA")

The search keywords were employed to differentiate the comparison between the conventional and advanced methods of biodiversity assessment techniques in special relation to eDNA. The literature was searched from 2010 to June 2024 for the present chapter, as no metabarcoding data were available before the search time (Keck et al., 2022).

6.3 Over view of environmental DNA research

eDNA is essentially defined as the total pool of DNA fragments isolated from environmental samples, including water, soil, and air (Pawlowski et al., 2020; Nagler et al., 2022). It is also derived from macroorganisms such as fish, vertebrates, birds, and mammals and is considered to be released in the form of epidermis, mucus, scales, feces, and gametes (Rodriguez-Ezpeleta et al., 2021). The use of eDNA in ecological assessments has been rapidly increasing, particularly in fisheries management (Ramirez-Amaro et al., 2022). eDNA surveys have been proposed as a tool to enhance fisheries assessments and conservation success by reducing systematic errors and providing information on species cooccurrence and relationships across ecological networks (Ramirez-Amaro et al., 2022). By analyzing DNA from environmental samples, researchers can determine whether specific species are present and, in some cases, even estimate their relative abundance based on the concentration of recovered DNA. It can detect, measure, and monitor the presence and abundance of various species, including plants and animals, without direct observation or capture of the organisms themselves (Johnson et al., 2023). eDNA is effective for monitoring aquatic and amphibian species, but it can also be applied in terrestrial environments. In terrestrial habitats, eDNA metabarcoding can be used to study plant community diversity. Researchers have historically employed plant community diversity as an indicator of habitat health, but traditional morphological studies often fail to capture the full spectrum of plant biodiversity due to physical limitations. Fahner et al.

(2016) adopted a metabarcoding approach to characterize vascular plant communities in a boreal wetland and to identify the ideal genetic markers for vascular plant metabarcoding applications. They discovered that by utilizing all four genes—rbcL, matK, ITS2, and tmL—the metabarcoding approach identified all 56 genera of vascular plants previously recorded in that habitat, along with 86 additional ones, resulting in a total of 142 genera (Ramirez-Amaro et al., 2022). These findings support eDNA metabarcoding as a high-fidelity method for studying vascular plant communities. Metabarcoding can also be employed in plant communities within temperate and tropical environments (Nagarajan et al., 2022). The application of eDNA metabarcoding is particularly crucial in conservation biology for detecting invasive species. Universal primers imply that the metabarcoding approach does not require prior knowledge of present taxa, enabling eDNA metabarcoding to detect the presence of new native species, whether they can be observed morphologically or visually. This application of eDNA metabarcoding has already been functionally utilized in both marine and freshwater environments (Borrell et al., 2017). eDNA also employed in aquaculture to discover the invasive aquatic species, like as the Asian carp, in North American rivers. The detection of invasive species in aquatic ports has become a high priority in ecological conservation efforts, given the increase in global trade through ocean ports, which has led to marine habitat invasion and the destruction of numerous marine habitats (Miralles et al., 2021). For habitat restoration initiatives, eDNA can measure the success of restoration interventions by tracking species' return to previously degraded habitats. Moreover, eDNA can assess how climate change impacts species distributions. By sampling various locations over time, scientists can monitor how species migrate or disappear in response to shifting environmental conditions, such as changing water temperatures or habitat loss. In polluted environments, eDNA can be utilized to monitor how pollutants affect species occurrence and community composition. For instance, the existence/nonexistence of certain aquatic species in polluted rivers can indicate water quality and overall ecosystem health (Albert et al., 2021) (Table 6.1).

There are several biases and limitations associated with marker gene selection. For example, 12S and 16S rRNA genes are highly conserved and effective for detecting vertebrates (such as amphibians, birds, and fish) and a few invertebrate species (Chan et al., 2022). However, they have limitations, including poor resolution at the species level in closely related taxa (e.g., some fish and amphibians). Additionally, amplification bias may exclude nontarget taxa or lead to preferential binding (Collins et al., 2019). These primers may also coamplify NUMTs (nuclear mitochondrial pseudogenes), resulting in false positives. The COI gene offers high species-level resolution for many invertebrates and fish and is widely used in barcoding (e.g., compatibility with the BOLD database) (Koroiva & Santana, 2022). However, it is less suitable for degraded eDNA due to typically longer amplicon sizes. COI may also fail to amplify taxa with low mitochondrial diversity (e.g., cnidarians) and can exhibit amplification biases due to mismatches at primer-binding sites (Clarke et al., 2017). In the case of Cyt B and D-loop genes, their strengths include usefulness in phylogeographic studies and providing species-level resolution in fish and mammals. However, Cyt B, although it evolves faster than 12S/16 S, may still fail to distinguish between closely related species (Chan et al., 2021). The D-loop region, while highly variable, often produces nonspecific or multiple bands, especially when using degraded samples (Mezhzherin et al., 2023). ITS regions (ITS1, ITS2, full ITS) are effective for detecting fungi, some mollusks, and protists, and offer high taxonomic resolution (Chan et al., 2021). Nevertheless, intragenomic variation can lead to inconsistent results. In addition, secondary structures and high GC content can reduce amplification efficiency (Bellemain et al., 2010). ITS regions also tend to show a bias toward abundant taxa. Nuclear markers (e.g., β-tubulin, SSU rDNA) are useful for detecting mammals and higher taxonomic groups. However, they have lower copy numbers, leading to reduced sensitivity in environmental samples. The SSU rDNA regions are often too long for eDNA metabarcoding, degrade faster, and their universal primers may compromise specificity (Chan et al., 2021). Lastly, ND genes (ND1, ND2, ND4, ND6) have high copy numbers and are useful for detecting vertebrate groups. However, high mutation rates can lead to primer mismatches, and there are limited reference sequences available in databases for certain taxa (Elyasigorji et al., 2023).

6.4 Environmental DNA versus traditional survey methods

In 21st century, an important issue is the persistent loss of biodiversity on Earth. Anthropogenic disturbances are causing populations of wild plants and animals to decline globally (Dirzo et al., 2014), and the rate of extinction of species is higher than it was before humans. The sustainability of our planet and human health are significantly impacted (Díaz et al., 2006). Despite the fact that many taxa and geographical locations lack complete or even described knowledge of biodiversity (Vié et al., 2009), a global political consensus exists to stop the current decline in biodiversity. To acquire accurate data on species distributions and population levels on an ecological and political time scale, biological monitoring is essential to all such conservation initiatives that protect biodiversity. Traditionally, species monitoring has depended on the physical identification of species through methods such as visual surveys and counting individuals in the field, based

TABLE 6.1 Targeted multispecies environmental DNA (*eDNA*) genetic markers sequence for different taxonomic groups.

Target group	Barcode region	Primer sequence	Amplicon size	Specificity	Efficiency	References
Algae, Seastar	12 S	112 F: 5'-GCACAACCGGGATCTTTACTTC-3' 113 R: 5'-AAATCCTCCTTCGAGCACTAAGTTT-3'	57	High	High	Ellis et al. (2022)
Amphibian	12 S	BuNe_SF: 5'-GTTTGCATGAGTATTRCGGATT-3' Bluegill_CytB_R: 5'-AAGTGTAAGGGCGAGTCTAGGG-3'	89	Moderate	Moderate	Spear et al. (2015)
Amphibian	16 S	C.car_12S_F: 5'-GTGGCATATAAATCTATGTC-3' Bluegill_CytB_R: 5'-ACACCTCCGAGTTTGTTTGGA-3'	90	Moderate	Moderate	Voros et al. (2017)
Arthropod	16 S	Cpa_CO1_F2: 5'-TCGGAATGATCTCTCATATTATCAGTC-3' BuNe_SR: 5'-ACCTAAGACACCAATAGCTAATATAGC-3'	93	High	High	Moushomi et al. (2019)
Bird	16 S	Cpa_COI_F: 5'-ACCTAATCTTTGGGGCCTGA-3' BUT1-R: 5'-ACGGCGGTTACAAGCACGGA-3'	95	High	High	Neice & McRae, (2021)
Bryzoa	16 S	Cpa_COI_F: 5'-GGTCAACAAATCATAAAGATATTGG-3' C.car_12S_R: 5'-AGAAGGAAGGAGGGAGGTAAAAGTCAG-3'	96	Moderate	Moderate	Kim et al. (2021)
Crustacean	16 S	Enj_Cytb_F: 5'-GGGATGATCAGATCTATAATGTGGTAGTTACAGC-3' Cpa_CO1_R2: 5'-AGTAGTATTAAAATTACGATCTGTTAGT-3'	100	High	High	Hauck et al. (2019)
Cnidaria	18 S	F-2317: 5'-TTACTACCCCCAGCTCTGCTTT-3' Cpa_COI_R: 5'-AGTCATTTACCAAATCCTCCTATCATAATA-3'	101	High	High	Takahashi et al. (2020)
Crustacean	CO1	Forward: 5'-AAAGTCTAACCTGCCCACTGAGTTA-3' EcCr_SR: 5'-CAGGAAACAGCTATGACAATCCCAGTTTGTGTCTC-3'	102	High	High	Wu et al. (2018)
Fish	CO1	FP104F: 5'-TACATTTAATGGAAAACAAGCATAAGCC-3' Eng_241R: 5'-CAGGAAACAGCTATGACCGTGAGTTCAAAGGTGT-3'	103	High	High	Ahn, (2020)
Fish	Cyt B	Man-F: 5'-AGCCCACGCAGGAGCATCAG-3' M13coraR: 5'-TAAACTTCAGGGTGACCAAAAAATCA-3'	146	Moderate	Moderate	Mahon et al. (2013)
Fish	D-Loop	NS-193-F: 5'-TGTAAAACGACGGCCAGTGGGTGGTTAAGAATAAACTAAG-3' qLC16S-R: 5'-TAACAAGAACGGATCAGACAAACAGA-3'	239	Moderate	Moderate	Stoeckle et al. (2018)
Fish	D-Loop region	Ofa_COI_F: 5'-TGTAAAACGACGGCCAGTGGTTAAGGGCCCAACTTTTAT-3' R.marina_16S_R: 5'-TAGGATTGGGTCTCCTCCCC-3'	244	Moderate	Moderate	Stoeckle et al. (2018)
Fish, Cnidaria	ITS	Ony-j_12S_F: 5'-CAGATATCGCAACCGCCTTT-3' R-2471: 5'-TCCAAAGTTTCATCATGCAGAGA-3'	251	Moderate	Moderate	Ogata et al. (2021)
Fish, Mollusk	ITS-1, ITS-2	PE1-CS1-ID50X: 5'-AGTACTTCACCTTCTGTTTCTGCATG-3' Reverse: 5'-TCGCTATRCATTATACAGCAGATACA-3'	270	High	High	Wilcox et al. (2020)
Fish, Mollusk	ITS2	Peixoto_Sal_2019-F: 5'-AGTCTCTCCCTGTATATCGTC-3' Reverse: 5'-TCTGGGTCTCCAAGCAGGTT-3'	271	High	High	Wilcox et al. (2020)

Fish, Mollusc	M13	Pflu_12S_F: 5'-CCACAGTGCTTCACCTTCTATTTCTA-3' Reverse: 5'-TGCAGATAAGAGGTTAGTAATGACGGTA-3'	281	High	High	Wilcox et al. (2020)
Fish, Mollusk	MCR	PnGenusF: 5'-GGGTTTTGGTAATTGACTTATTCCACT-3' Reverse: 5'-TGGACGAGAAGGAAGCTGTC-3'	290	High	High	Wilcox et al. (2020)
Fish, Mollusk	NADH	PpF: 5'-CACCAGACATAGCCTTCC-3' Reverse: 5'-TGGTTGATTTCGTATACCGACG-3'	290	High	High	Chen et al. (2020)
Fish, Sharks/ Rays, Mammals	NADH4	qLC16S-F: 5'-TGTAAAACGACGGCCAGTCGAGTAACTCACATTAACACAC-3' S. trutta insert reverse: 5'-TGTTGGGTTAACGGTTTGTGGTA-3'	299	High	High	Stoeckle et al. (2018)
Fish, Sharks/ Rays, Mammals	ND1	R. marina_16S_F: 5'-TGTAAAACGACGGCCAGTCGAGTAACTCACATTAATACTTCAC-3' S11V.2–588 R: 5'-TGTTTATACGAGGGAAGGCCATA-3'	303	High	High	Stoeckle et al. (2018)
Fish, Sharks/ Rays, Mammals	ND2	R. marina_16S_F: 5'-TGTAAAACGACGGCCAGTCGAGTAACTCATATTAATACTTCAC-3' Salmo_Mito-951R: 5'-TRTTATTCGTTTTCTAGAG-3'	310	High	High	Stoeckle et al. (2018)
Fish, Sharks/ Rays, Mammals	ND6; tRNA-Glu; cytB	RNS2: 5'-TGTAAAACGACGGCCAGTGGGCTCAAATTGATATTACA-3' SC-TM5_R: 5'-TTCTGTTGCTTGGAGTTGTA-3'	331	High	High	Stoeckle et al. (2018)
Mammal	Nuclear DNA	WC2302F: 5'-GTCATAGCCACCGCATTTGY-3' WC2302R: 5'-TTGCAGGCGTGAAGTTGTCT-3'	73–110	High	High	Ma et al. (2016)
Mammal	SSU(Small Subunit) rDNA	WC2302F: 5'-CGATAYCTCCACGCAAACG-3' WC2302R: 5'-TTGGTTGGTTCTAGCCGCA-3'	800–1000	Moderate	Low	Ma et al. (2016)
Mammal	β-Tubulin	westslope cutthroat-F: 5'-TTTCGGCCCTTTAGCCARCT-3' westslope cutthroat-R: 5'-TTTAAGCCTTTTCACTTAAAGGTCA-3'	87–88	High	Moderate	Ma et al. (2016)

ITS, Internal transcribed spacer.

on unique morphological characteristics. However, due to factors like phenotypic plasticity and closely related species that have remarkably similar juveniles, these methods occasionally fail to conduct effective and standardized surveys. As a result, several species databases contain incorrect information (Thomsen and Willerslev.,2015). In the absence of biological source material, eDNA metabarcoding is used for high-throughput amplicon sequencing to simultaneously identify multiple organisms using genetic material directly extracted from environmental samples (Taberlet et al., 2018; Eichmiller et al., 2016). According to Taberlet et al. (2018), the definition of DNA metabarcoding is "the automated identification of multiple species from a single bulk sample—containing entire organisms or from a single environmental sample containing degraded DNA." This methodology can be used for various target applications, such as species identification from eDNA and characterization of the diversity of a bulk sample—an environmental sample that contains organisms from various taxonomic groups under study (Taberlet et al., 2018).

In the environment, organisms release DNA through their skin, hair, gametes, urine, or feces (Pawlowski et al., 2020). Water, soil, sediment, air, or organic matter like gut contents, excrement, bird pellets, or even honey—which may include a range of plant pollen as well as DNA profiles of visiting pollinators—are among the media that can be examined for the identification of species (Harper et al., 2023). As demonstrated in lake sediments and Arctic permafrost, the released DNA remains in the medium and may persist for days to months, or even longer (Pascher et al., 2022). It makes it possible to identify species in a medium without the use of acoustic detection, sighting, or capture. eDNA metabarcoding has the potential to significantly lower sampling time, sample costs, and invasiveness, supporting ecosystem management and conservation (Thomsen and Willerslev, 2015). While at the same time, it increases the number of taxonomic groups that have to be surveyed. The number of papers concerning eDNA has increased exponentially in recent years due to the acknowledged potential of eDNA research (Jeunen et al., 2019). By using this technique, standard survey windows for migratory species like the great crested newt (GCN) could be extended. GCNs typically reside on land, hibernate from October to March, and breed in ponds and pools during their prime breeding season, which lasts from March to May. This creates a survey "window" where the GCN can be detected using eDNA analysis as well as traditional survey techniques (e.g., traps, torchlight surveys, and egg counts) (Thomsen et al.2012). When conventional sampling techniques are not as effective, one of the intents of eDNA analysis is that this survey window can be expanded to include the time that larvae and juveniles are present in ponds. Particularly when examining the distribution of rare or endangered species, eDNA analysis can offer significant time and cost savings over conventional survey approaches in terms of sample effort. According to research on invasive Asian carp in Chicago, Illinois, electrofishing at a site required 93 days of human effort to identify a single silver carp, whereas eDNA analysis required only 0.174 days of human effort to achieve a positive detection (Thomsen and Willerslev, 2015). The ability to identify live animals is the most significant advantage of traditional survey methods over eDNA methods. Furthermore, traditional methods can discriminate between the many life stages of the species of interest, whereas eDNA approaches cannot (e.g., in amphibians: eggs, larvae [tadpole], juveniles, and adults). eDNA methods cannot provide specific information on the location of a target species or on any microhabitat variation, in contrast to traditional approaches that allow you to know in "real time" where and when a target species was present. eDNA techniques should be an important tool to supplement the limited conservation resources, not to replace or ignore the knowledge and experience of experienced field ecologists and taxon specialists. Data on species presence and distribution can be gathered reasonably quickly and cheaply using eDNA analysis. This is especially useful if the sampling region is quite vast (Rees et al., 2014) (Fig. 6.1).

In biodiversity assessment, DNA-based methods have become increasingly vital for understanding ecological dynamics and species interactions (Bohmann et al., 2014). A standard workflow for these methods involves several critical steps, each with its own technical details and challenges. The first step is sample collection, where preservation methods play a crucial role. Samples can be preserved by freezing at −20°C or −80°C, or by using ethanol for immediate fixation (Hajibabaei et al., 2006). However, improper preservation can lead to DNA degradation, which poses a significant challenge in obtaining reliable results (Leray et al., 2019). Once samples are collected, the next step is filtration, where filter sizes typically range from 0.45 to 5 mm. This step is essential for concentrating DNA from environmental samples, such as water or soil (Bohmann et al., 2014). However, challenges such as filter clogging and loss of genetic material can occur, necessitating careful selection of filter types and sizes (Zhan et al., 2013). Following filtration, DNA extraction is performed, which can be hindered by polymerase chain reaction (PCR) inhibition due to the presence of contaminants like humic acids or polysaccharides (Kumar et al., 2015). It is crucial to use appropriate extraction kits and protocols to minimize these inhibitors and maximize yield (Baker et al., 2013). The amplification step, often using PCR, is another critical phase. Here, challenges such as nonspecific amplification and primer–dimer formation can affect the quality of the results (Meyer et al., 2019). Optimizing PCR conditions, including temperature and cycle number, is essential to mitigate these issues (Bohmann et al., 2014). Finally, the analysis phase involves sequencing and bioinformatics processing. This step can be complicated by the need for high-quality sequence data and the challenges of accurately identifying species

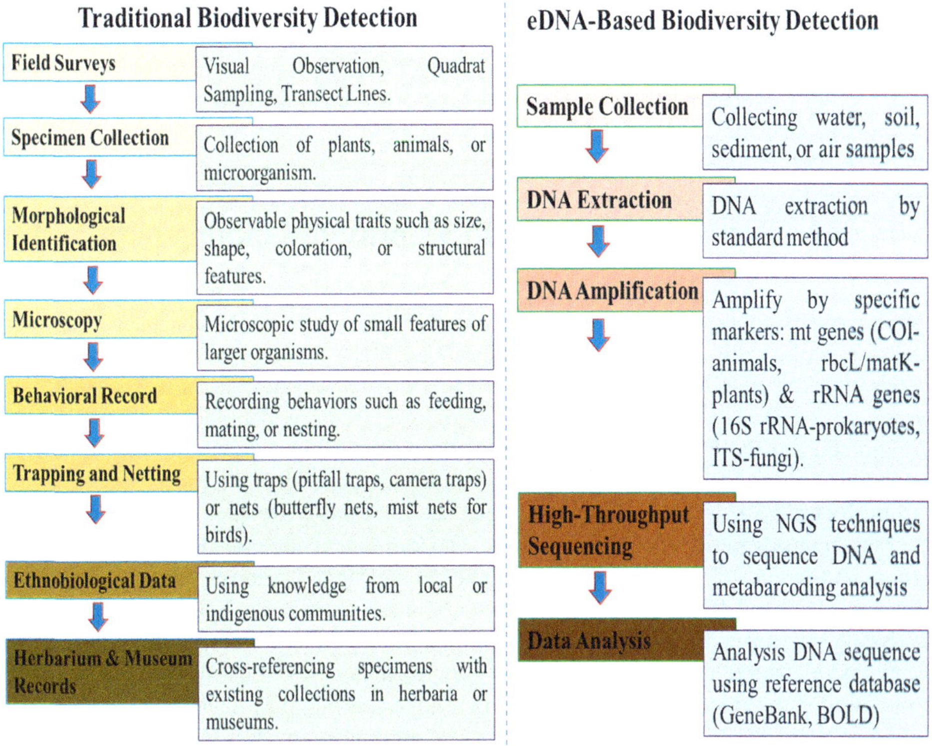

FIGURE 6.1 Traditional versus environmental DNA (*eDNA*) based biodiversity detection methods.

from genetic sequences (Elbrecht & Leese, 2015). In summary, while DNA-based methods for biodiversity assessment offer powerful tools, they are accompanied by various technical challenges at each stage of the workflow. Addressing these challenges through careful planning and execution is essential for obtaining reliable and meaningful results (Leray et al., 2019).

6.5 Comparisons of environmental DNA versus traditional method based on their work efficiency

eDNA methods offer significantly higher sensitivity and lower detection limits compared to traditional survey methods. Studies show that eDNA can detect as little as 0.5 copies/μL or 1 pg/L of target DNA, making it highly effective for identifying low-abundance or elusive species. Limits of detection and quantification can be as low as 1 pg/L or 0.1 pg/μL, depending on assay design and optimization (Wilcox et al., 2013). While eDNA is highly specific with validated primers, it remains susceptible to false positives from environmental contamination, DNA transport, or cross-amplification. Conversely, traditional methods rarely produce false positives but may result in misidentification or miss rare species due to low detectability and sampling limitations. False negatives in eDNA can arise from inhibitors, sample degradation, or insufficient replication but can be mitigated through strict protocols and replicate sampling. Traditional surveys are also prone to false negatives, especially for cryptic species, and often require more effort and cost. Overall, eDNA offers a more sensitive, cost-effective, and scalable tool for biodiversity monitoring when supported by robust assay validation and contamination control (Buxton et al., 2017; Roux et al., 2020; Fonseca et al., 2023; Chang et al., 2025).

In the present investigation, we reviewed more than 30 articles comparing eDNA and traditional survey methods, where eDNA was consistently found to be more sensitive, cost-effective, and capable of detecting a greater number of species overall (Smart et al., 2015; Lugg et al., 2018; Fediajevaite et al., 2021). Previous studies reported that eDNA detection probabilities for a freshwater amphibian species ranged from 0.29 to 1.0, compared to 0.01–0.26 for traditional trapping methods (Smart et al., 2015). In mammal monitoring, stream-based eDNA metabarcoding identified 25% more taxa than camera traps, at half the sampling cost. Similarly, other studies have shown that eDNA detected up to 32% more species than bottom trawling in fish monitoring. Notably, a 6-hour eDNA sampling period detected 112 marine fish species, compared to just 80 species recorded over 14 years of visual surveys (Lyet et al., 2021; Johnson et al., 2021; Yao et al., 2022). Regarding cost comparisons, it was shown that a single eDNA survey (including field and lab work) costs approximately AUD 395–482, whereas traditional electrofishing or trapping methods cost nearly AUD 730, representing up to 67% cost savings in favor of eDNA-based approaches (Lugg et al., 2018; Evans et al., 2017) (Table 6.2).

6.6 Integrating of environmental DNA to identify the biodiversity in different ecosystems

Traditional morphological-based biodiversity surveys may be replaced by molecular-based, environmental (e)DNA surveys. According to Taberlet et al. (2012), "eDNA" is simply the DNA found in environmental samples like water, silt, soil, or feces. Sequencing this eDNA allows identification of taxa without morphological identification across all life phases (Lisnerova et al., 2023). There are five essential steps in eDNA-based biodiversity surveys. Every step of the method should be modified to the target taxonomic group, the scientific investigation, and the sample type's intrinsic restrictions. To guarantee representative variation is achieved, the sample collection, processing, and DNA extraction are all crucial steps (Suresh et al., 2024). After extraction, each sample's target DNA is amplified by PCR using particular "tagged" primers, producing several identical DNA copies. The target gene area (also known as a genetic marker) that is conserved throughout the taxon group of interest is bound by the tagged primers, and the gene region is amplified throughout both the conserved and variable (i.e., species-specific) parts (Sun et al., 2023). Every amplified DNA sequence in that sample also receives a special tag particular to that sample during this procedure. This step significantly reduces the throughput time and cost of sample processing by allowing multiple samples to be merged and sequenced together. The samples undergo purification to get rid of any small nontarget sequences that may have remained after PCR amplification. Ineffective sequencing may occur if small nontarget sequences are not eliminated. A "DNA library" can be created by pooling the labeled samples at equal molar concentrations after PCR (Lilja et al., 2023). A next-generation sequencing (NGS) technology is used to sequence the DNA library. The processing of data and result visualization constitutes the last

TABLE 6.2 Metaanalyzes of quantified performance metrics comparing environmental DNA (*eDNA*) with traditional methods.

Metric	eDNA	Traditional survey
Sensitivity	Very high: LOD ≈ 0.5 copies/μL; LOQ≈1 pg/L	Moderate: detection depends on visibility, density, and effort
Specificity	High with validated primers; vulnerable to contamination and cross-amplification	High: direct observation or capture; misidentification possible
False positives	Possible due to contamination or DNA transport; mitigated by protocols	Rare
False negatives	Possible from inhibition, degradation, or poor sampling; mitigated with replication and controls	Common for elusive, rare, or behaviorally cryptic species
Detection limits	Extremely low thresholds possible with qPCR/ddPCR	Limited by physical detectability and sampling limitations
Detection probability	0.29–1.00	0.01–0.26
Number of species	25%–44% more	Baseline
Cost per survey	~AUD 395–482	~AUD 730
Taxa detected	Higher sensitivity	Misses some rare/elusive species

Source: Based on Wilcox et al. (2013), Smart et al. (2015), Evans et al. (2017), Buxton et al. (2017), Lugg et al. (2018), Roux et al. (2020), Fediajevaite et al. (2021), Lyet et al. (2021), Johnson et al. (2021), Yao et al. (2022), Fonseca et al. (2023), and Chang et al. (2025).

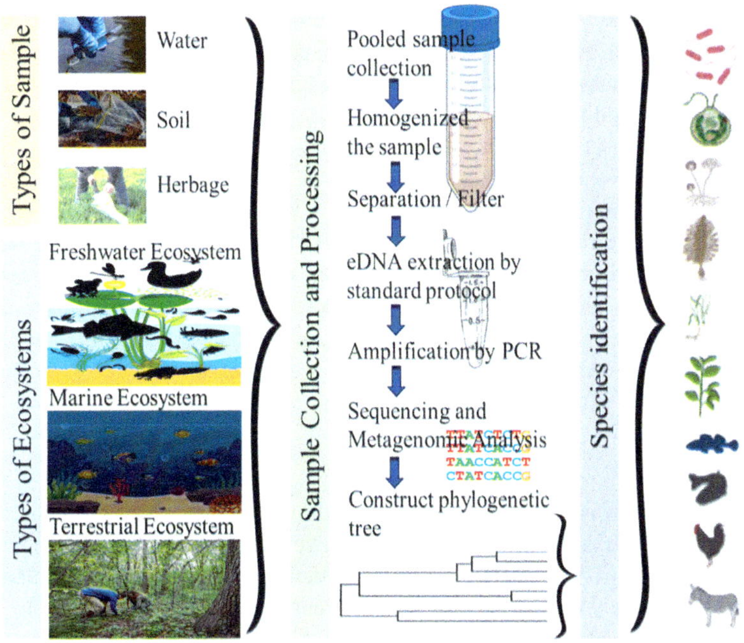

FIGURE 6.2 Schematic diagram of environmental sample collection from global ecosystem and biodiversity detection by environmental DNA (*eDNA*) extraction method and species identification. *PCR*, Polymerase chain reaction.

step of an eDNA survey. There are many software packages available right now, some designed especially for users who don't know much about programming (Shaw et al., 2016). Because of the abovementioning advantages of this process, it can be used for the identification of diversity in different ecosystems (Fig. 6.2).

6.6.1 Freshwater ecosystems

Freshwater ecosystems are the most vulnerable ecosystems worldwide Albert et al. (2021). Hence, surveying of freshwater ecosystems and organisms imposes much importance, but it is considered a challenging task, mainly because of two reasons. First, freshwater ecosystems are difficult to access because of the water current, turbidity, and sometimes navigation. Second, freshwater organisms are difficult to sample (Prié et al., 2018). For assessing the freshwater ecosystems, a large number of biotic indices have been created based on morphological identification of various groups of aquatic indicator organisms of different levels of organization, and have been developed in different countries (Lisnerova et al., 2023). For example, WFD requires ecological status assessment of surface waters to be based on "biological quality elements," which dependent on the hydrotype, include "phytoplankton," "diatoms," "aquatic flora," "macroalgae and angiosperms," "benthic invertebrate fauna," and "fish fauna." The ensuing lists of taxonomic groups and their abundance are used to calculate biotic indices. One of the earliest investigations to extract macroorganismal eDNA from freshwater concentrated on human, cow, pig, and sheep DNA as a way to identify the sources in surface water contaminated by feces (Thomsen and Willerslev, 2015). A later study in France successfully extracted eDNA from natural pond water samples belonging to American bullfrogs, an invasive amphibian species endemic to North America (Ficetola et al., 2008). Numerous follow-up investigations continued to concentrate on invasive species detection. Using the model species, the American bullfrog, (Dejean et al., 2012) showed that eDNA can provide better detection of invasive species than conventional approaches. The study conducted by Jerde et al. (2011) on two species of Asian carbs in river systems showed that the species' forefronts were closer to the invasion of upstream lake systems than conventional surveillance techniques had first detected. In a study that verified the existence of eDNA for low-abundance species in river systems as well, this strategy was later further enlarged and extended to additional related fish species (Antognazza et al., 2019; Takahara et al., 2013), which estimated the invasive bluegill sunfish's population on the Japanese mainland and nearby islands using eDNA. In this case, eDNA also outperformed conventional visual detections. The use of eDNA has recently been extended to crustaceans, snails, and invasive reptiles (Coghlan et al., 2021). Exploration of endangered species is another area of freshwater eDNA study. The majority of research in this area has concentrated on amphibian detection and even quantification (Biggs et al., 2015; Hata et al., 2022), which opens up a possible new range of possibilities for estimating relative abundance. Biomass of fish from freshwater has also been quantified, and species composition and communities have been examined. Additionally, eDNA has been shown to be useful for monitoring endangered freshwater insects, crustaceans, fish, and mammals (Coghlan et al., 2021). This method can also use high-throughput sequencing to account for entire lake faunas of fish and amphibians (Keskin et al., 2016). Despite the recent discovery by Turner et al. (2014) that aqueous macrobiological eDNA most likely resides within mitochondria or small cells, eDNA decay in freshwater beyond the detectability threshold occurs over days to weeks (Pilliod et al., 2014). In contrast, it has been demonstrated that eDNA can survive in soil for decades or even centuries after deposition (Lilja et al., 2023). eDNA degrades rapidly in freshwater ecosystems, making it more useful for conservation, as positive findings usually indicate the presence of species without interference from past populations. However, long-distance transport of eDNA from hundreds of meters to several kilometers has been reported in river systems and should be taken into consideration in flowing waters (Elberri et al., 2020) (Table 6.3).

6.6.2 Marine ecosystems

DNA sequencing from filtered seawater samples has been shown to be a useful method for studying microbial (prokaryotic and eukaryotic) biodiversity (Sun et al., 2023). However, it was recently demonstrated by Thomsen et al. (2012) that metabarcoding of eDNA from seawater samples can identify a rich marine fish fauna, and that this method can cover the fish diversity more effectively than any of the nine conventional methodologies employed in marine fish surveys. Additionally, Knudsen et al. (2019) demonstrated that whale eDNA can be found in seawater samples, and an experiment conducted in sea enclosures revealed that the eDNA signal is only visible within a few meters of the animals. Foremost indications that the detectable eDNA from macroorganisms can be found in marine water samples. Even small eDNA (100 bp) fragments in saltwater degrade beyond detectability in a matter of days. According to these findings, there is very little probability that eDNA will spread across great distances in marine environments. Initial research indicates promising prospects for future monitoring and management of marine biodiversity and resources, even though further research is required to confirm the eDNA approach in modern marine ecosystems (Mariani et al., 2021) (Table 6.3).

6.6.3 Terrestrial ecosystems

eDNA was first used to evaluate the variety of macroorganismal groups; it revealed DNA from both extinct and existing plants, birds, and mammals in sediments (Ariza et al., 2022; Valentin et al.,2018). Dry cave sediments then revealed DNA from extinct birds and mammals that lived in caves. Subsequent research revealed that DNA from invertebrates, such as fungi (Navarro–Noya et al. 2021) and insects (Aizpurua et al., 2017), is also preserved in ancient strata. Although the initial research on soil eDNA concentrated on previous deposits, there have been a number of new applications for current ecosystems. Both intracellular and extracellular DNA are included in total soil DNA, the latter of which likely makes up the majority (Lilja et al., 2023). Thus plant taxonomic diversity in a variety of terrestrial ecosystems has been effectively represented by metabarcoding eDNA from surface soil (Ariza et al., 2022). It has also been extensively shown that soil eDNA is applicable to fungi diversity analysis. For instance, rich and varied fungus communities were revealed by employing DNA metabarcoding from different forest soils (Sternhagen et al., 2020). Regarding animal eDNA in soil samples, Bienert et al. (2012) showed that NGS can be used to identify earthworm communities, a significant indicator group of ecosystem health and functioning, while researchers discovered that soil surface DNA reflects the relative biomass and overall taxonomic richness of individual vertebrate species. Based on the finding of DNA from keratinophilic and coprophilic fungi in ancient sediments, it has been hypothesized that sedimentary eDNA originates from hair, urine, feces, and epidermal cells (Makiola et al., 2019). The ability of NGS approaches to reconstruct paleo-ecosystems from permafrost samples has been shown to analyze recent environmental change in Greenland and reveal abundant diversity in previous flora. However, some findings suggest that physical remains like pollen and macrofossils are complementary rather than overlapping with sedimentary ancient DNA, and when combined, provide more detailed information on ancient plant communities than can be obtained by each method alone (Ariza et al., 2022) (Table 6.3).

6.7 Persistence of environmental DNA and the impact of environmental variables on its detection

Understanding eDNA persistence is crucial for accurate biodiversity assessments, especially in dynamic marine and lotic systems. Persistence affects both false negatives (due to low DNA density) and false positives (due to DNA transport).

TABLE 6.3 Environmental DNA (*eDNA*) study in different aquatic ecosystems.

Sampling site	Sample habitat	Geographical location	Taxon studied	Barcoding	Uses	References
Freshwater ecosystems	Yangtze River	China	Bacteria to vertebrates	Metabarcoding	Biomonitoring and management of the ecosystem	Zhang et al. (2024)
	Reservoir, Pond	Czech Republic	Parasites	Metabarcoding	Detection of freshwater myxozoan communities	Lisnerova et al. (2023)
	River	Japan	Unionid mussels	Species–specific	Monitoring declines of population	Hata et al. (2022)
	Lake	Canada	Freshwater Mussel	Metabarcoding	Freshwater mussel detection	Coghlan et al. (2021)
	River	Egypt	Sharptooth catfish	Species–specific	Tracking of catfish	Elberri et al. (2020)
	River	United Kingdom	Fish	Species–specific	Spawning distribution detection	Antognazza et al. (2019)
	Pond	France	Crayfishes	Species–specific	Detecting invasive crayfishes	Mauvisseau et al. (2018)
	Laboratory	United States	Freshwater mussel	Species–specific	Decay rates and eDNA shedding determination	Sansom and Sassoubre, (2017)
	Lake	Turkey	Fish	Metabarcoding	Rare and invasive freshwater fish identification	Keskin et al. (2016)
	Ponds	England, Wales, Scotland	Amphibians	Species–specific	Great crested newt monitoring	Biggs et al. (2015)
Marine ecosystems	Ocean	Ireland	Fish	Metabarcoding	Fish community investigation in the upper mesopelagic layer	Tibone et al. (2024)
	Peninsula	China	Fish	Species–specific	Biomass evaluation	Sun et al. (2023)
	Coast	Australia	Sessile benthic	Metabarcoding	Sessile benthic inspection	West et al. (2022)
	Island	France	Ray and Shark	Metabarcoding	Distribution and temporal variation of ray and shark	Mariani et al. (2021)
	Estuarie	Spain	European eel	Species–specific	Endangered eel monitoring	Cardas et al. (2020)
	Sea	Denmark	Fish	Species–specific	Fish species determination	Knudsen et al. (2019)
	Coastal	United States	Shark	Species–specific	Shark screening	Lafferty et al. (2018)
	Ocean	Azores	Ray	Species–specific	Survey of pelagic biodiversity	Gargan et al. (2017)
	Bay	United States	Vertebrates	Metabarcoding	Vertebrates' biodiversity assessment	Port et al. (2016)

(Continued)

TABLE 6.3 (Continued)

Sampling site	Sample habitat	Geographical location	Taxon studied	Barcoding	Uses	References
	Lagoon	Russia	Mollusk	Metabarcoding	Invasive species screening	Ardura et al. (2015)
Terrestrial ecosystems	Bay area	United States	Bat	Metabarcoding	Noninvasively monitoring the ecosystems	Suresh et al. (2024)
	Soil	Denmark	Earthworms	Metabarcoding	Earthworms' diversity surveillance	Lilja et al. (2023)
	Soil	Norway	Plant	Metabarcoding	Plant biodiversity assessment	Ariza et al. (2022)
	Soil	Mexico	Fungi	Metabarcoding	Fungal community comparison	Navarro–Noya et al. (2021)
	Soil	Costa Rica	Fungi	Metabarcoding	Fungal community surveillance	Sternhagen et al. (2020)
	Cropland	New Zealand	Fungi, bacteria, oomycetes	Metabarcoding	Plant pathogen screening	Makiola et al. (2019)
	Agricultural fields	United States	Bugs	Species–specific	Invasive exotic insect identification	Valentin et al. (2018)
	Agricultural fields	Europe	Insects	Metabarcoding	Prey detection	Aizpurua et al. (2017)
	Macadamia orchards	Australia	Arthropod	Metabarcoding	Dietary patterns and pest management of sympatric birds	Crisol–Martínez et al. (2016)
	Palm plantations	Malaysia	Insects	Metabarcoding	Biological impacts quantification of plantations	Edwards et al. (2014)

eDNA can travel long distances—up to 10 km downstream (Deiner et al., 2016)—and remains detectable despite tidal cycles (Kelly et al., 2018). Its persistence reflects the balance between DNA release and environmental degradation. Thus detection is inherently imperfect and influenced by species density and local conditions (Barnes and Turner, 2016; Seymour et al., 2018). Reliable monitoring requires accounting for these temporal and spatial factors.

Temperature influences aquatic species metabolism, physiology, and waste output, enhancing eDNA shedding through increased mucus and epithelial cell loss. It also boosts microbial activity, accelerating DNA release from sediments. These effects may lead to spatially uniform genetic signals. Predictive eDNA models should account for seasonal, longitudinal, and latitudinal temperature variations (Lacoursière-Roussel et al., 2016). Salinity and pH significantly affect eDNA stability in the environment. DNA degrades faster in acidic or highly basic environments, while neutral to slightly basic conditions promote stability. Salinity influences osmoregulation, energy expenditure, and key life stages, such as fertilization and larval growth, across different species (Bœuf and Payan, 2001). It alters hormonal control of metabolism and growth, with marine species typically growing faster at slightly lower salinities, and freshwater species showing the opposite trend. Fish, like salmon, undergo smoltification to adapt to saltwater. eDNA has proven effective in detecting marine biodiversity, though read abundance often poorly correlates with biomass estimates (Günther et al., 2018). Salinity-driven abiotic factors must be considered in both marine and freshwater eDNA applications (Çevik & Çevik, 2025). UV radiation accelerates eDNA degradation, especially at the water surface under strong sunlight. Daytime or open-area samples often yield less eDNA than shaded or nighttime samples. UVB causes direct DNA lesions like cyclobutane pyrimidine dimers and 6-4 photoproducts (6-4PPs), while UVA induces oxidative damage (Fraikin et al., 2024). However, studies show eDNA can persist even under high UVB exposure, suggesting microbial degradation may be more critical (Machler et al., 2018; Guthrie et al., 2024). Understanding UV impact is vital for optimizing eDNA-based biodiversity monitoring. Though UV affects eDNA stability, cumulative and long-term effects under varying environmental conditions remain underexplored, necessitating further research for refining detection strategies (Guthrie et al., 2024; Fraikin et al., 2024). Sediments are a major reservoir of eDNA in aquatic ecosystems, especially marine sediments, which hold an estimated 0.30–0.45 Gt of extracellular DNA (Dell'Anno & Danovaro, 2005). Over 90% of sediment DNA is extracellular, often adsorbed to particles, protecting it from degradation (Dell'Anno et al., 2002). eDNA concentrations vary by location and depth—higher at the surface and decreasing with depth, along with particle size. For example, Haihe River sediments contain ~96.8 $\mu g\ g^{-1}$ eDNA (Mao et al., 2014). Sediment-bound eDNA provides insights into both present and past biodiversity, helping reconstruct environmental and climatic histories, and predict future ecosystem dynamics (Ellegaard et al., 2020; Vuillemin et al., 2017).

6.8 Limitations of conventional and advanced biodiversity assessment methods

6.8.1 Limitations of conventional methods

Conventional methods of biodiversity assessment have been foundational in understanding the ecosystems, but it has several limitations, which restrict their effectiveness in highly dynamic ecosystems. Some key challenges associated with conventional biodiversity assessment techniques are:

- Survey-surveillance compilation, specimen collection, and observational methods require significant time and manpower-intensive. Diverse ecosystem services demand extensive effort to be adequately covered.
- Limited detection of cryptic and rare species: Cryptic species (those with similar physical appearances but different genetics) are difficult to distinguish without molecular tools. Rare or elusive species (e.g., nocturnal or deep-water organisms) are often missed using conventional methods.
- Species identification challenges: Accurate identification often requires taxonomic expertise, which is increasingly scarce. Morphological similarities between species can lead to misidentification. Specimen degradation (e.g., dried plant samples or damaged animal specimens) can make identification difficult.
- Ethical and ecological concerns: Methods such as trapping or specimen collection can harm organisms and disrupt ecosystems. Ethical concerns arise from invasive procedures, especially with endangered species.
- Habitat accessibility: Some habitats, like dense forests, deep oceans, or high-altitude regions, are difficult or dangerous to access. Seasonal and weather constraints may limit data collection in certain ecosystems.
- Snapshot nature of data: Conventional methods often provide a static snapshot of biodiversity at a specific time and location. This fails to capture dynamic changes, such as migrations, seasonal variations, or population shifts.
- Inability to assess microbial diversity: Traditional techniques focus on visible plants and animals, neglecting microorganisms that play critical ecological roles. Soil, water, and air microbiomes remain largely unexplored using conventional methods.

- Limited scalability: Scaling up field-based methods to monitor biodiversity across large landscapes is resource-intensive and impractical.
- Detection bias: Observer bias can affect results, especially in visual surveys. Some species are easier to detect because of their size, behavior, or habitat preferences, leading to skewed data.
- Lack of real-time monitoring: Conventional methods do not enable continuous or real-time monitoring, which is crucial for detecting rapid changes in ecosystems.
- Cost inefficiency: Fieldwork, specimen preservation, and taxonomic identification can be costly, particularly in remote regions.
- Lack of standardization: Protocols for data collection and analysis vary widely, making comparisons between studies difficult.
- Climate change and environmental dynamics: Traditional methods may not be sufficiently sensitive to detect early effects of climate change on species distributions or ecosystem dynamics.

6.8.2 Limitations of advanced methods

While advanced biodiversity assessment techniques, including eDNA, have revolutionized ecological monitoring, they come with their own set of limitations. These constraints can affect data reliability, accessibility, and applicability. Below are the general limitations of advanced techniques, with a focus on eDNA.

- Sensitivity to the ecosystems: eDNA is likely to degrade due to environmental factors such as pH, temperature, UV light, and microbial activity, making it challenging to extract DNA from certain ecosystems.
- Detection of false positives: Contamination during sampling or lab analysis can lead to the identification of species that do not exist. DNA from transported organisms (via water currents or bird droppings) can result in misleading presence data.
- Detection of false negatives: Low DNA concentration or uneven distribution of eDNA in the environment can result in missed species detections. Sampling methods may fail to capture DNA from rare or cryptic species.
- Difficulty in quantifying abundance: eDNA provides presence/absence data but struggles to estimate population sizes or biomass accurately. DNA concentration does not always correlate with the number of individuals, as it depends on shedding rates, organism size, and environmental factors.
- Taxonomic resolution limitations: eDNA relies on genetic databases for species identification, which are incomplete for many taxa, especially microorganisms, invertebrates, and rare species. Closely related species may have overlapping genetic markers, leading to identification challenges.
- High cost and specialization: Innovative molecular methods need specific equipment and skilled personnel. Costs associated with lab analysis, sequencing, and bioinformatics tools can be prohibitive, especially for large-scale studies or in low-resource settings.
- Lack of standardization: Protocols for eDNA collection, extraction, amplification, and sequencing vary across studies, making it difficult to compare results. No universal benchmarks exist for interpreting eDNA concentration or degradation rates across ecosystems.
- Limited applicability in complex ecosystems: Ecosystems with high biodiversity, such as tropical rainforests, may produce overlapping or mixed genetic signals, complicating species identification. In soil and sediment, DNA from multiple sources can accumulate, making it hard to distinguish between current and past biodiversity.
- No contextual ecological data: eDNA provides little to no behavioral, ecological, or life-stage information about the detected organisms. It cannot capture interactions like predator-prey relationships, symbiosis, or habitat use patterns.
- Dependence on bioinformatics: Data analysis requires advanced computational tools and software, which can introduce errors if not properly configured. Processing large datasets, such as those from high-throughput sequencing, can be time-intensive and computationally expensive.

6.9 Summary and future direction

While advanced techniques like eDNA offer significant advantages over conventional methods, their limitations highlight the need for careful application and interpretation. Combining eDNA with other techniques and addressing challenges in standardization, data interpretation, and cost will maximize their potential in biodiversity assessment and conservation.

Potential solutions to address limitations:

1. Improved protocols: Standardizing methods for sample collection, preservation, and analysis to reduce variability.

2. Advances in bioinformatics: Developing robust algorithms for taxonomic assignment and abundance estimation.
3. Expanding reference databases: Increasing genetic database coverage for underrepresented taxa.
4. Integration with conventional methods: Combining eDNA with traditional fieldwork for more comprehensive assessments.
5. Quality control measures: Implementing rigorous controls to minimize contamination and validate results.

Emerging trends and future directions:
Research is advancing to address these challenges and expand eDNA's utility:

1. Integrating eDNA with remote sensing: Combining satellite data with eDNA to map species distributions and ecosystem health on larger scales.
2. Long-term monitoring: Developing protocols for repeat sampling to track biodiversity trends over time, aiding climate change impact studies.
3. Improved analytical techniques: Focus on eRNA (environmental RNA) to differentiate between alive and deceased organisms in ecosystems. New bioinformatics tools to enhance taxonomic resolution and functional ecology insights.
4. Multiomics approaches: Using eDNA alongside metagenomics, metatranscriptomics, and proteomics to study ecosystem dynamics holistically.
5. Policy and management integration: Increasing use of eDNA data in shaping conservation policies, fisheries management, and invasive species control.

Funding

No funding was received to assist with the preparation of this manuscript.

Ethical approval

Not applicable.

Conflict of interest

The authors declare that they have no conflict of interest

Acknowledgments

The authors acknowledge Raja N. L. Khan Women's College (Autonomous), and Aliah University for supporting all sorts of library facilities.

References

Ahn, H. (2020). Variation of Japanese eel eDNA in sequentially changing conditions and in different sample volumes. *Journal of Fish Biology, 97*(4), 1238–1241.

Aizpurua, O., Budinski, I., Georgiakakis, P., Gopalakrishnan, S., Ibanez, C., Mata, V., Rebelo, H., Russo, D., Szodoray-Paradi, F., & Zhelyazkova, V. (2017). Agriculture shapes the trophic niche of a bat preying on multiple pest arthropods across Europe: Evidence from DNA metabarcoding. *Molecular Ecology, 27*(3), 815–825.

Albert, J. S., Destouni, G., Duke-Sylvester, S. M., Magurran, A. E., Oberdorff, T., Reis, R. E., & Ripple, W. J. (2021). Scientists' warning to humanity on the freshwater biodiversity crisis. *Ambio, 50*(1), 85–94.

Antognazza, C. M., Britton, J. R., Potter, C., Franklin, E., Hardouin, E. A., Gutmann Roberts, C., Aprahamian, M., & Andreou, D. (2019). Environmental DNA as a non–invasive sampling tool to detect the spawning distribution of European anadromous shads (*Alosa* spp.). *Aquatic Conservation, 29*(1), 148–152.

Ardura, A., Zaiko, A., Martinez, J. L., Samulioviene, A., Semenova, A., & Garcia–Vazquez, E. (2015). eDNA and specific primers for early detection of invasive species–A case study on the bivalve *Rangia cuneata,* currently spreading in Europe. *Marine Environmental Research, 112*, 48–55.

Ariza, M., Fouks, B., Mauvisseau, Q., Halvorsen, R., Alsos, I. G., & de Boer, H. J. (2022). Plant biodiversity assessment through soil eDNA reflects temporal and local diversity. *Methods in Ecology and Evolution, 14*(2), 415–430.

Bœuf, G., & Payan, P. (2001). How should salinity influence fish growth? *Comp Biochem Physiol Part C: Toxicol & Pharmacology, 130*(4), 411–423.

Baker, M. A., et al. (2013). DNA extraction from environmental samples: A review. *Molecular Ecology Resources.*

Balint, M., Pfenninger, M., Grossart, H., Taberlet, P., Vellend, M., Leibold, M. A., Englund, G., & Bowler, D. (2018). Environmental DNA time series in ecology. *Trends in Ecology & Evolution (Personal Edition), 33*, 945–957.

Barnes, M. A., & Turner, C. R. (2016). The ecology of environmental DNA and implications for conservation genetics. *Conserv Genet, 17*(1), 1–17.

Bellemain, E., Carlsen, T., Brochmann, C., Coissac, E., Taberlet, P., & Kauserud, H. (2010). ITS as an environmental DNA barcode for fungi: an in silico approach reveals potential PCR biases. *BMC microbiology, 10*(1), 189.

Beng, K. C., & Corlett, R. T. (2020). Applications of environmental DNA (eDNA) in ecology and conservation: Opportunities, challenges and prospects. *Biodiversity and Conservation, 29*, 2089–2121.

Bienert, F., De Danieli, S., Miquel, C., Coissac, E., Poillot, C., Brun, J. J., & Taberlet, P. (2012). Tracking earthworm communities from soil DNA. *Molecular Ecology, 21*(8), 2017–2030.

Biggs, J., Ewald, N., Valentini, A., Gaboriaud, C., Dejean, T., Griffiths, R. A., Foster, J., Wilkinson, J. W., Arnell, A., & Brotherton, P. (2015). Using eDNA to develop a national citizen science–Based monitoring programme for the great crested newt (*Triturus cristatus*). *Biological Conservation, 183*, 19–28.

Bohmann, K., et al. (2014). Environmental DNA for biodiversity monitoring: A review. *Molecular Ecology.*

Borrell, Y. J., Miralles, L., Do Huu, H., Mohammed-Geba, K., & Garcia-Vazquez, E. (2017). DNA in a bottle—Rapid metabarcoding survey for early alerts of invasive species in ports. *PLoS One, 12*(9) Article e0183347.

Brondízio, E. S., Settele, J., Díaz, S., & Ngo, H. T. (2019). Global assessment report on biodiversity and ecosystem services of the Intergovernmental Science-Policy Platform on Biodiversity and Ecosystem Services (IPBES). IPBES Secretariat. 1148p.

Buxton, A. S., Groombridge, J. J., & Griffiths, R. A. (2017). Is the detection of aquatic environmental DNA influenced by substrate type? *PLoS One, 12*(8), e0183371.

Cardas, J. B., Deconinck, D., Marquez, I., Torre, P. P., Garcia–Vazquez, E., & Machado–Schiaffino, G. (2020). New eDNA based tool applied to the specific detection and monitoring of the endangered European eel. *Biological Conservation, 250*, 108750.

Çevik, T., & Çevik, N. (2025). Environmental DNA (eDNA): A review of ecosystem biodiversity detection and applications. *Biodiversity and Conservation,* 1–37.

Chan, A. H. E., Chaisiri, K., Saralamba, S., Morand, S., & Thaenkham, U. (2021). Assessing the suitability of mitochondrial and nuclear DNA genetic markers for molecular systematics and species identification of helminths. *Parasites & Vectors, 14*(1), 233.

Chan, A. H. E., Saralamba, N., Saralamba, S., Ruangsittichai, J., & Thaenkham, U. (2022). The potential use of mitochondrial ribosomal genes (12S and 16S) in DNA barcoding and phylogenetic analysis of trematodes. *BMC Genomics, 23*(1), 104.

Chang, H., Ye, T., Xie, Z., & Liu, X. (2025). Application of environmental DNA in aquatic ecosystem monitoring: Opportunities, challenges and prospects. *Water (20734441),* (5), 17.

Chen, Z., Minamoto, T., Lin, L., & Gao, T. (2020). An optional low-cost method of extracting environmental DNA of macro-organisms from filter membranes in large scale eDNA surveys. *Pakistan Journal of Zoology* https://doi. org/10.17582/journ al. pjz/20190,11810,0108.

Clarke, L. J., Beard, J. M., Swadling, K. M., & Deagle, B. E. (2017). Effect of marker choice and thermal cycling protocol on zooplankton DNA metabarcoding studies. *Ecol evolution, 7*(3), 873–883.

Coghlan, S. A., Currier, C. A., Freeland, J., Morris, T. J., & Wilson, C. C. (2021). Community eDNA metabarcoding as a detection tool for documenting freshwater mussel (Unionidae) species assemblages. *Environ. DNA, 3*(6), 1172–1191.

Collins, R. A., Bakker, J., Wangensteen, O. S., et al. (2019). Non-specific amplification compromises environmental DNA metabarcoding with COI. *Methods Ecol Evolution, 10*(11), 1985–2001. https://doi.org/10.1111/2041-210X.13276 581.

Crisol–Martínez, E., Moreno–Moyano, L. T., Wormington, K. R., Brown, P. H., & Stanley, D. (2016). Using next–generation sequencing to contrast the diet and explore pest–reduction services of sympatric bird species in macadamia orchards in Australia. *PLoS One, 11*(3), e0150159.

Díaz, S., Fargione, J., Chapin III, F. S., & Tilman, D. (2006). Biodiversity loss threatens human well-being. *PLoS Biol, 4*(8), e277.

Deiner, K., Fronhofer, E. A., Mächler, E., Walser, J. C., & Altermatt, F. (2016). Environmental DNA reveals that rivers are conveyer belts of biodiversity information. *Nat Commun, 7*(1), 12544.

Deiner, K., Yamanaka, H., & Bernatchez, L. (2021). The future of biodiversity monitoring and conservation utilizing environmental DNA. *Environment DNA, 3*, 3–7.

Dejean, T., Valentini, A., Miquel, C., Taberlet, P., Bellemain, E., & &Miaud, C. (2012). Improved detection of an alien invasive species through environmental DNA barcoding: The example of the American bullfrog Lithobates catesbeianus. *Journal of applied ecology, 49*(4), 953–959.

Dell'Anno, A., & Danovaro, R. (2005). Extracellular DNA plays a key role in deep-sea ecosystem functioning. *Science, 309*(5744), 2179.

Dirzo, R., Young, H. S., Galetti, M., Ceballos, G., Isaac, N. J., & Collen, B. (2014). Defaunation in the anthropocene. *Science (New York, N.Y.), 345*(6195), 401–406.

Edwards, D. P., Magrach, A., Woodcock, P., Ji, Y., Lim, N. T.–L., Edwards, F. A., Larsen, T. H., Hsu, W. W., Benedick, S., & Khen, C. V. (2014). Selective–logging and oil palm: Multitaxon impacts, biodiversity indicators, and trade–Offs for conservation planning. *Ecological Applications: a Publication of the Ecological Society of America, 24*(8), 2029–2049.

Eichmiller, J. J., Miller, L. M., & Sorensen, P. W. (2016). Optimizing techniques to capture and extract environmental DNA for detection and quantification of fish. *Molecular Ecology Resources, 16*(1), 56–68.

Elberri, A. I., Galal–Khallaf, A., Gibreel, S. E., El–Sakhawy, S. F., El–Garawani, I., ElNabi, S. E.–S. H., & Mohammed–Geba, K. (2020). DNA and eDNA–based tracking of the North African sharptooth catfish *Clarias gariepinus*. *Molecular and Cellular Probes, 51*, 101535.

Elbrecht, V., & Leese, F. (2015). Can DNA-based biodiversity assessments be improved by using multiple markers? *Molecular Ecology Resources.*

Ellegaard, M., Clokie, M. R., Czypionka, T., et al. (2020). Dead or alive: sediment DNA archives as tools for tracking aquatic evolution and adaptation. *Commun Biol, 3*(1), 169.

Ellis, M. R., Clark, Z. S., Treml, E. A., Brown, M. S., Matthews, T. G., Pocklington, J. B., & Sherman, C. D. (2022). Detecting marine pests using environmental DNA and biophysical models. *Science of the Total Environment, 816*, 151666.

Elyasigorji, Z., Izadpanah, M., Hadi, F., & Zare, M. (2023). Mitochondrial genes as strong molecular markers for species identification. *The Nucleus, 66*(1), 81–93.

Evans, N. T., Shirey, P. D., Wieringa, J. G., Mahon, A. R., & Lamberti, G. A. (2017). Comparative cost and effort of fish distribution detection via environmental DNA analysis and electrofishing. *Fisheries, 42*(2), 90–99.

Fahner, N. A., Shokralla, S., Baird, D. J., & &Hajibabaei, M. (2016). Large-scale monitoring of plants through environmental DNA metabarcoding of soil: Recovery, resolution, and annotation of four DNA markers. *PLoS One, 11*(6) Article e0157505.

Fediajevaite, J., Priestley, V., Arnold, R., & Savolainen, V. (2021). Meta-analysis shows that environmental DNA outperforms traditional surveys, but warrants better reporting standards. *Ecology and Evolution, 11*(9), 4803–4815.

Fediajevaite, J., Priestley, V., Arnold, R., & Savolainen, V. (2021). Meta-analysis shows that environmental DNA outperforms traditional surveys, but warrants better reporting standards. In *Ecology and Evolution, 11*, John Wiley & Sons, Ltd, 4803–4815.

Ficetola, G. F., Miaud, C., Pompanon, F., & &Taberlet, P. (2008). Species detection using environmental DNA from water samples. *Biology Letters, 4*(4), 423–425.

Fonseca, V. G., Davison, P. I., Creach, V., Stone, D., Bass, D., & Tidbury, H. J. (2023). The application of eDNA for monitoring aquatic non-indigenous species: Practical and policy considerations. *Diversity, 15*(5), 631.

Fraikin, N., Couturier, A., & Lesterlin, C. (2024). The winding journey of conjugative plasmids toward a novel host cell. *Curr OpMicrobiology, 78*, 102449.

Günther, B., Knebelsberger, T., Neumann, H., Laakmann, S., & Martínez Arbizu, P. (2018). Metabarcoding of marine environmental DNA based on mitochondrial and nuclear genes. *Sci Rep, 8*(1), 14822.

Gargan, L. M., Morato, T., Pham, C. K., Finarelli, J. A., Carlsson, J. E., & Carlsson, J. (2017). Development of a sensitive detection method to survey pelagic biodiversity using eDNA and quantitative PCR: A case study of devil ray at seamounts. *Marine Biology, 164*(5), 1–9.

Guthrie, A. M., Cooper, C. E., Bateman, P. W., van der Heyde, M., Allentoft, M. E., & Nevill, P. (2024). A quantitative analysis of vertebrate environmental DNA degradation in soil in response to time, UV light, and temperature. *Environ DNA, 6*(4), e581.

Hajibabaei, M., et al. (2006). Environmental barcoding: A next-generation sequencing approach for biodiversity assessment. *Molecular Ecology*.

Halder, M., & Jha, S. (2023). The current status of population extinction and biodiversity crisis of medicinal plants. In *Medicinal plants: Biodiversity, biotechnology and conservation*. Springer Nature Singapore, 3–38.

Harper, L. R., Niemiller, M. L., Benito, J. B., Paddock, L. E., Knittle, E., Molano-Flores, B., & Davis, M. A. (2023). BeeDNA: Microfluidic environmental DNA metabarcoding as a tool for connecting plant and pollinator communities. *Environmental DNA, 5*(1), 191–211.

Hata, H., Ogasawara, K., & Yamashita, N. (2022). Population decline of an endangered unionid, *Pronodulariajapanensis*, in streams is revealed by eDNA and conventional monitoring approaches. *Hydrobiologia, 849*, 2635–2646.

Hauck, L. L., Weitemier, K. A., Penaluna, B. E., Garcia, T. S., & Cronn, R. (2019). Casting a broader net: Using microfluidic metagenomics to capture aquatic biodiversity data from diverse taxonomic targets. *Environmental DNA, 1*(3), 251–267.

Jerde, C. L., Mahon, A. R., Chadderton, W. L., & Lodge, D. M. (2011). "Sight-unseen" detection of rare aquatic species using environmental DNA. *Conservation Letters, 4*(2), 150–157.

Jeunen, G. J., Knapp, M., Spencer, H. G., Taylor, H. R., Lamare, M. D., Stat, M., & Gemmell, N. J. (2019). Species-level biodiversity assessment using marine environmental DNA metabarcoding requires protocol optimization and standardization. *Ecology and Evolution, 9*(3), 1323–1335.

Johnson, M. D., Fokar, M., Cox, R. D., & Barnes, M. A. (2021). Airborne environmental DNA metabarcoding detects more diversity, with less sampling effort, than a traditional plant community survey. *BMC Ecology and Evolution, 21*(1), 218.

Johnson, M. D., Freeland, J. R., Parducci, L., Evans, D. M., Meyer, R. S., Molano-Flores, B., & Davis, M. A. (2023). Environmental DNA as an emerging tool in botanical research. *American Journal of Botany, 110*(2), e16120.

Karim, F., Oyewande, A. A., Abdalla, L. F., Ehsanullah, R. C., & Khan, S. (2020). Social media use and its connection to mental health: A systematic review. *Cureus, 12*(6).

Keck, F., Blackman, R. C., Bossart, R., Brantschen, J., Couton, M., Hürlemann, S., Kirschner, D., Locher, N., Zhang, H., & Altermatt, F. (2022). Meta-analysis shows both congruence and complementarity of DNA and eDNA metabarcoding to traditional methods for biological community assessment. In *Mol. Ecol., 31*, John Wiley & Sons, Ltd, 1820–1835.

Kelly, R. P., Gallego, R., & Jacobs-Palmer, E. (2018). The effect of tides on nearshore environmental DNA. *PeerJ, 6*, e4521.

Keskin, E., Unal, E. M., & Atar, H. H. (2016). Detection of rare and invasive freshwater fish species using eDNA pyrosequencing: Lake Iznik ichthyofauna revised. *Biochemical Systematics and Ecology, 67*, 29–36.

Kim, P., Yoon, T. J., & Shin, S. (2021). Monthly eDNA monitoring of an invasive Bryozoan, Bugulina californica, in seawater using species-specific markers. *Animals, 11*(7), 1966.

Knudsen, S. W., Ebert, R. B., Hesselsøe, M., Kuntke, F., Hassingboe, J., Mortensen, P. B., Thomsen, P. F., Sigsgaard, E. E., Hansen, B. K., & Nielsen, E. E. (2019). Species–specific detection and quantification of environmental DNA from marine fishes in the Baltic Sea. *Journal of Experimental Marine Biology and Ecology. 510*, 31–45.

Koroiva, R., & Santana, D. J. (2022). Evaluation of partial 12S rRNA, 16S rRNA, COI and Cytb gene sequence datasets for potential single DNA barcode for hylids (Anura: Hylidae). *Anais da Academia Brasileira de Ciências, 94*, e20200825.

Kumar, S., et al. (2015). PCR inhibition in environmental samples: A review. *Molecular Ecology Resources*.

Lacoursière-Roussel, A., Rosabal, M., & Bernatchez, L. (2016). Estimating fish abundance and biomass from eDNA concentrations: variability among capture methods and environmental conditions. *Mol Ecol Resour, 16*(6), 1401–1414.

Lafferty, K. D., Benesh, K. C., Mahon, A. R., Jerde, C. L., & Lowe, C. G. (2018). Detecting southern California's white sharks with environmental DNA. *Frontiers in Marine Science, 5*, 355.

Leray, M., et al. (2019). A new perspective on the use of environmental DNA for biodiversity assessment. *Molecular Ecology*.

Lilja, M. A., Buivydaite, Z., Zervas, A., Krogh, P. H., Hansen, B. W., Winding, A., & Sapkota, R. (2023). Comparing earthworm biodiversity estimated by DNA metabarcoding and morphology–based approaches. *Applied Soil Ecology, 185*, 104798.

Lisnerova, M., Holzer, A., Blabolil, P., & Fiala, I. (2023). Evaluation and optimization of an eDNA metabarcoding assay for detection of freshwater myxozoan communities. *Environmental DNA, 5*(2), 312–325.

Lugg, W. H., Griffiths, J., van Rooyen, A. R., Weeks, A. R., & Tingley, R. (2018). Optimal survey designs for environmental DNA sampling. *Methods in Ecology and Evolution, 9*(4), 1049–1059.

Lyet, A., Pellissier, L., Valentini, A., Dejean, T., Hehmeyer, A., & Naidoo, R. (2021). eDNA sampled from stream networks correlates with camera trap detection rates of terrestrial mammals. *Scientific Reports, 11*(1), 11362.

Ma, H., Stewart, K., Lougheed, S., Zheng, J., Wang, Y., & Zhao, J. (2016). Characterization, optimization, and validation of environmental DNA (eDNA) markers to detect an endangered aquatic mammal. *Conservation Genetics Resources, 8*, 561–568.

Machler, E., Osathanunkul, M., & Altermatt, F. (2018). Shedding light on eDNA: neither natural levels of UV radiation nor the presence of a filter feeder affect eDNA-based detection of aquatic organisms. *PLoS one, 13*(4), e0195529.

Mahon, A. R., Jerde, C. L., Galaska, M., Bergner, J. L., Chadderton, W. L., Lodge, D. M., & Nico, L. G. (2013). Validation of eDNA surveillance sensitivity for detection of Asian carps in controlled and field experiments. *PLoS One, 8*(3), e58316.

Makiola, A., Dickie, I. A., Holdaway, R. J., Wood, J. R., Orwin, K. H., & Glare, T. R. (2019). Land use is a determinant of plant pathogen alpha–but not beta–diversity. *Molecular Ecology, 28*(16), 3786–3798.

Mao, D., Luo, Y., Mathieu, J., et al. (2014). Persistence of extracellular DNA in river sediment facilitates antibiotic resistance gene propagation. *Environ Sci & Technol, 48*(1), 71–78.

Mariani, S., Fernandez, C., Baillie, C., Magalon, H., & Jaquemet, S. (2021). Shark and ray diversity, abundance and temporal variation around an Indian Ocean Island, inferred by eDNA metabarcoding. *Conservation Science and Practice, 3*(6), e407.

Mauvisseau, Q., Coignet, A., Delaunay, C., Pinet, F., Bouchon, D., & Souty–Grosset, C. (2018). Environmental DNA as an efficient tool for detecting invasive crayfishes in freshwater ponds. *Hydrobiologia, 805*(1), 163–175.

Meyer, C. P., et al. (2019). Challenges and opportunities in the use of environmental DNA for biodiversity assessment. *Molecular Ecology*.

Mezhzherin, S. V., Morozov-Leonov, S. Y., & Tereshchenko, V. O. (2023). Transition bias and its compensation in the evolutionary lineage of the subfamily Murinae (Rodentia): Analysis of nuclear and mitochondrial DNA markers. *Cytology and Genetics, 57*(6), 550–555.

Miralles, L., Ibabe, A., González, M., García-Vázquez, E., & Borrell, Y. J. (2021). "If you know the enemy and know yourself": Addressing the problem of biological invasions in ports through a new NIS invasion threat score, routine monitoring, and preventive action plans. *Frontiers in Marine Science, 8*.

Moushomi, R., Wilgar, G., Carvalho, G., Creer, S., & Seymour, M. (2019). Environmental DNA size sorting and degradation experiment indicates the state of Daphnia magna mitochondrial and nuclear eDNA is subcellular. *Scientific Reports, 9*(1), 12500.

Nagarajan, R. P., Bedwell, M., Holmes, A. E., Sanches, T., Acuña, S., Baerwald, M., & &Schreier, A. (2022). Environmental DNA methods for ecological monitoring and biodiversityassessment in estuaries. *Estuaries and Coasts, 45*(7), 2254–2273.

Nagler, M., Podmirseg, S. M., Ascher-Jenull, J., Sint, D., & Traugott, M. (2022). Why eDNA fractions need consideration in biomonitoring. *Molecular Ecology Resources, 22*, 2458–2470.

Navarro–Noya, Y. E., Montoya–Ciriaco, N., Munoz ˜ Arenas, L. C., Hereira–Pacheco, S., Estrada–Torres, A., & Dendooven, L. (2021). Conversion of a high–altitude temperate forest for agriculture reduced alpha and beta diversity of the soil fungal communities as revealed by a metabarcoding analysis. *Frontiers in Microbiology, 12*, 667566.

Neice, A. A., & McRae, S. B. (2021). An eDNA diagnostic test to detect a rare, secretive marsh bird. *Global Ecology and Conservation, 27*, e01529.

Ogata, M., Masuda, R., Harino, H., Sakata, M. K., Hatakeyama, M., Yokoyama, K., & Minamoto, T. (2021). Environmental DNA preserved in marine sediment for detecting jellyfish blooms after a tsunami. *Scientific Reports, 11*(1), 16830.

Pascher, K., Švara, V., & &Jungmeier, M. (2022). Environmental DNA-based methods in biodiversity monitoring of protected areas: Application range, limitations, and needs. *Diversity, 14*(6), 463.

Pawlowski, J., Apothéloz-Perret-Gentil, L., & Altermatt, F. (2020). Environmental DNA: What's behind the term? Clarifying the terminology and recommendations for its future use in biomonitoring. *Molecular Ecology, 29*, 4258–4264.

Pfleger, M. O., Rider, S. J., Johnston, C. E., & Janosik, A. M. (2016). Saving the doomed: Using eDNA to aid in detection of rare sturgeon for conservation (Acipenseridae). *Global Ecology and Conservation, 8*, 99–107.

Pilliod, D. S., Goldberg, C. S., Arkle, R. S., & Waits, L. P. (2014). Factors influencing detection of eDNA from a stream-dwelling amphibian. *Molecular Ecology Resources, 14*(1), 109–116.

Port, J. A., O'Donnell, J. L., Romero–Maraccini, O. C., Leary, P. R., Litvin, S. Y., Nickols, K. J., Yamahara, K. M., & Kelly, R. P. (2016). Assessing vertebrate biodiversity in a kelp forest ecosystem using environmental DNA. *Molecular Ecology, 25*(2), 527–541.

Prié, V., Soler, J., Araujo, R., et al. (2018). Challenging exploration of troubled waters: a decade of surveys of the giant freshwater pearl mussel Margaritifera auricularia in Europe. *Hydrobiologia, 810*(1), 157–175.

Ramirez-Amaro, S., Bassitta, M., Picornell, A., Ramon, C., & Terrasa, B. (2022). Environmental DNA: State-of-the-art of its application for fisheries assessment in marine environments. *Frontiers in Marine Science, 9*, 1004674.

Ratnasingham, S., & Hebert, P. D. N. (2007). BOLD: The barcode of life data system (www. barcodinglife.org). *Molecular Ecology Notes, 7*(3), 355–364.

Rees, H. C., Maddison, B. C., Middleditch, D. J., Patmore, J. R., & Gough, K. C. (2014). The detection of aquatic animal species using environmental DNA–A review of eDNA as a survey tool in ecology. *Journal of Applied Ecology, 51*(5), 1450–1459.

Rodriguez-Ezpeleta, N., et al. (2021). Trade-offs between reducing complex terminology and producing accurate interpretations from environmental DNA: Comment on'Environmental DNA: What's behind the term?' by Pawlowski et al., (2020). *Molecular Ecology, 30*, 4601–4605.

Roux, L. M. D., Giblot-Ducray, D., Bott, N. J., Wiltshire, K. H., Deveney, M. R., Westfall, K. M., & Abbott, C. L. (2020). Analytical validation and field testing of a specific qPCR assay for environmental DNA detection of invasive European green crab *(Carcinus maenas). Environmental DNA, 2*(3), 309–320.

Ruppert, K. M., Kline, R. J., & Rahman, M. S. (2019). Past, present, and future perspectives of environmental DNA (eDNA) metabarcoding: A systematic review in methods, monitoring, and applications of global eDNA. *Global Ecology and Conservation, 17*, 1–29.

Sansom, B. J., & Sassoubre, L. M. (2017). Environmental DNA (eDNA) shedding and decay rates to model freshwater mussel eDNA transport in a river. *Environmental Science & Technology, 51*(24), 14244–14253.

Seymour, M., Durance, I., Cosby, B. J., et al. (2018). Acidity promotes degradation of multi-species environmental DNA in lotic mesocosms. *Commun Biol, 1*(1), 4.

Shaw, J. L., Weyrich, L., & Cooper, A. (2016). Using environmental (e) DNA sequencing for aquatic biodiversity surveys: A beginner's guide. *Marine and Freshwater Research, 68*(1), 20–33.

Smart, A. S., Tingley, R., Weeks, A. R., Van Rooyen, A. R., & McCarthy, M. A. (2015). Environmental DNA sampling is more sensitive than a traditional survey technique for detecting an aquatic invader. *Ecological Applications, 25*(7), 1944–1952.

Spear, S. F., Groves, J. D., Williams, L. A., & Waits, L. P. (2015). Using environmental DNA methods to improve detectability in a hellbender (Cryptobranchusalleganiensis) monitoring program. *Biological Conservation, 183*, 38–45.

Sternhagen, E. C., Black, K. L., Hartmann, E. D., Shivega, W. G., Johnson, P. G., McGlynn, R. D., Schmaltz, L. C., Asheim Keller, R. J., Vink, S. N., & Aldrich–Wolfe, L. (2020). Contrasting patterns of functional diversity in coffee root fungal communities associated with organic and conventionally managed fields. *Applied and Environmental Microbiology, 86*(11), e00052–00020.

Stoeckle, M. Y., Das Mishu, M., & &Charlop-Powers, Z. (2018). GoFish: A versatile nested PCR strategy for environmental DNA assays for marine vertebrates. *PLoS One, 13*(12), e0198717.

Sun, S., Lyu, D., Qian, T., Shan, X., & Wang, W. (2023). Evaluate the biomass of Fenneropenaeus chinensis from the Southern coast of Shandong Peninsula using eDNA. *Water, 15*(2), 342.

Suresh, V. M., Hébert, T., Verster, K. I., & Hadly, E. A. (2024). Temporal study of environmental DNA and acoustic data reveals coexistence of sympatric bat species in a North American Ecosystem. *Environmental DNA, 6*(6), e70035.

Taberlet, P., Prud'homme, S. M., Campione, E., Roy, J., Miquel, C., Shehzad, W., & &Coissac, E. (2012). Soil sampling and isolation of extracellular DNA from large amount of starting material suitable for metabarcoding studies. *Molecular Ecology, 21*(8), 1816–1820.

Taberlet, P., Bonin, A., Zinger, L., & &Coissac, E. (2018). *Environmental DNA: For biodiversity research and monitoring*. Oxford University Press.

Takahara, T., Minamoto, T., & Doi, H. (2013). Using environmental DNA to estimate the distribution of an invasive fish species in ponds. *PLoS One, 8*(2), e56584.

Takahashi, S., Sakata, M. K., Minamoto, T., & Masuda, R. (2020). Comparing the efficiency of open and enclosed filtration systems in environmental DNA quantification for fish and jellyfish. *PLoS One, 15*(4), e0231718.

Thomsen, P. F., & Willerslev, E. (2015). Environmental DNA–An emerging tool in conservation for monitoring past and present biodiversity. *Biological Conservation, 183*, 4–18.

Thomsen, P. F., Kielgast, J. O. S., Iversen, L. L., Wiuf, C., Rasmussen, M., Gilbert, M. T. P., & Willerslev, E. (2012). Monitoring endangered freshwater biodiversity using environmental DNA. *Molecular Ecology, 21*(11), 2565–2573.

Tibone, M., Cariou, T., O'Donnell, C., Stefanni, S., Aguzzi, J., O'Neill, B., & Mirimin, L. (2024). Towards the integration of environmental DNA analysis to profile the upper mesopelagic fish layer in the Northeast Atlantic Ocean. *ICES Journal of Marine Science, 81*(10), 2065–2078.

Turner, C. R., Barnes, M. A., Xu, C. C., Jones, S. E., Jerde, C. L., & Lodge, D. M. (2014). Particle size distribution and optimal capture of aqueous macrobial eDNA. *Methods Ecol Evolution, 5*(7), 676–684.

Valentin, R. E., Fonseca, D. M., Nielsen, A. L., Leskey, T. C., & Lockwood, J. L. (2018). Early detection of invasive exotic insect infestations using eDNA from crop surfaces. *Frontiers in Ecology and the Environment, 16*(5), 265–270.

Vié, J. C., Hilton-Taylor, C., & Stuart, S. N. (Eds.). (2009). *Wildlife in a changing world: an analysis of the 2008 IUCN Red List of threatened species*. IUCN.

Voros, J., Marton, O., Schmidt, B. R., Gál, J. T., & Jelić, D. (2017). Surveying Europe's only cave-dwelling chordate species (*Proteus anguinus*) using environmental DNA. *PLoS One, 12*(1), e0170945.

Vuillemin, A., Horn, F., Alawi, M., et al. (2017). Preservation and significance of extracellular DNA in ferruginous sediments from Lake Towuti, Indonesia. *Front Microbiology, 8*, 1440.

West, K. M., Adam, A. A., White, N., Robbins, W. D., Barrow, D., Lane, A. T., & Richards, Z. (2022). The applicability of eDNA metabarcoding approaches for sessile benthic surveying in the Kimberley region, north–western Australia. *Environ. DNA, 4*(1), 34–49.

Wheeler, Q. D., & Wilson, E. O. (2004). Taxonomy: Impediment or expedient? *Science (New York, N.Y.), 303*, 285.

Wilcox, T. M., McKelvey, K. S., Young, M. K., Jane, S. F., Lowe, W. H., Whiteley, A. R., & Schwartz, M. K. (2013). Robust detection of rare species using environmental DNA: The importance of primer specificity. *PLoS One, 8*(3), e59520.

Wilcox, T. M., McKelvey, K. S., Young, M. K., Engkjer, C., Lance, R. F., Lahr, A., & Schwartz, M. K. (2020). Parallel, targeted analysis of environmental samples via high-throughput quantitative PCR. *Environmental DNA, 2*(4), 544–553.

Wu, Q., Kawano, K., Uehara, Y., Okuda, N., Hongo, M., Tsuji, S., & Minamoto, T. (2018). Environmental DNA reveals nonmigratory individuals of *Palaemon paucidens* overwintering in Lake Biwa shallow waters. *Freshwater Science, 37*(2), 307–314.

Yang, J., Zhang, X., Jin, X., Seymour, M., Ritcher, C., Logares, R., Khim, J. S., & Klymus, K. (2021). Recent advances in environmental DNA-based biodiversity assessment and conservation. *Diversity and Distributions, 27*, 1876–1879.

Yao, M., Zhang, S., Lu, Q., Chen, X., Zhang, S. Y., Kong, Y., & Zhao, J. (2022). Fishing for fish environmental DNA: Ecological applications, methodological considerations, surveying designs, and ways forward. *Molecular Ecology, 31*(20), 5132–5164.

Zhan, A., et al. (2013). Optimizing the use of environmental DNA for biodiversity assessment. *Molecular Ecology*.

Zhang, Y., Qiu, Y., Liu, K., Zhong, W., Yang, J., Altermatt, F., & Zhang, X. (2024). Evaluating eDNA and eRNA metabarcoding for aquatic biodiversity assessment: From bacteria to vertebrates. *Environmental Science and Ecotechnology,* 100441.

Chapter 7

Insect diversity, distribution, and its functional roles in wetland ecosystems of India

Shipra Sonali[1,2], Ashis Debnath[1], Arnab Chattopadhyay[1], Inderjeet Tyagi[3], Vikas Kumar[3], and Devanshu Gupta[1]
[1]*Zoological Survey of India, Prani Vigyan Bhavan, New Alipore, Kolkata, West Bengal, India,* [2]*University of Calcutta, Kolkata, West Bengal, India,* [3]*Centre for DNA Taxonomy, Molecular Systematics Division, Zoological Survey of India, Kolkata, West Bengal, India*

7.1 Introduction

India, recognized globally as one of the 17 megadiverse nations, possesses a remarkable array of ecosystems, ranging from the towering Himalayas and arid deserts to extensive coastal regions, vital wetlands, and islands (Chandra & Gupta, 2022a). Within this rich biodiversity, aquatic insect diversity is both extensive and varied, reflecting the country's abundant freshwater habitats, including rivers, lakes, ponds, and streams. Freshwater ecosystems support a significant portion of India's fauna, with approximately 9.9% of the total faunal species residing in these habitats. Of the 10,168 freshwater animal species recorded in India, the phylum Arthropoda alone accounts for 6607 species, representing 61.4% of this freshwater faunal component (Chandra & Gupta, 2022a, 2022b).

Wetlands are critical components of these freshwater systems, characterized as dynamic and often transient landscape features. They undergo continuous natural processes such as sedimentation, peat accumulation, and soil formation, resulting in constant ecological succession. This inherent dynamism supports high biodiversity, facilitated by four key characteristics: (1) continuous transformation through ecological succession, enabling diverse species to thrive at different stages; (2) distinct zonation patterns, both within the ecosystem and along its edges, creating varied habitats; (3) regular fluctuations in water levels (seasonal or hydrological pulses), which enhance habitat diversity; and (4) their proximity to contrasting ecosystems, fostering unique ecological interactions (Maltby, 2009). Wetlands are ecologically vital ecosystems that perform a range of critical functions. They regulate water flow by absorbing excess rainfall and gradually releasing it during dry periods, helping to mitigate floods and maintain base flows. Acting as natural filters, wetlands improve water quality by trapping sediments and removing pollutants and excess nutrients through biological processes. They are also biodiversity hotspots, providing essential habitats for a wide range of species, including many that are endangered or endemic. Additionally, wetlands play a key role in climate regulation by storing large amounts of carbon, especially in peatlands, which are among the most efficient carbon sinks on the planet. Wetlands also help filter pollutants and impurities from water (Mishra et al., 2021). Beyond ecological functions, wetlands offer valuable ecosystem services such as supporting fisheries, agriculture, and tourism, while reducing economic costs related to water management and infrastructure (Bhowmik, 2020; Wantzen & Junk, 2000).

India, comprising only 2.4% of the global landmass, holds 4% of the world's freshwater resources (Ministry of Water Resources, Govt. of India). The country's inland wetlands—ranging from lakes and rivers to bogs and swamps—are estimated to cover over 10.56 million ha (6.62 million natural and 3.94 million artificial), according to the National Wetland Inventory and Assessment (MoEF&CC, Govt. of India). The Ministry of Environment, Forest and Climate Change (MoEF&CC) has identified 115 wetlands across 24 states and 2 Union Territories under the National Wetland Conservation Program (NWCP), with 89 of these designated as Ramsar wetland sites.

Aquatic insects are essential components of freshwater food webs, serving as crucial prey for fish, amphibians, and birds. Many species, including caddisflies, beetles, and fly larvae, also contribute significantly to decomposition and

Wetlands: Biodiversity, Health and their Role in Ecosystem Services. DOI: https://doi.org/10.1016/B978-0-443-34222-6.00007-3

nutrient cycling by breaking down organic matter, thereby helping to maintain water quality (Thorp & Covich, 2018). Furthermore, numerous aquatic insects—particularly mayflies and caddisflies—are sensitive to pollution, making their presence or absence reliable indicators of freshwater ecosystem health. Their diversity is directly influenced by ecological functions, physical habitat characteristics, and the availability of food resources (Wallace & Webster, 1996).

Global assessments of freshwater ecosystems, notably by the Intergovernmental Science-Policy Platform on Biodiversity and Ecosystem Services (IPBES, 2019, 2024), consistently highlight their severe endangerment due to human activities, including habitat destruction, pollution, overexploitation, and climate change impacts. Approximately one-third of the world's freshwater species are now at risk of extinction—a rate higher than that observed in marine or terrestrial ecosystems (Dudgeon et al., 2006; IPBES, 2019). This alarming decline in freshwater biodiversity, driven by anthropogenic pressures, necessitates comprehensive action to protect and restore these critical habitats and species, thereby safeguarding the vital ecosystem services they provide (Dudgeon et al., 2006; IPBES, 2019).

Given these pressing concerns, biodiversity assessments in wetlands must comprehensively account for all species that depend, directly or indirectly, on these habitats at any stage of their life cycles. Effective conservation of these species inherently requires the integrated conservation and management of the wetland habitats themselves. This chapter aims to summarize the diversity, distribution patterns, and functional roles of freshwater insects in India.

7.2 Aquatic insect diversity and distribution in India

An overview of aquatic insect diversity indicates that India harbors a significant share of global species richness across several major orders, although it still comprises a relatively small proportion of the global total. The combined global diversity of the surveyed aquatic insect orders—Ephemeroptera, Odonata, Plecoptera, Hemiptera, Coleoptera, Hymenoptera, Trichoptera, Lepidoptera, and Diptera—is estimated at approximately 105,324 species, of which India contributes 5138 species, accounting for approximately 4.88% (Table 7.1).

Among individual orders, India shows notably high representation in Trichoptera (1407 species; 8.14%), Diptera (1588; 10.05%), Lepidoptera (80; 10.81%), and Odonata (504; 7.88%), likely reflecting both comprehensive documentation efforts and the presence of favorable habitats or endemic lineages. Moderate representation is observed in Hemiptera (325; 6.85%) and Ephemeroptera (247; 6.67%), whereas lower proportions are seen in Coleoptera (831; 5.01%), Hymenoptera (10; 6.6%), and Plecoptera (146; 3.22%).

TABLE 7.1 Aquatic insect diversity in India and world.

N	Order	Species		
		World	India	Percentage
1.	Ephemeroptera[a]	3700	247	6.67
2.	Odonata[b]	6392	504	7.88
3.	Plecoptera[c,d]	4524	146	3.22
4.	Hemiptera[e]	4738	325	6.85
5.	Coleoptera	16,604	831	5.01
6.	Hymenoptera	150	10	6.6
7.	Trichoptera	17,279	1407	8.14
8.	Lepidoptera	740	80	10.81
9.	Diptera	51,197	1588	10.05
	Total	105,324	5138	4.88

[a] *Subramanian et al. (2024).*
[b] *Subramanian and Babu (2024).*
[c] *DeWalt et al. (2025).*
[d] *Chandra et al. (2019).*
[e] *Chandra, Gopi, et al. (2017), Chandra, Jaiswal, et al. (2017).*

These data underscore India's important, though unevenly documented, role in global aquatic insect biodiversity and reinforce the urgent need for continued taxonomic revision, ecological investigation, and targeted survey efforts—particularly for underrepresented groups and less-explored regions.

7.3 Ephemeroptera (mayflies)

Mayflies are among the most ecologically significant and commonly encountered groups of aquatic insects in India. Globally, the order Ephemeroptera comprises approximately 3700 species across 465 genera and 42 families. In India, 247 species have been recorded, representing 63 genera from 15 families within four suborders (Chandra & Gupta, 2022a, 2022b; Subramanian et al., 2024). Of these, an estimated 182 species are endemic to India, with high endemism particularly noted in the hill streams of the Western Ghats (Tamil Nadu, Kerala, Karnataka) as well as in Himachal Pradesh and Arunachal Pradesh (Subramanian et al., 2024). Mayflies are especially abundant in lotic (flowing water) habitats, where they play a vital role in aquatic food webs. Both larval and adult stages serve as important prey for a range of invertebrate and vertebrate predators. In addition to their ecological roles, mayflies are recognized as sensitive bioindicators of aquatic ecosystem health, with their assemblages often shifting in response to varying levels of pollution and human disturbance (Selvakumar, 2018).

7.4 Odonata (dragonflies and damselflies)

Odonates (Figs. 7.1 and 7.2) are abundant and ecologically important inhabitants of India's freshwater ecosystems, including rivers, ponds, marshes, and wetlands. These amphibiotic insects, with aquatic larvae and terrestrial adults, are represented globally by 6392 species across 685 genera and 46 families. In India, 504 species have been recorded, belonging to 157 genera and 18 families (Subramanian & Babu, 2024). The country's odonate diversity is particularly rich in the Northeast and the Western Ghats, the two recognized biodiversity hotspots. A total of 175 species are endemic to India, with the highest endemism reported from Kerala, followed by Tamil Nadu and Karnataka. Most endemic species exhibit restricted distributions, confined primarily to these biodiversity hotspots (Subramanian & Babu, 2024).

Ecologically, odonates play a significant role as predators in both their larval and adult stages. Larvae prey on aquatic organisms such as mosquito larvae, beetles, and tadpoles, while adults feed on mosquitoes, flies, termites, aphids, and small moths, thus serving as natural biocontrol agents for many pest species (Subramanian & Babu, 2018). Their dependence on specific aquatic habitats—ranging from lentic (standing water) to lotic (flowing water) systems—makes them particularly sensitive to environmental changes. Due to this habitat specificity, odonates are widely used as bioindicators for monitoring the ecological health of freshwater ecosystems (Bulánková, 1997; Chandra, Gupta, Gopi, et al., 2018; Chandra, Gupta, Kushwaha, et al., 2018).

FIGURE 7.1 Common Picture Wing, *Rhyothemis variegata* (Linnaeus, 1763), a dragonfly photographed in Damoh district, Madhya Pradesh, India.

FIGURE 7.2 Emerald spreadwing, *Lestes viridulus* (Rambur, 1842), a damselfly photographed in Jabalpur district, Madhya Pradesh, India.

7.5 Plecoptera (stoneflies)

Stoneflies, members of the order Plecoptera, are a small monophyletic group of hemimetabolous insects, globally represented by approximately 3984 described species across 303 genera and 16 families (DeWalt et al., 2019). In India, 146 species belonging to 27 genera and 8 families have been recorded, with 90 species reported as endemic (Chandra et al., 2019). These insects are primarily associated with high-elevation montane streams and rivers, where they play a vital ecological role. Alongside mayflies and caddisflies, stoneflies are among the most sensitive bioindicators of freshwater quality, making them crucial for assessing environmental health and the regional impacts of climate change.

Despite their ecological significance, species-level identification of stoneflies—particularly in their larval and female stages—remains challenging, thereby limiting the application of precise biotic indices in monitoring and conservation efforts (Gattolliat et al., 2016). Within India's 10 biogeographic zones, the Himalayan region harbors the highest diversity (87 species), followed by the Northeast (51 species), the Deccan Peninsula (18), the Western Ghats (11), and the Trans-Himalaya (6). Notably, there are no records of stoneflies from the Gangetic Plains, semiarid regions, deserts, Coastal areas, or Islands (Chandra et al., 2019).

Family-level diversity within Indian Plecoptera reveals Nemouridae as the most species-rich family, comprising 7 genera and 69 species. Other families include Perlidae (9 genera, 49 species), Peltoperlidae (2 genera, 8 species), Capniidae (1 genus, 7 species), Perlodidae (2 genera, 4 species), Chloroperlidae (3 genera, 4 species), Leuctridae (1 genus, 3 species), and Taeniopterygidae (2 genera, 2 species). This diversity underscores the ecological importance of stoneflies in India's freshwater ecosystems and highlights the need for enhanced taxonomic and ecological research.

7.6 Order hemiptera (aquatic bugs)

Hemiptera, commonly known as "aquatic bugs," represent a significant component of freshwater ecosystems, having secondarily adapted to aquatic life where both their nymphal and adult stages are fully aquatic. In India, this order is notably diverse, with 325 species distributed across 84 genera and 18 families under the suborder Heteroptera, as documented by Chandra, Gopi, et al. (2017), Chandra, Jaiswal, et al. (2017), and Praveen et al. (2024). These true bugs are categorized into three infraorders based on their ecological niches and habitat preferences: Nepomorpha, Gerromorpha, and Leptopodomorpha. Common examples such as water striders (Gerridae), water boatmen (Corixidae), and backswimmers (Notonectidae) are prevalent across various freshwater habitats in India. Aquatic and semiaquatic hemipterans primarily inhabit inland freshwater systems, though some species extend into intertidal brackish and marine environments. Their ecological roles are diverse, with most species acting as predators or scavengers. A notable exception is the family Corixidae, which consists of detritivores. Predatory species within families such as Belostomatidae, Aphelocheiridae, Naucoridae, Nepidae, and Gerridae are known to prey on other aquatic insects of similar size and even larger organisms, including tadpoles, fish, mollusks, and frogs, often immobilizing their prey (Basu & Subramanian, 2017). The taxonomic classification and diversity of families within the suborder Heteroptera are detailed as follows:

- **Infraorder Gerromorpha**: Key families include Gerridae (28 genera, 105 species), Hebridae (4 genera, 18 species), Hydrometridae (1 genus, 7 species), and Mesoveliidae (1 genus, 12 species).
- **Infraorder Nepomorpha**: This infraorder encompasses Belostomatidae (2 genera, 6 species), Nepidae (4 genera, 22 species), Notonectidae (4 genera, 32 species), Corixidae (7 genera, 19 species), and Micronectidae (1 genus, 22 species). Other families within Nepomorpha are Pleidae (6 genera, 13 species), Aphelocheiridae (1 genus, 5 species), Naucoridae (6 genera, 16 species), Ochteridae (1 genus, 2 species), Gelastocoridae (1 genus, 4 species), and Helotrephidae (7 genera, 14 species).
- **Infraorder Leptopodomorpha**: This group includes families such as Leptopodidae (3 genera, 7 species) and Saldidae (6 genera, 14 species).

7.7 Hymenoptera

While most Hymenoptera are terrestrial, a fascinating adaptation to aquatic environments exists among certain parasitoid wasps. These unique hymenopterans target the egg, larval, and pupal stages of various aquatic insects, including those from orders such as Heteroptera, Odonata, and even some Lepidoptera, Coleoptera, and Diptera. The female wasps of these parasitoid families exhibit remarkable behavior, actively swimming underwater to deposit their eggs into their aquatic insect hosts.

In India, species-level data on aquatic Hymenoptera remain notably limited. Confirmed records currently exist for only 10 species across 7 genera and 3 families, primarily from or near water bodies. These aquatic parasitic Hymenoptera include species from Braconidae and Ichneumonidae (Ichneumonoidea), Mymaridae, Eulophidae, and Trichogrammatidae (Chalcidoidea), Platygastridae (Platygastroidea), Figitidae (Cynipoidea), and Diapriidae (Diaprioidea). Beyond these egg parasitoids, some spider wasps within the family Pompilidae (Vespoidea) are also known to be aquatic (Rajmohana, 2017). Globally, the diversity of aquatic Hymenoptera is more recognized, with approximately 150 species from 11 families identified as aquatic (Bennett, 2008). This disparity highlights the need for more focused research and documentation of aquatic Hymenoptera in India.

7.8 Coleoptera (beetles)

Coleoptera (Fig. 7.3), commonly known as beetles, exhibit significant aquatic adaptations. Globally, 25 families across three suborders—Myxophaga, Adephaga, and Polyphaga—are recognized as water beetles (Chandra, Gopi, et al., 2017; Chandra, Jaiswal, et al., 2017; Freitag, 2015). The aquatic beetle fauna of India is remarkably diverse, comprising 831 species belonging to 140 genera across these 17 families and 3 suborders (Chandra, Gupta, Gopi, et al., 2018; Chandra, Gupta, Kushwaha, et al., 2018; Ghosh et al., 2022; Gupta et al., 2024, 2025; Sonali et al., 2024).

Suborder Adephaga: This suborder includes Gyrinidae (5 genera, 60 species), Haliplidae (1 genus, 11 species), Noteridae (3 genera, 12 species), and the highly speciose Dytiscidae (37 genera, 273 species).

Suborder Myxophaga: This suborder is represented by Hydroscaphidae (1 genus, 5 species) and Sphaeriusidae (1 genus, 2 species).

Suborder Polyphaga: This extensive group comprises several families:

FIGURE 7.3 *Regimbartia attenuata* (Fabricius, 1801), an aquatic beetle in the family Hydrophilidae (Coleoptera).

- Under Byrrhoidea: Elmidae (11 genera, 24 species), Dryopidae (6 genera, 10 species), and Psephenidae (10 genera, 28 species).
- Under Scirtoidea: Scirtidae (9 genera, 74 species).
- Under Hydrophiloidea: Hydrophilidae (44 genera, 260 species), Georissidae (1 genus, 7 species), Helophoridae (1 genus, 8 species), Epimetopidae (1 genus, 5 species), Hydrochidae (1 genus, 4 species), and Spercheidae (1 genus, 2 species).
- Under Staphylinoidea: Hydraenidae (7 genera, 46 species).

Various species of water beetles, including diving beetles (Dytiscidae) and water scavenger beetles (Hydrophilidae), are commonly found across diverse aquatic habitats such as ponds, lakes, and slow-moving streams. Aquatic beetles are vital components of aquatic ecosystems and essential to freshwater biodiversity (Akünal & Aslan, 2017). They play a significant role in nutrient recycling and contribute to the breakdown of organic matter within freshwater food webs (Akünal & Aslan, 2017). Furthermore, aquatic beetles are effective indicators of overall community patterns among freshwater macroinvertebrates, making them valuable for ecological assessments (Ortega et al., 2021). As one of the most diverse orders of aquatic macroinvertebrates, Coleoptera exhibit a wide morphological, functional, and ecological range; they are relatively easy to sample, and their taxonomy and biogeography are reasonably well-understood (Bilton et al., 2019).

7.9 Trichoptera (caddisflies)

Trichoptera, commonly known as caddisflies, are renowned for their aquatic larval stages, during which they construct characteristic protective cases from various organic materials such as plant debris or sand (Wiggins, 2004). These "underwater architects" are critical bioindicators, typically found in clean water habitats across India, notably in the Himalayan and Western Ghats regions, as well as in rivers, streams, and lakes. Caddisfly larvae are an important and beneficial component of the trophic dynamics and energy flow in both lentic (still water) and lotic (flowing water) environments. Their sensitivity to environmental changes makes them extensively utilized in biomonitoring surveys, alongside other benthic macroinvertebrates (Lenat, 1993).

Globally, the order Trichoptera encompasses over 17,279 extant species, inhabiting a diverse array of freshwater environments on all continents except Antarctica (Morse, 2023). In India, the documented diversity includes 27 families, comprising 102 genera, and a remarkable 1407 species (Chandra, Gopi, et al., 2017; Chandra, Jaiswal, et al., 2017; Pandher, 2024).

The family Leptoceridae stands out as the most speciose within India, with 13 genera and 296 species. Other significantly diverse families include Hydropsychidae (15 genera, 238 species), Philopotamidae (5 genera, 156 species), and Rhyacophilidae (2 genera, 193 species). Families with moderate diversity include Psychomyiidae (5 genera, 71 species), Lepidostomatidae (3 genera, 66 species), and Hydroptilidae (8 genera, 52 species). In contrast, several families are represented by a smaller number of genera and species, such as Stenopsychidae (1 genus, 14 species), Dipseudopsidae (2 genera, 13 species), Ecnomidae (1 genus, 9 species), Limnocentropodidae (1 genus, 3 species), and Odontoceridae (1 genus, 2 species).

Other notable families and their reported diversity include Xiphocentronidae (5 genera, 45 species), Calamoceratidae (2 genera, 6 species), Molannidae (2 genera, 8 species), Helicopsychidae (1 genus, 28 species), Apataniidae (4 genera, 22 species), Goeridae (2 genera, 31 species), Limnephilidae (7 genera, 44 species), Uenoidae (1 genus, 5 species), Brachycentridae (2 genera, 12 species), Phryganeidae (2 genera, 10 species), Phryganopsychidae (1 genus, 1 species), Glossosomatidae (3 genera, 26 species), and Hydrobiosidae (1 genus, 14 species). Additionally, a group classified as "Uncertain in superfamily Sericostomatoidea" comprises 4 genera and 6 species.

7.10 Lepidoptera (moths)

While the order Lepidoptera comprises approximately 157,424 described species globally (Nieukerken et al., 2011), only a small fraction, specifically 740 species, are known to exhibit aquatic adaptations. This aquatic evolution in lepidopteran caterpillars is understood to have occurred through processes of coevolution and radiation, allowing them to thrive in various aquatic habitats.

In India, the documented aquatic diversity of Lepidoptera is considerably limited, with only 80 species recorded. All of these species belong to the subfamily Acentropinae, which represents only 0.53% of the total Lepidopteran species documented in the country. Despite their low species count, the immature stages of aquatic Lepidoptera are considered important constituents within diverse aquatic food chains (Singh, 2017), highlighting their ecological role in freshwater ecosystems.

7.11 Order Diptera (true flies)

Diptera, commonly known as "true flies," constitute one of the world's most diverse insect groups, representing over 8.05% of known global biodiversity and 15.3% of global insect diversity (Chapman, 2009). This order demonstrates significant aquatic representation, with 40 families globally recognized as aquatic.

In India, 36 families, encompassing 151 genera and 1588 species of Diptera, are known to be aquatic or semiaquatic (Banerjee et al., 2017). The family Culicidae (mosquitoes) is the most prominent aquatic group in terms of generic abundance, with 22 genera. This is followed by Chironomidae (nonbiting midges) with 21 genera, Ephydridae (shore flies) with 15 genera, Limoniidae (limoniid crane flies) with 14 genera, and Ceratopogonidae (biting midges) with 12 genera (Banerjee et al., 2017). Larvae of these families, such as midge larvae, are commonly found in water bodies and play a crucial role in the organic breakdown process within aquatic ecosystems.

7.11.1 Aquatic insect diversity across Indian states and Union Territories

India, renowned for its vast terrestrial biodiversity, also harbors a significant and often underappreciated wealth of aquatic insect diversity. The provided table, focusing on four key orders—Ephemeroptera (Subramanian et al., 2024), Plecoptera (Chandra et al., 2019), Trichoptera (Pandher, 2024), and Odonata (Subramanian & Babu, 2024)—illustrates this rich, albeit unevenly documented, faunal tapestry.

The Himalayan and Western Ghats regions, particularly states like Meghalaya (454 total species) and Arunachal Pradesh (418 total species), emerge as clear hotspots for these aquatic groups, showcasing the influence of diverse freshwater habitats (Table 7.2). Conversely, numerous states in the central and western parts of India exhibit considerably lower recorded species counts. For Ephemeroptera and Plecoptera in many states, especially in the northern plains and some coastal regions (e.g., Chandigarh, Gujarat, Haryana, Lakshadweep, Punjab, Rajasthan, Telangana, Tripura, Uttar Pradesh, Puducherry), strongly indicates significant survey gaps for these orders, rather than a genuine absence of these insects (Table 7.2). As mayflies and stoneflies are often considered indicators of good water quality, their unrecorded status in vast areas suggests an incomplete understanding of freshwater ecosystem health in those regions.

Conversely, Odonata, despite some missing data, consistently show higher species counts across almost all states (Subramanian & Babu, 2024). This implies either more robust survey efforts for this charismatic order or a broader environmental tolerance that allows them to persist even where mayflies and stoneflies might be absent. The general trend of higher total species counts in states with diverse mountainous and forested landscapes (e.g., Meghalaya, Arunachal Pradesh, Uttarakhand) compared to more developed or arid regions reinforces the known association between habitat complexity and aquatic insect biodiversity (Table 7.2). This pattern, directly discernible from the quantitative data, points to inherent biases in current knowledge and priorities for future targeted faunistic research. The overall picture reveals that India's aquatic insect fauna, particularly in sensitive orders such as mayflies and stoneflies, serves as a crucial indicator of freshwater ecosystem health and underscores the pressing need for continued, expanded research to fully catalog and conserve this vital component of the nation's biodiversity.

7.12 Insect species diversity in Indian Ramsar wetlands

An assessment of documented species diversity across various Ramsar wetlands in India (Banerjee et al., 2023; Chandra et al., 2021), based on available data for specific insect orders, reveals substantial ecological variation. The Sundarban Wetland exhibits the highest recorded diversity, with 504 insect species, indicating its exceptional ecological richness. This is followed by the East Kolkata Wetlands (253 species) and the Bhoj Wetland (202 species), both of which reflect high biodiversity levels within the surveyed insect taxa (Fig. 7.4).

Other wetlands with notable species counts include Satkosia Gorge (119 species), Loktak Lake (118 species), Chilka Lake (99 species), and Renuka Wetland (89 species). Lonar Lake also presents a moderate level of diversity with 63 species recorded. Wetlands with intermediate diversity include Pong Dam Lake (58 species), Nawabganj Bird Sanctuary (54), Asan (48), Ansupa Lake (47), Vembannur Wetland Complex (46), Kabar Taal Lake (41), Nalsarovar (32), and Deepor Beel (28). Additional wetlands such as Sandi Bird Sanctuary (25), Nandur Madhameshwar and Sultanpur National Park (23 each), Sambhar Lake (22), Vembanad-Kol Wetland (21), and both Hirakud Reservoir and Thane Creek (20 each) demonstrate moderate insect species richness (Chandra et al., 2021; Banerjee et al., 2023).

TABLE 7.2 Species diversity for four key aquatic insect orders—Ephemeroptera (mayflies), Plecoptera (stoneflies), Trichoptera (caddisflies), and Odonata (dragonflies and damselflies).

Sl. no.	Name of state/UT	Ephemeroptera (mayflies)	Plecoptera (stoneflies)	Trichoptera (caddisflies)	Odonata (dragonflies and damselflies)	Total insect species
1	Meghalaya	14	26	254	160	454
2	Arunachal Pradesh	46	35	227	110	418
3	West Bengal	29	27	86	240	382
4	Uttarakhand	6	17	196	127	346
5	Assam	6	29	130	173	338
6	Tamil Nadu	72	15	88	147	322
7	Sikkim	6	8	206	82	302
8	Manipur	5	17	186	92	300
9	Kerala	38	8	59	178	283
10	Himachal Pradesh	38	25	97	122	282
11	Karnataka	44	10	77	141	272
12	Maharashtra	26	3	30	139	198
13	Jammu and Kashmir	34	7	52	67	160
14	Odisha	1	2	21	114	138
15	Madhya Pradesh	9	2	30	88	129
16	Nagaland	2	–	7	107	116
17	Andaman and Nicobar	3	–	22	80	105
18	Uttar Pradesh	4	–	20	70	94
19	Chhattisgarh	–	–	1	87	88
20	Goa	1	–	–	87	88
21	Mizoram	–	2	18	65	85
22	Bihar	5	–	25	53	83
23	Andhra Pradesh	4	–	–	72	76
24	Jharkhand	1	1	2	72	76
25	Tripura	–	–	–	76	76
26	Gujarat	–	–	–	63	63
27	Telangana	–	–	5	57	62
28	Rajasthan	1	–	6	51	58
29	Dadra and Nagar Haveli, Daman and Diu	–	–	–	51	51
30	Delhi	–	–	1	48	49
31	Punjab	–	–	4	41	45
32	Chandigarh	–	–	–	41	41

(Continued)

TABLE 7.2 (Continued)

Sl. no.	Name of state/UT	Ephemeroptera (mayflies)	Plecoptera (stoneflies)	Trichoptera (caddisflies)	Odonata (dragonflies and damselflies)	Total insect species
33	Puducherry	–	–	0	33	33
34	Haryana	–	–	–	26	26
35	Ladakh	8	2	–	9	19
36	Lakshadweep	–	–	–	17	17

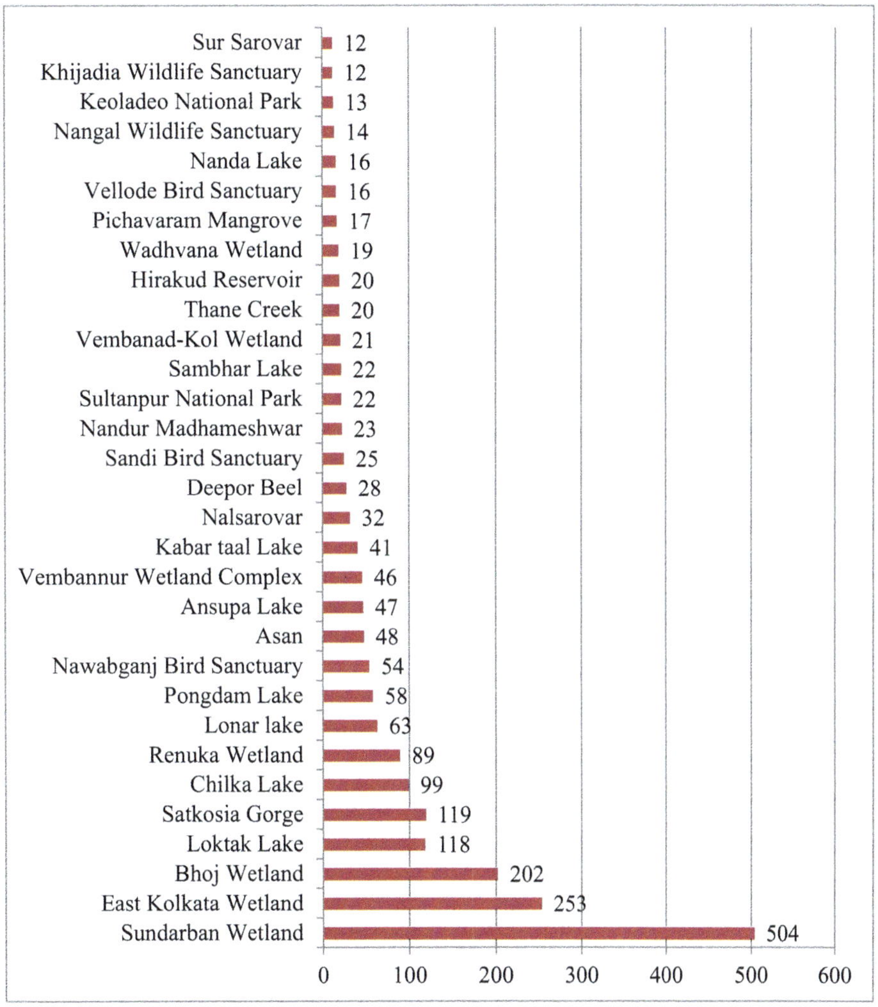

FIGURE 7.4 Graph showing the number of species in selected Ramsar wetland Sites in India.

Several sites report low species diversity, potentially indicating undersurveyed regions (Banerjee et al., 2023; Chandra et al., 2021). These include: Wadhvana Wetland (19 species), Pichavaram Mangrove (17), Vellode Bird Sanctuary and Nanda Lake (16 each), Nangal Wildlife Sanctuary (14), Keoladeo National Park (13), Khijadia Wildlife Sanctuary and Sur Sarovar (12 each), Point Calimere Wildlife and Bird Sanctuary (9), Keshopur-Miani Community Reserve (8), Surinsar-

Mansar Lake and Tsomoriri (6 each), Kanjli and Shallbugh Wetland Conservation Reserve (4 each), Chandertal Wetland (3), and five sites—Wular Lake, Sasthamkotta Lake, Hokersar Wetland, Koonthankulam Bird Sanctuary, and Samaspur Bird Sanctuary—with only a single recorded species each.

Notably, several Ramsar sites reported no species within the surveyed insect orders, suggesting either insufficient sampling or the absence of published data (Banerjee et al., 2023; Chandra et al., 2021). These include: Harike Lake, Ropar, Kolleru Lake, Ashtamudi Wetland, Bhitarkanika Mangroves, Rudrasagar Lake, Upper Ganga (Brijghat to Narora Stretch), Sarsai Nawar Jheel, Beas Conservation Reserve, Parvati Arga Bird Sanctuary, Saman Bird Sanctuary, Tso Kar, Thol Lake Wildlife Sanctuary, Haiderpur Wetland, Bhindawas Wildlife Sanctuary, Bakhira Wildlife Sanctuary, Pala Wetland, Tampara Lake, Chitrangudi Bird Sanctuary, Sirpur Wetland, Yashwant Sagar, Sakhya Sagar, Ranganathittu Bird Sanctuary, Gulf of Mannar Marine Biosphere Reserve, Vedanthangal Bird Sanctuary, Udhayamarthandapuram Bird Sanctuary, Pallikaranai Marsh Reserve Forest, Suchindram Theroor Wetland Complex, Vaduvur Bird Sanctuary, Kanjirankulam Bird Sanctuary, Karikili Bird Sanctuary, and Hygam Wetland Conservation Reserve. It is important to note that the lack of recorded species in this dataset does not necessarily imply an absence of insect fauna. Rather, it likely reflects incomplete survey efforts or the unavailability of published data for certain insect orders at these sites. Expanded and targeted biodiversity surveys are essential to fill these knowledge gaps and provide a more comprehensive understanding of insect diversity in India's wetland ecosystems.

7.13 Conclusion

India's wetland ecosystems, though historically underrepresented in biodiversity assessments, host a diverse assemblage of aquatic insects that play key functional roles in freshwater ecosystem dynamics. Contributing approximately 4.88% to global aquatic insect diversity, India stands out as a significant but unevenly explored hotspot.

Regions such as the Himalayas and Western Ghats, and wetlands like the Sundarbans (504 species), East Kolkata Wetlands (253), and Bhoj Wetland (202), support high species richness, reflecting the ecological value of complex habitats. Moderate diversity in Satkosia Gorge (119), Loktak Lake (118), and Chilika Lake (99) reinforces this trend. In contrast, low species counts in wetlands such as Sur Sarovar, Khijadia WLS, and Keoladeo NP (<15 species) likely reflect survey limitations rather than true faunal absence.

The uneven distribution highlights substantial gaps in surveys and data, particularly in central, arid, and undersampled regions. Sensitive bioindicator groups such as Plecoptera and Ephemeroptera are often missing, underscoring the need for targeted surveys, standardized monitoring, and taxonomic capacity building.

Given their roles in water quality assessment, nutrient cycling, and trophic dynamics, aquatic insects merit conservation focus beyond species inventories. In the face of increasing anthropogenic pressures—pollution, habitat loss, and climate change—integrative biodiversity assessments are critical for guiding freshwater ecosystem conservation and management in India (Richa et al., 2025).

References

Akünal, A., & Aslan, E. G. (2017). Ecological investigations on Hydrophilidae and Helophoridae (Coleoptera) specimens gathered from several water bodies of Western Turkey. *Knowledge Management of Aquatic Ecosystem, 418*, 43.

Banerjee, D., Bharti, D., Kumar, S., Mitra, A., Joshi, R., & Gupta, D. (2023). Faunal composition of Ramsar Wetlands from India: An analysis. *Records of the Zoological Survey of India, 123*(1S), 01–15. https://doi.org/10.26515/rzsi/v123/i1S/2023/172458.

Banerjee, D., Naskar, A., Sengupta, J., Hazra, S., Mridha, R., & Parui, P. (2017). Insecta: Diptera. In K. Chandra, K. C. Gopi, D. V. Rao, K. Valaharmati, & J. R. B. Alfred (Eds.). *Current status of freshwater faunal diversity in India* (pp. 467–479). Zoological Survey of India Published by the Director.

Basu, S., & Subramanian, K. A. (2017). Insecta: Hemiptera: (Aquatic Bugs). In K. Chandra, K. C. Gopi, D. V. Rao, K. Valaharmati, & J. R. B. Alfred (Eds.). *Current status of freshwater faunal diversity in India* (pp. 357–378). Zoological Survey of India Published by the Director.

Bennett, A. M. R. (2008). Global diversity of hymenopterans (Hymenoptera; Insecta) in freshwater. *Hydrobiologia, 595*(1), 529–534.

Bhowmik, S. (2020). Ecological and economic importance of wetlands and their vulnerability: A review. *Research Anthology on Ecosystem Conservation and Preserving Biodiversity,* 11–27.

Bilton, D. T., Ribera, I., & Short, A. E. Z. (2019). Water beetles as models in ecology and evolution. *Annual Review of Entomology, 64*(1), 359–377.

Bulánková, E. (1997). Dragonflies (Odonata) as bioindicators. *Biologia, Bratislava, 52*(2), 177–180.

Chandra, K., & Gupta, D. (2022a). Chapter 13: Biodiversity issues and challenges: Non-agricultural Insects. In S. Kaur, D. R. Batish, H. P. Singh, & R. Kohli (Eds.). *Biodiversity in India: Status, issues and challenges* (pp. 285–324). https://doi.org/10.1007/978-981-16-9777-7_13.

Chandra, K., & Gupta, D. (2022b). Chapter 18: Faunal diversity at ecosystem level in India: Dynamics and status. In S. Kaur, D. R. Batish, H. P. Singh, & R. Kohli (Eds.). *Biodiversity in India: Status, issues and challenges* (pp. 463–492). https://doi.org/10.1007/978-981-16-9777-7_18.

Chandra, K., Bharti, D., Kumar, S., Raghunathan, C., Gupta, D., Alfred, J. R. B., & Chowdhury, B. R. (2021). Faunal diversity in Ramsar Wetlands of India. In *Zoological survey of India and Wetland division, ministry of environment, forest and climate change*. Government of India, 292.

Chandra, K., Gopi, K. C., Rao, D. V., Subramanian, K. A., & Valarmathi, K. (2017). Current status on freshwater faunal diversity of India – An overview. In K. Chandra, K. C. Gopi, D. V. Rao, K. Valaharmati, & J. R. B. Alfred (Eds.). *Current Status of Freshwater Faunal Diversity in India* (pp. 1–25). Zoological Survey of India (Published by the Director.

Chandra, K., Gupta, D., & Ahmed, I. (2019). A catalogue of Indian stoneflies (Insecta: Plecoptera). *Zootaxa, 4646*(2), 201–235. https://doi.org/10.11646/zootaxa.4646.2.1.

Chandra, K., Gupta, D., Gopi, K. C., Tripathy, B., Kumar, V., Mandal, K., Kumar, H., & Saini, J. (2018). Faunal diversity of Indian Himalaya: An overview. In K. Chandra, D. Gupta, K. C. Gopi, B. Tripathy, & V. Kumar (Eds.). *Faunal Diversity of Indian Himalaya* (pp. 1–44). Kolkata: Zoological Survey of India Published by the Director.

Chandra, K., Gupta, D., Kushwaha, S., Das, P., & Ghosh, J. (2018). Arthropoda: Hexapoda. In K. Chandra, & C. Raghunathan (Eds.). *Faunal diversity of biogeographic zones: Islands of India* (pp. 247–320). Zoological Survey of India Published by the Director.

Chandra, K., Jaiswal, D., & Gupta, D. (2017). Insecta: Coleoptera (Aquatic Beetles). In K. Chandra, K. C. Gopi, D. V. Rao, K. Valaharmati, & J. R. B. Alfred (Eds.). *Current status of freshwater faunal diversity in India* (pp. 379–400). Zoological Survey of India Published by the Director.

Chapman, A. D. (2009). Numbers of living species in Australia and the World (PDF) (2nd edn.). Canberra: Australian Biological Resources Study, 1-80.

DeWalt, R. E., Maehr, M. D., Neu-Becker, U., & Stueber, G. (2019). Plecoptera species file online. Version 5.0/5.0.

DeWalt, R. E., Hopkins, H., Neu-Becker, U., & Stueber, G. (2025). Plecoptera Species File (version May 2025). In O. Bánki, Y. Roskov, M. Döring, G. Ower, D. R. Hernández Robles, C. A. Plata Corredor, T. Stjernegaard Jeppesen, A. Örn, T. Pape, D. Hobern, S. Garnett, H. Little, R. E. DeWalt, J. Miller, & T. Orrell, Catalogue of Life (Annual Checklist 2025). Catalogue of Life Foundation, Amsterdam, Netherlands. https://doi.org/10.48580/dgqdn-39r.

Dudgeon, D., Arthington, A. H., Gessner, M. O., Kawabata, Z., Knowler, D., Leveque, C., ... Stiassny, M. L. J. (2006). Freshwater biodiversity: A global crisis. *Bioscience, 56*(11), 922–933.

Elango, K., Vijayalakshmi, G., Arunkumar, P., Sobhana, E., & Sujithra, P. (2021). Aquatic insect's biodiversity: Importance and their conservation. *In: Biological Diversity: Current Status and Conservation Policies, 1*, 289–303. https://doi.org/10.26832/aesa-2021-bdcp-019.

Freitag, H. (2015). Clues to aquatic beetle research in Southeast Asia: A multitude of ecological niches, microhabitats and deduced field sampling techniques. *Entomology, Ornithology & Herpetology, 4*, 4.

Gattolliat, J. L., Vinson, G., Wyler, S., Pawlowski, J., & Sartori, M. (2016). Towards a comprehensive COI DNA barcode library for Swiss stoneflies (Insecta: Plecoptera) with special emphasis on the genus *Leuctra*. *Zoosymposia, 11*, 135–155.

Ghosh, J., Saini, J., Gupta, D., Ghosh, S. K., & Chandra, K. (2022). A catalogue of Indian hydraenidae (Insecta: Coleoptera). *Zootaxa, 5087*(4), 558–570. https://doi.org/10.11646/zootaxa.5087.4.4.

Gupta, D., Sonali, S., Ghosh, S. K., Das, P., Jaiswal, D., & Chandra, K. (2024). A catalogue of Indian hydrophiloidea (Insecta: Coleoptera). *Zootaxa, 5546*(1), 1–92. https://doi.org/10.11646/zootaxa.5546.1.1.

Gupta, D., Sonali, S., Wani, I., Ghosh, P., Kumar Ghosh, A., & Kumar Ghosh, S. (2025). A comprehensive taxonomic review of Indian Gyrinidae (Coleoptera): New combinations, geographic distribution, and annotated species checklist. *Journal of Natural History, 59*(17-20), 1267–1338. https://doi.org/10.1080/00222933.2025.2474742.

IPBES. (2019). The interlinkages among biodiversity, water, food and health: The IPBES nexus assessment. IPBES Secretariat.

IPBES. (2024). The interlinkages among biodiversity, water, food and health: The IPBES nexus assessment. IPBES Secretariat.

Lenat, D. R. (1993). A biotic index for the southeastern United States: Derivation and list of tolerance values, with criteria for assigning water-quality ratings. *Journal of the North American Benthological Society, 12*(3), 279–290.

Maltby, E. (2009). The changing wetland paradigm. In E. Maltby, & T. Barker (Vol. Eds.), *The wetlands handbook:* 2Set. John Wiley & Sons.

Mishra, A., Kumar, R. & Richa, R. (2021). Biodiversity conservation to mitigate the impact of climate change on agro-ecosystems. In: Biological Diversity: Current Status and Conservation Policies, Volume 1, Eds. Kumar., V., Kumar, S., Kamboj, N., Payum, T., Kumar, P. and Kumari, S. pp. 89-107. https://doi.org/10.26832/aesa2021-bdcp-06.

Morse, J. C. (2023). Trichoptera world checklist. Clemson University. Retrieved June 4, 2025, from https://trichopt.app.clemson.edu/.

Nieukerken, E. J. V., Kaila, L., Kitching, I. J., Kristensen, N. P., Lees, D. C., Minet, J., Mitter, C., Mutanen, M., Regier, J. C., Simonsen, T. J., Wahlberg, N., Yen, S.-H., Zahiri, R., Adamski, D., Baixeras, J., Bartsch, D., Bengtsson, B. A., Brown, J. W., Bucheli, S. R., ... Zwick., A. (2011). Order lepidoptera linnaeus, 1758. In Z.-Q. Zhang (Vol. Ed.), *Animal biodiversity: An outline of higher-level classification and survey of taxonomic richness: 3148*, (pp. 212–221). *Zootaxa*.

Ortega, J. C., Geijer, J., Bergsten, J., Heino, J., Herrmann, J., Johansson, F., & Bini, L. M. (2021). Spatio-temporal variation in water beetle assemblages across temperate freshwater ecosystems. *Science of the Total Environment, 792*, 148071.

Pandher, M. S. (2024). Fauna of India checklist: Arthropoda: Insecta: Trichoptera. Version 1.0. *Zoological Survey India*. https://doi.org/10.26515/Fauna/1/2023/Arthropoda:Insecta:Trichoptera.

Praveen, K., Hassan, M. E., Saha, P. C., Kushwaha, S., Dubey, A. K., Dash, S., Lyngdoh, J., Pal, A., Jehamalar, E. & Khanra, S. (2024). Checklist of Fauna of India: Arthropoda: Insecta: Hemiptera.Version 1.0, 2024, Zoological Survey of India.Published by the Director. https://doi.org/10.26515/Fauna/1/2023/Arthropoda:Insecta:Hemiptera.

Rajmohana, K. (2017). Insecta: Hymenoptera (Aquatic). In K. Chandra, K. C. Gopi, D. V. Rao, K. Valaharmati, & J. R. B. Alfred (Eds.). *Current status of freshwater faunal diversity in India* (pp. 487–499). Zoological Survey of India Published by the Director.

Richa, R., Gupta, D., Mukherjee, S., Duneria, K. & Makhdumi, P. M. A. (2025). Harnessing Agrivoltaics for Sustainable Agriculture, Indigenous Ecosystems and Biodiversity in the Himalayan and Gangetic Plains. *Uttar Pradesh Journal of Zoology, 46*(8), 240–250. https://doi.org/10.56557/upjoz/2025/v46i84911.

Selvakumar, C. (2018). Insecta: Ephemeroptera. In K. Chandra, D. Gupta, K. C. Gopi, B. Tripathy, & V. Kumar (Eds.). *Faunal diversity of Indian Himalaya* (pp. 219–226). Zoological Survey of India (Published by the Director, Zoological Survey of India, Kolkata).

Singh, N. (2017). Insecta: Lepidoptera. In K. Chandra, K. C. Gopi, D. V. Rao, K. Valaharmati, & J. R. B. Alfred (Eds.). *Current status of freshwater faunal diversity in India* (pp. 457–465). Zoological Survey of India Published by the Director.

Sonali, S., Shukla, S., & Gupta, D. (2024). An annotated checklist of Indian psephenidae (Coleoptera). *Halteres, 14*, 31–42. https://doi.org/10.5281/zenodo.14439998.

Subramanian, K. A., & Babu, R. (2018). Insecta: Odonata. In K. Chandra, D. Gupta, K. C. Gopi, B. Tripathy, & V. Kumar (Eds.). *Faunal diversity of Indian Himalaya* (pp. 227–240). Zoological Survey of India Published by the Director.

Subramanian, K. A., & Babu, R. (2024). *Fauna of India checklist: Arthropoda: Insecta: Odonata*. Version 1.0. *Zoological Survey India*. https://doi.org/10.26515/Fauna/1/2023/Arthropoda:Insecta:Odonata.

Subramanian, K. A., Vasanth, M., & Kubendran, T. (2024). *Fauna of India checklist: Arthropoda: Insecta: Ephemeroptera*. Version 1.0. *Zoological Survey India*. https://doi.org/10.26515/Fauna/1/2023/Arthropoda:Insecta:Ephmeroptera.

Thorp and covich's freshwater invertebrates. In (4th ed.). J. H. Thorp, & A. P. Covich (Vol. Eds.), *Ecology and general biology): 1*Academic Press.

Wallace, J. B., & Webster, J. R. (1996). The role of macroinvertebrates in stream ecosystem function. *Annual Review of Entomology, 41*(1), 115–139.

Wantzen, K. M., & Junk, W. J. (2000). The importance of stream-wetland-systems for biodiversity: A tropical perspective. In B. Gopal, W. J. Junk, & J. A. Davies (Eds.). *Biodiversity in wetlands: Assessment, function and conservation* (pp. 11–34). Backhuys.

Wiggins, G. B. (2004). *Caddisflies: The underwater architects*. University of Toronto Press.

Chapter 8

Migratory birds and their role in wetlands dynamics with special focus on the Indian wetlands/RAMSAR sites

Gopinathan Maheswaran
Zoological Survey of India, Kolkata, West Bengal, India

8.1 Introduction

Wetlands are biodiverse ecosystems that function as breeding, feeding, and resting sites for numerous wildlife species, notably migratory birds. These areas are crucial for the survival of migratory birds and, reciprocally, gain significant ecological benefits from the services those birds provide. As a key country along the Central Asian Flyway, India offers important wetland habitats to more than 370 species of migratory birds. With 80 wetlands listed as Ramsar Sites, India bears considerable responsibility for safeguarding wetland health and biodiversity at both national and global scales. Wetlands rank among the planet's most productive ecosystems, supporting a wide array of plant and animal life. Migratory birds are among their most prominent inhabitants and visitors, undertaking annual journeys of thousands of kilometers that link different continents and ecosystems. In India, wetlands—including Ramsar-designated sites—are vital for these migrants, and conversely, the birds' presence and activities affect wetland condition, nutrient dynamics, and ecological equilibrium.

Balachandran et al. (2018) note that approximately 12% (165 species) of the 1395 species recorded on the Indian subcontinent are winter migrants, the majority originating from the Palearctic region beyond the Himalaya. Many of these species breed in northern Asia (Russia, Mongolia, and China) and Central Asian countries (Kazakhstan, Turkmenistan, Uzbekistan, Kyrgyzstan, Tajikistan), while a minority arrive from as far as western Europe. A large number of waterbirds are migratory (roughly 360 species), encompassing Anatidae (ducks, geese, swans), Charadriiformes (about 130 shorebirds), Podicipedidae (grebes), Phoenicopteridae (flamingos), Ciconiidae (storks), Threskiornithidae (ibises, spoonbills), Ardeidae (bitterns, herons, egrets), Pelecanidae (pelicans), Rallidae (rails) and Gruidae (cranes) (Baker et al., 2004). Across India's wetlands, 243 bird taxa, including wetland-dependent species, have been recorded, representing 23 families across 9 orders. Of the 315 identified Indian wetland species, about 130 (c. 42%) are migrants, 173 are residents, and a few have uncertain status.

Each year, following the start of winter across much of the Northern Hemisphere, many bird species move southward to spend the colder months in milder regions. After overwintering, most return to their breeding grounds in spring (Alerstam, 1990; Berthold, 2001; Newton, 2003; Somveille et al., 2013).

Migratory bird populations confront numerous threats, including pollution, habitat loss, climate change, illegal hunting, urban expansion, disease, collisions with wind turbines, oil spills, radiation, artificial lighting, and other anthropogenic pressures; these threats are driving declines in species diversity and abundance globally (Baker et al., 2004; Biber & Salathe, 1991; Trouwborst, 2012).

India hosts many extensive waterbodies, inland and coastal wetlands, marshes, ephemeral slushy areas, and ponds that attract migratory birds (Islam & Rahmani, 2004; Kumar, 2015). These wetlands span from high-altitude trans-Himalayan lakes such as Tso Moriri, Pangong Tso, Tso Kar, and Chander Taal to coastal wetlands, including salt marshes (from Point Calimere in Tamil Nadu to the Rann of Kutch in Gujarat) and mangrove forests (the Sundarbans in West Bengal). Many of these sites draw large numbers of migratory birds because they provide rich resources through primary production. Several wetlands are recognized as "Wetlands of International Importance" for their unique ecosystems, exceptional conservation value, and critical ecosystem services. Currently, India has around 1264 wetlands, including 91 Ramsar sites, 120

Wetlands: Biodiversity, Health and their Role in Ecosystem Services. DOI: https://doi.org/10.1016/B978-0-443-34222-6.00017-6

significant wetlands, and 1053 other wetlands (https://rsis.ramsar.org/). Indian Ramsar sites cover 13,599.51 km^2. India acceded to the Ramsar Convention on Wetlands on February 1, 1982.

A majority of these wetlands are located in the north-west Himalaya and the Gangetic plains. Well-known wetlands that attract large numbers of wintering migrants include Pong Dam in Himachal Pradesh (Kumar & Paliwal, 2015; Kumar, 2011), Asan Conservation Reserve in Uttarakhand, Wular Lake in Jammu and Kashmir, Tso Kar Wetland Complex in Ladakh, Harike Lake in Punjab, Bhindawas Wildlife Sanctuary and Sultanpur National Park in Haryana, Sur Sarovar Bird Sanctuary in Uttar Pradesh, Keoladeo National Park and Sambhar Lake in Rajasthan, Bhoj Wetland in Madhya Pradesh, Kabartal Wetland in Bihar, Deepor Beel in Assam, Loktak Lake in Manipur, Bhitarkanika Mangroves and Hirakud Reservoir in Odisha, East Kolkata Wetlands and Sundarbans in West Bengal, Kolleru Lake in Andhra Pradesh, Point Calimere Bird Sanctuary and Pichavaram Mangrove in Tamil Nadu, Ashtamudi Wetland and Sasthamkotta Lake in Kerala, Nanda Lake in Goa, Thane Creek in Maharashtra, and Nalsarovar Bird Sanctuary, Khijadia Wildlife Sanctuary and Wadhwana Wetland in Gujarat. Additionally, numerous bird sanctuaries with wetland habitats host considerable numbers of migratory species, for example, Patna Bird Sanctuary, Nawabganj Bird Sanctuary, Lakh Bahosi Sanctuary, and Sandi Bird Sanctuary in Uttar Pradesh, Kanwar Lake Bird Sanctuary in Bihar, Khichan Bird Sanctuary, Pulicat Lake on the Tamil Nadu–Andhra Pradesh border, D'Ering Wildlife Sanctuary in Arunachal Pradesh, the Rann of Kutch in Gujarat, Kumarakom Bird Sanctuary in Kerala, and Ranganathittu Bird Sanctuary in Karnataka.

Although the roles of birds in freshwater and wetland ecosystems are acknowledged, much remains to be studied, especially in the Indian context. Waterbirds provide a range of ecosystem services, yet our knowledge of their functional roles is still limited. Kear (1990) highlights the cultural importance of Anatidae (ducks) and other waterbirds in many regions. Ducks have significant recreational and commercial value for both rural and urban communities (van Kooten, Withey & Wong, 2011) and can generate income. Over the past decades, human fascination with large congregations of waterfowl—during migration or at wintering sites—has motivated local communities to protect and conserve them, contributing to the establishment of the Ramsar Convention (1971) and organizations such as The Wildfowl & Wetlands Trust (United Kingdom) and Wetlands International (originally The International Waterfowl Research Bureau), as well as conservation initiatives in India. Ducks, swans, cranes, geese, waders, flamingos, and other waterbirds often serve as flagship species for broader conservation campaigns, increasing public awareness and promoting actions to protect species and their habitats.

8.2 Migration and flyways: India's strategic position

Migratory birds embark on seasonal movements between breeding and wintering areas in response to environmental cues like temperature, food availability, and day length. India lies within the Central Asian Flyway, which stretches from Arctic Russia to the Indian Ocean. This flyway supports over 600 bird species, many of which rely on Indian wetlands during parts of their annual cycle. Most migratory species seek refuge at major Ramsar sites in India, especially during onward and return migration, and many resident bird species also make these sites their home (Table 8.1).

8.3 Broad ecological roles of migratory birds in wetlands

8.3.1 Nutrient cycling

Migratory birds add nutrients to wetlands through guano, which is rich in nitrogen and phosphorus. This input boosts aquatic productivity, encouraging algal growth and supporting food webs that include invertebrates and fish.

8.3.2 Seed dispersal

Numerous migrants, such as ducks and waders, consume fruits and seeds, transporting them in their droppings. This behavior aids vegetation regeneration and increases plant diversity in wetland habitats.

8.3.3 Trophic interactions

As both predators and prey, migratory birds regulate populations of aquatic organisms, including mollusks, crustaceans, and insects. Their presence shapes the composition and functioning of aquatic communities.

TABLE 8.1 Important Ramsar sites in India with representative bird species.

Ramsar sites in India (State/UT)	Representative bird species
Kolleru Lake (Andhra Pradesh)	Spot-billed pelican, Painted stork, Little egret, Eurasian coot, Northern shoveler
Deepor Beel (Assam)	Spot-billed pelican, Pheasant-tailed jacana, Painted stork, Indian pond heron, Garganey
Kanwar/Kabartal (Kanwar Taal) (Bihar)	Bar-headed goose, Northern pintail, Black-headed ibis, Common teal, Gray heron
Nagi Bird Sanctuary (Bihar)	Spot-billed duck, Painted stork, Asian openbill, Little cormorant, Black-headed ibis
Nakti Bird Sanctuary (Bihar)	Resident waterfowl and waders (e.g., Little grebe, Common teal, Little egret, Indian pond heron)
Nalsarovar (Gujarat)	Greater flamingo, Greylag goose, Common pochard, Northern shoveler, Indian spot-billed duck
Wadhvana Wetland (Gujarat)	Dalmatian pelican, Pallas's fish-eagle (occasional), Common pochard, Ferruginous duck
Thol Lake Wildlife Sanctuary (Gujarat)	Eurasian coot, Common teal, Bar-headed goose (migratory), Little ringed plover
Khijadiya Wildlife Sanctuary (Gujarat)	Black-necked stork, Painted stork, Common teal, Little cormorant, Marsh sandpiper
Sultanpur National Park (Haryana)	Black-tailed godwit, Common teal, Ruddy shelduck, Gray heron, Common pochard
Bhindawas Wildlife Sanctuary (Haryana)	Little egret, Common teal, Eurasian coot, Indian spot-billed duck
Chandra Taal/Chandertal Wetland (Himachal Pradesh)	Brown-headed gull, Bar-headed goose (nearby areas), Horned lark, Himalayan snowcock (high-altitude species)
Pong Dam/Beas Conservation Reserve (Himachal Pradesh)	Great cormorant, Bar-headed goose, Ruddy shelduck, Common pochard, Gray heron
Renuka Lake (Himachal Pradesh)	Little cormorant, Little egret, Kingfisher, Common teal
Hokersar Wetland (Jammu and Kashmir)	Common pochard, Ruddy shelduck, Northern pintail, Gray heron, Common teal
Hygam Wetland Conservation Reserve (Jammu and Kashmir)	Migratory ducks and waders (e.g., Gadwall, Northern pintail, Common teal, Coot)
Shallabugh Wetland (Jammu and Kashmir)	Common pochard, Tufted duck, Gadwall, Several waders
Mansar-Surinsar Lakes (Jammu and Kashmir)	Little cormorant, Indian spot-billed duck, Common teal, Gray heron
Wular Lake (Jammu and Kashmir)	Common pochard, Ferruginous duck, Great cormorant, Eurasian teal
Udhwa Lake (Jharkhand)	Mixed waterfowl: Little cormorant, Little egret, Common teal, Asian openbill
Ranganathittu Bird Sanctuary (Karnataka)	Painted stork, Asian openbill, Woolly-necked stork, Indian pond heron, Pied kingfisher
Ankasamudra Bird Conservation Reserve (Karnataka)	Resident and migratory waterfowl (e.g., Spot-billed duck, Little cormorant, Whiskered tern)
Aghanashini Estuary (Karnataka)	Indian river tern, Painted stork, Indian cormorant, Black-headed ibis, Various waders
Magadi Kere Conservation Reserve (Karnataka)	Resident ducks and waders (e.g., Common teal, Little egret, Sandpipers)
Nanda Lake (Goa)	Lesser whistling-duck, Little cormorant, Indian pond heron, Black-headed ibis, Kingfisher

(Continued)

TABLE 8.1 (Continued)

Ramsar sites in India (State/UT)	Representative bird species
Koonthankulam Bird Sanctuary (Tamil Nadu)	Painted stork, Black-headed ibis, Asian openbill, Woolly-necked stork, Little egret
Vedanthangal Bird Sanctuary (Tamil Nadu)	Little cormorant, Indian pond heron, Spot-billed pelican (occasionally nearby), Asian openbill
Point Calimere WLS/Bird Sanctuary (Tamil Nadu)	Greater flamingo, Black-headed ibis, Painted stork, Eurasian curlew, Kentish plover
Vembannur Wetland Complex (Tamil Nadu)	Resident waterfowl and waders (e.g., Little egret, Black-winged stilt, Common sandpiper)
Vellode Bird Sanctuary (Tamil Nadu)	Spot-billed pelican, Little cormorant, Common teal, Black-winged stilt
Udhayamarthandapuram Bird Sanctuary (Tamil Nadu)	Little egret, Little cormorant, Openbill stork, Pond herons
Tampara Lake (Odisha)	Little cormorant, Spot-billed duck, Intermediate egret, Kingfisher
Hirakud Reservoir (Odisha)	Bar-headed goose (migratory), Common pochard, Ferruginous duck, Gray heron
Chilika Lake (Odisha)	Greater flamingo, Eurasian spoonbill, Bar-headed goose, Spot-billed pelican, Black-tailed godwit
Bhitarkanika Mangroves/Bhitarkanika WLS (Odisha)	Black-headed ibis, Lesser adjutant, White-bellied sea-eagle, Herons & kingfishers
Pulicat Lake (Andhra Pradesh/Tamil Nadu)	Terek sandpiper, Whimbrel, Indian river tern, Greater sand plover, Little stint
Keoladeo National Park/Bharatpur (Rajasthan)	Sarus crane, Bar-headed goose, Eurasian spoonbill, Northern pintail, Pallas's fish-eagle (occasional)
Sambhar Salt Lake (Rajasthan)	Greater flamingo, Lesser flamingo, Pied avocet, Kentish plover, Black-winged stilt
Khichan (Phalodi) (Rajasthan)	Demoiselle crane (large winter flocks), Sociable lapwing (occasionally), Local passerines
Menar (Udaipur) (Rajasthan)	Local and migratory waterbirds (e.g., Egrets, Herons, Ducks), and the resident village bird community species
Soor Sarovar/Sur Sarovar (Uttar Pradesh)	Bar-headed goose, Pied kingfisher, Little cormorant, Common teal, Indian spot-billed duck
Nawabganj Bird Sanctuary (Unnao?/Uttar Pradesh)	Little cormorant, Painted stork (seasonal), Asian openbill, Kingfishers
Sirpur Lake (Madhya Pradesh)	Common teal, Little cormorant, Pond heron, Gray heron, Purple moorhen
Sakhya Sagar/Other MP wetlands	Local waterfowl (e.g., Common pochard, Little grebe, Little egret)
Ashtamudi Wetland (Kerala)	Little egret, Black-winged stilt, Common greenshank, Terek sandpiper
Vembanad-Kol Wetland (Kerala)	Spot-billed pelican, Painted stork, Black-headed ibis, Eurasian teal
Sasthamkotta Lake (Kerala)	Resident freshwater species: Kingfisher, Little cormorant, Pond herons
Kanjirankulam Bird Sanctuary (Tamil Nadu)	Resident waterfowl and migratory waders (e.g., Painted stork, Little egret)
Loktak Lake (Manipur)	Ferruginous pochard, Bar-headed goose (on migration), Spot-billed pelican (occasional), Little egret
Keibul Lamjao/Loktak complex (Manipur)	Lesser whistling-duck, Spot-billed duck, Clamorous reed warbler (marsh species)
Bhoj Wetland/Upper Lake (Bhopal) (Madhya Pradesh)	Lesser whistling-duck, Little grebe, Common teal, Little cormorant
Tawa Reservoir (Madhya Pradesh)	Resident and migratory waterfowl (e.g., Common pochard, Little cormorant, Herons)

(*Continued*)

TABLE 8.1 (Continued)

Ramsar sites in India (State/UT)	Representative bird species
Hirakud Reservoir (duplicate earlier)	–
Sundarbans Wetland (West Bengal)	White-bellied sea-eagle, Black-headed ibis (mangrove edges), Kingfishers, Herons
East Calcutta Wetlands (West Bengal)	Herons and egrets, Kingfisher, Little cormorant, Common coot
Chakratirtha/other Bengal wetlands	Local waterbirds and waders
Harike Wetland (Punjab)	Ferruginous duck, Common pochard, Northern shoveler, Gray heron
Pallikaranai Marsh Reserve Forest (Tamil Nadu)	Little egret, Indian pond heron, Kingfisher, Black-headed ibis (nearby)
Gulf of Mannar (Tamil Nadu)	Coastal/wetland birds: Terns, Kingfishers, Herons, Egrets
Pichavaram Mangrove (Tamil Nadu)	Herons and egrets, Kingfishers, Terns, Brahminy kite
Vellayani-Karakulam Lake complex (Kerala)	Local resident ducks and waders, Little cormorant
Coringa Wildlife Sanctuary/Godavari mangroves (Andhra Pradesh)	Indian river tern, Black-headed ibis, Herons, Kingfishers
Khecheopalri/Khecheopalli Lake (Sikkim)	High-altitude forest and wetland species: Himalayan swiftlet, White-throated laughingthrush (local forest birds)
Telineelapuram?/other Andhra wetlands	Local waterbirds (e.g., egrets, herons, ducks)
Kadalundi-Binangad (Kerala)	Coastal waders, Terns, Herons
Nanjarayan Bird Sanctuary (Tamil Nadu)	Resident waterbirds: Little cormorant, Pond herons, Kingfisher
Kazhuveli Bird Sanctuary (Tamil Nadu)	Migratory waders and waterfowl: Sandpipers, Little egret, Black-winged stilt
Sakkarakottai Bird Sanctuary (Tamil Nadu)	Local waders and waterfowl: Little egret, Common sandpiper, Little cormorant
Therthangal Bird Sanctuary (Tamil Nadu)	Resident and migratory waterbirds (egrets, cormorants, ducks)

8.3.4 Sediment and habitat structuring

Species like dabbling ducks stir surface sediments while feeding, which can enhance oxygenation and modify nutrient dynamics. Their foraging also creates microhabitats that benefit other wetland species.

8.4 Ramsar wetlands in India: Avian refuges and ecological hubs

India's Ramsar sites (Fig. 8.1) encompass diverse wetland types—from freshwater lakes and mangroves to saline lagoons. Tamil Nadu and Uttar Pradesh have the majority of the Ramsar sites in India (Fig. 8.2). Notable examples include:

- Keoladeo National Park (Rajasthan): Formerly a duck-hunting area, now a refuge for over 350 bird species, including Greylag Geese and Eurasian Spoonbills.
- Chilika Lake (Odisha): Hosts more than a million birds annually, including Lesser Flamingos, gulls, and Black-tailed Godwits.
- Kolleru Lake (Andhra Pradesh): Supports Spot-billed Pelicans, Painted Storks, and migratory ducks.
- Sambhar Lake (Rajasthan): A saline wetland important for flamingos and avocets.
- Point Calimere (Tamil Nadu): Attracts a variety of shorebirds, including threatened species such as the Spoon-billed Sandpiper.

8.4.1 Biodiversity of other taxa in wetlands where migratory birds reside

In many respects, waterbirds and their activities positively influence aquatic biodiversity by acting as dispersal agents that move plant and animal propagules, thereby helping maintain diversity in isolated or temporary wetlands. This multiscale service supports biodiversity at local, regional, and even continental levels.

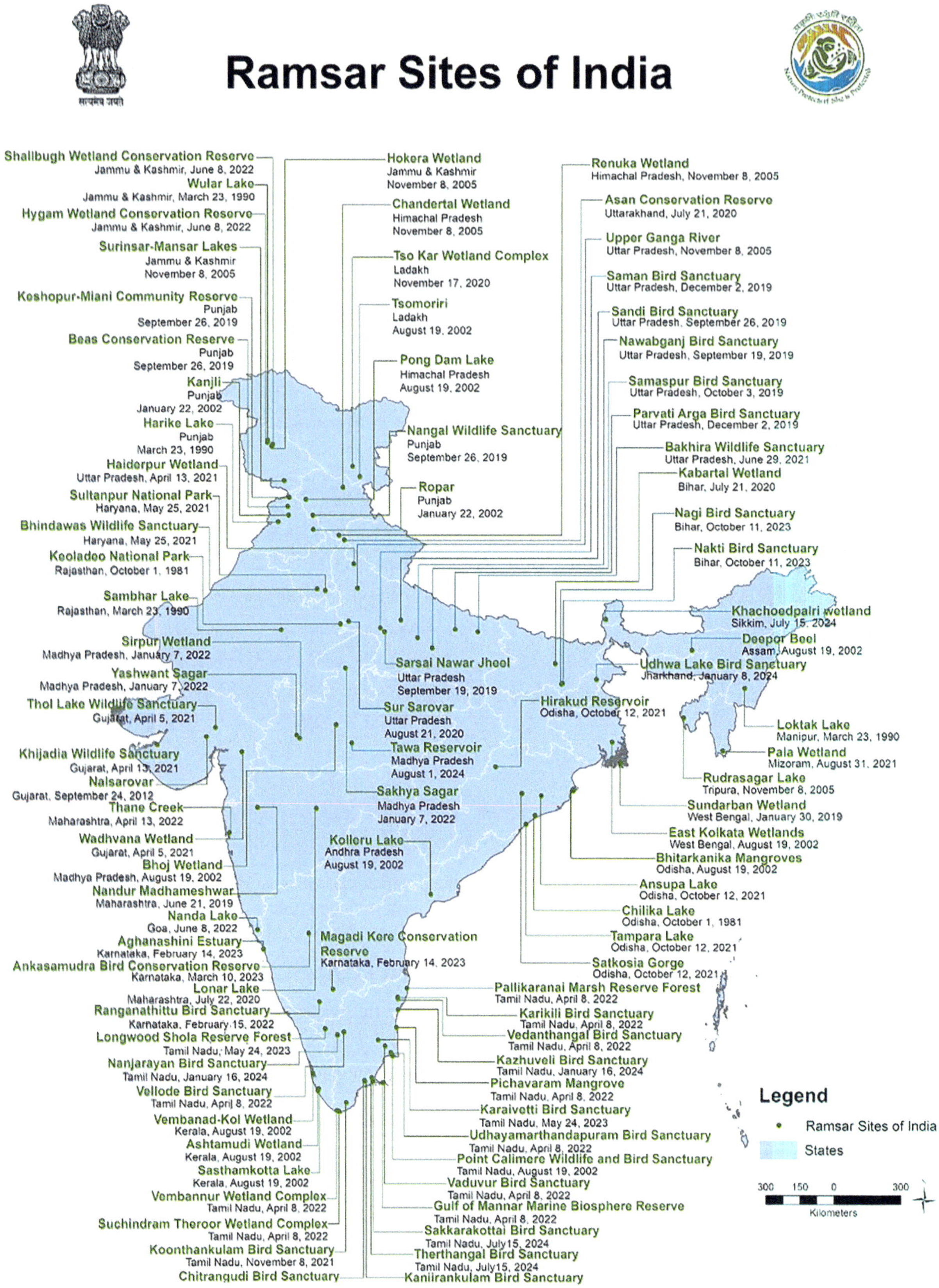

FIGURE 8.1 Ramsar sites in India. Map lines delineate study areas and do not necessarily depict accepted national boundaries. Courtesy: MoEFCC, New Delhi.

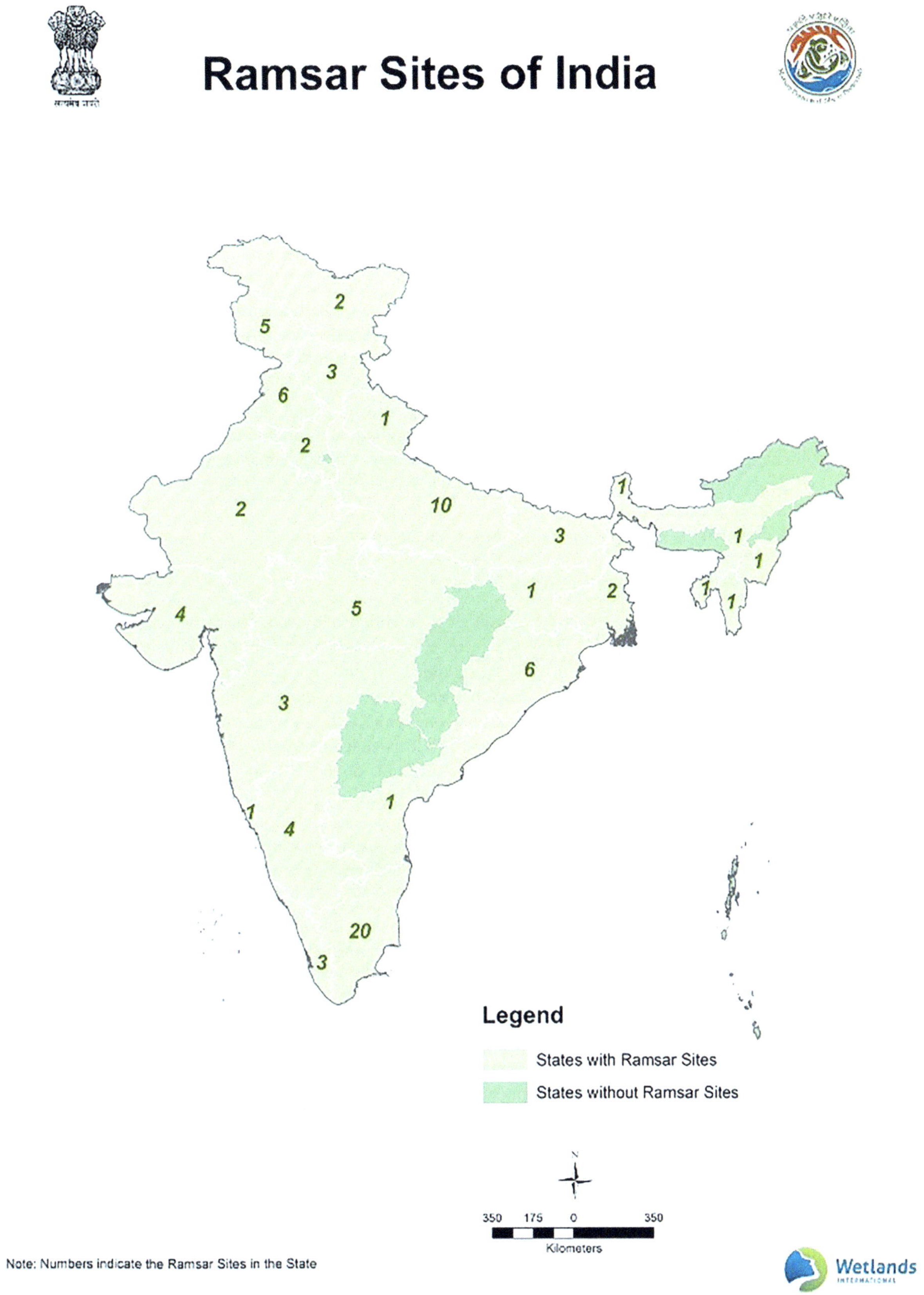

FIGURE 8.2 State-wise Ramsar sites in India. Map lines delineate study areas and do not necessarily depict accepted national boundaries. Courtesy: MoEFCC, New Delhi.

Jasmin et al. (2008) report that geese can help maintain grasslands in a diverse state by moderating interspecific competition and stimulating primary production (Bazely & Jefferies, 1985; Cargill & Jefferies, 1984). Likewise, submerged macrophyte diversity tends to be higher where waterfowl occur (Hidding et al., 2010), which in turn boosts wetland productivity (Nolet, 2004). Although no direct Indian studies exist, grazing by geese and wigeon *Anas penelope* on seagrasses has produced long-term positive effects by reducing mud accumulation in grazed beds elsewhere (Nacken & Reise, 2000). Conversely, overgrazing by geese can cause soil erosion, increased salinity, and diminished soil nitrogen retention, among other negative impacts (Buckeridge & Jefferies, 2007; Gauthier et al., 2006; Iacobelli & Jefferies, 1991). Geese populations may also facilitate the spread of invasive grasses that can variably benefit or harm local bird species, even when geese preferentially graze on those same grasses (Best & Arcese, 2009; Isaac-Renton et al., 2010).

Waterbirds can also offer protective benefits to other taxa. For instance, Canada geese *Branta canadensis* fiercely defend their nests, and neighboring nesting species often experience higher breeding success (Allard & Gilchrist, 2002). This protective effect can influence community composition; in some Swedish archipelagos, islets with nesting Canada geese exhibit higher species richness of nesting birds compared with nearby islets without geese (Fabricius & Norgren, 1987). Mann et al. (1999) proposed that shorebird predation on herbivorous invertebrates in shallow ponds allows charophyte species (e.g., *Tolypella glomerata*), which lack defenses against herbivory, to persist.

Waterbirds themselves host diverse communities of organisms. For example, the geomorphology of Pulicat Lake, Andhra Pradesh (not a Ramsar site) contributes to its high avian and aquatic biodiversity (Kumar et al., 2010). Chacko et al. (1953) documented 59 phytoplankton and 23 zooplankton species, while Krishnan and Sampath (1973) recorded 16 phytoplankton and 35 zooplankton species. They also reported 65 fish species, including three elasmobranchs and 62 teleosts. Along the higher shorelines, littoral forests occur (Champion & Seth, 1968), dominated by pioneering sand binders like *Spinifex littoralis* and *Ipomoea biloba*, followed by secondary colonizers.

8.4.2 Migratory waterbirds as bioindicators

When assemblages of bird species reflect the overall ecological condition of a waterbody, they can function as low-cost bioindicators—particularly useful in protected areas where monitoring budgets are limited. Pulicat Lake Bird Sanctuary (Andhra Pradesh, not a Ramsar site) illustrates this: surveys recorded 161 bird species across 19 orders, 52 families, and 114 genera (Alam et al., 2023). Among these, 85 species were observed in adjoining waterbodies—2 near threatened and 83 least concern; 3 species are in schedule I, 81 in schedule II, and one is unprotected. From various study sites around Pulicat, 144 species were recorded—10 near threatened, 1 vulnerable, and 133 least concern; 18 are schedule I, 125 schedule II, and one unprotected. Sixty-eight species were common to both areas. Overall, of the 161 species, 11 are near threatened, 1 is vulnerable, and 149 are least concern per the IUCN. These species span 19 of India's 26 bird orders, with Charadriiformes (shorebirds) being the most numerous, followed by Passeriformes and Pelecaniformes. Such richness largely reflects the diverse invertebrate communities that sustain these birds. Moreover, waterfowl behavior—such as the pair formation of Greater Flamingoes before migration or the foraging patterns of waders and long-legged birds as water levels change—can indicate nutrient status, as observed at Pulicat and in boreal lakes elsewhere (Nummi et al., 2000; Pöysä et al., 2001).

Pulicat Lake harbors around 500 macrofaunal species, including 168 fish species (Chacko et al., 1953; Raj, 2006; Selvanathan & Kaliyamurthy, 1972), 12 penaeid prawn species (Chacko et al., 1953), 36 crab species (Kannan & Pandiyan, 2012), and 19 mollusk species (Thangavelu & Sanjeevaraj, 1985). The edible oyster (*Crassostrea madrasensis*) is widespread, though Raj (2006) reported its decline in Pulicat. Surveys list 57 algal species across five families (Raghavaiah, 2007), about 14 algal species and 88 zooplankton species that constitute major food for flamingos (Ramesh & Ramachandran, 2005). These micro- and macrofauna underpin the avian diversity and support submerged macrophyte propagation. Fluctuations in duck and coot numbers can reliably signal changes in submerged macrophyte abundance (Wicker & Endres, 1995), which in turn positively influence aquatic biodiversity (Carpenter & Lodge, 1986). The majority of the migratory birds come to India using the Central Asian Flyway (Fig. 8.3), especially during peak migratory season.

8.4.3 Migratory waterbirds' role in disease surveillance

Pong Dam in Himachal Pradesh is a staging ground for Bar-headed Geese, with over 50,000 individuals arriving annually before moving further south in winter. Pong, one of the largest man-made wetlands in northern India, serves as a preferred wintering area for this species. During the H5N1 outbreak in India, Pong was among the first northern wetlands affected, and the bar-headed goose—an endangered visitor—was severely impacted. As poultry farming expands in areas adjacent to natural wetlands frequented by migratory birds each winter, the risk of disease transmission from domestic birds to wild populations increases. Given India's large poultry and egg production industry, the threat of avian influenza to wild birds is substantial.

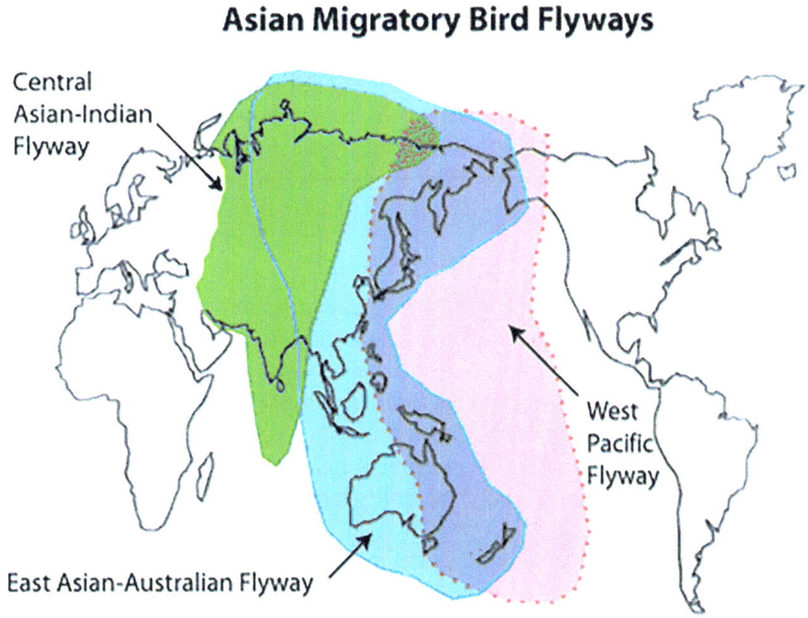

FIGURE 8.3 Major migratory bird flyways, including the Central Asian-Indian flyway (highlighted in *green*), bring in majority of the migratory birds to India. Map lines delineate study areas and do not necessarily depict accepted national boundaries.
Courtesy: US Fish and Wildlife Service, Public domain, via Wikimedia Commons (https://commons.wikimedia.org/wiki/File:Central_Asian_Flyway_Map.png)

Waterbirds can host and spread a broad range of pathogens, including avian-endemic and zoonotic agents (e.g., Hubalek, 2004; Wobeser, 1997). Although negative impacts are documented elsewhere, relatively few detailed studies have been conducted in India. Migratory wild ducks have been effectively used to monitor temporal variation in avian influenza virus subtypes (Wallensten et al., 2007), providing a tool to anticipate outbreaks of highly pathogenic strains that can cause major economic losses in poultry and potentially trigger human epidemics or pandemics (Munster et al., 2005). Captive or resident wild ducks that interact with migratory birds can function as sentinels for local pathogen dynamics across a range of infections (e.g., Farnsworth et al., 2012; Rollo et al., 2012; Ziegler et al., 2010).

8.4.4 Migratory waterbirds as scavengers

The Greater Adjutant Stork *Leptoptilos dubius*, now globally threatened, was historically associated with disposing of human corpses in India (del Hoyo et al., 1992); this practice ceased with changing regulations and a steep decline in stork numbers. Today, adjutants forage at open landfills, feeding largely on garbage in Assam. Many other waterbirds are facultative scavengers—gulls, storks, and ibises commonly aggregate at rubbish dumps to feed on human refuse. Although these birds can have negative impacts (e.g., depositing guano) (Whelan et al., 2008), they provide benefits to surrounding communities by accelerating decomposition and reducing odors associated with biodegradable waste. In Varanasi, offerings left in the Ganges—flowers and puffed rice—are consumed by gulls and terns, preventing accumulation and maintaining the sanctity of the riverfront.

8.4.5 Migratory birds' role in wetlands' nutrient and biogeochemical cycling

The use of bird guano as fertilizer has a long history worldwide, including in India—its agricultural importance has been reviewed by Hutchinson (1950) and Whelan et al. (2008). When waterbirds congregate or establish colonies, they alter nutrient flows. At roosting or breeding sites, their excreta can deliver sufficient nutrients to shift the trophic status of wetlands. For instance, Kolleru Lake—a Ramsar site and bird sanctuary in coastal Andhra Pradesh (16°37'N 081°12'E), situated between the Godavari and Krishna river basins—functions as a natural eutrophic lake and flood-balancing reservoir. It provides refuge to resident and migratory waterbirds, including the vulnerable Gray Pelican (*Pelecanus philippensis*). During rainy seasons, overflow water rich in guano reaches surrounding agricultural fields and enhances paddy and seasonal crop growth.

In Hungarian soda pans, nutrient inputs from geese and other waterbirds supply up to 50% of carbon and 70% of phosphorus, underpinning high production of fairy shrimps and other crustaceans (Boros et al., 2008a, b). Waterbirds such as gulls, storks, geese, swans, and some shorebirds commonly feed across aquatic and terrestrial habitats, thereby mediating cross-ecosystem transfers of energy and nutrients. Breeding colonies in forested areas contribute substantial nutrient subsidies through their droppings (Kameda et al., 2006). These cross-habitat subsidies remain poorly understood but are an expanding area of research (Baxter et al., 2004; Burdon & Harding, 2008).

8.4.6 Migratory birds' role in supporting local livelihoods

Deepor Beel Sanctuary (4000 ha, 26°08'N 091°39'E)—a Ramsar site in Assam—functions as a permanent freshwater lake formed from an old Brahmaputra channel, serving both ecological and practical roles for Guwahati as a major stormwater storage basin. As a staging site along the East-Asian Australasian flyway, it supports globally threatened species such as the Spot-billed Pelican *P. philippensis*, Lesser Adjutant Stork *Leptoptilos javanicus*, Greater Adjutant Stork, and the Critically Endangered Baer's Pochard *Aythya baeri*. More than 50 fish species provide livelihoods for many nearby villages; nymphaea nuts and flowers, ornamental fish, medicinal plants, and seeds of the giant water lily Euryale ferox generate significant local income, while commercially valuable orchids occur in adjacent forests.

The East Calcutta Wetlands (12,500 ha, 22°27'N 088°27'E) in West Bengal are a globally recognized example of multifunctional wetlands. Local resource-recovery practices have historically spared Calcutta (Kolkata) the cost of constructing and operating wastewater treatment plants. The wetland treats city sewage and reuses the effluent for pisciculture and agriculture, recovering nutrients efficiently: water flows through fish ponds covering roughly 4000 ha that act as solar reactors, completing many biochemical processes. During winter, numerous migratory waterbirds visit the wetlands for weeks, depositing guano that further enriches the system while relying on it for food.

Chilika Lake—a Ramsar site on India's east coast in Odisha (between 19°28' and 19°54'N and 85°05'–85°38'E)—hosts millions of migratory birds feeding on a diverse diet that includes around 300 fish species (Mahapatro et al., 2016). Chilika's sediments contain about 126 molluscan species across 56 families and multiple orders/classes. Mollusks play a crucial role linking benthic and pelagic food webs, recycling nutrients and serving as prey for fish, shellfish, birds, and humans.

8.4.7 Migratory birds' role as pest control agents

Waterbirds consume aquatic insect larvae and can reduce abundances of Chironomidae (nonbiting midges) (Rodriguez-Perez, Green, & Figuerola, 2007; Sánchez et al., 2006). Chironomids can be pests in ricefields, wastewater wetlands, and other habitats (Ali, 1996). However, comparable studies are scarce in India. Waterbirds also prey on mosquito larvae—an important pest and disease vector—with ducklings consuming large proportions of mosquito larvae (Miles et al., 2002) and terns and shorebirds also feeding on them (Tucakov & Puzovic, 2006; Fonteneau et al., 2009). No studies are known to quantify the impacts of waterbirds on biting dipteran populations in India.

Waterbirds help biologically control snails that damage rice (Yusa, Sugiura, & Wada, 2006). Teo (2001) found that 5–10 domestic ducks per hectare reduced golden apple snail *Pomacea canaliculata* densities by over 80%. In northern Europe, domestic and wild ducks can reduce slug populations, including the invasive Spanish slug Arion vulgaris (Grimm, Paill, & Kaiser, 2000; Speiser et al., 2001). Storks, ibises, egrets, and gulls forage across aquatic and terrestrial environments, and likely help control agricultural pests. Wading birds prey on invasive red-swamp crayfish *Procambarus clarkii*, a pest in ricefields that damages seeds, seedlings, and dykes (Anastacio, Parente, & Correia, 2005; Negro et al., 2000; Tablado et al., 2010). Herons also feed on tadpole shrimps Triops spp., pests of rice (Hafner & Britton, 1983).

Introduced fish species often harm water quality by destroying macrophytes, increasing phosphorus loads, or preying on large zooplankton (Søndergaard et al., 2008; Starling et al., 2002; Weber & Brown, 2009). Waterbirds can be effective predators on fish, especially in shallow waters (Britton & Moser, 1982; Winfield, 1990; Gawlik, 2002; Steinmetz et al., 2003). Their presence may also produce indirect effects—fish alter behavior and growth due to predation risk (Allouche & Saudin, 2001; Wacht-Katz et al., 2010), avoiding surface and shallow littoral zones by day (Gliwicz & Jachner, 1992; Lewin et al., 2004; Mehner et al., 2007). Such fear-induced changes can cascade through food webs—for example, algal bands in tropical streams form where catfish avoid herons and kingfishers (Power et al., 1989).

Mass fish kills associated with heat waves, droughts, or pollution severely degrade water quality (Hoyer et al., 2009; Kushlan, 1974). Piscivorous birds can mitigate such events by reducing fish populations as water levels fall (Gawlik, 2002; Kushlan, 1976), thereby preventing synchronized fish mortality and removing nutrients that would otherwise fuel toxic cyanobacterial blooms, anoxia, and disease from decomposing fish. In Everglades experiments, birds removed all fish from 10-cm-deep ponds within 12 days and over 50% from 28-cm-deep ponds (Gawlik, 2002).

Biomanipulation schemes often aim to improve water quality by manipulating fish communities (Jeppesen et al., 2007). Many shallow lakes might avoid costly remediation thanks to waterbirds' direct and indirect regulation of fish. For instance, cormorants have been linked to increased water clarity via control of benthivorous fish (Dirksen et al., 1995; Leah et al., 1980). Yet quantified assessments of birds' effects on fish remain rare, even in perceived conflict situations, due to limited information on whether bird predation is additive to other mortality sources (Harris et al., 2008). A detailed Scottish fishery study estimated cormorants removed 40% of the rainbow trout *Oncorhynchus mykiss* catch and 16 times the brown trout Salmo trutta catch (Stewart et al., 2005). Anatidae consume weed seeds in ricefields and can reduce weed seedbanks—wintering ducks in flooded ricefields have lowered subsequent weed biomass by over 50% (van Groenigen et al., 2003). Ducks typically arrive after harvest, so they do not directly cause preharvest grain loss. However, as dispersal agents, ducks can also spread weeds (Brochet et al., 2009). Future research should quantify the costs and benefits of waterbird granivory in rice and other crops. Beyond direct predation, waterbirds may control pests indirectly by transmitting parasites that reduce pest abundance—for example, terns host trematodes that use mosquitoes as intermediate hosts (Fonteneau et al., 2009), and cestodes using piscivorous birds as final hosts can reduce introduced cyprinids (Kennedy et al., 2001). Many helminth life cycles remain unknown (Lefebvre et al., 2009), so birds likely carry more pest-affecting parasites than currently documented (see also https://web2.uconn.edu/tapeworm/index.php).

8.4.8 Migratory birds' role in ecosystem engineering

The role of waterbirds as ecosystem engineers has been underappreciated but warrants more research. Large birds like pelicans and flamingos exert strong bioturbation when feeding, substantially altering sediment distribution (Glassom & Branch, 1997; Rodríguez-Pérez & Green, 2006; Sandsten & Klaassen, 2008; Scott et al., 2012). Mukherjee and Borad (2001) note that waterbirds can influence reservoir eutrophication depending on where they defecate: feeding in water and excreting on land can delay eutrophication, while feeding on land and excreting into water accelerates it. In Pandloli reservoirs, egrets and herons were estimated to remove 13.47 t of N and 3.98 t of P annually in their food, demonstrating the potential importance of waterbirds in eutrophication processes in Matar taluka, Kheda district, Gujarat. Colonial waterbirds also markedly affect vegetation at colony sites (Kolb, Jerling, & Hambäck, 2010; Telfair & Bister, 2004), for example, opening dense reedbeds and increasing habitat heterogeneity that benefits other species.

8.4.9 Migratory birds' role in dispersal of seeds, invertebrates, and nonpathogenic microbes

Terrestrial birds are well known for dispersing seeds and are major providers of this ecosystem service (Sekercioglu, 2006). About 36% of 135 extant terrestrial bird families are largely frugivorous (Herrera, 2002), though this excludes Anatidae, rails, shorebirds, and other aquatic birds that disperse propagules of nonfruiting plants (Green et al., 2002). Historically, dispersal of aquatic plants by waterfowl received little attention, but recent progress over the past decade has improved our understanding. For instance, nonanecdotal studies of dispersal by waterbirds from Africa or Asia were scarce until recently (Aoyama et al., 2012 for seabirds). Waterbirds are now recognized as vectors for a broad suite of aquatic invertebrates—including mollusks, bryozoans, crustaceans, dipterans, rotifers and annelids (Frisch et al., 2007; Green & Figuerola, 2005). These organisms cannot disperse between lakes and rivers on their own or effectively by wind, making waterbirds a primary means of gene flow among populations (Figuerola et al., 2005). Waterbirds also transport phytoplankton, diatoms, ciliates, and spores of bacteria, archaea, and fungi (Brito-Echeverria et al., 2009; Figuerola & Green, 2002; Green et al., 2008; Schlichting, 1960; Thornton, 1971). Nonetheless, the relative importance of bird-mediated dispersal compared to other modes such as wind remains poorly quantified globally, including in India (Wilkinson et al., 2012).

Anatidae and other waterbirds are crucial for colonization and recovery of new and restored wetlands by aquatic plants and animals. Thousands of plant and invertebrate species likely benefit from dispersal via waterbirds through colonization of suitable habitats, gene flow, enhanced germination, and escape from high-mortality areas (Bilton et al., 2001; Brochet et al., 2009; Frisch et al., 2007). In India, the invasive water hyacinth *Eichhornia crassipes* in Brahmaputra floodplains has altered freshwater plant communities and expanded rapidly—waterbirds contribute to seed dispersal—reducing light penetration, lowering productivity, and causing loss of native species (Lahon et al., 2023).

8.5 Conservation policies and initiatives in India

India's commitment to conserving migratory birds and wetlands is reflected in its legislative and international engagements:

- **Wildlife Protection Act (1972):** Provides legal cover for protected species, including migratory birds.
- **Wetlands (Conservation and Management) Rules, 2017:** Emphasize conservation through wise use and regulated development.

- **Ramsar Convention on Wetlands (1971):** India is a contracting party, committed to the wise use of wetlands.
- **Convention on Migratory Species (CMS):** Promotes global cooperation in safeguarding migratory species.

8.5.1 Wetland conservation policies in India as initiated by the Ministry of Environment, Forest and Climate Change

8.5.1.1 Institutional framework in India

The Ministry of Environment, Forest and Climate Change (MoEFCC) serves as the primary authority responsible for developing and implementing policies related to the conservation and sustainable management of wetlands across India. A dedicated Wetlands Division within the Ministry oversees and coordinates the government's various initiatives in this domain.

8.5.1.2 Policy framework by the Ministry of Environment, Forest and Climate Change

India's commitment to wetland conservation is anchored in the National Environment Policy (2006), which recognizes wetlands as integral components of the country's freshwater resources. The policy emphasizes their inclusion in development planning, promotion of sustainable management practices, encouragement of ecotourism, and the establishment of appropriate regulatory mechanisms.

Several legal instruments strengthen the national regulatory framework for wetlands. These include the Indian Forest Act (1927), Wildlife (Protection) Act (1972), Indian Fisheries Act (1897), Water (Prevention and Control of Pollution) Act (1974), and the Biological Diversity Act (2002). Together, they provide the foundation for the protection and sustainable use of wetlands within forests, protected areas, and other ecologically significant landscapes.

Further, the National Wildlife Action Plan (2017–31) identifies inland aquatic ecosystems as a key priority area, recommending the establishment of a national wetlands mission and a comprehensive biodiversity register. Similarly, the National Action Plan on Climate Change integrates wetland conservation into its National Water Mission and Green India Mission, acknowledging their role in climate resilience and ecosystem stability.

8.5.2 Integrated wetland management in India

To assist states in implementing conservation and management strategies, the MoEFCC launched a national program for wetlands in 1986. This initiative evolved into the National Plan for Conservation of Aquatic Ecosystems (NPCA), which currently covers both rural and urban wetlands.

Wetlands are protected under several central legislations, including the Forest (Conservation) Act (1980) and the Wildlife (Protection) Act (1972). In addition, the Wetlands (Conservation and Management) Rules, 2017, formulated under the Environment (Protection) Act, 1986, provide a structured regulatory framework. These rules mandate the creation of State and Union Territory Wetland Authorities, which function as the primary policy and regulatory bodies for wetland management at the regional level.

8.5.2.1 Indian wetland inventory and data system

A national wetlands inventory has been prepared using remote sensing technology by the Space Applications Centre. The resulting National Wetland Atlas provides spatial data at a scale of 1:50,000, supporting evidence-based decision-making and policy development.

8.5.2.2 Knowledge and research initiatives on wetlands in India

In 2021 the Ministry launched a National Wetlands Portal, designed as a comprehensive knowledge platform for policymakers, site managers, researchers, students, and citizens. The portal offers information on wetland resources, health cards for over 600 sites, capacity-building materials, awareness campaigns, and e-learning resources to promote informed participation in wetland management.

8.5.2.3 Participatory management: a society approach

To encourage community participation, the MoEFCC introduced Mission Sahbhagita in May 2022. The initiative adopts a "whole-of-government, whole-of-society" approach, engaging local communities, citizens, and key stakeholders in wetland conservation and sustainable use.

Building on this participatory model, the Ministry launched the "Save Wetlands" Campaign in 2023 to raise public awareness, expand the network of wetland mitras (friends of wetlands), and promote citizen partnerships. The campaign aligns with the Mission LiFE (Lifestyle for Environment) movement, which promotes environmentally responsible behavior among citizens.

In June 2023, the MoEFCC introduced the Amrit Dharohar Initiative, focusing on the conservation and sustainable utilization of Ramsar sites. This three-year program targets four key components—species and habitat conservation, wetland-based livelihoods, nature tourism, and carbon sequestration—to enhance the ecological and socio-economic value of India's wetlands.

8.5.2.4 Communication and outreach

The Ministry conducts extensive outreach activities to foster awareness about the ecological and economic importance of wetlands. Annual celebrations of World Wetlands Day are held at national and state levels, complemented by publications such as the "Anup" newsletter, which provides updates on national and regional conservation efforts. The Ministry also supports the development of Wetland Interpretation Centers, designed as hubs for education, communication, and public engagement.

8.5.2.5 Private sector and partnership for wetland management and conservation

Recognizing the role of businesses in biodiversity protection, the MoEFCC has promoted private-sector collaboration in wetland management. Through the Indian Business and Biodiversity Initiative (IBBI), developed jointly with the Confederation of Indian Industry (CII), companies are encouraged to commit to sustainable practices. In 2022, the Ministry launched the India Wetland Coalition (IWC) to further strengthen partnerships that address nature loss and promote climate resilience through corporate engagement.

8.5.2.6 Capacity building and technical support

The Ministry places strong emphasis on developing the capacity of wetland managers and stakeholders. Under Mission Sahbhagita, regional workshops and training programs are organized to share best practices and address management challenges. The National Wetlands Portal hosts online training courses, manuals, and operational guidelines that assist local authorities in implementing regulatory and conservation measures effectively.

8.6 Recommendations and the way forward

To ensure the sustainable coexistence of wetlands and migratory birds, the following actions are recommended:

- **Strengthen monitoring and research**, including satellite tagging and long-term population studies.
- **Restore degraded wetlands** through hydrological rehabilitation and native vegetation planting.
- **Enhance community participation** in bird conservation through ecotourism and livelihood incentives.
- **Foster transboundary cooperation**, particularly among Central Asian Flyway countries, to protect interconnected habitats.

8.7 Conclusion

Migratory birds are more than seasonal visitors; they are integral to the structure and function of wetland ecosystems. Indian wetlands, particularly Ramsar-designated sites, provide indispensable services to these birds, while the birds reciprocate through ecological contributions that sustain wetland productivity and biodiversity. However, research authenticating such information is scanty and study properly. Therefore the need of the hour in the Indian research community is to pay attention to the beneficial roles migratory waterbirds play in ecosystem functioning and their well-being. Continued protection, restoration, and research into this dynamic relationship are crucial for the future of both wetlands and migratory species in India and beyond.

References

Alam, I., Majumder, A., Naskar, A., & Maheswaran, G. (2023). Bird observations from Pulicate Lake Bird Sanctuary and the adjoining areas in India. *Records of the Zoological Survey of India, 123*(2), 233–249. https://doi.org/10.26515/rzsi/v123/i2/2023/172562.

Alerstam, T. (1990). *Bird Migration*. Cambridge University Press.

Ali, A. (1996). A concise review of chironomid midges (Diptera: Chironomidae) as pests and their management. *Journal of Vector Ecology, 21*, 105–121.

Allard, K., & Gilchrist, H. G. (2002). Kleptoparasitism of herring gulls taking eider eggs by Canada Geese. *Waterbirds, 25*, 235–238.

Allouche, S., & Saudin, P. (2001). Effects of avian predation threat, water flow and cover on growth and habitat use by chub, *Leuciscus cephalus*, in an experimental stream. *Oikos, 94*, 481–492.

Anastacio, P. M., Parente, V. S., & Correia, A. M. (2005). Crayfish effects on seeds and seedlings: Identification and quantification of damage. *Freshwater Biology, 50*, 697–704.

Aoyama, Y., Kawakami, K., & Chiba, S. (2012). Seabirds as adhesive seed dispersers of alien and native plants in the oceanic Ogasawara Islands, Japan. *Biodiversity and Conservation, 21*, 2787–2801.

Baker, A. J., Gonzalez, P. M., Piersma, T., Niles, L. J., deLima Serraro do Nascimento, I., Atkinson, P. W., Clark, N. A., Minton, C. D. T., Peck, M. K., & Aarts, G. (2004). Rapid population decline in Red Knots: Fitness consequences of decreased refuelling rates and late arrival in Delaware Bay. In: *Proceedings of the Royal Society of London. B. 271*, 875–882.

Balachandran, S., Sinha, T., & Manakadan, R. (2018). *Indian bird migration atlas*. Bombay natural history society, Mumbai and Oxford University Press pp. xvi + 216.

Baxter, C. V., Fausch, K. D., Muramaki, M., & Chapman, P. L. (2004). Fish invasion restructures stream and forest food webs by interrupting reciprocal prey subsidies. *Ecology, 85*, 2656–2663.

Bazely, D. R., & Jefferies, R. L. (1985). Goose feces - a Source of nitrogen for plant-growth in a grazed salt-marsh. *Journal of Applied Ecology, 22*, 693–703.

Berthold, P. (2001). *Bird migration. A general survey*. Oxford University Press.

Best, R. J., & Arcese, P. (2009). Exotic herbivores directly facilitate the exotic grasses they graze: Mechanisms for an unexpected positive feedback between invaders. *Oecologia, 159*, 139–150.

Biber, J.-P., & Salathe, T. (1991). Threats to migratory birds. In T. Salathe (Ed.). *Conserving migratory birds* (pp. 17–35). International Council for Bird Preservation (Technical Publication No. 12).

Bilton, D. T., Freeland, J. R., & Okamura, B. (2001). Dispersal in freshwater invertebrates. *Annual Review of Ecology and Systematics, 32*, 159–181.

Boros, E., Nagy, T., Pigniczki, C., Kotyman, L., Balogh, K. V., & Voros, L. (2008b). The effect of aquatic birds on the nutrient load and water quality of soda pans in hungary. *Acta Zoologica Academiae Scientiarum Hungaricae, 54*, 207–224.

Boros, Forro, E., l, Gere, G., Kiss, O., Voros, L., & Andrikovics, S. (2008a). The role of aquatic birds in the regulation of trophic relationships of continental soda pans in hungary. *Acta Zoologica Academiae Scientiarum Hungaricae, 54*, 189–206.

Brito-Echeverria, J., Lopez-Lopez, A., Yarza, P., Anton, J., & Rossello-Mora, R. (2009). Occurrence of *Halococcus* spp. in the nostrils salt glands of the seabird *Calonectris diomedea. Extremophiles: Life Under Extreme Conditions, 13*, 557–565.

Britton, R. H., & Moser, M. E. (1982). Size specific predation by herons and its effect on the sex-ratio of natural-populations of the mosquito fish gambusia-affinis baird and girard. *Oecologia, 53*, 146–151.

Brochet, A. L., Guillemain, M., Fritz, H., Gauthier-Clerc, M., & Green, A. J. (2009). The role of migratory ducks in the long-distance dispersal of native plants and the spread of exotic plants in Europe. *Ecography, 32*, 919–928.

Buckeridge, K. M., & Jefferies, R. L. (2007). Vegetation loss alters soil nitrogen dynamics in an Arctic salt marsh. *Journal of Ecology, 95*, 283–293.

Burdon, F. J., & Harding, J. S. (2008). The linkage between riparian predators and aquatic insects across a streamresource spectrum. *Freshwater Biology, 53*, 330–346.

Cargill, S. M., & Jefferies, R. L. (1984). The Effects Of Grazing By Lesser Snow Geese On The Vegetation Of A Sub-arctic Salt-marsh. *Journal of Applied Ecology, 21*, 669–686.

Carpenter, S. R., & Lodge, D. M. (1986). Effects of submersed macrophytes on ecosystem processes. *Aquatic Botany, 26*, 341–370.

Chacko, P. I., Abraham, J. G., & Andal, R. (1953). Report on survey of the flora, fauna and fisheries of the Pulicat Lake, Madras State, India, 1951–52 (Fresh Water Fisheries Biological Station, Madras, No. 8).

Champion, H. G., & Seth, S. K. (1968). *A revised survey of forest types of India*. Government of India Press.

Dirksen, S., Boudewijn, T. J., Noordhuis, R., & Marteijn, E. C. L. (1995). Cormorants *Phalacrocorax carbo sinensis* in shallow eutrophic fresh-water lakes - prey choice and fish consumption in the non-breeding period and effects of large-scale fish removal. *Ardea, 83*, 167–184.

Fabricius, E., & Norgren, H. (1987). *Lär känna kandagåsen* [Get to know the Canada Goose; in Swedish]. Svenska Jägareförbundet [Swedish Sportsmens`Association]. *Stockholm*, 58.

Farnsworth, M. L., Miller, R. S., Pedersen, K., Lutman, M. W., Swafford, S. R., Riggs, P. D., & Webb, C. T. (2012). Environmental and demographic determinants of avian influenza viruses in waterfowl across the contiguous United States. *PLoS One, 7*.

Figuerola, J., & Green, A. J. (2002). Dispersal of aquatic organisms by waterbirds: A review of past research and priorities for future studies. *Freshwater Biology, 47*, 483–494.

Figuerola, J., Green, A. J., & Michot, T. C. (2005). Invertebrate eggs can fly: Evidence of waterfowl-mediated gene flow in aquatic invertebrates. *American Naturalist, 165*, 274–280.

Fonteneau, F., Paillisson, J. M., Kinsella, J. M., Latraube, F., & Marion, l (2009). First description of the helminth community in the Whiskered Tern *Chlidonias hybrida* (Lariforma: Sternidae) in France. *Revue D Ecologie-La Terre Et La Vie, 64*, 79–84.

Frisch, D., Green, A. J., & Figuerola, J. (2007). High dispersal capacity of a broad spectrum of aquatic invertebrates via waterbirds. *Aquatic Sciences, 69*, 568–574.

Gauthier, G., Giroux, J. F., & Lochefort, l (2006). The impact of goose grazing on arctic and temperate wetlands. *Acta Zoologica Sinica, 52*, 108–111.

Gawlik, D. E. (2002). The effects of prey availability on the numerical response of wading birds. *Ecological Monographs, 72*, 329–346.

Glassom, D., & Branch, G. M. (1997). Impact of predation by greater flamingos *Phoenicopterus ruber* on the meiofauna, microflora, and sediment properties of two southern African lagoons. *Marine Ecology Progress Series, 150*, 1–10.

Gliwicz, Z. M., & Jachner, A. (1992). Diel migrations of juvenile fish - A ghost of predation past or present. *Archiv für Hydrobiologie, 124*, 385–410.

Green, A. J., & Figuerola, J. (2005). Recent advances in the study of long-distance dispersal of aquatic invertebrates via birds. *Diversity and Distributions, 11*, 149–156.

Green, A. J., Figuerola, J., & Sánchez, M. I. (2002). Implications of waterbird ecology for the dispersal of aquatic organisms. *Acta Oecologica-International Journal of Ecology, 23*, 177–189.

Green, Aa. J., Jenkins, K. M., Bell, D., Morris, P. J., & Kingsford, R. T. (2008). The potential role of waterbirds in dispersing invertebrates and plants in arid Australia. *Freshwater Biology, 53*, 380–392.

Grimm, B., Paill, W., & Kaiser, H. (2000). The "Spanish slug": autecology, predators and wild plants as food plants. *Förderungsdienst, 48*, 11–16.

Van Groenigen, J. W., Burns, E. G., Eadie, J. M., Horwath, W. R., & Van Kessel, C. (2003). Effects of foraging waterfowl in winter flooded rice fields on weed stress and residue decomposition. *Agriculture, Ecosystems & Environment, 95*, 289–296.

Hafner, H., & Britton, R. H. (1983). Changes of foraging sites by nesting little egrets (*Egretta garzetta* L.) in relation to food supply. *Colonial Waterbirds, 6*, 24–30.

Harris, C. M., Calladine, J. R., Wernham, C. V., & Park, K. J. (2008). Impacts of piscivorous birds on salmonid populations and game fisheries in Scotland: A review. *Wildlife Biology, 14*, 395–411.

Herrera, C. M. (2002). In C. M. Herrera and, & O. Pellmyr (Eds.). *Seed dispersal by vertebrates. In* Plant–animal interactions: an evolutionary approach (pp. 185–208). Blackwell Publishing By.

Hidding, B., Nolet, B. A., De Boer, T., De Vries, P. P., & Klaassen, M. (2010). Above- and below-ground vertebrate herbivory may each favour a different subordinate species in an aquatic plant community. *Oecologia, 162*, 199–208.

Hoyer, M. V., Watson, D. L., Willis, D. J., & Canfield, D. E., Jr. (2009). Fish kills in Florida's canals, creeks/rivers, and ponds/lakes. *Journal of Aquatic Plant Management, 47*, 53–56.

del Hoyo, J., Elliot, A., & Christie, D. (1992). Handbook of the birds of the world. Vol. 1: Ostrich to ducks. Lynx Edicions.

Hubalek, Z. (2004). An annotated checklist of pathogenic microorganisms associated with migratory birds. *Journal of Wildlife Diseases, 40*, 639–659.

Hutchinson, G. E. (1950). *Survey of Contemporary Knowledge of Biochemistry. 3. The Biogeochemistry of Vertebrate Excretion. 96.* Bulletin of the American Museum of Natural History.

Iacobelli, A., & Jefferies, R. L. (1991). Inverse salinity gradients in coastal marshes and the death of stands of salix - the effects of grubbing by geese. *Journal of Ecology, 79*, 61–73.

Isaac-Renton, M., Bennett, J. R., Best, R. J., & Arcese, P. (2010). Effects of introduced Canada geese (*Branta canadensis*) on native plant communities of the Southern Gulf Islands, British Columbia. *Ecoscience, 17*, 394–399.

Islam, M. Z., & Rahmani, A. R. (2004). *Important bird areas in India: Priority sites for conservation.* Indian Bird Conservation Network: Bombay Natural History Society and BirdLife International Pp. xviii + 1133.

Jasmin, J. N., Rochefort, L., & Gauthier, G. (2008). Goose grazing influences the fine-scale structure of a bryophyte community in arctic wetlands. *Polar Biology, 31*, 1043–1049.

Jeppesen, E., Meerhoff, M., Jacobsen, B. A., Hansen, R. S., Søndergaard, M., Jensen, J. P., Lauridsen, T. L., Mazzeo, N., & Branco, C. W. C. (2007). Restoration of shallow lakes by nutrient control and biomanipulationthe successful strategy varies with lake size and climate. *Hydrobiologia, 581*, 269–285.

Kameda, K., Koba, K., Hobara, S., Osono, T., & Terai, M. (2006). Pattern of natural N-15 abundance in lakeside forest ecosystem affected by cormorant-derived nitrogen. *Hydrobiologia, 567*, 69–86.

Kannan, V., & Pandiyan, J. (2012). Shorebirds (Charadriidae) of Pulicat Lake, India with special reference to conservation. *World Journal of Zoology, 7*(3), 178–191.

Kear, J. (1990). *Man and wildfowl.* T & A.D. Poyser.

Kennedy, C. R., Shears, P. C., & Shears, J. A. (2001). Long-term dynamics of Ligula intestinalis and roach Rutilus rutilus: A study of three epizootic cycles over thirty-one years. *Parasitology, 123*, 257–269.

Kolb, G. S., Jerling, L., & Hambäck, P. A. (2010). The impact of cormorants on plant-arthropod food webs on their nesting Islands. *Ecosystems, 13*, 353–366.

Van Kooten, G. C., Withey, P., & Wong, l D. (2011). Bioeconomic modeling of wetlands and waterfowl in western Canada: Accounting for amenity values. *Canadian Journal of Agricultural Economics-Revue Canadienne d´Agroeconomie, 59*, 167–183.

Krishnan, P., & Sampath, V. (1973). *A survey of Pulicat Lake and its fisheries.* Department of Fisheries, Government of Tamil Nadu.

Kumar, A. (2011). Pong dam wetlands and zoological survey of India. *Current Science, 85*(10), 1398.

Kumar, A. (2015). Migratory waterbirds of Himachal Pradesh: Status and conservation. In G. V. Gopi, & S. A. Hussain (Vol. Eds.), *Waterbirds of India, ENVIS Bulletin: Wildlife & Protected Areas: Vol. 16*, (pp. 258–267). pp. 367, Wildlife Institute of India, Dehradun-248001, India.

Kumar, A., & Paliwal, R. (2015). Galloanserae and aquatic neoaves of pong dam wetland, Himachal Pradesh: Status and conservation issues. In M. Rawat, (Ed.). *Aquatic ecosystem: Biodiversity, ecology and conservation* (pp. 307–321). c Springer India. https://doi.org/10.1007/978-81-322-2178-4_18 pp. xi+333.

Kumar, N. V. N., Nagarjuna, A., & Reddy, D. C. (2010). Ecoresiliency and remediation strategies for biodiversity conservation of aquatic and avifauna of Pulicat brackish water lagoon. *World Journal of Fish and Marine Sciences, 2*(5), 389–400.

Kushlan, J. A. (1974). Effects of a natural fish kill on water-quality, plankton, and fish population of a pond in big cypress swamp, florida. *Transactions of the American Fisheries Society, 103*, 235–243.

Kushlan, J. A. (1976). Wading bird predation in a seasonally fluctuating pond. *The Auk, 93*, 464–476.

Lahon, D., Sahariah, D., Debnath, J., Nath, N., Meraj, G., Farooq, M., Kanga, S., Singh, S. K., & Chand, K. (2023). Growth of water hyacinth biomass and its impact on the floristic composition of aquatic plants in a wetland ecosystem of the Brahmaputra floodplain of Assam, India. *PeerJ, 3*, 11. https://doi.org/10.7717/peerj.14811 e14811.

Leah, R. T., Moss, B., & Forrest, D. E. (1980). The role of predation in causing major changes in the limnology of a hyper-Eutrophic Lake. *Internationale Revue der Gesamten Hydrobiologie, 65*, 223–247.

Lefebvre, F., Georgiev, B. B., Bray, R. A., & Littlewood, D. T. J. (2009). Developing a dedicated cestode life cycle database: Lessons from the hymenolepidids. *Helminthologia, 46*, 21–27.

Lewin, W. C., Okun, N., & Mehner, T. (2004). Determinants of the distribution of juvenile fish in the littoral area of a shallow lake. *Freshwater Biology, 49*, 410–424.

Mahapatro, D., Panigrahy, R., Panda, S., & Mishra, R. K. (2016). Malcofaunal diversity of chilika Lake, Odisha, India. *Proceedings of the Zoological Society*. https://doi.org/10.1007/s12595-016-0204-9.

Mann, H., Proctor, V. W., & Taylor, A. S. (1999). Towards a biogeography of north American charophytes. *Australian Journal of Botany, 47*, 445–458.

Mehner, T., Kasprzak, P., & Holker, F. (2007). Exploring ultimate hypotheses to predict diel vertical migrations in coregonid fish. *Canadian Journal of Fisheries and Aquatic Sciences, 64*, 874–886.

Miles, A. K., Lawler, S. P., Dritz, D., & Spring, S. (2002). Effects of mosquito larvicide on mallard ducklings and prey. *Wildlife Society Bulletin, 30*, 675–682.

Mukherjee, A., & Borad, C. K. (2001). Effects of waterbirds on water quality. *Hydrobiologia, 464*, 201–205.

Munster, V. J., Wallensten, A., Baas, C., Rimmelzwaan, G. F., Schutten, M., Olsen, B., Osterhaus, A. D. M. E., & Fouchier, R. A. M. (2005). Mallards and highly pathogenic avian influenza ancestral viruses, northern Europe. *Emerging Infectious Diseases, 11*, 1545–1551.

Nacken, N., & Reise, K. (2000). Effects of herbivorous birds on intertidal seagrass beds in the northern Wadden Sea. *Helgoland Marine Research, 54*, 87–94.

Negro, J. J., Tella, J. L., Blanco, G., Forero, M. G., & Garrido-Fernández, J. (2000). Diet explains interpopulation variation of plasma carotenoids and skin pigmentation in nestling white storks. *Physiological and Biochemical Zoology, 73*, 97–101.

Newton, I. (2003). Geographical patterns in bird migration. In P. Berthold, E. Gwinner, & E. Sonnenschein (Eds.). *Avian Migration*Springerhttps://doi.org/10.1007/978-3-662-05957-9_14.

Nolet, B. A. (2004). Overcompensation and grazing optimisation in a swan-pondweed system? *Freshwater Biology, 49*, 1391–1399.

Nummi, P., Sjöberg, K., Pöysä, H., & Elmberg, J. (2000). Individual foraging behaviour indicates resource limitation: An experiment with mallard ducklings. *Canadian Journal of Zoology, 78*, 1891–1895.

Power, M. E., Dudley, T. L., & Cooper, S. D. (1989). Grazing catfish, fishing birds, and attached algae in a panamanian stream. *Environmental Biology of Fishes, 26*, 285–294.

Pöysä, H., Sjöberg, K., Elmberg, J., & Nummi, P. (2001). Pair formation among experimentally introduced mallards *Anas platyrhynchos* reflects habitat quality. *Annales Zoologici Fennici, 38*, 179–184.

Raghavaiah, P. S. (2007). Wetland birds of Pulicat Lake, India: Seasonal patterns and habitat preferences [Doctoral dissertation, Pondicherry University].

Raj, P. J. S. (2006). *Macro fauna of Pulicat Lake NBA Bulletin. No. 6*. National Biodiversity Authority.

Ramesh, D. A., & Ramachandran, S. (2005). Factors influencing flamingo (*Phoenicopterus roseuis*) distribution in the Pulicat Lagoon ecosystem, India. *Wetlands Ecology and Management, 13*, 69–72. https://doi.org/10.1007/s11273-003-9072-y.

Rodriguez-Perez, H., Green, A. J., & Figuerola, J. (2007). Effects of greater flamingo *Phoenicopterus ruber* on macrophytes, chironomids and turbidity in natural marshes in Doñana, SW Spain. *Fundamental and Applied Limnology, 170*, 167–175.

Rodríguez-Pérez, H., & Green, A. J. (2006). Waterbird impacts on widgeongrass *Ruppia maritima* in a Mediterranean wetland: Comparing bird groups and seasonal effects. *Oikos, 112*, 525–534.

Rollo, S. N., Ferro, P. J., Peterson, M. J., Ward, M. P., Ballard, B. M., & Lupiani, B. (2012). Role of nonmigratory mottled ducks (Anas fulvigula) as sentinels for avian influenza surveillance. *Journal of Zoo and Wildlife Medicine, 43*, 168–170.

Sandsten, H., & Klaassen, M. (2008). Swan foraging shapes spatial distribution of two submerged plants, favouring the preferred prey species. *Oecologia, 156*, 569–576.

Schlichting, H. E. J. (1960). The role of waterfowl in the dispersal of algae. *Transactions of the American Microscopical Society, 79*, 160–166.

Scott, J. J., Renaut, R. W., & Owen, R. B. (2012). Impacts of flamingos on saline lake margin and shallow lacustrine sediments in the Kenya Rift Valley. *Sedimentary Geology, 277–278*, 32–51.

Sekercioglu, C. H. (2006). Increasing awareness of avian ecological function. *Trends in Ecology & Evolution, 21*, 464–471.

Selvanathan, M., & Kaliyamurthy, M. (1972). New records of fish from the Pulicat Lake. *Records of the Zoological Survey of India, 67*, 367–371. https://doi.org/10.26515/rzsi/v67/i1-4/1972/161484.

Somveille, M., Manica, A., Butchart, S. H. M., & Rodrigues, A. S. L. (2013). Mapping global diversity patterns for migratory birds. *PLoS One, 8*(8), e70907. https://doi.org/10.1371/journal.pone.0070907.

Speiser, B., Glen, D., Piggot, S., Ester, A., Davies, K., Castillejo, J., & Coupland, J. (2001). *Slug damage and control of slugs in horticultural crops.* FiBL-Merkblatt. Research Institute of Organic Agriculture (FiBL). http://orgprints.org/515/1/Slugcontrol.pdf.

Starling, F., Lazzaro, X., Cavalcanti, C., & Moreira, R. (2002). Contribution of omnivorous tilapia to eutrophication of a shallow tropical reservoir: Evidence from a fish kill. *Freshwater Biology, 47*, 2443–2452.

Steinmetz, J., Kohler, S. L., & Soluk, D. A. (2003). Birds are overlooked top predators in aquatic food webs. *Ecology, 84*, 1324–1328.

Stewart, D. C., Middlemas, S. J., Gardiner, W. R., Mackay, S., & Armstrong, J. D. (2005). Diet and prey selection of cormorants (*Phalacrocorax carbo*) at Loch Leven, a major stocked trout fishery. *Journal of Zoology, 267*, 191–201.

Sánchez, M. I., Green, A. J., & Alejandre, R. (2006). Shorebird predation affects density, biomass, and size distribution of benthic chironomids in salt pans: An exclosure experiment. *Journal of the North American Benthological Society, 25*, 9–18.

Søndergaard, M., Liboriussen, L., Pedersen, A. R., & Jeppesen, E. (2008). Lake restoration by fish removal: Short- and long-term effects in 36 danish Lakes. *Ecosystems, 11*, 1291–1305.

Tablado, Z., Tella, J. L., Sanchez-Zapata, J. A., & Hiraldo, F. (2010). The paradox of the long-term positive effects of a north american crayfish on a European community of predators. *Conservation Biology, 24*, 1230–1238.

Telfair, P. C., & Bister, T. J. (2004). Long-term breeding success of the cattle egret in Texas. *Waterbirds, 27*, 69–78.

Teo, S. S. (2001). Evaluation of different duck varieties for the control of the golden apple snail (*Pomacea canaliculata*) in transplanted and direct seeded rice. *Crop Protection, 20*, 599–604.

Thangavelu, R., & Sanjeevaraj, P. J. (1985). Exploitation of shell-deposits in Pulicat Lake. *Journal of the Marine Biological Association of India, 27*(1-2), 124–128.

Thornton, M. L. (1971). Potential for long-range dispersal of aquatic phycomycetes by internal transport in birds. *Transactions of the British Mycological Society, 57*, 49.

Trouwborst, A. (2012). Transboundary wildlife conservation in a changing climate: Adaptation of the bonn convention on migratory species and its daughter instruments to climate change. *Diversity, 4*, 258–300. https://doi.org/10.3390/d4030258.

Wacht-Katz, M., Abramsky, Z., Kotler, B., Altstein, O., & Rosenzweig, M. L. (2010). Playing the waiting game: Predator and prey in a test environment. *Evolutionary Ecology Research, 12*, 793–801.

Weber, M. J., & Brown, M. L. (2009). Effects of common carp on aquatic ecosystems 80 years after "carp as a dominant": ecological insights for fisheries management. *Reviews in Fisheries Science, 17*, 524–537.

Wobeser, G. (1997). *Diseases of wild waterfowl* (2nd edition.). Plenum Press 324 pp.

Wicker, A.M., Endres, K.M. Relationship between waterfowl and American coot abundance with submersed macrophytic vegetation in Currituck Sound, North Carolina. Estuaries 18, 1995, 428–431. https://doi.org/10.2307/1352325.

Wallensten, A., Munster, V. J., Latorre-Margalef, N., Brytting, M., Elmberg, J., Fouchier, R.A.M., Fransson, T., Haemig, P.D., Karlsson, M., Lundkvist, Å., Osterhaus, A.D.M.E., Stervander, M., Waldenström, J., Olsen, B. Surveillance of Influenza Virus A in Migratory Waterfowl in Northern Europe. Emerging Infectious Diseases. 13(3), 2007, 404-411.

Whelan, C.J., Wenny, D.G. and Marquis, R.J. Ecosystem Services Provided by Birds. Annals of the New York Academy of Sciences, 1134, 2008, 25-60. https://doi.org/10.1196/annals.1439.003.

Yusa, Y., Sugiura, N., & Wada, T. (2006). Predatory potential of freshwater animals on an invasive agricultural pest, the apple snail *Pomacea canaliculata* (Gastropoda: Ampullariidae), in southern Japan. *Biological Invasions, 8*, 137–147.

Ziegler, U., Seidowski, D., Globig, A., Fereidouni, S. R., Ulrich, R. G., & Groschup, M. H. (2010). Sentinel birds in wild-bird resting sites as potential indicators for West Nile virus infections in Germany. *Archives of Virology, 155*, 965–969.

Chapter 9

Impact of water and soil pollution on the wetland ecological health and dynamics

Yogesh Chandra Nainwal[1] **and Bhaskar Dutt Nainwal**[2]

[1]*Department of Chemistry, Government Post Graduate College Maldevta (Raipur), Dehradun, Uttarakhand, India,* [2]*Department of Chemistry, Doon University, Dehradun, Uttarakhand, India*

9.1 Introduction

A wetland is an area where water covers the soil, either seasonally or permanently. They are characterized by the presence of water at or near the surface, and the water table is usually at or near the land surface for much of the year. Estuarine ecosystems (Mitsch & Gosselink, 2015) serve as habitats for numerous species, including rare and endangered ones. Wetlands are ecosystems that can be utilized to remove pollutants from water bodies. Wetlands are among the most biologically diverse ecosystems, supporting a wide variety of plant and animal species (Reddy & DeLaune, 2008). Wetlands act as natural filters, trapping sediment and filtering pollutants from water that flows through them. They contribute to better water quality by filtering out excess nutrients, heavy metals, and various other contaminants.

Wetlands function like natural sponges, soaking up and storing excess water during heavy rains or storms, which helps lower the risk of flooding in downstream areas. They help manage water flow and minimize the impacts of flooding. Wetlands store significant amounts of carbon, both in their plant biomass and in the soils. They play a crucial role in the global carbon cycle by sequestering carbon dioxide from the atmosphere. Wetlands stabilize shorelines and reduce erosion by absorbing energy from waves and currents.

Coastal wetlands are among the most productive ecosystems, yet some of them are at-risk environments (Navarro et al., 2021; Wang et al., 2016). Coastal wetlands are found in the area where freshwater and marine water generally intermix (Navarro & Rodríguez-Santalla, 2023; Turner & Ellis, 2010). They play a vital role in shielding coastlines from storm surges and wave action, while also significantly contributing to carbon sequestration. In addition to naturally removing pollutants from water bodies (Nainwal, 2024a), wetlands also offer recreational activities such as birdwatching, fishing, and boating. They also hold cultural importance for many communities, serving as sites for traditional practices and spiritual ceremonies. Besides, wetlands are essential ecosystems that provide numerous ecological services, making them vital to both wildlife and human communities. Protecting and restoring wetlands (Zedler & Kercher, 2005) is crucial for maintaining biodiversity, improving water quality, and mitigating the impacts of climate change.

Water pollution in wetlands results from various contaminants, including industrial effluents, agricultural runoff, and municipal waste. These contaminants add an overload of nutrients like nitrogen and phosphorus, heavy metals, pesticides, and organic matter to wetlands. Excess nutrients lead to eutrophication, causing algal blooms and subsequent oxygen depletion, which can suffocate aquatic life like fish and amphibians. Heavy metals and pesticides generally build up in sediments and water bodies, poisoning aquatic life and disrupting the food webs within wetland ecosystems. Pollution can affect water flow and drainage patterns, altering the hydrology critical for wetland plant communities and wildlife.

The cumulative effects of pollution can lead to declines in native species diversity and the expansion of invasive species (Nainwal, 2024b). Soil pollution in wetlands arises from contaminated sediments, often resulting from historical industrial activities, mining, or improper waste disposal. Pollutants present in soils can seep into nearby water bodies, impacting aquatic organisms either directly or indirectly through the food chain. Polluted soil reduces habitat quality for wetland plants and animals, affecting their reproductive success and overall population dynamics. Soil pollution can persist for decades, continuing to impact wetland ecosystems long after initial contamination.

Wetlands: Biodiversity, Health and their Role in Ecosystem Services. DOI: https://doi.org/10.1016/B978-0-443-34222-6.00012-7

Wetlands provide crucial ecosystem services like water purification (Hallin et al., 2015; Maiga et al., 2017; Sumiahadi & Acar, 2018), flood control, and carbon sequestration. Pollution diminishes these services, affecting surrounding ecosystems and human communities. Pollution disrupts the intricate food chain within wetlands, affecting predator–prey dynamics and nutrient cycling. Pollution can lead to declines in native species, including plants, fish, birds, and amphibians, impacting biodiversity and resilience. Healthy wetlands play a vital role in climate resilience. Pollution reduces wetlands' resilience to the effects of climate change, including rising sea levels and extreme weather events.

Water and soil pollution pose significant threats to wetland ecosystems, impacting their ecological health, biodiversity, and ability to provide essential services. Addressing these challenges requires integrated management approaches that prioritize pollution prevention, habitat restoration, and sustainable land use practices.

9.2 Types, sources, and impact of water and soil pollution

Water and soil pollution can originate from both point and nonpoint sources. Point sources of water and soil pollution are specific, identifiable locations from which pollutants are discharged directly into the wetland. These sources typically have a distinct point of entry into the wetland area, like industrial discharge from factories, refineries, and manufacturing facilities, effluents from the sewage system, direct discharge points from which pollutants like chemicals, heavy metals, or pathogens, and accidental spills of oil or chemicals from storage tanks or transportation vessels enter the wetland area.

Point sources are often regulated and monitored by environmental authorities due to their identifiable nature. They can have significant local impacts on wetland health dynamics depending on the volume and nature of pollutants discharged. Nonpoint sources of pollution refer to diffuse sources that are not easily traced to a single point of discharge. These sources are characterized by runoff or seepage that carries pollutants into a broad area like agricultural runoff where excess fertilizers, pesticides, and animal waste washed off from fields reach into nearby streams or rivers, urban runoff in which storm water carrying pollutants such as oil, heavy metals (from vehicles), litter, and chemicals (from roads, lawns, and roofs) reach into drainage systems and water bodies, construction sites where sediments, debris, and chemicals from construction activities enter wetland area during rainfall or erosion events.

Nonpoint sources are challenging to regulate and control due to their diffuse nature and diverse origins. They can contribute to cumulative impacts on water quality, soil texture, and ecosystem health over large areas (Carpenter et al., 1998). Best management practices such as vegetated buffers, erosion control measures, and proper waste management are typically employed to mitigate nonpoint sources of pollution (Wang et al., 2016).

Point sources are identifiable and have specific discharge points, while nonpoint sources are diffuse and originate from multiple, often unpredictable locations. Point sources are subject to stricter regulatory controls and monitoring requirements; in contrast, managing nonpoint sources typically involves comprehensive watershed-level approaches and the implementation of voluntary practices. Both point and nonpoint sources can negatively affect wetland health and damage aquatic ecosystems, although the degree and intensity of impacts depend upon the types and amounts of pollutants present. Addressing pollution through effective management practices is crucial for preserving the integrity of wetland ecosystems and ensuring their vital contributions to ecological balance and human well-being (Flynn et al., 1995; UNESCO, 1971; White & Kaplan, 2017). Tables 9.1–9.3 effectively illustrate the types and sources of water and soil pollution, their respective impacts on the wetland ecosystem, and relevant pollution statistics.

TABLE 9.1 Water pollutant sources and their percent contribution.

Types of pollutants	Source of pollution	Percentage contribution (%)
Nutrients (like nitrogen, phosphorus)	Agricultural runoff	62
Pathogens (microbes)	Sewage	12
Heavy metals (lead, Mercury)	Industrial effluents	21
Microplastics	Urban runoff	10

TABLE 9.2 Soil pollutant, sources, and their percent contribution.

Types of pollutants	Source of pollution	Percentage contribution
Pesticides and herbicides	Agricultural activities	High
Heavy metals	Industrial activities	As per the region
Hydrocarbons (oil, fuels)	Industrial waste	As per the location
General waste (plastics, chemicals)	Urbanization, industries	Varies by location

TABLE 9.3 Pollution statistics.

Statistic	Water pollution	Soil pollution
Global impact	More than two billion people lack safe drinking water	35% of global soils are degraded
Health impact	Two million deaths annually from water-related diseases	Significant impact on food security
US pollution	45% of rivers/lakes are polluted	Varies, but significant

9.3 Effect of pollution on wetland ecosystem

Pollution significantly affects various dynamics of wetland ecosystems, triggering a chain of ecological effects (Jones & Bockstael, 2000). The term **"dynamics"** of wetlands reflects their biological, physical, and chemical interactions that occur within a wetland ecosystem. These dynamics determine how wetlands operate, respond, and adapt to internal and external factors. Pollution disrupts the natural rhythm of these finely balanced ecosystems, reducing the wetland's resilience, productivity, and ecological functioning. Thus gaining insight into wetland dynamics is crucial for preserving their ecological health and the valuable services they offer to both the environment and human populations. The polluted runoff containing heavy metals, nutrients, sediments and oil, blocks soil pores, declining ability of soil to absorb water and groundwater recharging, thereby disrupting not only the hydrological dynamics of wetlands but severely affects the biological, chemical and nutrient dynamics of wetland by accelerating the phenomenon of loss of biodiversity, habitat degradation, disruption of food web, alteration in microbial communities, eutrophication, change in pH and formation of harmful compounds.

Contaminants can damage habitats, rendering them uninhabitable for numerous species and leading to a decline in biodiversity. Altered conditions may favor invasive species, allowing them to outcompete native species and disrupt ecological equilibrium. Excess nutrients from agricultural runoff can lead to algal blooms, which deplete oxygen levels in the water and harm aquatic life (Vitousek et al., 1997). Heavy metals, pesticides, and other toxic substances can accumulate in organisms, posing risks to both wildlife and humans. Contaminated water can result in polluted soil, disrupting nutrient cycles and hindering plant growth (Fig. 9.1).

Polluted water can lead to the contamination of wetland soils, affecting their ability to support plant life and altering nutrient cycling (Table 9.4). Soil pollution can alter soil microbial communities, impacting decomposition and nutrient availability (Fig. 9.2). Pollution can disrupt natural hydrology, changing water levels and flow patterns that are crucial for wetland functions. Runoff can increase sedimentation rates, burying plant life and interfering with natural filtration. Healthy wetlands are significant carbon sinks. Pollution can reduce their capacity to sequester carbon, contributing to climate change. Contaminated wetlands can lead to the spread of diseases among wildlife and humans, especially in communities that rely on these ecosystems for water and food. A decline in the health of fish and plant life can threaten local food supplies and endanger people's livelihoods. Contaminated wetlands may struggle to recover from disturbances, leaving them more susceptible to climate change and other environmental stresses.

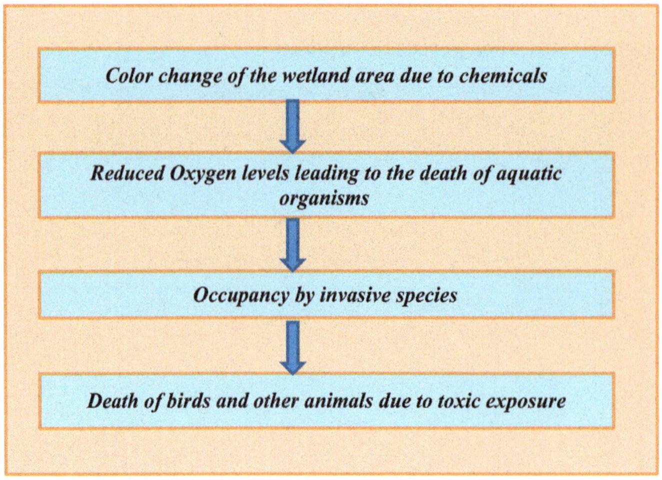

FIGURE 9.1 Impact of polluted water on wetland ecology.

TABLE 9.4 Effect of pollution on the wetland ecosystem.

Types of pollution	Effect on wetland ecosystem	Examples
Nutrient pollution	Algal blooms reduce oxygen levels and lead to a loss of biodiversity	Excess fertilizer runoff
Chemical pollution	Toxicity to flora and fauna, disruption of food chains	Pesticides, heavy metals
Sediment pollution	Habitat alteration, reduced light penetration, and smothering	Erosion, construction runoff
Plastic pollution	Ingestion by wildlife, habitat degradation	Microplastic litter
Thermal pollution	Altered species composition, increased respiration rates	Industrial discharge
Oil pollution	Toxic effects on organisms, degradation of habitat	Oil spills

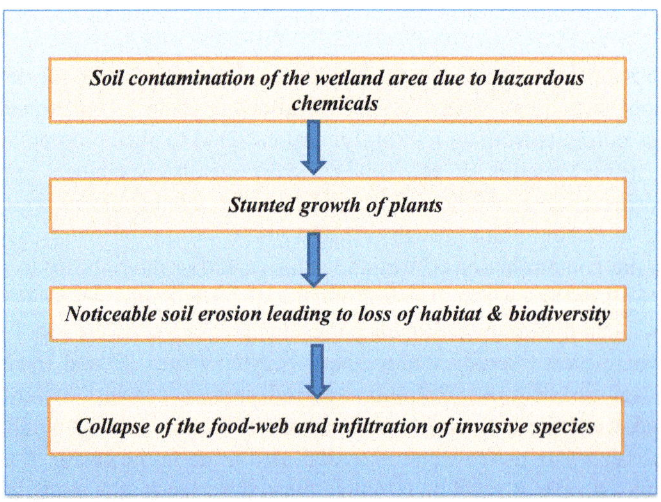

FIGURE 9.2 Impact of soil pollution on wetland ecology.

Wetlands provide essential services like flood control, water purification, and recreational opportunities. Pollution diminishes these benefits, affecting local economies and community well-being. Pollution poses severe threats to the health of wetland ecosystems, highlighting the need for comprehensive management strategies. Effective pollution control, habitat restoration, and community engagement are crucial for safeguarding these vital ecosystems and the services they provide. Addressing pollution in wetlands is essential for preserving biodiversity, protecting human health, and maintaining ecological integrity. The table well summarizes the effect of pollution on the ecosystem.

9.4 Cumulative impacts on flora and fauna in wetland ecology

Cumulative impacts refer to the combined and synergistic effects of multiple stressors or activities on an ecosystem over time. These impacts can arise from various sources, such as pollution, habitat destruction, climate change, invasive species, and resource extraction. Unlike direct impacts that result from a single action or event, cumulative impacts consider the aggregate effects of numerous influences, which can lead to significant alterations in wetland ecosystem structure, function, and health (Fig. 9.3).

Sensitive plant species are often specialized and adapted to specific environmental conditions within wetlands. Wetland plants have developed unique adaptations that allow them to thrive in a harsh environment characterized by waterlogging, low oxygen (anoxic) soils, high salinity, flooding, and contamination due to human activities. One of their key survival strategies is the development of aerenchyma, a specialized tissue that facilitates oxygen transport from aerial parts to submerged roots. Many species develop adventitious roots, which aid in plant stability and help absorb oxygen from above the waterline. In salty environments, such as those inhabited by mangroves and salt marsh plants, excess salt is either expelled through specialized glands or isolated within vacuoles to protect cellular functions. To cope with prolonged flooding, some plants switch to anaerobic respiration, allowing them to generate energy without oxygen. They also produce antioxidants and phenolic compounds to combat oxidative stress and pollution. Structural features such as thick cuticles, waxy coatings, and altered leaf shapes help regulate water retention and improve gas exchange. Furthermore, certain wetland species exhibit vivipary, in which seeds germinate while still attached to the parent plant, thereby improving survival in unstable habitats. All these adaptations not only support the survival of individual plants but also contribute to the strength and stability of wetland ecosystems under both natural and human-induced stress.

Habitat degradation and competition from invasive species are critical threats to their survival. Habitat degradation may result from pollution, altered hydrology, agricultural expansion, industrialization, and urbanization. Degraded habitats may no longer provide the necessary conditions (e.g., water levels, soil quality) for sensitive species to thrive. Contaminated soils can impair plant growth and nutrient uptake, impacting sensitive species that rely on specific soil conditions. Changes in vegetation can lead to altered light, temperature, and moisture levels, further stressing the sensitive plants.

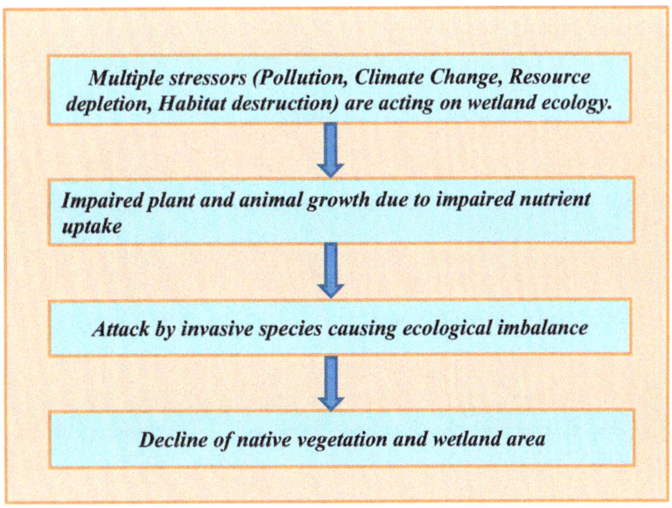

FIGURE 9.3 Cumulative impact of stressors on wetland ecology.

Invasive species often compete with native plants for essential resources such as water, light, and nutrients, leading to declines in sensitive populations. Invasive species grow quickly, thrive in disturbed environments, are easily adaptable to new environments, and have high reproductive rates. Invasive species can alter the ecological balance, changing interactions among plant species and disrupting pollination and seed dispersal processes. Some invasive plants can physically change habitats (e.g., by creating dense thickets), making them unsuitable for sensitive species that require open or specific conditions. The decline of sensitive plant species due to habitat degradation and competition from invasive species underscores the importance of conservation efforts. Protecting and restoring wetland habitats, controlling invasive species, and adopting sustainable land use practices are essential for conserving these vulnerable plant communities and supporting the overall health of the wetland ecosystem. Table 9.5 well explains the cumulative impacts on flora and fauna in wetland ecology.

9.5 Wetland restoration strategies

Phytoremediation is one of the best techniques used for wetland restoration in which plants absorb, degrade, or stabilize contaminants in soil and water (Herath & Vithanage, 2015). It leverages the natural abilities of certain plant species to remediate polluted environments. Wetlands play a vital role in filtering pollutants and improving water quality (McCutcheon & Jørgensen, 2008). Phytoremediation can enhance the ecological health of degraded wetlands by addressing contamination issues while promoting biodiversity (Zhang et al., 2010).

The application of exact phytoremediation mechanisms depends on the types of contaminants and the availability of specific plant species in the area where the technique is to be applied. If the localities are rich in heavy metals like arsenic, lead, cadmium, and uranium, the mechanism of Phytoextraction, also called Phytoaccumulation, can be applied, in which plants absorb these metals through their roots and transport them to their stem and leaves, for example, the Sunflower plant. The plant tissues can now be harvested to obtain these heavy metals. Plants like Vetiver grass can immobilize contaminants in the soil, preventing their migration to groundwater and reducing their bioavailability (phytostabilization) by secreting specific substances through their roots.

Grasses and shrubs are commonly used species that can tolerate heavy metals and stabilize contaminated soils. Certain plants (e.g., Poplar tree) produce specific enzymes that can break down contaminants through metabolic processes, converting them into less harmful substances (Phytodegradation). This process is effective for organic pollutants, including pesticides and hydrocarbons (Schnoor et al., 1995). Microorganisms associated with plant roots degrade contaminants, a process enhanced by plant root exudates (Rhizodegradation). This process is beneficial in the remediation of organic pollutants. Phytoremediation requires selecting plants that survive and thrive in contaminated conditions. Fast-growing plant species can facilitate quicker remediation. Deep-rooted plants can access deeper contaminants and improve soil structure. Planting native species aids in restoring the natural ecosystem and provides vital support for local wildlife. The plants like Cattails (*Typha* spp.) are efficient at removing nutrients and can thrive in wet environments. The plant Bulrush (*Schoenoplectus* spp.) is effective at stabilizing heavy metals. The plant Willow (*Salix* spp.) grows rapidly and is highly effective in enhancing water quality.

TABLE 9.5 Cumulative impacts on flora and fauna in wetland ecology.

Impact type	Flora effects	Fauna effects
Habitat loss	Reduced plant diversity, loss of native species	Decline in species populations, reduced breeding sites
Nutrient overload	Algal blooms, decreased light penetration	Fish kills, changes in food availability
Pollution	Toxicity leading to plant mortality	Bioaccumulation of toxins, reproductive issues
Invasive species	Competition with native plants, habitat alteration	Predation on native species, reduced biodiversity
Climate change	Altered plant community dynamics, Stress on species	Migration patterns disrupted, increased mortality rates
Water quality degradation	Reduced growth and reproduction	Habitat fragmentation, loss of aquatic life

Disposal of exhausted (harvested) plants after phytoremediation is of utmost importance to manage the contaminants they've absorbed. Because these plants can release pollutants back into the wetland environment after they die and decompose, contamination may recur. The application of an effective pollutant management strategy relies on the type of contaminant, the plant species, and the local regulatory framework. For instance, incinerating used plant biomass is an effective method of dealing with organic pollutants, but it can result in the release of harmful gases. Composting harvested plants is a suitable approach for areas with a lower level of organic and nutrient pollution. Heavy metals from exhausted plant biomass can be recovered using chemical and thermal extraction methods. Disposal of harmful and radioactive pollutants can be done by landfilling harvested plant biomass in a safe, secure location. Encapsulation or solidification (embedding plant material in concrete or any other solid surface to prevent leaching) can also be followed for toxic and radioactive substances (Fig. 9.4).

Phytoremediation can take longer than other remediation methods, requiring patience and long-term commitment. Careful monitoring is necessary to avoid bioaccumulation in food webs. Variability in site conditions can affect the success of the strategy, requiring adaptive management approaches. Phytoremediation is a promising strategy for wetland restoration, utilizing the natural capabilities of plants to remediate contaminated environments (McCutcheon & Schnoor, 2004; Susarla et al., 2002). By carefully selecting appropriate species and implementing a well-designed management plan, phytoremediation can effectively improve wetland health, enhance biodiversity, and restore essential ecosystem services. This sustainable approach not only addresses pollution but also fosters resilience in wetland ecosystems.

The bioremediation technique involves the use of microorganisms, plants, or fungi to degrade, remove, or stabilize contaminants from the environment (Chen et al., 2021). It is an environment-friendly approach that harnesses natural processes to clean up polluted sites. Wetlands are critical ecosystems for water purification, biodiversity, and climate regulation (Euliss et al., 2006; Smith & Hargreaves, 2015). Bioremediation can effectively restore the ecological health of degraded wetlands by addressing various types of contamination (Gibert et al., 2005). Indigenous microorganisms break down contaminants naturally over time (Kugler et al., 2022; Kumar et al., 2023), requiring minimal intervention. Some microbial strains are quite effective at degrading particular pollutants, such as hydrocarbons, heavy metals, or nutrients (Hazen, 2010). Certain fungi can break down complex organic pollutants and heavy metals through enzymatic processes, thereby enhancing contaminant degradation in wetland soils (Fungal bioremediation).

Factors such as temperature, moisture, and pH can influence microbial and plant activity. Bioremediation may take longer than other methods, necessitating patience and ongoing commitment. Some remediation processes can create intermediate byproducts that may require further assessment and management. Bioremediation is a viable and sustainable strategy for wetland restoration, leveraging natural biological processes to remediate contaminated environments. By selecting appropriate microorganisms, plants, and fungi and implementing a tailored management plan, bioremediation can effectively restore the ecological integrity of wetlands, improve water quality, and enhance biodiversity. This approach not only addresses pollution but also promotes long-term ecological resilience in these critical ecosystems.

Aeration is also a good wetland restoration technique in which the oxygen content is increased in water or soil (Allen et al., 2023; Wallace et al., 2007). Increased dissolved oxygen helps in the remediation of pollutants from the wetland (Chen et al., 2011; Dong et al., 2012). Aeration enhances microbial activity, promotes plant growth (Butterworth et al., 2013; Uggetti et al., 2016), and improves overall ecosystem health by controlling nutrient levels (Garcia Chance & White, 2018;

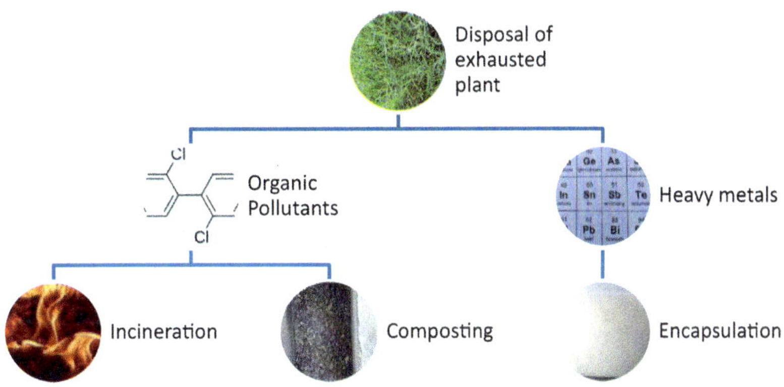

FIGURE 9.4 Pollutant management through Phytoremediation.

Nivala et al., 2019). Aeration can help mitigate problems associated with low oxygen levels, such as organic matter buildup and the release of harmful gases (e.g., methane). It can facilitate the breakdown of pollutants, enhance nutrient cycling, and support the health of wetland flora and fauna (Arias et al., 2015; Chyan et al., 2016; Lin et al., 2015). Aerators (e.g., paddle wheels, aeration fountains) are used to agitate water, increasing oxygen exchange. The technique restores natural hydrological patterns to promote water movement and oxygenation.

Planting certain species that promote aeration through their root structures can enhance oxygen diffusion into the soil. Aerobic microorganisms can thrive in environments with higher oxygen levels and aid in the breakdown of organic matter. Incorporation of organic materials (e.g., compost) can improve soil structure and aeration capacity. Factors such as temperature, seasonality, and weather can affect the efficiency of aeration methods.

Mechanical aeration can be costly, requiring consideration of budget and resource allocation. Too much aeration can disrupt anaerobic processes that some species and nutrient cycles rely on, so a balanced approach is necessary. Aeration is a valuable strategy for wetland restoration that can enhance ecological health by improving oxygen availability, promoting the breakdown of organic matter, and supporting plant and microbial life.

By implementing appropriate aeration techniques tailored to specific site conditions, restoration efforts can significantly enhance the functioning and resilience of wetland ecosystems, ultimately leading to improved water quality and biodiversity. Apart from this, some other strategies depicted below in Table 9.6 can be adopted to improve wetland health dynamics.

Sedimentation also plays a crucial role in wetland restoration (O'Brien et al., 2004; Rodríguez-Santalla & Navarro, 2021). It affects the physical, chemical, and biological processes of the ecosystem. Sedimentation promotes soil formation, water retention, maintenance of water levels, nutrient recycling, flood control, increased biodiversity, decreased soil erosion, and enhanced carbon sequestration. Besides, excessive sedimentation has drawbacks, as it reduces light availability for underwater vegetation, thereby hindering their growth (Rodríguez-Santalla et al., 2021). The process of sedimentation must be carried out very carefully, as excessive sedimentation can lead to a significant reduction in the area's biodiversity.

TABLE 9.6 Wetland restoration strategies.

Restoration strategy	Description	Benefits	Process limitations
Reforestation	Plantation of native trees and shrubs	Biodiversity enhancement and soil stabilization	It takes a long time for trees. Sometimes, reforestation may lead to disruption of the hydrological conditions of the area
Hydrological restoration	Reestablishing natural water flow	Improves water quality and restores aquatic habitats	Advanced engineering is required for prolonged sustainability
Invasive species management	Removal of nonnative species	Protects native biodiversity and improves ecosystem health	Continuous vigilance and hard work are required for the thorough removal of invasive species
Soil amendments	Adding organic matter or nutrients	Enhancement in soil fertility and promotion of plant growth	Excess nutrients may cause alterations in the microbial communities of native soil
Habitat creation	Constructing new wetland areas	Increased habitat availability for wildlife	It requires high labor and cost
Monitoring and adaptive management	Promoting sedimentation in wetland areas in a controlled way	Increased biodiversity, water retention, water level management, soil formation, decreased soil erosion, flood control, habitat creation, and thriving carbon sequestration	It needs prolonged funding and process expertise for sustainability
Community engagement	Involving local stakeholders	Fosters public awareness	It entirely depends on the community's interests, which may differ

9.6 Conclusion

Water and soil pollution significantly deteriorate the ecological health and dynamics of wetlands. The introduction of pollutants disrupts nutrient cycles, degrades habitat quality, diminishes biodiversity, and leads to declines in both plant and animal species. Contaminated water sources can create toxic environments, impacting species and the overall functionality of the ecosystem. Furthermore, the resilience of wetlands to climate change and other environmental stressors is compromised, frightening their ability to provide essential services such as water filtration, flood control, and carbon sequestration. Therefore mitigating pollution through reliable and effective management practices is essential to preserving the integrity of wetland ecosystems, ensuring their vital contributions to environmental quality and human welfare.

Declarations

Competing interest: The authors declare no competing interests.

Acknowledgment: The authors are highly grateful to Prof. Vinod Prakash Aggarawal, Principal, Government Post Graduate College Maldevta, Raipur (Dehradun), and Prof. Surekha Dangwal, Vice Chancellor, Doon University Dehradun, for providing their full support to complete this study.

Funding: No funding was received for this study.

References

Allen, D. J., Farrell, M., Huang, J., Nair, R., & Mashtare, M. L. (2023). Artificial aeration of an overloaded constructed wetland improves hypoxia but does not ameliorate high nitrogen loads. *Journal of Environmental Management, 326*, 116625. https://doi.org/10.1016/j.jenvman.2022.116625.

Arias, C. A., van Oirschot, D., Kilian, R., & Brix, H. (2015). Design and performance evaluation of a highly loaded aerated treatment wetland managing effluents from a food processing industry in Denmark. *Water Practice and Technology, 10*(4), 644–651.

Butterworth, E., Dotro, G., Jones, M., & Nivala, J. (2013). Effect of artificial aeration on tertiary nitrification in a full-scale subsurface horizontal flow constructed wetland. *Ecological Engineering, 54*, 236–244.

Carpenter, S. R., Caraco, N. F., Correll, D. L., Howarth, R. W., Sharpley, A. N., & Smith, V. H. (1998). Nonpoint pollution of surface waters with phosphorus and nitrogen. *Ecological Applications, 8*(3), 559–568.

Chen, J., Li, X., Jia, W., Zhou, Q., & Liu, Y. (2021). Promotion of bioremediation performance in constructed wetland microcosms for acid mine drainage treatment by using organic substrates and supplementing domestic wastewater and plant litter broth. *Journal of Hazardous Materials, 404*, 124125. https://doi.org/10.1016/j.jhazmat.2020.124125.

Chen, Y., Wen, Y., Cheng, J., Li, Y., & Zhou, Q. (2011). Effects of dissolved oxygen on extracellular enzymes activities and transformation of carbon sources from plant biomass: Implications for denitrification in constructed wetlands. *Bioresource Technology, 102*(3), 2433–2440. https://doi.org/10.1016/j.biortech.2010.10.122.

Chyan, J. M., Lin, C. J., Lin, Y. C., & Chou, Y. A. (2016). Improving the removal performance of pollutants by artificial aeration and flow rectification in a free water surface constructed wetland. *International Biodeterioration & Biodegradation, 113*, 146–154.

Dong, H., Qiang, Z., Li, T., & Jin, H. (2012). Effect of artificial aeration on the performance of vertical-flow constructed wetland treating heavily polluted river water. *Journal of Environmental Sciences, 24*(4), 596–601. https://doi.org/10.1016/S1001-0742(11)60804-8.

Euliss, N. H., Smith, L. M., Wilcox, D. A., & Mushet, D. M. (2006). Wetland restoration: How do we define success? *Wetlands, 26*(4), 665–678.

Flynn, K. M., McKee, K. L., & Mendelssohn, I. A. (1995). Recovery of freshwater marsh vegetation after a saltwater intrusion event. *Oecologia, 103*, 63–72.

Garcia Chance, L. M., & White, S. A. (2018). Aeration and plant coverage influence floating treatment wetland remediation efficacy. *Ecological Engineering*. https://doi.org/10.1016/j.ecoleng.2018.07.011.

Gibert, O., De Pablo, J., Cortina, J. L., & Ayora, C. (2005). Municipal compost-based mixture for acid mine drainage bioremediation: Metal retention mechanisms. *Applied Geochemistry, 20*(9), 1648–1657. https://doi.org/10.1016/j.apgeochem.2005.04.012.

Hallin, S., Hellman, M., Choudhury, M. I., & Ecke, F. (2015). Relative importance of plant uptake and plant-associated denitrification for the removal of nitrogen from mine drainage in subarctic wetland. *Water Research, 85*, 377–383. https://doi.org/10.1016/j.watres.2015.08.060.

Hazen, T. C. (2010). In situ groundwater bioremediation. In K. N. Timmis (Ed.). *Handbook of hydrocarbon and lipid microbiology, part 24* (pp. 2584–2596). Springer-Verlag ISBN: 978-3-540-77587-4.

Herath, I., & Vithanage, M. (2015). Phytoremediation in constructed wetlands. In Environmental Applications of Nanomaterials. https://doi.org/10.1007/978-3-319-10969-5_21.

Jones, K. B., & Bockstael, N. E. (2000). Spatial dynamics of wetland ecosystems. In *Wetlands: A Threatened Ecosystem*. Springer, 95–115.

Kugler, A., Brigmon, R. L., Friedman, A., et al. (2022). Bioremediation of copper in sediments from a constructed wetland ex-situ with the novel bacterium *Cupriavidus basilensis* SRS. *Scientific Reports, 12*(1), 17615. https://doi.org/10.1038/s41598-022-20930-0.

Kumar, V., Bera, T., Roy, S., et al. (2023). Investigating bioremediation capabilities of a constructed wetland through spatial successional study of the sediment microbiome. *npj Clean Water, 6*, 8. https://doi.org/10.1038/s41545-023-00225-1.

Lin, J. L., Tu, Y. T., Chiang, P. C., Chen, C. L., & Wang, Y. T. (2015). Using aerated gravel-packed contact bed and constructed wetland system for polluted river water purification: A case study in Taiwan. *Journal of Hydrology, 525*, 400–408. https://doi.org/10.1016/j.jhydrol.2015.03.049.

Maiga, Y., Nikiema, J., & Cissé, G. (2017). Constructed wetlands. In Water and sanitation for the 21st century: Health and microbiological aspects of excreta and wastewater management. https://doi.org/10.14321/waterpathogens.66.

McCutcheon, S. C., & Jørgensen, S. E. (2008). Phytoremediation. *Ecological Engineering,* 2751–2766.

McCutcheon, S. C., & Schnoor, J. L. (2004). *Phytoremediation: Transformation and control of contaminants.* John Wiley & Sons.

Mitsch, W. J., & Gosselink, J. G. (2015). *Wetlands* (5th ed.). John Wiley & Sons.

Nainwal, Y. C. (2024a). Cost analysis in terms of wastewater treatment and sustainable development. In *Water, the environment, and the sustainable development goals.* Elsevier, 209–220. https://doi.org/10.1016/B978-0-443-15354-9.00020-7.

Nainwal, Y. C. (2024b). Detailed overview of techniques used across the globe for the remediation of pollutants from polluted water. In *Environmental metagenomics, water quality and suggested remediation measures of polluted waters: A combined approach.* Elsevier, 49–60. https://doi.org/10.1016/B978-0-443-13659-7.00015-1.

Navarro, N., Abad, M., Bonnail, E., & Izquierdo, T. (2021). The arid coastal wetlands of Northern Chile: Towards an integrated management of highly threatened systems. *Journal of Marine Science and Engineering, 9*(9), 948. https://doi.org/10.3390/jmse9090948.

Navarro, N., & Rodríguez-Santalla, I. (2023). Coastal wetlands. *Journal of Marine Science and Engineering, 11*, 767. https://doi.org/10.3390/jmse11040767.

Nivala, J., Abdallat, G., Aubron, T., et al. (2019). Vertical flow constructed wetlands for decentralized wastewater treatment in Jordan: Optimization of total nitrogen removal. *Science of the Total Environment, 671*, 495–504.

O'Brien, J. M., Rejmánková, E., Weaver, K., & Smith, J. (2004). The role of sediment in wetland restoration. *Ecological Engineering, 22*(4), 329–339.

Reddy, K. R., & DeLaune, R. D. (2008). *Biogeochemistry of Wetlands: Science and applications.* CRC Press.

Rodríguez-Santalla, I., Díez-Martínez, A., & Navarro, N. (2021). Vulnerability analysis of the Riumar Dune Field in El Garxal Coastal Wetland (Ebro Delta, Spain). *Journal of Marine Science and Engineering, 9*(6), 601. https://doi.org/10.3390/jmse9060601.

Rodríguez-Santalla, I., & Navarro, N. (2021). Main threats in Mediterranean coastal wetlands: The Ebro Delta case. *Journal of Marine Science and Engineering, 9*(11), 1190. https://doi.org/10.3390/jmse9111190.

Schnoor, J. L., Licht, L. A., McCutcheon, S. C., Wolfe, N. L., & Carreira, L. H. (1995). Phytoremediation of organic and nutrient contaminants. *Environmental Science & Technology, 29*(7), 318–323. https://doi.org/10.1021/es00007a747.

Smith, R. D., & Hargreaves, B. R. (2015). Effects of water quality on wetland vegetation. *Wetland Ecology and Management, 23*(3), 341–354.

Sumiahadi, A., & Acar, R. (2018). A review of phytoremediation technology: Heavy metals uptake by plants. *IOP Conference Series: Earth and Environmental Science, 142*, 012023. https://doi.org/10.1088/1755-1315/142/1/012023.

Susarla, S., Medina, V. F., & McCutcheon, S. C. (2002). Phytoremediation: An ecological solution to organic chemical contamination. *Ecological Engineering, 18*(5), 647–658. https://doi.org/10.1016/S0925-8574(02)00026-5.

Turner, R. E., & Ellis, M. (2010). *Coastal Wetlands: A global perspective.* University of California Press.

Uggetti, E., Hughes-Riley, T., Morris, R. H., & Jefferson, B. (2016). Intermittent aeration to improve wastewater treatment efficiency in a pilot-scale constructed wetland. *Science of the Total Environment, 559*, 212–217.

UNESCO. (1971). *Convention on Wetlands of International Importance Especially as Waterfowl Habitat.* United Nations Educational, Scientific and Cultural Organisation.

Vitousek, P. M., Aber, J. D., Howarth, R. W., Likens, G. E., Matson, P. A., Schindler, D. W., Schlesinger, W. H., & Tilman, D. G. (1997). Human alteration of the global nitrogen cycle: Causes and consequences. *Ecological Applications, 7*(3), 737–750.

Wallace, S. D., Liner, M. O., Redmon, D., & Hildebrand, M. (2007). Oxygen transfer efficiency in aerated subsurface flow wetlands. In Proceedings of the 2nd International Symposium on Wetland Pollutant Dynamics and Control, University of Tartu, Estonia, 16–20 September 2007.

Wang, X., Wang, W., & Tong, C. (2016). A review on the impact of typhoons and hurricanes on coastal wetland ecosystems. *Acta Ecologica Sinica, 36*(1), 23–29.

White, E., & Kaplan, D. (2017). Restore or retreat? Saltwater intrusion and water management in coastal wetlands. *Ecosystem Health and Sustainability, 3*(1), e01258.

Zedler, J. B., & Kercher, S. (2005). Wetland resources: Status, trends, ecosystem services, and restoration. *Annual Review of Environment and Resources, 30*, 39–74.

Zhang, B. Y., Zheng, J. S., & Sharp, R. G. (2010). Phytoremediation in engineered wetlands: Mechanism and applications. *Procedia Environmental Science, 2*, 1325 1325.

Chapter 10

Physicochemical dynamics, species richness, and conservation challenges of wetlands in the Indian Himalaya

Prakash Chhetri, Rajesh Joshi, Lakpa Tamang, Kailash Singh Gaira, Sivaganga Kandoth, Sushmita Chettri, and Jullen Subba

GB Pant National Institute of Himalayan Environment, Sikkim Regional Centre, Gangtok, Sikkim, India

10.1 Introduction

Wetlands represent ecotonal zones between terrestrial and aquatic ecosystems and support significant components of global biodiversity (Ghermandi et al., 2008). Wetlands are defined as areas of marsh, fen, peatland, or water, whether natural or artificial, permanent or temporary, with water that is static or flowing, fresh, brackish, or saline, including areas of marine water where the depth at low tide does not exceed 6 m (Ramsar, 2016). Wetlands occur across the globe, ranging from coastal mangroves to high-altitude wetlands situated above 3000 m elevation. Although global estimates of wetland extent remain uncertain, reports suggest that approximately 12–16 million km^2 of the Earth's surface is covered by inland and coastal wetlands (Davidson et al., 2023), of which Asia accounts for the largest share (31.8%) (Sharma et al., 2020). In India, the total extent of wetlands (area >2.5 ha) is estimated at 15.98 Mha, constituting about 4.8% of the country's total geographical area (EnviStats India, 2022). The Indian Himalayan Region (IHR) harbors a substantial number of wetlands along with rich associated biodiversity. Wetlands located at elevations above 3000 m above mean sea level in the IHR are referred to as high-altitude wetlands (HAWs). The extent of HAWs (wetland area >2.5 ha) in the IHR is approximately 1.25 lakh ha. Himalayan wetlands possess exceptional ecological significance and are particularly recognized for their spiritual and cultural values (Sharma et al., 2020).

Water has long been regarded as a source of life and is considered sacred in many cultures worldwide. Wetlands typically occupy lowlands or natural depressions and are among the most productive ecosystems on Earth (Ghermandi et al., 2008), providing numerous ecosystem services to human society (Russi et al., 2013). Key provisioning services include freshwater supply, food, fuelwood, fodder, and other nontimber forest products, while regulating and cultural services such as tourism, flood control, water purification, climate change mitigation, and habitat provision for unique flora and fauna are also notable. Wetland plants constitute a fundamental component of wetland ecosystems and their catchment areas, reinforcing many of these ecosystem functions.

Owing to its diverse topography and climatic conditions, India's wetlands support approximately one-fifth of the country's known biodiversity (Murthy et al., 2013). The IHR, spanning thirteen mountain states and union territories, represents a repository of immense ecological wealth and biological diversity. Among the region's varied ecosystems, wetlands are of critical importance due to their unique environmental characteristics and their role in sustaining diverse floral and faunal assemblages. Wetland plants contribute substantially to ecological processes such as nutrient cycling, water purification, carbon sequestration, erosion control, and habitat provision, while also serving as indicators of wetland health and regulators of local microclimates. In the Indian Himalayan wetlands, plant diversity is particularly significant due to the fragility and distinctiveness of high-altitude and montane ecosystems. The IHR encompasses an extensive, ecologically complex landscape along India's northern frontier, where wetlands, especially those at high elevations, serve as vital ecotones between land and water systems. These wetlands act as biodiversity reservoirs, regulate regional hydrology, support endemic and specialized flora, and sustain traditional knowledge systems of indigenous communities.

Wetlands: Biodiversity, Health and their Role in Ecosystem Services. DOI: https://doi.org/10.1016/B978-0-443-34222-6.00019-X

Wetlands such as Tsomgo (Sikkim), Pangong Tso (Ladakh), Tso Moriri (Ladakh), Devikund (Himachal Pradesh), Brahma Tal (Uttarakhand), and Sangestar (Arunachal Pradesh) are not only ecologically fragile but also harbor a wide array of ethnobotanically important and medicinal plant species. Owing to steep environmental gradients and relative geographic isolation, many plant species inhabiting these wetlands exhibit high levels of endemism and ecological specialization, rendering them particularly vulnerable to disturbance. Consequently, their conservation has emerged as a crucial concern. This chapter examines in detail the diversity of plant species in Himalayan wetlands and evaluates their conservation status in the context of increasing anthropogenic pressures and natural threats.

10.2 Physicochemical status of wetlands in relation to plant growth

The physicochemical characteristics of wetland water and the associated floral composition are primarily governed by the nature of local bedrock and prevailing climatic conditions. The presence of high-altitude lakes reflects the availability of soil moisture capable of sustaining diverse vegetation under harsh climatic environments. Macrophytes are widely used as bioindicators to assess nutrient variability in high-altitude lakes (Lacoul & Freedman, 2006). Vegetation density within the lake catchment plays a critical role in regulating nutrient inflow and influencing aquatic productivity and plant growth.

Assessments of physicochemical parameters of Himalayan wetlands reveal considerable spatial variability, both horizontally and along altitudinal gradients. Studies from the western Himalaya report higher concentrations of total dissolved solids, pH, and electrical conductivity in lake waters compared to those of the central and eastern Himalaya (Bhat et al., 2011). These differences are largely attributed to the weathering of exposed parent materials, variations in rainfall patterns, and differences in vegetation cover. Dense vegetation in the eastern Himalaya moderates surface runoff and enhances nutrient retention within soils, thereby promoting plant growth. Additionally, the presence of grasses along lake peripheries reduces sediment inflow, a process more pronounced in the eastern Himalaya than in the cold, arid landscapes of the western Himalayan region.

Dissolved oxygen (DO) is a key determinant of water quality and is essential for the balanced functioning of aquatic ecosystems. DO concentrations ranging from 1.6 to 6.17 mg L^{-1} have been reported in several Himalayan wetlands, including Wular, Mansar, and Surinsar lakes, exceeding the Bureau of Indian Standards (BIS) limits for drinking water (Bhat & Pandit, 2014; Kumar et al., 2006; Singh & Jain, 2013). Increasing recreational activities in and around high-altitude lakes contribute to nutrient enrichment, which in turn influences DO dynamics in lake waters. Elevated nutrient inputs often promote algal blooms, leading to oxygen depletion and ecological imbalance.

Temperature plays a pivotal role in plant growth and significantly influences the physicochemical properties of wetland ecosystems. Recent studies from the Himalaya indicate widespread glacier retreat, resulting in the formation of new glacial lakes and the shrinkage of existing ones. Such alterations in high-altitude hydrology are likely to have profound implications for ecosystem functioning. Furthermore, the presence of microplastics in lake ecosystems, coupled with rising temperatures, may exacerbate environmental stress in these fragile systems (Neelavannan et al., 2022).

Climatic variability and increasing anthropogenic pressures have collectively altered the physicochemical composition of Himalayan wetland catchments. Changes in the interaction among abiotic factors are likely to modify regional hydro-ecological processes. Therefore continuous and long-term monitoring of Himalayan lakes, which constitute the base flow of many perennial Asian rivers, is essential for understanding ecosystem dynamics and ensuring their sustainable management.

10.3 Species richness and composition

Plant species richness and community composition in the IHR are largely governed by longitudinal and altitudinal gradients. The Eastern Himalaya, particularly Sikkim and Arunachal Pradesh, receives heavy monsoonal rainfall and is characterized by a warm, humid climate that supports tropical evergreen, subtropical, and montane rainforest ecosystems. As a result, the Eastern Himalaya (Fig. 10.1) exhibits exceptionally high plant diversity and endemism, particularly among orchids, rhododendrons, and bamboo species.

In contrast, the western Himalaya, including Jammu and Kashmir, Ladakh, Himachal Pradesh, and Uttarakhand, experiences comparatively lower precipitation and a drier climate, supporting predominantly temperate and alpine vegetation types. Subtropical pine forests, oak forests, and deodar stands are characteristic of this region. The cold desert zones of the north-western IHR support sparse vegetation cover (Fig. 10.2); however, these areas exhibit a high degree of endemism among cold-adapted and stress-tolerant plant species.

A greater number of studies on wetland vegetation have been conducted in the western Himalaya compared to the Eastern Himalaya. Lake catchments, owing to higher moisture availability, generally support greater and more distinct

FIGURE 10.1 Moist alpine scrub at Tsomgo high-altitude wetland, Sikkim, Eastern Himalaya.

FIGURE 10.2 Dry alpine vegetation in the catchment of Tso Moriri, Ladakh, western Himalaya.

plant diversity than the surrounding terrestrial landscapes. Species richness and diversity indices vary markedly with altitude, ecological conditions, and levels of anthropogenic disturbance (Table 10.1).

Lowland tropical wetlands along the fringes of the IHR support the highest plant species richness and diversity, whereas high-altitude alpine wetlands are relatively species-poor but ecologically specialized. Tropical and low-altitude wetlands exhibit higher richness and diversity, while alpine and high-altitude wetlands show reduced diversity due to harsh climatic conditions and shorter growing seasons. Notably, certain lakes, such as Nigeen Lake (Jammu and Kashmir), display low species richness but high evenness, resulting in elevated Simpson and Shannon diversity indices (Fig. 10.3A and B).

Eastern Himalayan wetlands show more stable and higher diversity indices (2.16 for Khecheopalri, 2.89 for Loktak, and 3.21 for Deeporbeel Lake) as compared to wide variations in western Himalaya, such as 1.3 for Tso Moriri to 3.04 for Nigeen Lake. The Eastern Himalayan region has consistency in higher Simpson values, suggesting lesser disturbances and a stable ecosystem, whereas the western Himalaya has extremes, suggesting higher evenness in Renuka Lake (0.99) and survival of hardy species in Tso Moriri (0.65). Alpine wetland in eastern Himalaya retains moderate Richness (e.g., Pangtenga Tso: 93) due to high humidity and less anthropogenic pressure as compared to the wetlands having cold stress and anthropogenic pressure, including tourism (e.g., Tso moriri: 45).

Poaceae and Cyperaceae are found to be dominant families across wetlands of IHR with *Phragmites australis*, Phragmites karka, *Echinochloa stagnina, Carex* spp, *Acorus calamus, Schoenoplectus lacustris*, as dominant species (Table 10.2). Tropical wetlands in eastern Himalaya are further enriched by floating and free-floating species (*Nelumbo nucifera, Pontederia crassipes, Pistia stratiotes*) reflecting warmer and nutrient-rich wetlands, whereas at higher altitudes, lake catchments are mostly dominated by the species of family Asteraceae, Ericaceae, and Rosaceae, while the most dominating genera are *Rhododendron*, *Primula*, *Potentilla*, *Carex*, *Juncus*, *Saussurea*, etc. Similarly, in the western Himalaya, *Ipomoea aquatica, A. calamus*, *Typha angustifolia*, *Centella asiatica*, *Alternanthera sessilis*, *Bacopa monnieri*, *Colocasia esculenta*, etc., are the common vegetation in the tropical and subtropical wetlands. Asteraceae, Rosaceae, and

TABLE 10.1 Selected wetlands across Indian Himalayan Region (*IHR*) and their phytosciological attributes.

Sl. no.	Wetland site	Location	Altitude (m)	S[a]	H'	D	References
1	Loktak	Manipur	768	312	2.89	0.87	Bassi et al. (2014), Devi et al. (2024), Mao et al. (2024)
2	Deeporbeel	Assam	55	150	3.21	0.91	MoEF&CC (2017a), Mao et al. (2024)
3	Tso Moriri	Ladakh	4522	45	1.32	0.65	Rawat (2017)
4	Chander Tal	Himachal Pradesh	4250	148	1.76	0.73	Singh and Hajra (1996)
5	Wular Lake	Jammu & Kashmir	1580	100	2.45	0.82	IUCN (2023), Mao et al. (2024)
6	Nainital Lake	Uttarakhand	1938	65	2.12	0.79	Upreti et al. (2019), Bargali et al. (2013)
7	Asan Barrage	Uttarakhand	420	281	–	–	Mao et al. (2024)
8	Renuka Lake	Himachal Pradesh	660	407	2.4	0.99	Choden et al. (2022)
9	Khecheopalri Wetland	Sikkim	1700	112	2.16	–	Rai, Bhutia, et al. (2025), Rai, Das, et al. (2025)
10	Nigeen Lake	Jammu and Kashmir	1584	35	3.04	0.94	Saima and Kamili (2016)
11	Pangteng Tso	Arunachal Pradesh	3912	93	–	–	Kanwal (2021)

[a]*S, Species Richness, H', Shannon Wiener Index, D, Simpson Index.*

Ranunculaceae are the most dominant families in the higher elevation wetlands of the western Himalaya. *Lentopodium, Aster, Artemisia, Tanacetum, Ranunculus, Clematis, Potentilla, Rosa*, etc., are the most dominant genus in higher arid region of western Himalaya (Bargali et al., 2013; Bassi et al., 2014; IUCN, 2023; Kanwal, 2021; Mao et al., 2024; MoEF&CC, 2017a, b; Rai, Bhutia, et al., 2025; Rai, Das, et al., 2025; Sharma et al., 2019; Singh & Hajra, 1996; Upreti et al., 2019).

10.4 Indicator species of wetlands in the Indian Himalayan Region

Indicator species are the plant taxa whose presence, absence, or abundance reflects a specific environmental condition, ecological process, or habitat type (Lindenmayer et al., 2000). IHR wetlands have several plant species that serve as bioindicators of wetland health, water quality, and habitat integrity (Table 10.3). Sensitive indicator species play a significant ecological role. The information presented in Table 10.3 highlights different plant species as ecological indicators for assessing water quality, wetland types, and land degradation across the IHR. For water quality, species such as *P. crassipes, P. stratiotes, Azolla pinnata, Hydrilla verticillata, Lemna minor,* and *Ceratophyllum demersum* indicate eutrophication or nutrient enrichment, especially at low to mid elevations of the eastern Himalaya, while the presence of *Vallisneria spiralis* reflects stable and unpolluted aquatic systems in the lower Himalayan foothills. Wetland types are also distinguished by characteristic vegetation, such as *Carex* and *Juncus* species, that dominate moist alpine meadows across the IHR. *Hippuris vulgaris, Carex moorcroftii*, and *Polygonum alpinum* signify high-altitude cold freshwater systems of the western Himalaya; *Sphagnum* species mark peat-forming bogs in acidic and anoxic wetlands of the eastern Himalaya; and *Trapa natans* indicates organic-rich shallow marshes in Assam's floodplains. In terms of land degradation, the occurrence of rare and threatened species such as *Saussurea obvallata, Aconitum heterophyllum*, and *Valeriana jatamansi* highlights endangered habitats and is used for identifying conservation priority habitats. Overall, these plants serve as reliable bioindicators, linking vegetation patterns to ecological conditions and environmental stress across the wetlands of IHR (Bassi et al., 2014; IUCN, 2025; Keddy, 2010; Mishra et al., 2023; MoEF&CC, 2017a, b; Rawat, 2017).

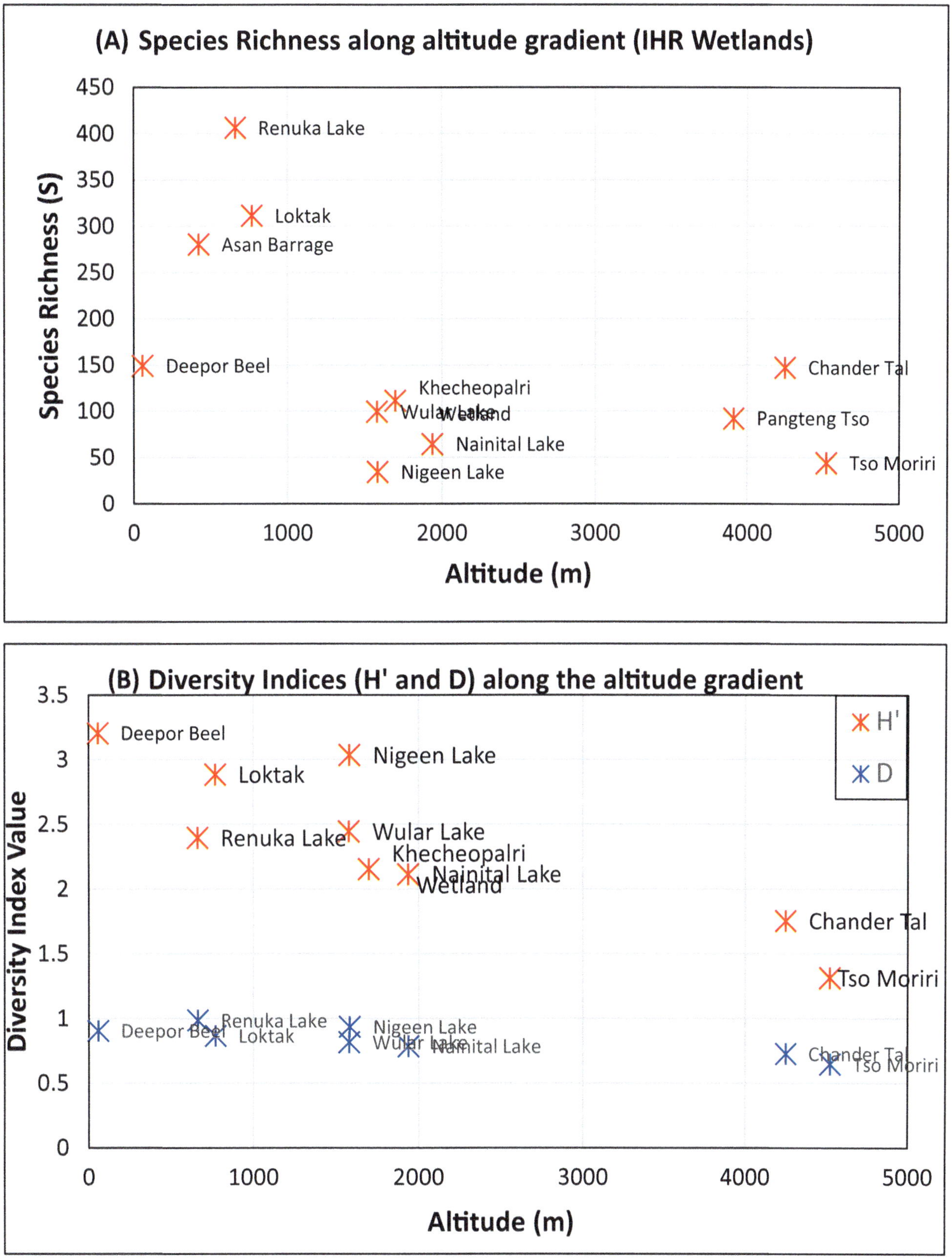

FIGURE 10.3 (A) Species Richness (*S*) and (B) Diversity Indices along the elevation gradient of different wetlands across Indian Himalayan Region (*IHR*).

TABLE 10.2 Dominant plant families and species in Indian Himalayan Region (*IHR*) wetlands.

Region	Wetlands	Dominant Families	Key species	Growth Forms
Eastern Himalaya	Loktak Lake, Deeporbeel, Khecheopalri Lake, Pangteng Tso	Poaceae, Cyperaceae, Araceae, Nymphaeaceae, Nelumbonaceae, Hydrocharitaceae, Pontederiaceae, Lemnaceae, Salviniaceae	*Phragmites karka, Echinochloa stagnina, Schoenoplectus lacustris, Cyperus iria, Colocasia esculenta, Acorus calamus, Nelumbo nucifera, Nymphaea nouchali, Hydrilla verticillata, Vallisneria spiralis, Pontederia crassipes, Pistia stratiotes, Lemna minor, Azolla pinnata*	Emergent grasses, sedges, rooted floating-leaved plants, free-floating aquatics, submerged hydrophytes
Western Himalaya	Wular Lake, Nigeen Lake, Chandratal, Renuka Lake, Nainital Lake, Tso Moriri	Poaceae, Cyperaceae, Potamogetonaceae, Hydrocharitaceae, Ceratophyllaceae, Characeae, Nymphaeaceae, Typhaceae	*Phragmites australis, Arundo donax, Carex* spp., *Scirpus* spp., *Potamogeton crispus, Potamogeton lucens, H. verticillata, V. spiralis, Ceratophyllum demersum, Chara vulgaris, Nymphaea pubescens, Typha latifolia*	Submerged macrophytes dominant, emergent sedges and reeds, floating-leaved lilies, some marsh plants

TABLE 10.3 Notable indicator species in Indian Himalayan Region (*IHR*).

Category	Indicative role	Indicator (Absence or presence)	Region of occurrence
Water quality	Eutrophication/nutrient load	*Pontederia crassipes, Pistia stratiotes, Azolla pinnata, Hydrilla verticillata, Lemna minor, Ceratophyllum demersum*	Low to mid elevations (mostly in eastern Himalaya)
Water quality	Stable, unpolluted aquatic system	*Vallisneria spiralis*	Lower Himalayan foothills
Wetland type	Moist alpine meadows	*Carex* sp., *Juncus* sp.	Entire IHR
Wetland type	High-altitude cold freshwater system	*Hippuris vulgaris, Carex moorcroftii, Polygonum alpinum*	High-altitude wetlands of western Himalaya
Wetland type	Peat-forming bogs, acidic and anoxic wetlands	*Sphagnum* spp.	Eastern Himalaya
Wetland type	Organic-rich shallow marshes	*Trapa natans*	Floodplains of Assam
Land degradation	Endangered habitat	*Saussurea obvallata, Aconitum heterophyllum, Valeriana jatamansi*, etc.	High-altitude region

10.5 Threat and conservation challenges

The Himalaya, with its exceptionally high species endemism, coupled with high anthropogenic pressure, has been identified as a significant global biodiversity hotspot (Wambulwa et al., 2021). IHR is represented by 8700 taxa of angiosperms, 51 taxa of gymnosperms, 766 taxa of pteridophytes, and 1955 taxa of bryophytes, out of which 12% are known to be endemic to the Himalaya and the same number are point endemic and under the threat category (Singh & Pusalkar, 2020).

In the Ladakh region of the Himalaya, 25 species from high-altitude wetlands are classified into various threat categories, of which 21 are native to the Himalayan region (Bhojak et al., 2024). Among these, *Arnebia euchroma* and *Ephedra gerardiana* are classified as Endangered (EN), while *Physochlaina praealta* is listed as Vulnerable (VU). Similarly, from Himachal Pradesh (Chandra Tal and Suraj Tal), six species have been reported as threatened, including two endemics (*Eritrichium nanum* and *Ranunculus trivedii*) and 17 native taxa (Dey et al., 2021). Within this group, *Aconitum violaceum*, *Rheum spiciforme*, and *Rhodiola heterodonta* are categorized as Vulnerable (VU); *A. euchroma* and *Saussurea gossypiphora* as Critically Endangered (CR); and *Papaver guilelmi-waldemarii* as Endangered (EN). Thirty-four species of high-value medicinal plants within the catchment of three different rivers within the watersheds in Nandadevi Biosphere Reserve have been identified under different threat categories. Among the 34 threatened species, *Angelica glauca, Dactylorhiza hatagirea,* and *Picrorhiza kurrooa* are endangered (EN), *A. violaceum* and *E. gerardiana* are vulnerable, and *Saussurea costus* is CR as per IUCN classification of threat categories (Mishra et al., 2023).

Pangteng Tso in Aruancahal Pradesh, located in the eastern Himalaya, stores 60 native species in the Himalayan region, while 15 species are endemic to the eastern Himalayan region. High-value medicinal plants from Pangteng Tso fall under different threat categories, such as *P. kurrooa, Platanthera bhutanica, Ponerorchis chusua,* are recorded under Appendix-II of CITES. *Abies densa* and *Rhododendron anthopogon* were categorized as near threatened and endangered threat categories, respectively, as per CAMP, 2003 (Kanwal, 2021). The other prominent species of eastern Himalaya, that are under threat category, occurring in the high moisture content areas and wetlands are *Coptis teeta*, *P. kurrooa, Allium wallichii, Podophyllum hexandrum, A. heterophyllum* (Endangered), *Nardostachys jatamansi* (CR), *Polygonatum verticillatum*, *Polygonatum cirrhifolium*, *Fritillaria cirrhosa*, etc. (Vulnerable). Most of these species occur in the HAWs of Sikkim (Tsomgo and Hanspokhari wetland complex) and Arunachal Pradesh.

Wetlands are the most threatened ecosystem, degrading three times faster than forests (Convention on Wetlands, 2021). Most of the wetland plants are under different threat categories due to various threats associated with the wetland ecosystem. The threats, both manmade and natural, impact the species or the habitat. Key threats and conservation challenges to the wetland ecosystem include:

- Climate change-induced alterations, such as shifting snowlines and glacial retreat, constrain the available habitats and accelerate biodiversity loss.
- Habitat degradation and fragmentation, siltation, and desertification are processes that are being further intensified by the impacts of regional climate change.
- Expansion of grazing areas and uncontrolled livestock movement exert severe pressure on wetland catchments, resulting in vegetation loss, soil compaction, and altered hydrological regimes.
- Unsustainable extraction of forest resources, and illegal collection and trafficking of wild medicinal and edible plants from these wetlands make species vulnerable.
- Unregulated and poorly managed tourism and rapid developmental interventions contribute to species loss, water and soil pollution, thereby reducing ecosystem resilience. In addition, species inhabiting these wetlands are increasingly vulnerable to illegal collection, overharvesting, and habitat destruction.
- Nutrient loading leading to eutrophication is one of the major threats to the tropical and subtropical wetlands in IHR, leading to the expansion of algae and invasive species like *P. crassipes* and *P. stratiotes*.

These combined pressures not only threaten species persistence but also undermine the ecological services that HAWs provide to downstream communities. Given the scale and intensity of these challenges, there is an urgent need for a comprehensive, science-based management framework, supported by systematic monitoring and up-to-date baseline information, to guide conservation planning and policy interventions in the region.

10.6 Conclusion and recommendations

Himalayan wetlands, covering about 1.25 lakh ha of high-altitude area, are ecologically vital yet highly vulnerable systems. Strong east-west contrasts exist, with eastern Himalayan wetlands exhibiting higher species richness and ecosystem stability (Shannon index up to 3.21) due to humid climate and dense vegetation, while western Himalayan wetlands show lower diversity (as low as 1.32) and greater physicochemical variability under cold and arid conditions. Species richness declines sharply with altitude, from >400 species in low- and mid-altitude wetlands to <50 species in cold arid alpine lakes, reflecting strong climatic filtering. Physicochemical data indicate generally high DO (1.6–6.17 mg L^{-1}), but increasing nutrient loading, eutrophication, and invasive macrophytes are degrading water quality in accessible wetlands. The dominance of Poaceae and Cyperaceae and the presence of well-defined indicator species provide reliable measures of wetland health and disturbance. However, wetlands are degrading three times faster than forests, with numerous endemic and medicinal plant species now categorized as CR, Endangered, or Vulnerable.

To address these challenges, long-term monitoring of physicochemical parameters, biodiversity, and catchment processes is essential. In situ conservation of endemic and indicator species, catchment restoration, and eutrophication control should be prioritized. Regulation of tourism through carrying capacity assessments, invasive species management, and sustainable grazing practices is crucial for maintaining ecosystem health. Integrating wetland management into regional climate adaptation plans, supported by legal frameworks and participatory governance, will enhance resilience. HAWs are degrading rapidly and require science-based, multistakeholder management to safeguard their biodiversity, hydrological functions, and socio-ecological benefits under accelerating climate and anthropogenic pressures.

References

Bargali, K., Bisht, P., Khan, A., & Rawat, Y. S. (2013). Diversity and regeneration status of tree species at Nainital Catchment, Uttarakhand, India. *International Journal of Biodiversity and conservation, 5*(5), 270–280.

Bassi, N., Kumar, M. D., Sharma, A., & Pardha-Saradhi, P. (2014). Status of wetlands in India: A review of extent, ecosystem benefits, threats and management strategies. *Journal of hydrology: Regional studies, 2*, 1–19.

Bhat, F. A., Yousuf, A. R., Aftab, A., Arshid, J., Mahdi, M. D., & Balkhi, M. H. (2011). Ecology and biodiversity in Pangong Tso (lake) and its inlet stream in Ladakh, India. *International Journal of Biodiversity and Conservation, 3*(10), 501–511.

Bhat, S. A., & Pandit, A. K. (2014). Surface water quality assessment of Wular Lake, a Ramsar site in Kashmir Himalaya, using discriminant analysis and WQI. *Journal of Ecosystems*, 724728.

Bhojak, P., Dey, D., Sekar, K. C., Thapliyal, N., & Danu, H. (2024). Floristic diversity in high-altitude wetlands of Ladakh Union Territory, Trans Himalaya, India. *Check List, 20*(4), 915–939.

Choden, Y., Sharma, M. P., Pandey, G., Gupta, S. K., & Dema, K. D. (2022). Ecological health assessment of Renuka Lake, Himachal Pradesh, India. *Nature Environment and Pollution Technology, 21*(1), 167–174.

Convention on Wetlands. (2021). *Global wetland outlook: Special edition 2021*. Secretariat of the Convention on Wetlands.

Davidson, N. C., Finlayson, C. M., & Fluet-Chouinard, E. (2023). The extent and distribution of the world's wetlands. In *Ramsar wetlands*. Elsevier, 91–114.

Devi, R. S., Satapathy, K. B., & Kumar, S. (2024). Loktak lake of the state Manipur: A review to educate the intellectuals and awareness on its biowealth. *Educational Administration: Theory and Practice, 30*(5), 6936–6947.

Dey, D., Bhojak, P., Sekar, K. C., & Arya, D. (2021). An annotated checklist of vascular plants in and around two major high-altitude wetlands of Lahaul-Spiti, Himachal Pradesh, India. *Check List, 17*(6), 1715–1730.

EnviStats India. (2022). EnviStats India, Vol:II Environment Accounts. Social Statistics Division, National Statistical Office, Ministry of Statistics and Programme Implementation. Govt. of India.

Ghermandi, A., Van den Bergh, J.C., Brander, L.M., De Groot, H.L., & Nunes, P.A. (2008). The economic value of wetland conservation and creation: A meta-analysis. Fondazione Eni Enrico Mattei Working Papers. 238.

IUCN. (2023). *Wetland ecosystems of Kashmir Himalaya: Biodiversity, conservation and management*. International Union for Conservation of Nature.

IUCN. (2025). *IUCN Red List Version 2025-1*. International Union for Conservation of Nature.

Kanwal, K. S. (2021). Floral biodiversity of Pangateng Tso (PTSO) high-altitude wetland of Eastern Himalaya. *Species,* 431–440.

Keddy, P. A. (2010). *Wetland ecology: Principles and conservation*. Cambridge University Press.

Kumar, V., Rai, S. P., & Singh, O. (2006). Water quantity and quality of Mansar Lake located in the Himalayan Foothills, India. *Lake and Reservoir Management, 22*(3), 191–198.

Lacoul, P., & Freedman, B. (2006). Relationships between aquatic plants and environmental factors along a steep Himalayan altitudinal gradient. *Aquatic Botany, 84*(1), 3–16.

Lindenmayer, D. B., Margules, C. R., & Botkin, D. B. (2000). Indicators of biodiversity for ecologically sustainable forest management. *Conservation Biology,* 941–950.

Mao, A. A., Vivek, C. P., Agrawala, D. K., Dash, S. S., Vivek, C. P., Mukherjee, S., Jayanthi, J., & Jalal, J. S. (2024). Compendium on phytodiversity of Ramsar sites in India. *Botanical Survey of India,* 1–732.

Mishra, A. P., Kumar, A., & Yadav, S. N. (2023). Ecology and conservation of threatened medicinal plants in the Trans-Himalayan region of Nanda Devi Biosphere Reserve, Western Himalaya. *Trees, Forests and People, 14*, 100451.

MoEF&CC. (2017a). Management Plant of Deepor Beel Ramsar Site, Assam (2017-2027). Ministry of Environment, Forest and Climate Change, Government of India.

MoEF&CC. (2017b). Wetlands (Conservation and Management) Rules, 2017. Ministry of Environment, Forest and Climate Change, Government of India. Accessed July 21, 2025.

Murthy, T.V.R., Patel, J.G., Panigrahy, S., & Parihar, J.S. (2013). National wetland atlas: Wetlands of international importance under Ramsar convention. Space Applications Centre (ISRO), Ahmedabad, India, 230.

Neelavannan, K., Sen, I. S., Lone, A. M., & Gopinath, K. (2022). Microplastics in the high-altitude Himalayas: Assessment of microplastic contamination in freshwater lake sediments, Northwest Himalaya (India). *Chemosphere, 290*, 13335.

Rai, P., Das, D. J., & Meena, D. K. (2025). *A note on Sikkim's first Ramsar site: Sacred Khecheopalri Lake where biodiversity, spirituality and culture converge*. Ela Journal of Forestry and Wildlife, 1715.

Rai, S. C., Bhutia, P. L., & Sharma, E. (2025). Floral diversity and conservation status of Khecheopalri wetland, Sikkim Himalaya. *Mountain Research and Development,* 67–79.

Ramsar, C. (2016). *An introduction to the Ramsar convention on wetlands* (7th ed.). Secretariat of the Convention on Wetlands.

Rawat, G. S. (2017). The Himalayan Vegetation along horizontal and vertical gradients. In *Bird migration across the Himalayas: In Wetland functioning amidst mountains and glaciers*. Cambridge University Press, 189.

Russi, D., ten Brink, P., Farmer, A., Badura, T., Coates, D., Förster, J., Kumar, R., & Davidson, N. (2013). *The economics of ecosystems and biodiversity for water and wetlands. 78*, IEEP, 118p.

Saima, B. M. S., & Kamili, A. N. (2016). Assessment of floristic diversity on the bank of nigeen lake, kashmir valley. *Biodiversity Management & Forestry Journals, 5*(1), 2.

Sharma, E., Chettri, N., & Tsering, K. (2019). *Biodiversity and ecosystem services of Himalayan wetlands*. ICIMOD.

Sharma, S., Gosavi, V. E., Kanwal, K. S., Agnihotri, V., Chandra, S. K., Arya, S. C., & Rawal, R. S. (2020). *High altitude wetlands in the Indian Himalaya: Conservation and management*. GB Pant National Institute of Himalayan Environment.

Singh, D., & Pusalkar, P. (2020). Floristic diversity of the Indian Himalaya. In *Biodiversity of the Himalaya: Jammu and Kashmir State*. Springer.

Singh, D. K., & Hajra, P. K. (1996). Floristic diversity. *Changing perspectives of biodiversity status in the Himalaya,* 23–38.

Singh, O., & Jain, C. K. (2013). Assessment of water quality and eutrophication of lakes. *Journal of Environmental Nanotechnology. 2*, 46–52.

Upreti, B. M., Pandey, N. C., Chopra, N., & Tewari, L. M. (2019). Species diversity among two forests of nainital, Kumaun Himalaya. *International Journal for Research in Applied Science & Engineering Technology (IJRASET),* 45.

Wambulwa, M. C., Milne, R., Wu, Z. Y., Spicer, R. A., Provan, J., Luo, Y. H., Zhu, G. F., Wang, W. T., Wang, H., Gao, L. M., & Li, D. Z. (2021). Spatiotemporal maintenance of flora in the Himalaya biodiversity hotspot: Current knowledge and future perspectives. *Ecology and Evolution, 11*(16), 10794–10812.

Chapter 11

Floral borders and reduced agronomic inputs in paddy fields: enhancing biodiversity and crop health in wetland ecosystems

Manikandan P.[1], Girija D.S.[1], Saravanaraman M.[1], Bryony K. Willcox[2], Rengalakshmi R.[1], and Deepa Senapathi[3]

[1]*Division of Ecotechnology, M. S. Swaminathan Research Foundation, Chennai, Tamil Nadu, India,* [2]*Centre for Agri-Environmental Research, School of Agriculture, Policy and Development, University of Reading, Reading, Berkshire, United Kingdom,* [3]*Department of Sustainable Land Management, University of Reading, Reading, Berkshire, United Kingdom*

11.1 Introduction

Wetlands are ecologically significant ecosystems with unique hydrology, soil properties, and biodiversity, occurring where land meets water. They include marshes, swamps, bogs, and estuaries and are defined by the Ramsar Convention as areas of marsh, fen, peatland, or water, whether natural or artificial, permanent or temporary, with fresh, brackish, or salt water. Wetlands cover about 6% of the Earth's surface (12.1 million km^2). In India, they account for 4.7% of the total area, with over 75% being inland and 25% coastal. Wetlands provide essential ecosystem services such as water purification, groundwater recharge, carbon sequestration, and flood mitigation. They act as biodiversity hotspots, supporting diverse flora and fauna, including endangered species. Additionally, they store up to 35% of global terrestrial carbon. Socioeconomically, they support livelihoods through fishing, agriculture, and ecotourism while holding cultural and educational significance. However, wetlands face threats from urbanization, pollution, water overextraction, and climate change, necessitating urgent conservation and sustainable management (Finlayson et al., 2018; Prasad et al., 2002; Secretariat, Ramsar Convention, 2016).

Paddy, or rice (*Oryza sativa*), is a vital staple crop that feeds over half the world's population and plays a key role in food security. It is especially important in Asia, where over 90% of global rice is produced and consumed. Paddy cultivation is central to the cultural and economic fabric of many nations, supporting millions of small-scale farmers. Rice occupies about 162 million hectares worldwide, yielding over 756 million tons annually. India, the second-largest producer after China, cultivates paddy on 43 million hectares, contributing 24% of global production. Wetlands, including irrigated and rainfed lowlands, are crucial for paddy cultivation, as rice thrives in submerged conditions. About 55% of global rice is grown in wetland ecosystems, mainly in tropical and subtropical regions. In India, 40%–45% of paddy cultivation occurs in wetlands, especially in West Bengal, Bihar, Tamil Nadu, and Andhra Pradesh. Beyond agriculture, paddy fields aid groundwater recharge, soil fertility, and carbon sequestration. However, water scarcity, greenhouse gas emissions, and soil degradation threaten its sustainability, highlighting the need for innovative practices (Bouman et al., 2007; FAO, 2021; Ministry of Water Resources (MoWR), 2014).

The extensive use of synthetic fertilizers, pesticides, and herbicides in agriculture significantly impacts wetland ecosystem services. These chemicals, transported through runoff and leaching, contaminate wetlands, causing nutrient enrichment, eutrophication, and biodiversity loss. Excess nitrogen and phosphorus from fertilizers trigger algal blooms, depleting oxygen, harming aquatic fauna, and disrupting food webs. Pesticides like organophosphates and neonicotinoids affect nontarget organisms, including amphibians, fish, and pollinators, reducing biodiversity and altering ecosystems. Wetlands' essential services, such as water purification and groundwater recharge, are impaired as agrochemical residues

Wetlands: Biodiversity, Health and their Role in Ecosystem Services. DOI: https://doi.org/10.1016/B978-0-443-34222-6.00009-7

change soil and water chemistry, reducing microbial diversity. Additionally, bioaccumulation of toxins in wetland species threatens higher trophic levels, including birds and mammals. These effects weaken wetlands' ability to provide flood mitigation, carbon sequestration, and habitat, impacting communities that depend on them. Addressing these challenges requires integrated pest and nutrient management, organic farming, and stronger wetland conservation policies (Ghosh & Das, 2020; Schäfer et al., 2016; Tilman et al., 2002).

Paddy cultivation in India relies heavily on synthetic fertilizers and pesticides to sustain high yields amid rising food demand. India consumes about 27 million tons of fertilizers annually, with paddy using a significant share, especially nitrogen-based fertilizers like urea. Pesticides such as carbamates, organophosphates, and pyrethroids are widely used to control pests such as stem borers and leafhoppers. However, excessive agrochemical use impacts wetlands, as paddy fields, primarily irrigated lowlands, facilitate runoff. Fertilizer leachates cause nutrient enrichment, leading to eutrophication, algal blooms, and hypoxic conditions that harm aquatic life. Pesticides further contaminate water bodies, causing toxic bioaccumulation and disrupting trophic dynamics. Such contamination weakens wetland services like water purification and habitat provision, reducing biodiversity. Additionally, overuse of agrochemicals alters wetland soil properties, depletes microbial diversity, and reduces carbon sequestration, thereby exacerbating climate change. These disruptions highlight the urgency of sustainable practices, including integrated pest management (IPM) and organic farming, to mitigate the environmental impacts of chemical-intensive paddy cultivation (Das et al., 2019; Gupta & Seth, 2007; Singh et al., 2020).

Reducing agrochemical use in paddy cultivation is vital for sustainable and healthy ecosystems. IPM, combining biological control, cultural practices, and minimal chemical use, is highly effective. Biopesticides like *Trichoderma*, *Metarhizium anisopliae*, and neem-based products provide ecofriendly alternatives. Organic fertilizers such as vermicompost, green manure, and farmyard manure enhance soil fertility and microbial activity while reducing chemical dependence. Techniques like the system of rice intensification and direct-seeded rice lower water and nutrient needs, cutting agrochemical use. Precision farming, including soil testing and variable-rate applications, ensures that nutrients are applied efficiently, minimizing waste and runoff. Reduced chemical inputs improve soil health, microbial diversity, and water quality in wetlands. Healthier soil strengthens root systems and natural pest resistance, decreasing chemical dependency. Lower agrochemical use also protects aquatic biodiversity by preventing eutrophication and toxic contamination. Additionally, sustainable practices enhance carbon sequestration in paddy fields, aiding climate change mitigation and preserving ecosystem services like water purification and habitat provision (Choudhury & Kennedy, 2005; Uphoff et al., 2018; Yadav et al., 2021).

Paddy cultivation is predominantly practiced as a monocropping system in wetland ecosystems, which, while ensuring food security, poses significant challenges to biodiversity. Monocropping simplifies the landscape, reducing habitat heterogeneity and adversely affecting the diversity of flora and fauna, including beneficial organisms such as pollinators, predators, and decomposers. This loss of biodiversity weakens ecosystem resilience, making paddy fields more vulnerable to pest outbreaks and reducing natural pest control mechanisms. Introducing crop diversity, such as intercropping and crop rotation, enhances habitat complexity, fostering biodiversity and ecological balance. One innovative approach to promoting biodiversity in mono-cropped paddy systems is the use of floral borders-rows of flowering plants along field margins. These borders attract and sustain beneficial arthropods, including pollinators and natural enemies of pests, by providing nectar, pollen, and shelter. For instance, planting marigolds or sunflowers as borders can attract predatory insects like ladybird beetles and parasitoids, which help control paddy pests such as leafhoppers and stem borers. This biological pest control reduces reliance on synthetic pesticides, mitigating their harmful effects on nontarget organisms and wetland ecosystems. Furthermore, limited chemical input, combined with floral borders, enhances soil health by maintaining microbial diversity and reducing the accumulation of toxic residues. This integrated approach not only increases the ecological health of paddy fields but also contributes to higher productivity and sustainability by leveraging natural ecosystem services, such as nutrient cycling and water purification, essential for maintaining the balance of wetland ecosystems (Anjum et al., 2020; Gurr et al., 2004; Landis et al., 2000).

11.2 Methods

11.2.1 Study area

This study was conducted during the 2019–20 samba paddy season (September 2019–January 2020) in the Puducherry region of India. The research focused on the local long-duration rice variety, White Ponni, which is traditionally cultivated during this season. The study area is a coastal plain with predominantly flat terrain, where agriculture is a major land use. Key crops grown in the region include rice, sugarcane, banana, vegetables, and coconut, interspersed with patches of seminatural vegetation. The region experiences a tropical climate with distinct seasons—a dry spell from January to June and a monsoon season spanning October to December. The average annual rainfall for the Puducherry district was recorded at 1320 mm (IMD (Indian Meteorological Department), 2020).

11.2.2 Field experiment design

The study integrated farm-level interventions and landscape assessments to evaluate the influence of different agricultural input management strategies and floral border crops on pest populations and natural enemy communities. Twelve smallholder farms were selected for the study, each spanning less than 2 ha. To minimize spatial dependence, we maintained a minimum distance of 1 km between farms, except for one pair.

Each farm was sampled intensively using six fixed quadrats representing edge and center zones, yielding 72 spatial sampling units across the 12 farms. These quadrats were observed at three critical crop stages: early, mid, and late season, leading to a total of 216 observations for pest and natural enemy analysis. This nested sampling design provided sufficient replication at the observational level, helping to enhance the robustness of statistical analyzes despite the limited number of farm-level replicates.

The farms were assigned to one of two agricultural input management approaches, with six farms per treatment group:

1. **High-input system**: Characterized by the intensive use of synthetic fertilizers and pesticides for crop nutrition and pest management. Farmers applied insecticides such as chlorpyrifos and Cartap hydrochloride, alongside fertilizers including mono-ammonium phosphate, di-ammonium phosphate, urea, and NPK (19:19:19) complex fertilizers. Fertilizer applications were divided into two phases—75% before transplanting and the remaining amount as a top dressing during the tillering or flowering stage, depending on crop growth.
2. **Low-input system**: A collaborative effort between researchers and farmers to develop a reduced-input approach, replacing synthetic inputs with organic alternatives. This included foliar applications of Panchakavya (a fermented mixture of cow urine, ghee, dung, milk, curd, ripe bananas, and jaggery) every 15 days from the active tillering to flowering stage. Pest control relied on DasaKavya (a fermented plant extract) and neem seed oil, serving as alternatives to synthetic pesticides.

Both treatments included a basal application of farmyard manure (2 t ha^{-1}) before the first plowing, sourced directly from the respective farms.

11.2.3 Floral border crop treatment

To assess the effect of floral border crops on pest dynamics, natural enemies, and crop yield, three farms from each input system (high and low) were selected to have black gram (*Vigna mungo*) planted along the field bunds. This species was chosen as it provides nectar and pollen resources that attract beneficial arthropods. The remaining three farms per input type had fallow bunds without additional floral resources.

11.2.4 Pest and natural enemy sampling

Surveys were conducted at three key growth stages of the rice crop: early, mid, and late season, to monitor major pests and natural enemy populations.

The major pests, such as yellow stem borer (YSB), Leaf Folder, gall midge, brown plant hopper (BPH), green leaf hopper (GLH), were surveyed using six quadrats (75 cm^2) per field. The quadrats were equally distributed between the field edge (≤5 m from the boundary) and the field center (≥10 m from the boundary), with a minimum spacing of 10 m between quadrats. Within each quadrat, rice plants were examined for YSB, leaf folder, and gall midge by counting the number of tillers affected and total number of tillers, number of BPH and GLH per plant, and number of beneficial insects were counted per quadrat. The counts were pooled to obtain per-field estimates by location (edge vs center) and sampling date. Captured insects were stored in 70% ethanol for later identification. The collected specimens were identified to the species level where possible, with unidentified specimens categorized into morpho-species. Identification was performed in the laboratory by a trained entomologist using morphological keys. The analysis focused on invertebrates known for their role in natural pest regulation, including both generalist and specialist predators or parasitoids targeting major insect pests of paddy. Specimens were identified using standard morphological keys by a trained entomologist. Representative voucher specimens were retained in the entomology lab at MSSRF for reference.

11.2.5 Yield measurement

Yield data were obtained through collaboration with farmers. Each farmer harvested their respective fields, and the total grain yield was recorded after threshing, cleaning, drying, and weighing. The final yield was expressed in kg per hectare (kg ha^{-1}) for comparison across study sites.

11.2.6 Data analysis

All statistical analyzes were conducted using R software (version 4.0), (R Core Team, 2019) to evaluate the effects of chemical input levels and floral borders on pest abundance, natural enemy populations, and biodiversity indices in paddy fields. Descriptive statistics (mean±SD) were calculated for pest count, natural enemy count, and biodiversity indices. Data visualization was performed using box plots. To assess the impact of chemical input (High/Low/Organic) and floral borders (Present/Absent) on pest suppression and natural enemies, a two-way ANOVA was applied. Spatial differences (field center vs border) were analyzed using a paired t-test (for normal data) or Wilcoxon signed-rank test (for nonnormal data). Pearson's or Spearman's correlation tested the relationship between pest count and natural enemies. Biodiversity was assessed using Shannon-Weiner and Simpson's Diversity Indices, with one-way ANOVA or Kruskal-Wallis test to compare treatments. All analyses were conducted at α=0.05, and visualized using ggplot2 in R.

11.2.7 Result and discussion

11.2.7.1 Diversity and abundance of insects in paddy cultivated under wetland conditions

A total of 38 insect species were recorded in the paddy fields, comprising 23 pest species, 7 natural enemy species, and 8 general insect species. The insect pests belonged to five orders: Lepidoptera (9 species), Hemiptera (6 species), Coleoptera (3 species), Diptera (3 species), and Orthoptera (2 species). Among these, *Scirpophaga incertulas* (YSB), *Nilaparvata lugens* (BPH), *Orseolia oryzae* (gall midge), and *Cnaphalocrocis medinalis* (rice leaf folder) were dominant pests (Table 11.1), aligning with previous studies highlighting their significant impact on paddy ecosystems (Heong, Escalada, et al., 2015; Horgan et al., 2017). The presence of multiple planthoppers and stink bugs suggests a high risk of sap-feeding damage and virus transmission, as reported in rice-growing regions (Caballero et al., 2020).

The natural enemy complex included three species of Coccinellid beetles (*Harmonia axyridis, Coccinella transversalis*, and *Menochilus sexmaculatus*), one predatory ground beetle (*Ophionea nigrofasciata*), one lynx spider (*Oxyopes javanus*), and two Odonates (*Calopteryx* sp. and *Sympetrum flaveolum*) (Table 11.1). These predators play a crucial role in suppressing pest populations, as documented in rice ecosystems (Gurr et al., 2016). Coccinellid beetles are well-known aphid predators, while odonates help control dipteran pests (Yu et al., 2021). The presence of these species suggests the potential for conservation biocontrol in reducing pest populations naturally.

Eight generalist insect species, including *Acheta domesticus* (cricket), *Musca domestica* (housefly), *Baetis* sp. (mayfly), and *Culex* sp. (mosquito), were also recorded (Table 11.1). The occurrence of *Lasius niger* (ant) and *Vespa germanica* (wasp) suggests potential interactions in pest regulation through predation or competition (Landis et al., 2000).

Across the four different treatments (Low input, Low input+Floral border, High input, and High input+Field border), pest species dominated the insect community. Low-input fields recorded a higher abundance of major rice pests, with *S. incertulas* (35 individuals), *C. medinalis* (33), and *N. lugens* (30) contributing to 10.26%, 9.68%, and 8.80% of the total community composition, respectively. However, in low-input+Floral border fields, these numbers declined, with *S. incertulas* (18 individuals) and *N. lugens* (13) contributing to 5.81% and 4.19% of the insect community. The reduction in pest numbers suggests that floral borders positively influenced pest suppression, likely by supporting higher populations of natural enemies (Bianchi et al., 2006).

In high-input fields, *S. incertulas* (40 individuals, 13.07%) and *C. medinalis* (38, 12.42%) were more abundant, possibly due to the impact of chemical inputs disrupting natural enemy populations (Heong, Escalada, et al., 2015). The High input +Field border treatment showed intermediate pest levels, indicating that border cropping may mitigate pest populations even in high-input systems.

The introduction of floral and field borders significantly enhanced beneficial insect populations. *M. sexmaculatus* and *Calopteryx* sp. were most abundant in Low input+Floral border fields (25 and 29 individuals, contributing to 8.06% and 9.35% of the community composition, respectively). *C. transversalis* was absent in Low input and High input fields but recorded in Low input+Floral border (14 individuals, 4.52%) and High input+Field border (9 individuals, 2.91%), suggesting that border cropping facilitated Coccinellid abundance (Landis et al., 2000). Spider populations, represented by *O. javanus*, increased in diversified habitats, with 21 individuals (6.77%) in Low input + Floral border and 16 (5.18%) in High input + Field border (Gurr et al., 2016).

TABLE 11.1 Abundance and community composition of insect species in paddy fields.

S. no	Scientific name	Type (Pest/ natural enemy/ general insect)	Low input		Low input+floral border		High input		High input+Field border	
			Abundance	Community composition	Abundance	Community composition	Abundance	Community composition	Abundance	Community composition
1	*Scirpophaga incertulas*	Pest	35	10.26	18	5.81	40	13.07	27	8.74
2	*Nilaparvata lugens*	Pest	30	8.80	13	4.19	34	11.11	24	7.77
3	*Orseolia oryzae*	Pest	19	5.57	9	2.90	22	7.19	15	4.85
4	*Spodoptera litura*	Pest	0	0.00	4	1.29	0	0.00	5	1.62
5	*Dicladispa armigera*	Pest	9	2.64	8	2.58	10	3.27	9	2.91
6	*Scotinophora lurida*	Pest	7	2.05	2	0.65	9	2.94	5	1.62
7	*Sogatella furcifera*	Pest	3	0.88	1	0.32	3	0.98	2	0.65
8	*Nephotettix virescens*	Pest	12	3.52	7	2.26	15	4.90	13	4.21
9	*Melanitis leda ismene*	Pest	4	1.17	0	0.00	4	1.31	2	0.65
10	*Schistocerca americana*	Pest	7	2.05	6	1.94	6	1.96	7	2.27
11	*Psalis pennatula*	Pest	0	0.00	2	0.65	0	0.00	3	0.97
12	*Amsacta albistriga*	Pest	0	0.00	1	0.32	0	0.00	2	0.65
13	*Cnaphalocrocis medinalis*	Pest	33	9.68	19	6.13	38	12.42	27	8.74
14	*Archips semiferanus*	Pest	0	0.00	1	0.32	0	0.00	0	0.00

(Continued)

TABLE 11.1 (Continued)

S. no	Scientific name	Type (Pest/ natural enemy/ general insect)	Low input		Low input+floral border		High input		High input+Field border	
			Abundance	Community composition	Abundance	Community composition	Abundance	Community composition	Abundance	Community composition
15	*Anoplocnemis curvipes*	Pest	0	0.00	2	0.65	0	0.00	1	0.32
16	*Melanagromyza obtusa*	Pest	0	0.00	3	0.97	0	0.00	1	0.32
17	*Leptocorisa oratorius*	Pest	6	1.76	3	0.97	7	2.29	5	1.62
18	*Leptocorisa acuta*	Pest	8	2.35	5	1.61	7	2.29	6	1.94
19	*Melanitis ismene*	Pest	7	2.05	6	1.94	11	3.59	7	2.27
20	*Hesperia comma*	Pest	3	0.88	0	0.00	3	0.98	1	0.32
21	*Sitophilus oryzae*	Pest	4	1.17	2	0.65	5	1.63	2	0.65
22	*Nezara viridula*	Pest	0	0.00	0	0.00	0	0.00	1	0.32
23	*Pelopidas mathias*	Pest	1	0.29	0	0.00	2	0.65	0	0.00
24	*Harmonia axyridis*	Natural enemy	18	5.28	19	6.13	15	4.90	16	5.18
25	*Coccinella transversalis*	Natural enemy	0	0.00	14	4.52	0	0.00	9	2.91
26	*Menochilus sexmaculatus*	Natural enemy	20	5.87	25	8.06	13	4.25	21	6.80
27	*Calopteryx* sp.	Natural enemy	20	5.87	29	9.35	12	3.92	18	5.83
28	*Sympetrum flaveolum*	Natural enemy	11	3.23	16	5.16	5	1.63	7	2.27
29	*Ophionea nigrofasciata*	Natural enemy	5	1.47	10	3.23	4	1.31	7	2.27

30	*Oxyopes javanus*	Natural enemy	13	3.81	21	6.77	7	2.29	16	5.18
31	*Acheta domesticus*	General insect	8	2.35	5	1.61	0	0.00	0	0.00
32	*Musca domestica*	General insect	10	2.93	17	5.48	8	2.61	13	4.21
33	*Baetis* sp.	General insect	14	4.11	5	1.61	3	0.98	10	3.24
34	*Culex* sp.	General insect	15	4.40	12	3.87	14	4.58	13	4.21
35	*Chrysochus cobaltinus*	General insect	7	2.05	7	2.26	6	1.96	5	1.62
36	*Lasius niger*	General insect	6	1.76	8	2.58	2	0.65	5	1.62
37	*Vespa germanica*	General insect	4	1.17	7	2.26	1	0.33	3	0.97
38	*Hydrophilus triangularis*	General insect	2	0.59	3	0.97	0	0.00	1	0.32

Generalist insect species such as houseflies (*M. domestica*), ants (*L. niger*), and mosquitoes (*Culex* sp.) were found across treatments. Houseflies and mosquitoes were more abundant in Low input + Floral border and High input + Field border, while *L. niger* was highest in Low input+Floral border (8 individuals, 2.58%), suggesting that floral diversity may provide nesting resources. The presence of ants and wasps suggests potential mutualistic interactions influencing pest regulation.

These results indicate that habitat diversification through floral and field borders significantly alters pest-natural enemy dynamics in rice ecosystems. The observed reduction in major pests in Low input+Floral border treatments aligns with studies emphasizing noncrop vegetation's role in enhancing biological control (Landis et al., 2000; Bianchi et al., 2006). Increased natural enemy populations in diversified habitats further support findings on conservation biological control (Gurr et al., 2016).

While the study included only three farms per treatment group, which limits statistical power at the farm scale, the intensive within-farm sampling design helped enhance observational replication. Specifically, six fixed quadrats were monitored per farm across three key crop stages, yielding 216 total observations. This nested approach improved the reliability of ecological inferences at the field level. However, we acknowledge that spatial autocorrelation cannot be entirely ruled out, and future research should increase the number of independent farm replicates and consider spatially explicit models to more broadly validate treatment-level effects.

Conversely, high-input fields exhibited increased pest abundance, likely due to the disruption of natural enemies by pesticide use, reinforcing concerns that excessive chemical inputs can cause secondary pest outbreaks and disrupt ecological balance (Heong, Escalada, et al., 2015). The presence of generalist insects across all treatments highlights the broader ecological roles of nonpest insect species in agroecosystems. These findings suggest that IPM strategies incorporating habitat diversification can help regulate pest populations naturally while maintaining ecosystem stability. Future research should assess the long-term impacts of habitat manipulation on biodiversity and pest suppression in rice cultivation.

The Shannon-Wiener Index (H') values suggest that Low input+Floral border (H'=3.21) and High input+Field border (H'=3.21) treatments supported higher species diversity compared to High input (H'=2.97). A higher H' value indicates a more diverse insect community, which is beneficial for ecological balance, reducing the dominance of specific pest species, and promoting natural enemy populations (Bianchi et al., 2006). Lower Simpson's Index (D) values indicate greater species richness and even distribution, whereas higher D values suggest dominance by fewer species. The High input treatment (D=0.0678) had the highest dominance, suggesting that certain pest species were more prevalent under intensive input conditions, possibly due to reduced natural enemy populations. The Low input+Floral border (D=0.0486) and High input+Field border (D=0.0492) treatments had the lowest dominance, implying that these conditions favored a more balanced insect community, likely due to additional habitat and resources for natural enemies (Tscharntke et al., 2016). Evenness (E) measures how uniformly individuals are distributed among species. The Low input treatment (E=0.9201) had the highest evenness, suggesting a well-balanced insect community with no single species dominating. Evenness was slightly lower in the High input treatment (E=0.8909) (Table 11.2), indicating a less balanced species distribution, likely due to an increase in dominant pest populations (Karp et al., 2018). The inclusion of floral and field borders significantly influenced insect dynamics: Low input+Floral border fields demonstrated reduced pest abundance and increased natural enemy diversity, reinforcing the ecological benefits of border cropping. High-input fields showed higher pest abundance but lower natural enemy populations, likely due to chemical inputs disrupting natural enemy populations. High input+Field border fields provided an intermediate effect, promoting some level of natural enemy augmentation but not as effectively as the floral border treatment.

11.2.8 Effect of inputs and floral borders on major pests of paddy

The presence of floral borders led to a marked reduction in major rice pests. The population of YSB declined significantly from 24.67 (Low Input) to 15.00 (Low Input+Flower Border) and from 18.33 (High Input) to 12.00 (High Input+Flower Border). Gall midge populations also decreased from 11.00 to 6.00 under Low Input conditions and from 15.00 to 12.00 under High Input conditions with floral borders. Leaf folder and BPH populations exhibited a similar decline, with BPH numbers dropping from 21.33 (High Input) to 15.00 (High Input+Flower Border) (Table 11.3; Fig. 11.1). These reductions suggest that floral borders provide refuge and resources for natural enemies that suppress pest populations, a trend consistent with previous research on habitat diversification (Bianchi et al., 2006; Gurr et al., 2017; Lu et al., 2014). The highest median abundance of YSB was observed in Low Input treatments (24–27 individuals per replication), while the lowest was in High Input+Flower Border treatments (10–13 individuals per replication). This suggests that while high-input practices such as fertilizers and pesticides reduce pest pressure, the addition of floral borders further enhances pest

TABLE 11.2 Diversity indices of insects in paddy fields under different treatments.

Treatment	Shannon-Wiener Index (*H'*)	Simpson's Index (*D*)	Evenness Index (*E*)
Low input	3.13	0.0536	0.9201
Low input+Floral border	3.21	0.0486	0.9113
High input	2.97	0.0678	0.8909
High input+Field border	3.21	0.0492	0.9028

suppression, likely by attracting parasitoids such as *Trichogramma* spp. and predators like spiders and mirid bugs (Gurr et al., 2016). Leaf folder populations showed a similar decline, with median values dropping from 6 to 10 individuals in Low Input treatments to 3–6 individuals in Low Input+Flower Border treatments, an effect linked to increased natural enemy abundance, particularly coccinellids and spiders (Heong et al., 1991). Gall midge infestation was highest in Low Input fields (8–16 individuals per replication) and lowest in High Input+Flower Border fields (11–13 individuals per replication). The additional reduction in floral border treatments suggests an ecological advantage in which alternative nectar sources improve parasitoid survival and enhance biological control (Gurr et al., 2017). BPH populations peaked in High Input fields (19–25 individuals per replication) but were significantly suppressed in floral border treatments (14–16 individuals per replication in High Input+Flower Border treatments) (Table 11.3; Fig. 11.1), supporting previous findings that excessive nitrogen fertilization promotes sap-feeding insect outbreaks while habitat diversification suppresses them through increased predator populations (Lu et al., 2007; Settle et al., 1996). GLH populations showed minimal variation across treatments, indicating that other environmental or plant-related factors, such as varietal resistance and plant health, may play a larger role in regulating their abundance.

The introduction of floral and field borders significantly enhanced the presence of beneficial insects, particularly predatory species such as coccinellids, dragonflies, and spiders. *M. sexmaculatus* and *Calopteryx* sp. were most abundant in Low Input+Floral Border fields, with 25 and 29 individuals, respectively, contributing to 8.06% and 9.35% of the community composition. *C. transversalis* was absent in Low Input and High Input fields but recorded in Low Input+Floral Border (14 individuals, 4.52%) and High Input+Field Border (9 individuals, 2.91%), suggesting that border cropping facilitated coccinellid abundance (Landis et al., 2000). Spider populations, represented by *O. javanus*, increased in diversified habitats, with 21 individuals (6.77%) in Low Input+Floral Border and 16 (5.18%) in High Input+Field Border (Gurr et al., 2016). The spider population increased from 39.33 (Low Input) to 59.00 (Low Input+Flower Border) and from 21.00 (High Input) to 34.00 (High Input+Flower Border) (Table 11.3; Fig. 11.2), underscoring the role of floral resources in sustaining predatory arthropods (Tscharntke et al., 2016). Similarly, coccinellid beetles, which contribute to aphid and planthopper control, increased from 21.33 to 34.00 under Low Input conditions and from 27.00 to 30.00 under High Input conditions. Mirid bug populations also rose significantly, from 15.00 to 27.00 (Low Input) and from 8.00 to 10.00 (High Input) (Table 11.3; Fig. 11.2), reinforcing their role in biological control strategies (Gurr et al., 2016). Generalist insect species such as houseflies (*M. domestica*), ants (*L. niger*), and mosquitoes (*Culex* sp.) were recorded across treatments, with ants showing the highest abundance in Low Input+Floral Border fields, indicating that floral diversity may provide nesting resources.

The results of the ANOVA analysis revealed significant variations in pest populations across different treatments, except for the GLH, which did not show significant differences. The presence of floral borders in paddy fields substantially influenced pest abundance, as indicated by significant *F*-values for YSB (F=27.96, P<.001), leaf folder (F=15.46, P=.001), gall midge (F=6.72, P=.014), and BPH (F=21.78, P<.001) (Table 11.4). The significant reduction in pest populations in treatments with floral borders aligns with studies demonstrating that diversified cropping systems suppress pest outbreaks by enhancing natural enemy populations (Bianchi et al., 2006). The abundance of natural enemies, including spiders, coccinellids, and mirid bugs, varied significantly across treatments. The Low Input+Flower Border treatment recorded the highest spider abundance (Mean±SD: 59.0±8.7), followed by the High Input+Flower Border treatment (34.0±3.6), while the High Input treatment had the lowest spider population (21.0±1.0). A similar trend was observed in coccinellids, with the Low Input+Flower Border treatment exhibiting the highest population (34.0±2.6), followed by the High Input+Flower Border treatment (30.0±1.7). The High Input treatment had a slightly higher coccinellid population (27.0±1.0) compared to the Low Input treatment (21.3±3.2) (Table 11.4). The abundance of mirid

TABLE 11.3 Descriptive statistics of insects in paddy fields under different treatments.

Common name	Type	Low Input		Low Input+Flower Border		High Input		High Input+Flower Border	
		Mean	SD	Mean	SD	Mean	SD	Mean	SD
Yellow stem borer	Pest	24.67	2.08	15.00	2.00	18.33	1.15	12.00	1.73
Leaf folder	Pest	8.00	2.00	4.00	1.73	12.00	1.00	10.33	1.15
Gall midge	Pest	11.00	4.36	6.00	1.00	15.00	2.00	12.00	1.00
Brown plant hopper	Pest	12.00	1.00	9.33	1.53	21.33	3.21	15.00	1.00
Green leaf hopper	Pest	5.33	0.58	3.33	0.58	5.00	1.00	4.00	1.73
Spiders	Natural enemy	39.33	3.79	59.00	8.54	21.00	1.00	34.00	3.61
Coccinellids	Natural enemy	21.33	3.21	34.00	2.65	27.00	1.00	30.00	1.73
Mirid bug	Natural enemy	15.00	1.00	27.00	2.00	8.00	1.00	10.00	1.73

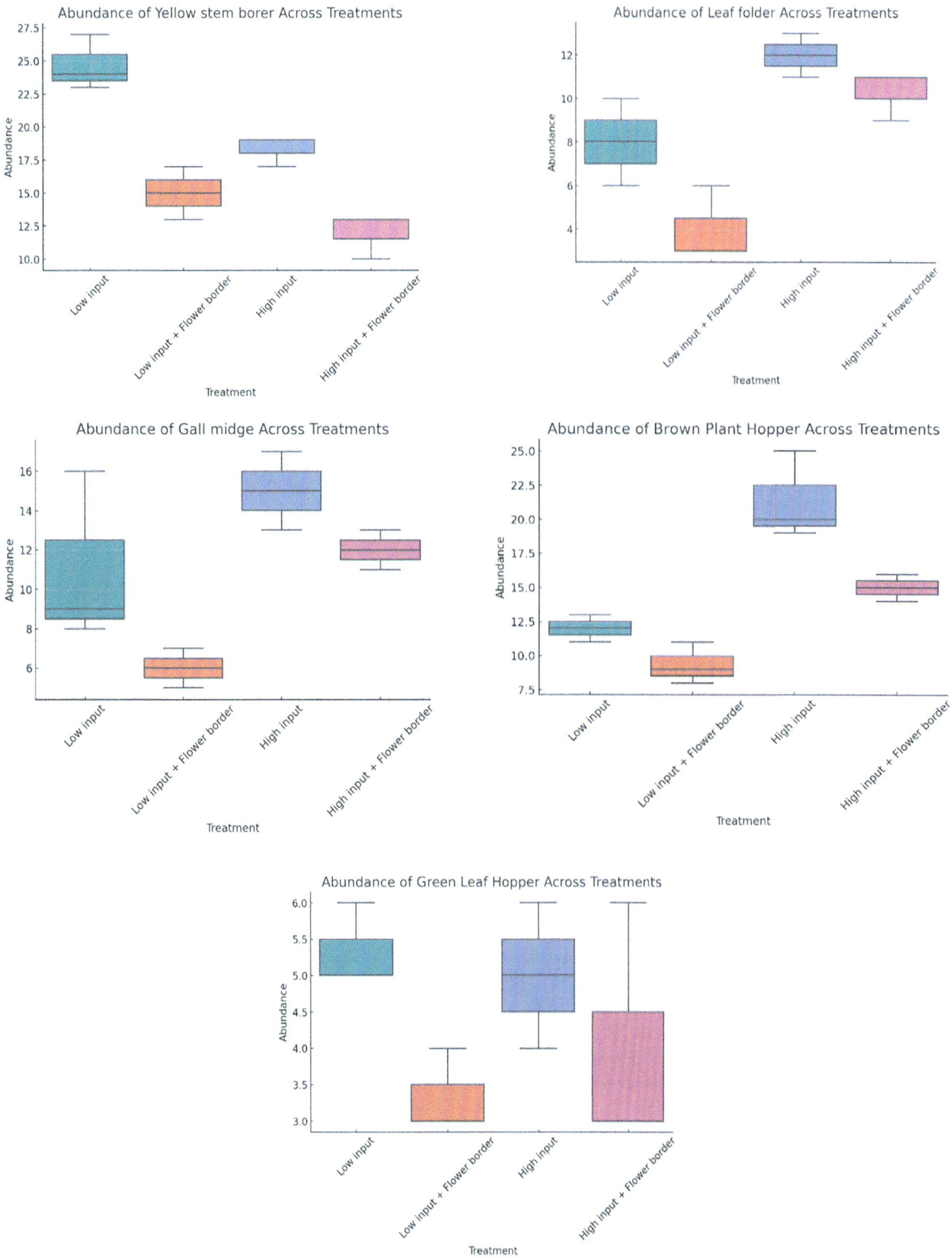

FIGURE 11.1 Effect of inputs and floral borders on the abundance of major insect pests in paddy fields in the wetland ecosystem.

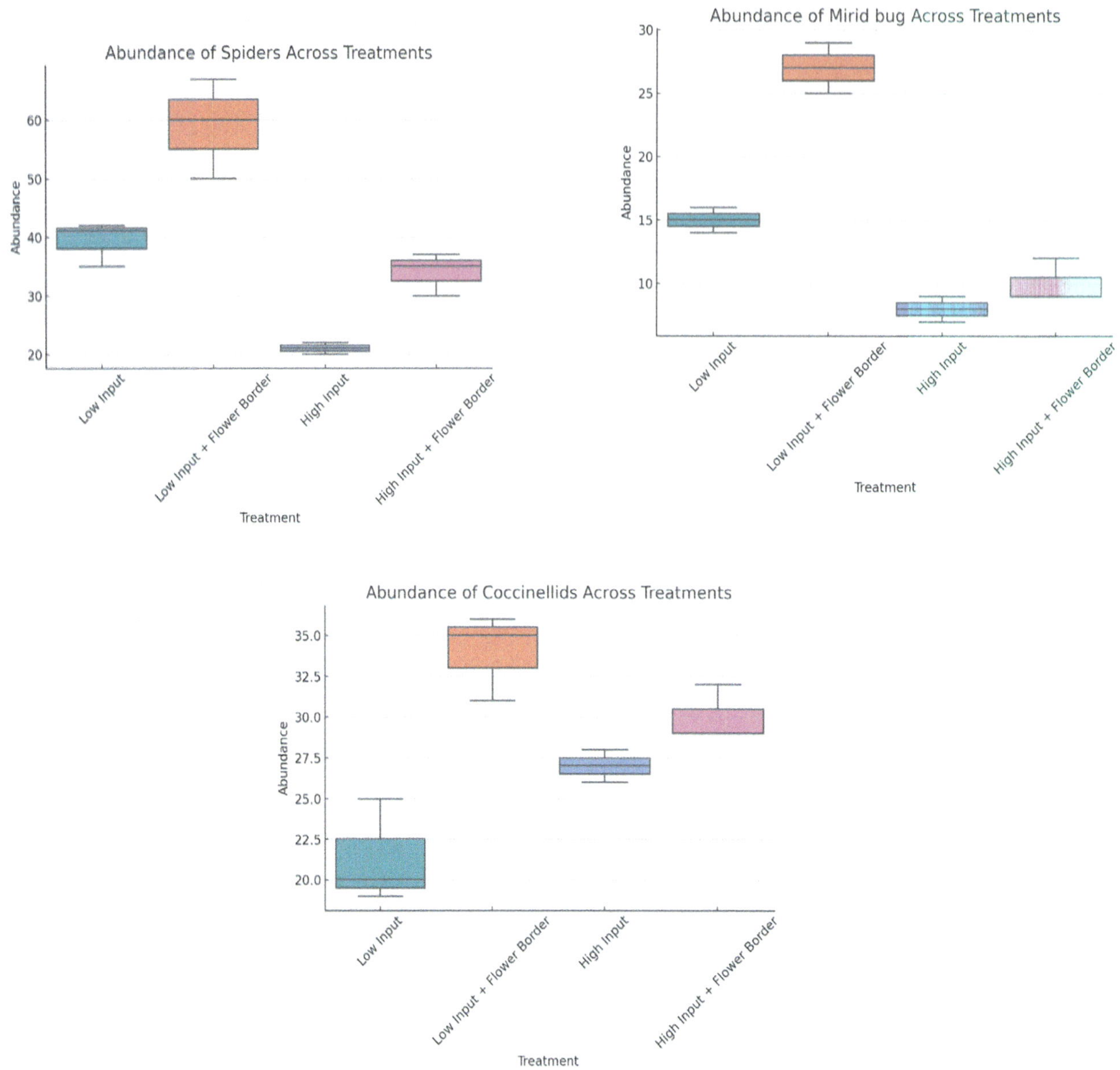

FIGURE 11.2 Effect of inputs and floral borders on the abundance of natural enemies of major insect pests in paddy fields in wetland ecosystem.

bugs was also significantly influenced by the treatments, with the Low Input+Flower Border treatment supporting the highest population (27.0±2.0), while the High Input treatment recorded the lowest (8.0±1.0). The High Input+Flower Border treatment exhibited a moderate increase in mirid bug abundance (10.0±1.7) compared to the High Input treatment.

The presence of floral borders significantly influenced the abundance of key natural enemies, including spiders (F=29.5, P<.001), coccinellids (F=16.02, P=.001), and mirid bugs (F=96.89, P<.001). The substantial increase in spider populations in floral border treatments suggests that these habitats provide shelter, alternative prey, and favorable microclimatic conditions for predatory arthropods (Schmidt-Entling & Döbeli, 2009). Similarly, the significant rise in coccinellid and mirid bug populations underscores the importance of floral diversity in sustaining predatory insects that contribute to biological control. The positive impact of floral borders on mirid bug populations is particularly noteworthy given their role in suppressing BPH populations. The observed increase in mirid bug numbers suggests that floral borders enhance their survival and reproduction, reducing the need for chemical insecticides (Gurr et al., 2017). To better interpret the ecological roles of natural enemies observed in the study, we compiled a supplementary table (Table S1) that outlines the type and key functional traits (e.g., feeding guild, prey type, and habitat use) of dominant arthropods, particularly spiders, coccinellids, and mirid bugs. For instance, *O. javanus* (lynx spider) is a generalist predator known to hunt actively

TABLE 11.4 ANOVA for major pests and natural enemies in paddy fields of the wetland ecosystem.

Pest/Natural enemy	Sum of squares (treatment)	df (treatment)	*F*-value	*P*-value
Yellow stem borer	265.67	3	27.96	.0001
Leaf folder	108.25	3	15.46	.0011
Gall midge	126.00	3	6.72	.0141
Brown plant hopper	239.58	3	21.78	.0003
Green leaf hopper	7.58	3	2.17	.1699
Spiders	2242.00	3	29.50	.0001
Coccinellids	256.25	3	16.02	.0010
Mirid bug	654.00	3	96.89	.0000

on plant foliage and suppress various lepidopteran and homopteran pests. Ladybird beetles such as *C. transversalis* and *M. sexmaculatus* are specialized in feeding on soft-bodied insects like aphids and hoppers, contributing to population regulation of *N. lugens* and *Nephotettix virescens*. Mirid bugs, as omnivorous predators, play a critical role in suppressing planthoppers while coexisting with crop plants. These functional insights reinforce the importance of floral border habitats in sustaining predator diversity and improving pest suppression services.

Although the current study focused primarily on pest dynamics and biodiversity, several qualitative observations were made during field visits related to soil health and field conditions. Farmers practicing low-input methods, particularly in combination with floral borders, reported improved soil friability and reduced surface crusting, especially after rainfall. Field staff also observed reduced compaction and easier root penetration in these plots. These perceived improvements could be linked to the application of organic inputs such as Panchakavya and farmyard manure, which are known to stimulate microbial activity and improve soil structure. However, no direct measurements of microbial biomass, nutrient cycling, or water quality were taken. Future studies should incorporate soil biochemical assays and water quality monitoring to better understand the broader ecosystem service benefits of reduced-input, biodiversity-enhancing practices.

These findings demonstrate that incorporating floral borders in paddy ecosystems effectively enhances natural enemy abundance while suppressing pest populations. The balance between pests and their natural enemies is crucial for sustainable pest management, reducing dependency on chemical pesticides that can disrupt ecosystem services. The significant variations observed across treatments highlight the potential of floral borders as a habitat management strategy to mitigate pest outbreaks in rice fields. Future studies should focus on identifying specific floral species that optimize natural enemy conservation while minimizing any potential trade-offs in agronomic performance. Additionally, long-term monitoring of biodiversity shifts in floral border fields could provide further insights into their role in ecological intensification strategies for rice cultivation.

11.2.9 Relationship between the major pests and natural enemies among the different treatments

The correlation analysis between pest populations and natural enemy abundance in the paddy ecosystem provides insights into the role of biological control in pest management. Negative correlation values indicate that an increase in natural enemies is associated with a decline in pest populations.

Spiders showed strong negative correlations with leaf folder (−0.82), gall midge (−0.84), and BPH (−0.84), suggesting their significant role in reducing these pests (Table 11.5). Previous studies confirm that spiders effectively control leaf folders and hoppers in rice fields (Schmidt-Entling & Chatterjee, 2020; Taher et al., 2018). The weaker correlation between spiders and YSB (−0.09) indicates that borers, which feed inside plant stems, are not their primary prey (Khan et al., 2021).

Coccinellids (ladybird beetles) exhibited moderate to weak negative correlations with pest populations, with the strongest observed against YSB (−0.80) (Table 11.5). This suggests a role in controlling stem borer larvae, though their primary prey are soft-bodied insects like aphids and whiteflies (Hodek et al., 2012). Their weaker correlations with gall

TABLE 11.5 Correlation between major insect pests and natural enemies.

Pest/Natural enemy	Yellow stem borer	Leaf folder	Gall midge	Brown plant hopper	Green leaf hopper
Spiders	−0.09	−0.82	−0.84	−0.84	−0.48
Coccinellids	−0.80	−0.45	−0.30	−0.26	−0.65
Mirid Bug	−0.04	−0.89	−0.81	−0.77	−0.42

midge (−0.30) and BPH (−0.26) suggest that other natural enemies, such as mirid bugs, may play a greater role in suppressing these pests (Kumar et al., 2019).

Mirid bugs showed strong negative correlations with leaf folder (−0.89), gall midge (−0.81), and BPH (−0.77), reinforcing their importance in pest suppression, particularly against planthoppers (Heong, Cheng, et al., 2015). Their weak correlation with YSB (−0.04) suggests they do not significantly prey on borers due to their concealed feeding behavior.

These findings highlight the critical role of natural enemies in maintaining lower pest populations, emphasizing the need for conservation biological control. Spiders and mirid bugs were particularly effective in suppressing key paddy pests. Enhancing their populations through habitat manipulation, such as floral borders, can improve pest control (Landis et al., 2000; Gurr et al., 2017). The weaker correlations of coccinellids with gall midge and BPH suggest that integrating multiple natural enemies is essential for effective pest management. These results align with studies advocating diversified farming systems to enhance natural enemy abundance and reduce pest outbreaks (Tscharntke et al., 2016).

11.2.10 Pest suppression efficiency in paddy fields

The study demonstrates that integrating floral border crops in both low-input and high-input paddy cultivation significantly enhanced pest suppression and natural enemy abundance. The pest suppression efficiency (PSE) analysis confirmed that floral borders effectively reduced pest populations while boosting beneficial arthropods.

In low-input fields, suppression efficiencies were highest for leaf folder (50.00%) and gall midge (45.45%), followed by YSB (39.19%) and GLH (37.50%). BPH showed a moderate decline of 22.22%, indicating that floral interventions effectively mitigate key paddy pests (Table 11.6). These findings align with research showing that diversified habitats enhance pest suppression by attracting natural enemies (Gurr et al., 2017; Zhang et al., 2020).

In high-input fields, pest suppression was evident but slightly lower. YSB suppression was 34.55%, while BPH, gall midge, and leaf folder reductions were 29.69%, 20.00%, and 13.89%, respectively. The lower suppression rates in high-input fields may result from increased fertilizers and pesticides, which can alter pest dynamics and predator-prey interactions (Chaplin-Kramer et al., 2011).

Floral borders significantly increased key natural enemies like spiders, coccinellids, and mirid bugs. In low-input systems, spiders increased by 50.00%, coccinellids by 59.38%, and mirid bugs by 80.00%, highlighting their role in pest suppression. Similar trends were observed in high-input fields, with increases of 61.90% (spiders), 11.11% (coccinellids), and 25.00% (mirid bugs) (Table 11.6). The greater impact of floral borders under low-input conditions aligns with previous studies, suggesting that reduced chemical inputs favor predator populations (Landis et al., 2000; Tschumi et al., 2016).

Habitat diversification through floral borders creates a favorable environment for natural enemies, enhancing their predatory potential. The significant rise in spider and mirid bug populations indicates their key role in suppressing pests like BPH and gall midge (Heong, Escalada, et al., 2015).

The stronger effects observed in low-input fields suggest that habitat manipulation is a viable alternative to chemical control, particularly in sustainable and organic farming (Gurr et al., 2016). Floral borders help maintain predator-prey balance, supporting conservation biological control in rice ecosystems.

11.2.11 Effect of floral borders and input levels on pest-to-defender ratios

The study revealed that integrating floral borders into paddy fields significantly influenced pest-to-defender ratios across treatments and spatial locations (center vs edge).

TABLE 11.6 Effect of input and floral border on pest suppression efficiency (*PSE*).

Pest	PSE (Low Input vs Low Input +Flower Border) (%)	PSE (High Input vs High Input +Flower Border) (%)
Yellow stem borer	39.19	34.55
Leaf folder	50.00	13.89
Gall midge	45.45	20.00
Brown plant hopper	22.22	29.69
Green leaf hopper	37.50	20.00

In low-input fields without floral borders, pest-to-defender ratios were higher at the center for YSB (0.471), gall midge (0.265), and BPH (0.176) compared to the edge, indicating weaker biological control within the field interior (Table 11.7). The edge had slightly lower ratios (YSB: 0.333, gall midge: 0.103, BPH: 0.179), likely due to natural enemy spillover from adjacent landscapes (Landis et al., 2000a). With floral borders, pest-to-defender ratios at the edge decreased substantially, particularly for YSB (0.195), leaf folder (0.049), and BPH (0.171), suggesting improved natural enemy recruitment and retention (Gurr et al., 2017).

In high-input fields without floral borders, pest-to-defender ratios were already lower due to pesticide use. However, floral borders further reduced these ratios, especially at the edge, where values were lowest (YSB: 0.071, leaf folder: 0.014, gall midge: 0.014, BPH: 0.043), indicating strong natural enemy activity (Tscharntke et al., 2016).

Across all treatments, pest-to-defender ratios were consistently lower at the edge, particularly in floral border treatments, suggesting enhanced predatory arthropod activity with potential spillover into field interiors (Chaplin-Kramer et al., 2011). The most notable reductions in high-input+flower border fields were seen in YSB (0.514 to 0.200 at the center, 0.195 to 0.071 at the edge), gall midge (0.162 to 0.089 at the center, 0.171 to 0.014 at the edge), and BPH (0.162 to 0.133 at the center, 0.171 to 0.043 at the edge) (Table 11.7).

The lower pest-to-defender ratios in floral border treatments highlight their role in enhancing biological control, reducing reliance on chemical inputs (Gurr et al., 2016). The edge effect suggests that natural enemies establish more readily near floral habitats, requiring further research on optimizing their movement into field interiors.

11.2.12 Effect of floral borders and input levels on yield of paddy fields in wetland ecosystem

The yield of paddy varied significantly among treatments, with the highest yield observed in the High input+Flower border treatment (4069±95 kg ha^{-1}), followed by High input (3643±85 kg ha^{-1}), Low input+Flower border (2746±70 kg ha^{-1}), and the lowest yield recorded in the Low input treatment (2117±105 kg ha^{-1}) (Fig. 11.3). The analysis revealed that integrating floral borders with different input levels positively impacted paddy yield, demonstrating the potential of ecological intensification strategies in rice cultivation.

The inclusion of flower borders in both low-input and high-input systems led to increased yields, likely due to improved pollination services, higher natural enemy populations, and better pest regulation. Studies show that floral borders enhance the diversity and abundance of spiders, coccinellids, and mirid bugs, contributing to biological pest suppression (Gurr et al., 2017; Tschumi et al., 2016). Similar results have been observed in other agroecosystems, where habitat manipulation through floral strips reduced pest infestation and increased crop yields (Landis et al., 2000). Although yield gains were evident in both low and high-input systems with floral borders, a comprehensive cost-benefit analysis is essential to assess the economic viability of this approach for smallholder adoption. Future research should integrate input costs, labor, and market returns.

The high-input treatments resulted in significantly higher yields than low-input systems, indicating the role of fertilizer and pest management interventions in crop productivity. However, in high-input systems, the addition of floral borders increased yield by 11.7%, suggesting that ecological approaches can complement conventional agronomic practices to enhance sustainable productivity.

In contrast, low-input systems had lower overall yields, but adding flower borders led to a 29.8% increase, highlighting the importance of biodiversity-driven pest regulation in resource-limited farming. Research supports that low-input

TABLE 11.7 Effect of input and floral border on Pest to defender ratio.

Intervention		Pest to defender ratio				
		Yellow stem borer	Leaf folder	Gall midge	Brown plant hopper	Green leaf hopper
Low input	Center	0.471	0.118	0.265	0.176	0.088
	Edge	0.333	0.128	0.103	0.179	0.051
Low Input+Flower Border	Center	0.514	0.189	0.162	0.162	0.108
	Edge	0.195	0.049	0.171	0.171	0.049
High Input	Center	0.195	0.049	0.098	0.122	0.073
	Edge	0.229	0.083	0.083	0.146	0.063
High Input+Flower Border	Center	0.200	0.067	0.089	0.133	0.044
	Edge	0.071	0.014	0.014	0.043	0.000

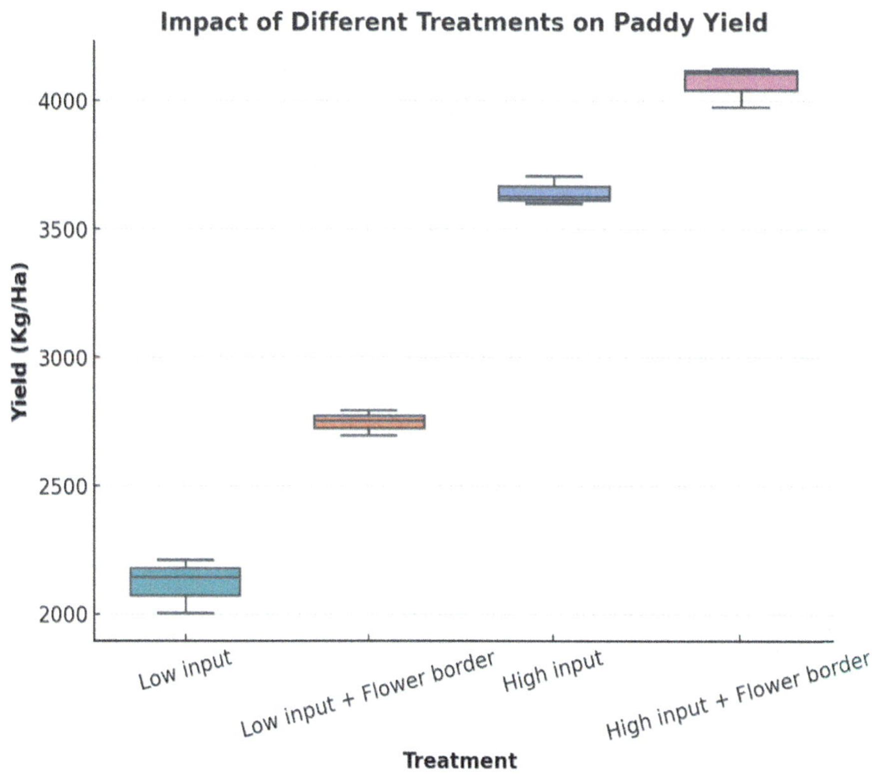

FIGURE 11.3 Effect of inputs and floral borders on paddy yield in the wetland ecosystem.

farming combined with agroecological interventions helps sustain yields while reducing chemical input dependency (Chaplin-Kramer et al., 2011).

These findings suggest that integrating floral borders in rice cultivation enhances ecosystem services, pest regulation, and yield, particularly in low-input farming systems. This aligns with global efforts to promote sustainable intensification and reduce agriculture's ecological footprint (Tilman et al., 2011). Future studies should assess long-term biodiversity benefits and economic feasibility to encourage broader adoption among farmers.

Appendix A Supplementary material

Supplementary material related to this chapter can be found online at doi:10.1016/B978-0-443-34222-6.00009-7.

Reference

Anjum, S. A., Ullah, R., Ashraf, U., et al. (2020). Agricultural biodiversity and ecological intensification for pest management in rice agroecosystems: A review. *Agriculture, Ecosystems & Environment, 298*, 106968.

Bianchi, F. J., Booij, C. J., & Tscharntke, T. (2006). Sustainable pest regulation in agricultural landscapes: A review on landscape composition, biodiversity and natural pest control. *Proceedings of the Royal Society B: Biological Sciences, 273*(1595), 1715–1727.

Bouman, B. A. M., Humphreys, E., Tuong, T. P., & Barker, R. (2007). Rice and water. *Advances in Agronomy, 92*, 187–237.

Caballero, J., Barrion, A., & Carino, F. (2020). Insect pests and their natural enemies in rice ecosystems: A comprehensive review. *Journal of Applied Entomology, 144*(3), 215–230.

Chaplin-Kramer, R., O'Rourke, M. E., Blitzer, E. J., & Kremen, C. (2011). A meta-analysis of crop pest and natural enemy response to landscape complexity. *Ecology Letters, 14*(9), 922–932.

Choudhury, A. T. M. A., & Kennedy, I. R. (2005). Nitrogen fertilizer losses from rice soils and control of environmental pollution problems. *Communications in Soil Science and Plant Analysis, 36*(11–12), 1625–1639.

Das, S., Chowdhury, R., & Ganguly, A. (2019). Impacts of pesticide use in paddy fields on wetland health: A case study from India. *Ecotoxicology and Environmental Safety, 176*, 177–184.

FAO (2021). *Rice market monitor*. Food and Agriculture Organization of the United Nations.

Finlayson, C. M., Davidson, N. C., & Stroud, D. A. (2018). The Ramsar convention and world wetlands: The role of the convention in the conservation and wise use of wetlands. *Marine and Freshwater Research, 69*(1), 6–18.

Ghosh, S., & Das, S. (2020). Impacts of agrochemical pollutants on wetlands: A review. *Ecotoxicology and Environmental Safety, 190*, 110114.

Gupta, P. K., & Seth, A. (2007). A review of agrochemical use and its effects on wetland ecosystems in India. *Current Science, 93*(4), 564–570.

Gurr, G. M., Lu, Z., Zheng, X., Xu, H., Zhu, P., Chen, G., Yao, X., Cheng, J., & Catindig, J. L. (2016). Multi-country evidence that crop diversification promotes ecological intensification of agriculture. *Nature Plants, 2*(3), 16014.

Gurr, G. M., Wratten, S. D., & Landis, D. A. (2017). Habitat management to suppress insect pests. *Annual Review of Entomology, 62*, 91–109.

Gurr, G. M., Wratten, S. D., & Luna, J. M. (2004). Multi-function agricultural biodiversity: Pest management and other benefits. *Basic and Applied Ecology, 5*(2), 107–116.

Heong, K. L., Aquino, G. B., & Barrion, A. T. (1991). Arthropod community structures of rice ecosystems in the philippines. *Environmental Entomology, 20*(3), 547–554.

Heong, K. L., Cheng, J. A., & Escalada, M. M. (2015). *Rice planthoppers: Ecology, management, socioeconomics and policy*. Springer.

Heong, K. L., Escalada, M. M., & Horgan, F. G. (2015). Ecology and management of rice insect pests. *Agricultural Systems, 138*, 1–12.

Hodek, I., van Emden, H. F., & Honek, A. (2012). *Ecology and behaviour of the ladybird beetles (Coccinellidae)*. Wiley-Blackwell.

Horgan, F. G., Stuart, A. M., & Gurr, G. M. (2017). Integrated pest management for sustainable rice production. *Pest Management Science, 73*(4), 751–758.

IMD (Indian Meteorological Department). (2020). Observed Rainfall Trend Over States. https://imdpune.gov.in/hydrology/rainfall%20variability%20page/rainfall%20trend.html.

Karp, D. S., Chaplin-Kramer, R., Meehan, T. D., Martin, E. A., DeClerck, F., Grab, H., & Zou, Y. (2018). Crop pests and predators exhibit inconsistent responses to surrounding landscape composition. *Proceedings of the National Academy of Sciences, 115*(33), E7863–E7870.

Khan, M. A., Jaleel, W., Aftab, M., & Ahmad, S. (2021). Biological control of insect pests in rice. *Journal of Asia-Pacific Entomology, 24*(1), 73–85.

Kumar, R., Pandey, S. K., & Singh, S. P. (2019). Role of natural enemies in the management of insect pests of rice. *Indian Journal of Entomology, 81*(4), 586–592.

Landis, D. A., Wratten, S. D., & Gurr, G. M. (2000). Habitat management to conserve natural enemies of arthropod pests in agriculture. *Annual Review of Entomology, 45*(1), 175–201.

Lu, Y., Wu, K., Jiang, Y., Xia, B., Li, P., Feng, H., & Wyckhuys, K. A. G. (2014). Widespread adoption of Bt cotton and insecticide decrease promotes biocontrol services. *Nature, 510*(7503), 110–113.

Lu, Z., Heong, K. L., Yu, X., Hu, C., & Cheng, J. (2007). Effects of nitrogen fertilizer on herbivores and its stimulation to major insect pests in rice. *Rice Science, 14*(1), 56–66.

Ministry of Water Resources (MoWR), Government of India. (2014). Status of Wetlands in India. Ministry of Jal Shakti.

Prasad, S. N., Ramachandra, T. V., Ahalya, N., Sengupta, T., Alakananda, M., & Thirumala, S. (2002). Conservation of Wetlands in India: A review. *Tropical Ecology, 43*(1), 173–186.

R. Core Team. 2019. "R: A Language and Environment for Statistical Computing." Vienna, Austria. https://www.R-project.org/.

Schäfer, R. B., von der Ohe, P. C., Rasmussen, J., et al. (2016). Thresholds for the effects of pesticides on invertebrate communities and leaf litter breakdown in european streams. *Environmental Toxicology and Chemistry, 35*(2), 190–202.

Schmidt-Entling, M. H., & Chatterjee, S. (2020). Predation pressure in rice fields: Spiders as natural enemies of pests. *Agricultural and Forest Entomology, 22*(1), 37–46.

Schmidt-Entling, M. H., & Döbeli, J. (2009). Sown wildflower areas to enhance spiders in arable fields. *Agriculture, Ecosystems & Environment, 133*(1–2), 19–22.

Secretariat, Ramsar Convention. (2016). An introduction to the ramsar convention on wetlands. Ramsar Convention Bureau.

Settle, W. H., Ariawan, H., Astuti, E. T., Cahyana, W., Hakim, A. L., Hindayana, D., & Lestari, A. S. (1996). Managing tropical rice pests through conservation of generalist natural enemies and alternative prey. *Ecology, 77*(7), 1975–1988.

Singh, R. P., Tripathi, R., & Prasad, S. N. (2020). Agrochemical contamination in indian rice fields: Challenges for wetland ecosystems. *Environmental Monitoring and Assessment, 192*(10), 637.

Taher, M., Ohtsuka, K., & Takabayashi, J. (2018). The role of spiders in rice pest management: A case study in Southeast Asia. *Journal of Economic Entomology, 111*(2), 584–592.

Tilman, D., Balzer, C., Hill, J., & Befort, B. L. (2011). Global food demand and the sustainable intensification of agriculture. *Proceedings of the National Academy of Sciences, 108*(50), 20260–20264.

Tilman, D., Cassman, K. G., Matson, P. A., Naylor, R., & Polasky, S. (2002). Agricultural sustainability and intensive production practices. *Nature, 418*(6898), 671–677.

Tscharntke, T., Karp, D. S., Chaplin-Kramer, R., Batáry, P., DeClerck, F., Gratton, C., & Martin, E. A. (2016). When natural habitat fails to enhance biological pest control–five hypotheses. *Biological Conservation, 204*, 449–458.

Tschumi, M., Albrecht, M., Entling, M. H., & Jacot, K. (2016). High effectiveness of tailored flower strips in reducing pests and crop plant damage. *Proceedings of the Royal Society B: Biological Sciences, 283*(1823), 20152529.

Uphoff, N., Kassam, A., & Harwood, R. (2018). *System of rice intensification (SRI): Advancements and future perspectives*. Springer.

Yadav, S. S., Kumar, R., Yadav, K. S., & Pal, S. (2021). Sustainable paddy cultivation: Practices and perspectives to minimize agrochemical use. *Journal of Cleaner Production, 282*, 124545.

Yu, X., Chen, X., & Zhou, H. (2021). Odonates as bioindicators for pest control in paddy ecosystems. *Ecological Applications, 31*(6), e02354.

Zhang, W., Lu, Y., Werf, W., Huang, J., Wu, F., Zhou, K., & Ouyang, F. (2020). Deciphering the mechanisms of natural enemy enhancement by floral diversification in agricultural landscapes: A meta-analysis. *Ecology Letters, 23*(6), 888–901.

Chapter 12

Wetlands of Himachal Pradesh: biodiversity reservoirs in the Northwest Himalaya

Kranti Thakur[1], Rajni Dhalaria[2], Sarita Pathania[3], Jagdeep Verma[4], and Deepak Pathania[5]

[1]*Department of Botany, Shoolini Institute of Life Sciences and Business Management, Solan, Himachal Pradesh,* [2]*Department of Biosciences, Minerva P.G. College of Arts, Sciences and Commerce, Kangra, Himachal Pradesh, India,* [3]*Department of Zoology, Sardar Patel University, Mandi, Himachal Pradesh,* [4]*Department of Botany, Sardar Patel University, Mandi, Himachal Pradesh, India,* [5]*Department of Environmental Sciences, Central University of Jammu, Jammu, Jammu and Kashmir, India*

12.1 Introduction

Wetlands are land areas characterized by the presence of water, either permanently or seasonally. They encompass diverse ecosystems such as swamps, fens, marshes, bogs, and pocosins (Balwan & Kour, 2021). Due to their transitional (ecotonal) nature, wetlands exhibit unique characteristics influenced by seasonal variability, making the delineation of their boundaries challenging (Arya et al., 2020). These ecosystems offer critical habitats for a broad range of flora and fauna, including reptiles, amphibians, and invertebrates (Mitsch & Gosselink, 2000), functioning as "organic supermarkets" that provide food and shelter. Biodiversity in wetlands is strongly influenced by water quality and surrounding vegetation (Buckton, 2007).

Historically, wetlands have been central to human civilization, offering essential services such as freshwater provisioning, flood regulation, carbon sequestration, climate moderation, recreational opportunities, and biodiversity conservation (Alikhani et al., 2021) (Fig. 12.1). Despite their ecological importance, wetlands worldwide face increasing threats from anthropogenic activities and climate change (Zhang et al., 2024), necessitating effective and adaptive management strategies.

In view of conserving and making wise use of wetlands and their resources, an intergovernmental treaty (Ramsar Convention) is already in place. India became a party to it in the year 1982. Wetlands cover around 4.1 million hectares (excluding irrigated agricultural lands, rivers, and streams) of land area in India, of which 1.5 million hectares are natural and 2.6 million ha are man-made (Shan et al., 2021). Ninety-One (91) wetlands in India are of international importance (Ramsar Wetlands), which include lakes, marshes, river stretches, mangrove swamps, lagoons, estuaries, and Himalayan high-altitude wetlands. As many as fourteen such wetlands are located in the northwest Himalaya. The Union Territories of Jammu and Kashmir and Ladakh have five and two Ramsar sites, respectively, whereas Himachal Pradesh has three, and Uttarakhand has one. Despite the ecological significance of Himalayan wetlands, information regarding their distribution and biodiversity remains fragmented. In this chapter, an attempt has been made to compile the scattered information to make it available for various stakeholders (local communities, Government agencies, researchers, etc.) at a single place. It will help devise and implement sustainable management strategies for the state's ecologically fragile ecosystems.

12.2 Major wetlands in Himachal Pradesh

Himachal Pradesh is located in the lap of the Western Himalaya between 30°22' and 30°12'N latitudes and 75°47' and 79°04'E longitudes, and this North Indian state harbours approximately 2000 wetlands of varying sizes. Of these, 958 are larger than 500 m^2 (Bhambri et al., 2018). These are distributed across an altitudinal range of 450–5093 m above mean sea

Wetlands: Biodiversity, Health and their Role in Ecosystem Services. DOI: https://doi.org/10.1016/B978-0-443-34222-6.00003-6

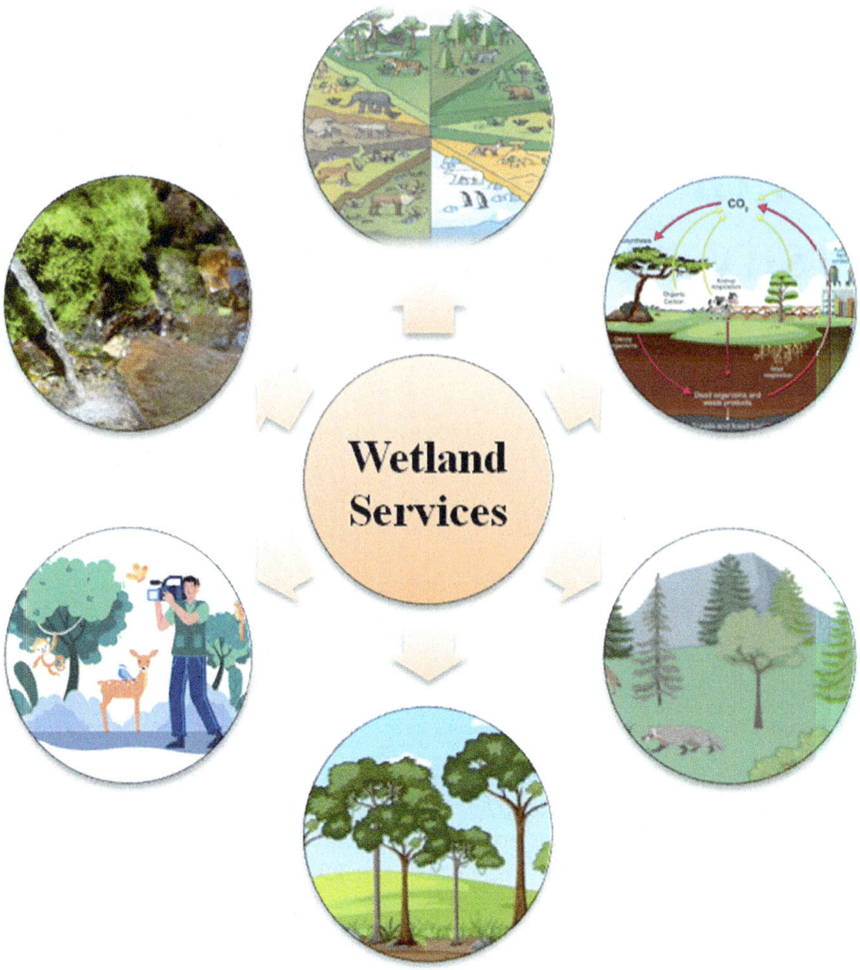

FIGURE 12.1 Important services provided by the wetlands.

level (amsl) and encompass subtropical, temperate, and alpine zones (Kohli, 2015; Sharma & Walia, 2014). Many of these wetlands serve as vital sources of livelihood for the local community and have significant biological, esthetic, and tourism value (Kohli, 2015).

Twenty-seven natural wetlands of considerably larger size, covering an area of 0.15 Km^2, and five man-made wetlands covering an area of 7.12 Km^2 occur in the state. Presently, the state has three major wetlands that have been declared as Ramsar sites. These are Pong Dam wetland, Renuka Lake, and Chandertal (Pathania & Kumar, 2017). Besides these, Rewalsar and Khajjiar lakes have been identified as wetlands of national importance (Attri, 2016; Kashyap et al., 2016). This section presents an overview of these and other key wetlands in the state, focusing on their geographic location and biodiversity profiles. Fig. 12.2 denotes the district-wise location of wetlands covered under the scope of the present paper.

12.2.1 Pong Dam wetland

This wetland, also known as the Maharana Pratap Sagar Reservoir, is situated in the Kangra district of Himachal Pradesh at 30°58'N latitude and 75°50'E longitude (Kumar et al., 2005). Spanning an impressive 156.62 Km^2, it is one of the largest man-made wetlands in Northern India (Malik & Rai, 2019). Designated as a Ramsar site in 2002, this low-altitude wetland (335–430 m amsl) falls within the subtropical zone and receives an average annual rainfall of around 1780 mm (Malik & Rai, 2019). Further, it serves as an important breeding ground and transitory resting site for many migrating waterbirds travelling from the Trans-Himalayan regions (Dhadwal, 2011).

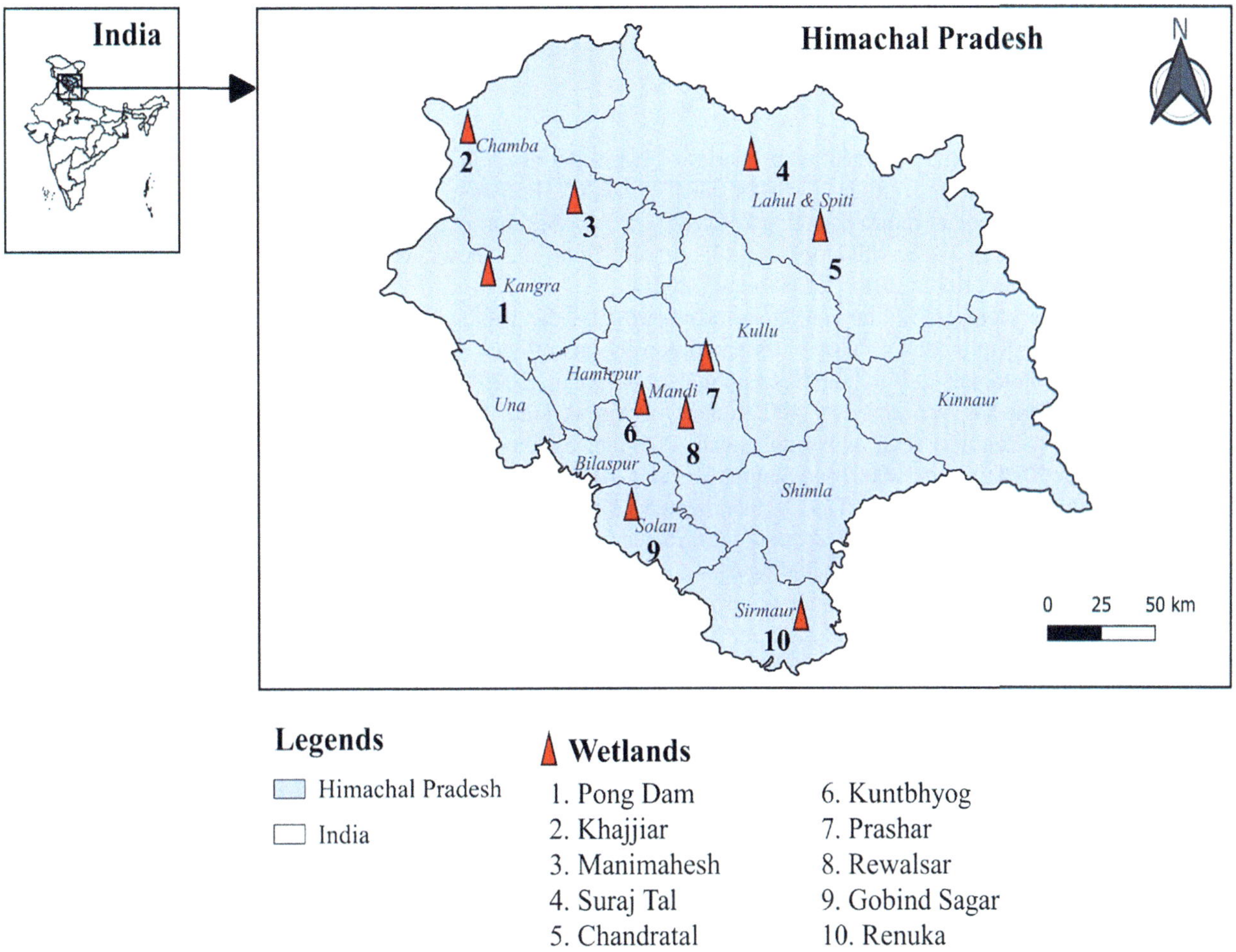

FIGURE 12.2 District-wise location of some important wetlands in Himachal Pradesh, India. Map lines delineate study areas and do not necessarily depict accepted national boundaries.

The wetland is a well-known habitat for a variety of animals, including species of fish, amphibians, reptiles, and mammals. Gaston and Pandey (1987) reported red-necked grebe (*Podiceps grisegena*) and Pandey (1989) documented 220 species of visitor and resident birds from this wetland. A huge population of Bar-headed Goose was observed by Tak et al. (2001). Jindal, and Singh, & Sharma (2013) identified 304 species of birds belonging to 52 families, of which 70 species were aquatic birds. The dominant bird species include the Himalayan barred owlet (*Glaucidium cuculoides*), Himalayan griffon vulture (*Gyps himalayensis*), hill myna (*Gracula religiosa*), etc. Pong Dam ecosystem is also an important habitat for some threatened species such as *Neophron percnopterus*, *Limosa limosa*, *Sterna acuticauda*, *Aquila heliaca*, *Aquila clanga* (*Clanga clanga*), *Haliaeetus leucoryphus,* and *Rynchops albicollis* (IUCN red list, 2025; Jindal, & Singh, & Sharma, 2013). Kumar and Paliwal (2015) documented 165 avian species from this wetland. Recently, Thakur, Radha, et al. (2022) documented 225 bird species under 150 genera of 54 families from this wetland.

In addition to avifaunal diversity, the wetlands support diverse mammal species such as nilgai, sambar, barking deer, wild boar, clawless otter, leopard, porcupine, and jackal, which have been reported from this wetland (Kohli, 2015). Kumar et al. (2005) identified 17 species of zooplankton comprising Cladocera, Rotifera, Protozoa, and Copepoda, with almost equal species proportions. This wetland also supports a diverse array of flora, including aquatic plants, grasses, and shrubs (Malik & Rai, 2019). Srivastava and Gupta (2004) and Kumar et al. (2005) reported more than 40 algal species, which belong to the classes Chlorophyceae, Bacillariophyceae, and Cyanophyceae. Thakur, Yardi, et al. (2022) provided a list of 22 wild edible plants occurring in this area, some of which (*Syzygium cumini, Dalbergia sissoo, Mangifera indica, Morus alba, Morus nigra, Ficus carica*, etc.) also act as attractants for many bird species. Shrubs (*Murraya koenigii*, *Ziziphus* spp., *Carissa opaca*, *Adhatoda vasica*, *Euphorbia* spp.), grasses (*Cymbopogon* spp., *Arundinaria* spp.), and climbers (*Bauhinia vahlii*, *Loranthus* spp.) are also of common occurrence (Kohli, 2015). According to Anuradha et al. (2023), habitat destruction, overfishing, tourist activities, and pollution are posing threats to this wetland.

12.2.2 Renuka wetland

It is a natural wetland situated in the Sirmaur district of Himachal Pradesh at 30°36′N and 77°27′E, covering an area of approximately 0.2 Km^2 at an altitude of 650 m amsl (Jindal & Prajapat, 2005). The areas experience a subtropical monsoon climate, with annual rainfall ranging between 279 and 728 mm (Choden et al., 2022; Yadav et al., 2021). The lake is nestled between two steep hill slopes, and its detached portion, known as Parasuram Tal, follows the abandoned course of the Giri River (Kohli, 2015; Yadav et al., 2021).

Due to its unique biodiversity and ecological characteristics, the wetland was designated as a Ramsar site in 2005 (Yadav et al., 2021). It provides a range of essential ecosystem services, including nutrient cycling, water purification, habitat for diverse biota, and cultural and recreational value (Gaur, 2020).

A fauna assessment has recorded over 400 animals in this wetland. Insect diversity includes more than 50 butterfly species belonging to families such as Nymphalidae, Pieridae, Papilionidae, and Satyridae (Parui & Mukherjee, 2000; Sharma & Kumar, 2015). Singh (2008) and Sengupta et al. (2019) reported the occurrence of more than 110 butterflies and dipteran flies, while Ghosh and Hegde (2013) documented 13 species of aquatic beetles distributed across 10 genera and 3 families (Dytiscidae, Gyrinidae, and Hydrophilidae). Recently, Banerjee et al. (2023) studied the faunal composition of this wetland and recorded 563 species, including 6 protozoans, 282 invertebrates, 26 fishes, eight amphibians, 114 reptiles, 103 birds, and 24 mammals. The wetland is also rich in plant species, where *Shorea robusta, Terminalia bellirica, Terminalia chebula, Cassia fistula, Toona ciliata, Bauhinia variegata, D. sissoo*, etc. represent the dominant trees (Kohli, 2015). Subramani et al. (2007) documented 395 plant species belonging to 316 genera and 115 families. Phytoplanktonic diversity is quite rich with Bacillariophyceae (diatoms) being the most dominant (56.38%) followed by green (25.17%) and blue-green (14.22%) algae (Jindal & Prajapat, 2005). Srivastava and Gupta (2009) reported the occurrence of 35 algal species across 22 genera and 12 families in this wetland.

12.2.3 Chandertal wetland

Chandertal is one of the most attractive natural high-altitude wetlands in Himachal Pradesh. It is located in Lahul and Spiti at 32°28'N and 77°37'E, covering an area of ~0.49 Km^2, at an altitude of 4420 m amsl (Rout et al., 2020). The region is characterized by an alpine climate with annual rainfall of around 750 mm. The lake derives its name, Chandertal or "Lake of the moon," from its distinct crescent shape (Singh & Thakur, 2019).

It is situated between a low ridge and the main Kunzam range; the lake drains into the Chandra River (Sharma & Samant, 2019; Singh & Thakur, 2019). In recognition of its exceptional ecological significance, Chandertal was designated as a Ramsar site in 2005 (Solanki & Ankanksha, 2021).

This wetland hosts a number of bird species, including the Himalayan snow cock (*Tetraogallus himalayensis*), magpie robin (*Copsychus saularis*), blue rock thrush (*Monticola solitarius*), common teal (*Anas crecca*), black-winged stilt (*Himantopus himantopus*), Kentish plover (*Charadrius alexandrinus*), greenshank (*Tringa nebularia*), common sandpiper (*Tringa hypoleucos*), and hoopoe (*Upupa epops*) (Sharma et al., 2012). Rana et al. (2014) listed 41 bird species belonging to 35 genera of 12 families from this wetland. Four species (*Gallinago solitaria, Calidris minuta, Charadrius mongolus, H. himantopus*) of the order Charadriiformes were documented by Singh and Thakur (2021a) and Thakur and Mehta (2015) confirmed the occurrence of garganey (*Anas querquedula*) for the first time in this wetland. This wetland is also a breeding spot for *Tadorna ferruginea* and *Anas strepera* (Mehta & Thakur, 2017; Thakur & Mehta, 2016).

Plumpley and Sharma (1990) reported some mammal species such as the Himalayan fox (*Vulpes vulpes*), snow leopard (*Panthera uncia*), lynx (*Felis lynx*), and wolf (*Canis lupus*) from within the lake catchment area. Later, Singh and Thakur (2019) enlisted 11 mammalian species under 6 families (Canidae, Felidae, Moschidae, Bovidae, Muridae, and Sciuridae). Six species of Rhopalocera (Lepidoptera) belonging to the families Nymphalidae, Pieridae, Lycaenidae, and Satyridae have also been documented (Singh & Thakur, 2021b). Banerjee et al. (2023) listed 107 faunal species from this wetland, including 15 protozoa, 22 invertebrates, 53 birds, and 17 mammals.

Due to the high altitude and harsh conditions, tree vegetation is absent. However, the herbaceous flora is remarkably diverse. The dominant species are *Aconitum violaceum, Anemone rivularis, Potentilla atrosanguinea, Pedicularis bicornuta, Gentianella moorcroftiana, Saxifraga brunonis, Juncus bufonius, Carex nigerrima, Potamogeton nodosus, Gentiana aquatica, Triglochin palustris*, etc. (Dey et al., 2017; Sharma et al., 2012). According to Dey et al. (2021), Asteraceae, Cyperaceae, Ranunculaceae, and Poaceae are the most representative families in this area. The grasses frequently met belong to the genus *Poa* and *Agropyron*.

12.2.4 Khajjiar wetland

Khajjiar wetland, located in the Chamba district at 32°32'N and 76°03'E at an elevation of 1920 m amsl, covers an area of approximately 6 km^2 (Singh & Banyal, 2014; Thakur, Radha, et al., 2022). It features a temperate to alpine climate, receiving around 800 mm of annual rainfall. Because of its scenic beauty, where a central water body is surrounded by gently sloping meadows and dense deodar (*Cedrus deodara*) forests. This distinctive landscape has earned Khajjiar the moniker "Mini-Switzerland" (Singh & Banyal, 2014).

The wetlands support a rich diversity of avifauna and mammals. Prominent bird species include the gray-headed woodpecker, Asian barred owlet, wedge-tailed green pigeon, koklass pheasant, kalij pheasant, Himalayan monal, black eagle, magpies, jungle fowl, parrots and partridges, whereas Himalayan weasel, serow, rhesus monkey, Hanuman langur, and Himalayan fox, etc. represent dominant mammals (Attri, 2016; Kohli, 2015).

Singh and Banyal (2014) studied the Lepidoptera fauna of Khajjiar Lake and reported the occurrence of 49 butterfly species under 41 genera of 10 families. Besides, six species of Herpetofauna, including two amphibians and four reptiles, were also recorded. Attri (2016) also identified some butterfly species, including *Gonepteryx rhamni*, *Pieris canidia*, *Parnassius hardwickii*, *Papilio polyctor*, and *Papilio protenor*. As many as 46 angiosperm and 3 gymnosperm species have been identified from this area, with *Pinus roxburghii*, *Quercus* spp., and *C. deodara* representing the dominant trees (Attri, 2016).

12.2.5 Rewalsar wetland

It is situated in the Middle Siwalik hills (1360 m amsl) in the Mandi district (31°37'N latitude, 76°49'E longitude). The lake has a very small surface area (about 0.3 Km^2), and the region is characterized by a subtropical monsoonal climate having an average annual rainfall of about 1136 mm (Gaury et al., 2018; Kumari & Kumar, 2020).

Despite its modest size, the wetland supports diverse wildlife. The surrounding forest patch and lake ecosystem provide habitat for numerous species, including barking deer, jackal, leopard, snakes and lizards, mongoose, porcupine, monkey, Indian coots, Kalij pheasant, white-breasted kingfisher, brown coot, etc. Some fish species such as *Rasbora rasbora*, *Puntius sarana*, *Carassius auratus*, *Cyprinus carpio*, *Channa puntius*, *Cyathus striatus*, and *Channa marulius* have also been recorded. Jindal and Thakur (2013) documented some Rotifera, Cladocera, and Copepoda from the Rewalsar wetland. Dominant flora includes Chir, Kachnar, Khair, Shisham, and species of *Nymphaea*, *Ceratophyllum*, *Trapa*, and *Phragmites*. Jindal and Thakur (2013) reported many species of Bacillariophyceae, Chlorophyceae, Cyanophyceae, Euglenophyceae, and Cryptophyceae from the Rewalsar wetland. Owing to its geographical position and religious significance, this lake attracts thousands of tourists every year (Gaury et al., 2018).

12.2.6 Prashar Lake

It is a small lake (0.02 Km^2) located in the Mandi district (31°37′N latitude, 77°48′E longitude) at an altitude of 2730 m amsl. The lake is surrounded by snow during winter and is characterized by an extremely cold and dry climate. Kumar et al. (2023) and Singh et al. (2014) documented 95 bird species from the area belonging to 74 genera and 30 families, of which 79 species were migrant and 16 were the resident ones. Additionally, thirteen (13) mammalian species have also been recorded from Prashar wetlands and the surrounding area (Singh et al., 2014). Of these, *Semnopithecus ajax* is endangered and *Ursus thibetanus* is vulnerable, whereas *Panthera pardus* and *Naemorhedus goral* are in the vulnerable and near threatened categories, respectively (IUCN Red List, 2025). Singh et al. (2015) identified three reptiles, one amphibian, and one fish species. Jindal et al. (2014) recorded 67 phytoplankton species from this wetland, and Sharma and Kumari (2018) identified more than 30 species belonging to 27 genera of zooplankton. Prashar Lake holds immense cultural and spiritual value for local communities. Traditional beliefs and practices associated with the lake contribute to biodiversity conservation. Sharma and Kumari (2018) documented 25 traditional medicinal plants used by locals for treating various ailments.

12.2.7 Kuntbhyog Lake

This lake is also situated in Mandi district (31°37′N latitude, 76°49′E longitude) at an altitude of 1750 m amsl. Surrounded by mountains and woodlands, it is a small spring-fed lake with an area of 0.03 Km^2 (Kumar et al., 2023). Jindal and Thakur (2014) investigated the hydrobiology and productivity of this lake and reported the occurrence of 30 phytoplankton and 18 zooplankton species. *Brachionus*, *Cyclops*, *Keratella*, and *Daphnia* were the dominant zooplankton genera, whereas the phytoplanktons chiefly belonged to class Bacillariophyceae, Chlorophyceae, Cyanophyceae, and

Euglenophyceae, with *Navicula, Nitzschia, Cymbella, Closterium, Cosmarium, Scenedesmus,* and *Oscillatoria* as the dominant genera.

12.2.8 Suraj Tal Lake

Suraj Tal is located in the Lahaul and Spiti district and positioned right below the Bara-lacha-la pass (32°45′N latitude, 77°23′E longitude) at an altitude of nearly 4800 m amsl. The Bhaga River originates from Suraj Tal, which converges with the Chandra River downstream near Tandi to form the Chandrabhaga River (Dey et al., 2021).

Due to its extreme alpine environment, biodiversity in and around Suraj Tal is relatively limited. However, several cold-adapted species have been reported. Kumar (1992) recorded some phytoplankton species belonging to Chlorophyceae, Cyanophyceae, and Bacillariophyceae, and some insect and crustacean species from this wetland.

Singh et al. (2014) identified 20 Cyanophycean species from this area belonging mainly to the genera *Cyanobium, Geitlerinema, Gloeocapsopsis, Leptolyngbya, Limnothrix, Synechocystis, Nostoc, Nodularia, Planktothrix, Pseudanabaena,* and *Phormidium.*

12.2.9 Gobind Sagar Lake

Gobind Sagar Lake is a large man-made reservoir located across the Bilaspur and Una districts of Himachal Pradesh, situated at approximately 31°24′N and 76°26′E. Created by the Bhakra Nangal Dam on the Sutlej River, the lake covers a surface area of about 12 km², making it one of the largest reservoirs in India (Sharma & Walia, 2014; Singh et al., 2018). More than 50 fish species have been reported from Gobind Sagar Lake (Jhingran & Sehgal, 1978; Lal & Dua, 2018; Rao Rama & Natrajan, 1979; Sharma, 2018). The wetland also supports more than 25 butterfly species (Sharma et al., 2020). According to Sarvade et al. (2016), this wetland witnesses the dominance of tropical and subtropical trees like *P. roxburghii, Acacia catechu, D. sissoo, Terminalia anogeissiana, Terminalia bellerica, T. chebula, S. cumini, Lannea coromandelica, Albizia chinensis, Albizia lebbeck*, etc.

12.2.10 Manimahesh Lake

This high-altitude lake is located in the Chamba district (32°23′N latitude, 76°38′E longitude) of the state at an altitude of 4088 m amsl (Suman et al., 2015). It lies at the base of Mount Kailash (Manimahesh Kailash), which holds deep religious and spiritual significance for Hindus. The lake is snow-bound for much of the year and is accessible primarily during the summer pilgrimage season (Sharma et al., 2012).

Attri et al. (2010) reported the occurrence of mammal species like the Himalayan languor (*Presbytis entellus*), red-faced monkey (*Macaca mulatta*), goral (*N. goral*), Himalayan fox (*Vulpes bengalensis*), etc. from this wetland. Some bird species including Monal (*Lophophorus impejanus*), western tragopan (*Tragopan melanocephalus*), koklass pheasant (*Pucrasia macrolopha*), White-crested Kalij (*Lophura leucomelana*), Himalayan Snow Cock (*T. himalayensis*), chukar (*Alectoris chukar*) have also been identified. Additionally, Sharma et al. (2012) reported that the main trees found in the lake's catchment area are Bhojpatra (*Betula utilis*), Rai (*Picea smithiana*), poplar (*Populus ciliata*), Buransh (*Rhododendron campanulatum*), and Diyar (*C. deodara*). The area is also rich in medicinal plant diversity; the notable among these were Patlehan (*Heracleum candicans*), Dhoop (*Jurinea macrocephala*), Guchhi (*Morchella esculenta*), Karu (*Picrorhiza kurroa*), Bankakri (*Podophyllum hexandrum*), Mushkbala (*Valeriana wallichii*), Banfsha (*Viola odorata*), Ratanjot (*Arnebia benthamii*), Dudali (*Tragopogon pratense*), Malora (*Rumex acetosa*), etc.

12.3 Importance of wetlands

Wetlands are among the most productive ecosystems on Earth, providing a multitude of ecological and societal benefits. They are extensively used for drinking water supply, irrigation, fisheries, and recreational activities. In addition to provisioning services, wetlands deliver vital ecological functions such as biodiversity conservation, carbon sequestration, flood attenuation, nutrient cycling, and groundwater recharge (Fig. 12.3). They possess distinctive ecological characteristics that offer a wide range of benefits and services to humanity (Arya et al., 2020; Prasad et al., 2002). Therefore they play an important role in regulating and maintaining the ecological balance.

12.3.1 Reservoirs of biodiversity

Wetlands host a wide array of floral, faunal, and microbial communities due to their unique hydrological and ecological conditions. These biotic components interact to maintain the functional integrity and equilibrium of the ecosystem.

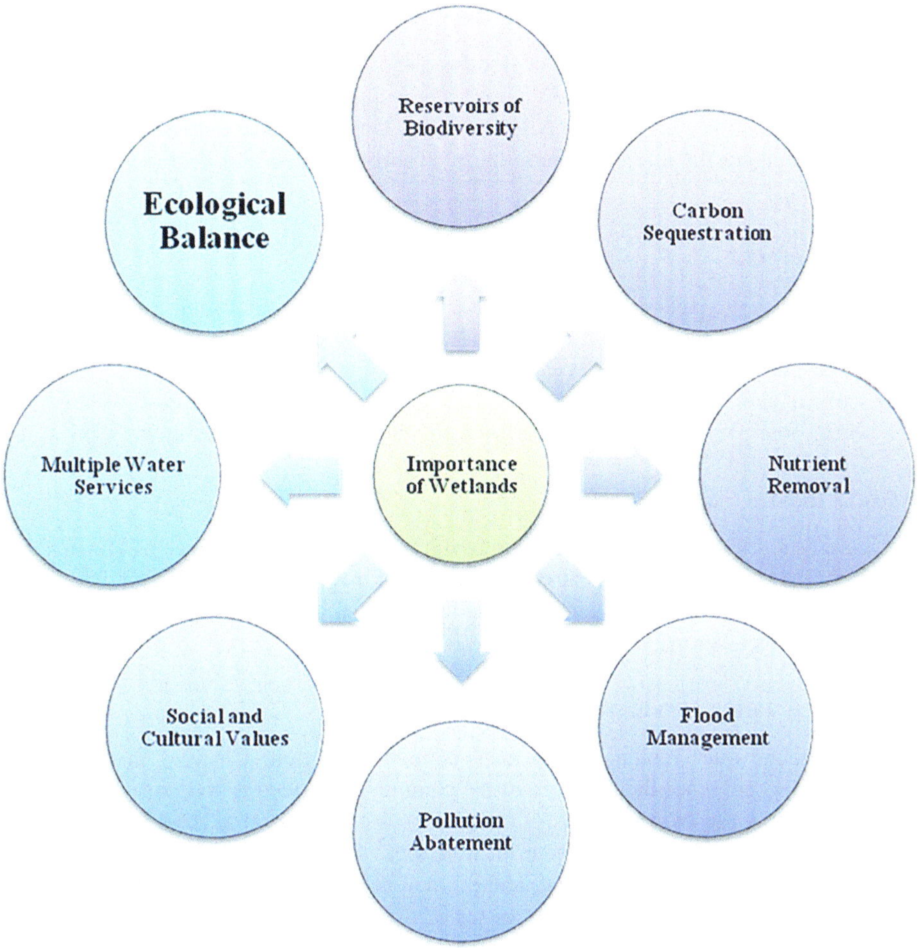

FIGURE 12.3 Multiple benefits and services provided by wetlands.

Biodiversity in wetlands governs essential biogeochemical processes, contributing to structural and functional dynamics over time. Additionally, large wetland complexes provide critical habitats for migratory bird species, underscoring their role in sustaining avian diversity across flyways.

12.3.2 Carbon sequestration

Wetlands act as substantial carbon sinks, storing organic carbon in the biomass of aquatic plants, animals, and microbes. Therefore they have a significant impact on the carbon cycle (Arya et al., 2020). Some amount of carbon infiltrates wetlands through sedimentation and as dissolved form from surface and groundwater sources (Pant et al., 2003).

12.3.3 Nutrient removal

Nutrient removal is an important ecosystem service provided by the wetlands, which results in a reduction of nutrient load in rivers and oceans. As a result, the natural habitats of the river and ocean are well protected. Wetlands exhibit the highest nitrogen retention capacity among aquatic systems, followed by lakes and rivers (Saunders & Kalff, 2001). Mechanisms such as plant uptake, sedimentation, and microbial denitrification drive the removal and retention of nitrogen and other nutrients.

12.3.4 Flood management

Many wetlands are capable of storing excess water or floodwater to safeguard the downstream area from flooding (Bassi et al., 2014). They play an important role in maintaining water quality, promoting the growth of aquatic plants and animals, and ensuring a continuous flow of water downstream. During flood conditions, wetlands trap nutrients and suspended particles. They mitigate the effects of flooding by absorbing water and decreasing the velocity of floodwater (Mitsch & Gosselink, 2015).

12.3.5 Pollution abatement

Economically and ecologically, wetlands offer a cost-effective strategy for controlling both point and nonpoint source pollution (Bassi et al., 2014). However, increasing anthropogenic pressures such as land drainage and agricultural runoff. They are threatening the integrity of natural wetlands. Elevated nutrient inputs can lead to eutrophication, habitat degradation, and loss of recreational value (Arya et al., 2020). Moreover, wetland plants such as *Typha*, *Phragmites*, *Eichhornia*, *Azolla*, and *Lemna* play an important role in mitigating heavy metal pollution (Dhir et al., 2009).

12.3.6 Social and cultural values

Numerous wetlands hold deep-rooted cultural and religious significance. In regions such as Himachal Pradesh, lakes like Chandertal, Manimahesh, Parashar, and Renuka attract thousands of pilgrims annually. Ritual practices include holy dips and the collection of water for ceremonial use, emphasizing the spiritual and socio-cultural connections that communities maintain with these ecosystems (Sharma et al., 2012).

12.3.7 Multiple-use of water services

Wetlands contribute to groundwater replenishment and sediment retention. Changes in the physical and chemical characteristics of wetlands also influence the structure of the community inside the wetlands. The biodiversity in wetlands plays a crucial role in regulating the biogeochemistry of the ecosystem (Sims et al., 2013). Apart from this, avifauna in particular depend heavily on wetlands, which offer critical services for feeding, breeding, roosting, premigration needs (molting, fat deposits), migration, staging, wintering, and protection from breaching predators. So many bird species are ecologically dependent on wetlands for at least some stages of their life cycle.

12.4 Threats to wetlands and their conservation

Wetland loss and degradation are global concerns driven by factors such as anthropogenic and environmental pressures. Some of the primary threats to these sensitive ecosystems are illustrated in Fig. 12.4 and briefly discussed next.

12.4.1 Industrial and agricultural activities

Large-scale conversion of forests to agricultural land has caused significant ecological disruption, including biodiversity loss, nutrient runoff, nonland degradation, and microclimatic alterations (Arya et al., 2020). Intensive farming practices increase the water demand for irrigation, which has a negative impact on wetland ecosystems. Declining water levels, in turn, disrupt hydrological regimes, degrade wetland vegetation, and diminish critical habitat for wildlife (Gupta et al., 2020).

12.4.2 Land use and land use changes

Land use and land cover alteration represent one of the most widespread drivers of environmental transformation. These changes are often driven by human needs and expansion (Harikrishna et al., 2013). Urban development mainly alters the hydrology and sedimentation regimes. Changes in land use also accelerate sedimentation rates, hence modifying the fundamental processes and functions of the ecosystem. Excessive nutrient concentrations from runoff in agricultural areas can lead to eutrophication (Arya et al., 2020).

12.4.3 Climate change

Climate change has an impact on a wetland environment by increasing temperatures, changing hydrological patterns, and affecting the ecosystem's biogeochemistry (Erwin, 2009). Changes in the frequency and intensity of rainfall, coupled with

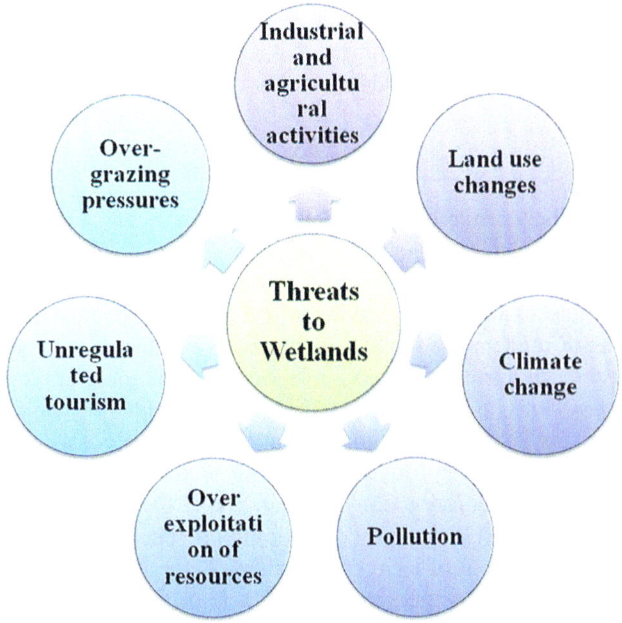

FIGURE 12.4 Major factors threatening wetlands and their biodiversity.

rising temperatures and extreme events such as droughts and floods, affect wetland structure and function. Temperature changes can result in modification of species composition, problems of algal blooms, fish-killing, and dead zone creation in the upper surface of water bodies, which will adversely affect the metabolism and output of wetlands (Gopal et al., 2010).

12.4.4 Pollution

Pollution of both soil and water in wetlands is a major threat, as it occurs through anthropogenic activities such as poor waste disposal and frequent use of pesticides and herbicides (Kingsford et al., 2016). The untreated water from industries, domestic wastewater, and the discharge of various nutrients and toxic substances into water bodies deteriorate the aquatic environment (Include reference).

12.4.5 Overexploitation of wetland resources

Wetlands harbor specialized plant communities that support primary productivity and system resilience. However, population pressures drive excessive fishing, hunting, and harvesting practices, threatening native biodiversity and reducing the capacity of wetlands to recover from disturbances (Gupta et al., 2020).

12.4.6 Unregulated tourism

More than 70% of Ramsar Sites worldwide provide recreational services. Even small-sized wetlands often attract significant tourist activity and offer different leisure activities such as photography and bird watching. The government of Himachal Pradesh has also recently decided to develop two islands in Govind Sagar Lake into major tourist destinations. Usually, the carelessness of tourists in terms of littering, playing high-volume music, and disturbing local flora and fauna can significantly affect the wetland biodiversity. Sharma et al. (2012) reported that unregulated camping near Manimahesh Lake caused habitat degradation. Vehicular emissions further compound environmental stress in these fragile areas.

12.4.7 Overgrazing pressures in catchment areas

Several indigenous communities in the Himalayas such as "Gaddis" and "Gujjars" used to graze their sheep, goats, and cattle near the marshes and wet grasslands. However, repetitive grazing in the same catchment areas can lead to habitat degradation. In high-altitude wetlands, where the growing season is short and conditions are extreme, the impacts of overgrazing are especially severe. This pressure hampers the regeneration of plant communities and disturbs critical nesting and breeding sites for wetland-dependent birds.

12.5 Conservation

Wetlands are distinctive ecosystems between terrestrial and aquatic environments (ecotones). Despite their high value and indispensable functions, many wetlands are deteriorating due to neglect and skewed, unplanned priorities. Their proper management can help attain some of the United Nations Sustainable Development Goals (SDGs), such as the availability and sustainable management of water for all (Goal 6), taking urgent actions to combat climate change and its impact (Goal 13), and strengthening the means of implementation and revitalizing the global partnership for sustainable development (Goal 17). It will further ensure the achievement of the Convention on Biological Diversity (CBD) 2050 long-term goals to protect and restore natural ecosystems, and prosper together with the whole of nature.

In India, many initiatives have been taken at the Government level for the conservation of wetlands. Indian Wildlife (Protection) Act (1972) establishes a regulatory framework for all wetlands located within forests and the designated protected areas. The National Wetlands Conservation Program (NWCP) was launched in 1987 by the Ministry of Environment, Forests and Climate Change (Government of India) to prevent the nation's wetlands from deteriorating and to warrant their judicious usage for the benefit of local communities. Later, the Ministry notified Wetlands (Conservation and Management) Rules (2017) for the conservation and management of wetlands in India. These rules shifted the focus of wetlands management from the central government to the state governments. In 2023, the Ministry launched the Save Wetlands Campaign to strengthen citizens' roles in wetlands management by aligning it with the Mission Lifestyle for Environment. The Ministry launched another program (Amrit Dharohar Initiative) in the same year to promote the unique conservation values of the Ramsar Sites through its four components dealing with (1) species and habitat conservation, (2) wetlands livelihoods, (3) nature tourism, and (4) wetlands carbon. In Himachal Pradesh, the State Wetland Authority (HPSWA) was constituted in 2017 with the aim of acting as a nodal agency for coordinating the Wetland Conservation Program with the active participation of various stakeholders. However, we have to understand that conservation is not the sole responsibility of any Government; every citizen should strive for the conservation of natural habitats, including wetlands. Here we suggest some measures for the protection of wetlands in the state:

1. The management plans developed by the Government should cover multiple aspects of a wetland and its relationship with its catchment area (integrated management plan).
2. The health and status of wetlands should be continuously monitored so as to assess potential threats well in time. For example, regular monitoring of water quality parameters to assess pollution levels.
3. Periodical cleaning of open drains and selective harvesting of macrophytes can be done to diminish nutrient levels in the lake sediments.
4. Reforestation activities should be enhanced, and grazing should be regulated. The same patches in the catchment areas should not be used for this purpose every year. This will help check the risk of heavy soil erosion and aid in better plant regeneration.
5. Local inhabitants and tourists should cooperate with the administration to achieve conservation goals by reducing the negative impacts of anthropogenic pressures. They should be sensitized to the importance of wetlands, the threats they face, and sustainable utilization practices.

12.6 Conclusion

No doubt that wetlands are unique ecosystems, which are ecologically sensitive and biodiversity-rich. Besides performing several essential functions such as water provision and regulation, carbon sequestration, climate moderating, biodiversity preservation, and recreational activities wetlands also possess significant cultural and social significance. India has eighty-five Ramsar wetlands but the information available on their distribution and diversity profiles is incomplete. Many large to small-sized wetlands are under stress because of activities like excessive deforestation, overgrazing, uncontrolled agricultural practices, expanding urbanization, industrialization, and unregulated tourism, as evidenced by their shrinking area. Despite many management plans devised by the Government (NWCP, framing of wetlands conservation and management rules, Save Wetlands Campaign, Amrit Dharohar Initiative, and constitution of state wetland authorities), no significant change has been noticed in the conservation and restoration of wetlands. This calls for honest and active participation of various stakeholders (local communities, Government agencies, researchers, etc.) toward sustainable management and conservation of these fragile ecosystems. This will help attain the United Nations SDGs and the CBD 2050 long-term goals pertaining to water availability and management.

References

Alikhani, S., Nummi, P., & Ojala, A. (2021). Urban wetlands: A review on ecological and cultural values. *Water, 13*(22), 1–47.

Anuradha, Thakur, M. L., & Banyal, H. S. (2023). Status and habitat use pattern of birds in Pong Dam Lake Wildlife Sanctuary, Himachal Pradesh. *Biological Forum, 15*(2), 1308–1319.

Arya, A. K., Joshi, K. K., Bachheti, A., & Deepti (2020). A review on distribution and importance of wetlands in the perspective of India. *Journal of Applied and Natural Science, 12*(4), 10–20.

Attri, P. K., Santvan, V. K., & Thakur, M. (2010). Assessment of natural resources for conservation of wetland in district Chamba–A case study of Manimehesh Lake. *Journal of Advanced Laboratory Research in Biology, 1*(1), 18–22.

Attri, P. K. (2016). Assessment of biodiversity conservation and restoration of Khajjiar wetland, Chamba (HP). *EPH-International Journal of Agriculture and Environmental Research, 2*(2), 17–24.

Balwan, W. K., & Kour, S. (2021). Wetland-an ecological boon for the environment. *East African Scholars Journal of Agriculture and Life Sciences, 4*(3), 38–48.

Banerjee, D., Bharti, D., Kumar, S., Mitra, A., Joshi, R., & Gupta, D. (2023). Faunal composition of Ramsar wetlands from India: An analysis. *Records of the Zoological Survey of India, 123*(1S), 01–15.

Bassi, N., Kumar, M. D., Sharma, A., & Pardha-Saradhi, P. (2014). Status of wetlands in India: A review of extent, ecosystem benefits, threats and management strategies. *Journal of Hydrology: Regional Studies, 2*, 1–19.

Bhambri, R., Misra, A., Kumar, A., Gupta, A. K., Verma, A., & Tiwari, S. K. (2018). Glacier Lake inventory of Himachal Pradesh. *Himalayan Geology, 39*(1), 1–32.

Buckton, S. (2007). Managing wetlands for sustainable livelihoods at KoshiTappu. *Danphe, 16*(1), 12–13.

Choden, Y., Sharma, M. P., Pandey, G., Gupta, S. K., & Dema, K. D. (2022). Ecological health assessment of Renuka Lake, Himachal Pradesh, India. *Nature Environment and Pollution Technology, 21*(1), 167–174.

Dey, D., Bhojak, P., Sekar, K. C., & Arya, D. (2021). An annotated checklist of vascular plants in and around two major high-altitude wetlands of Lahaul-Spiti, Himachal Pradesh, India. *Check List, 17*(6), 1715–1730.

Dey, D., Bhojak, P., & Sekar, K. C. (2017). Floristic diversity in Chandra Tal, Himachal Pradesh. *ENVIS Newsletter on Himalayan Ecology, 14*(4), 7.

Dhadwal, S. D. (2011). *Pong and its birds*. Wild Wings.

Dhir, B., Sharmila, P., & Saradhi, P. P. (2009). Potential of aquatic macrophytes for removing contaminants from the environment. *Critical Reviews in Environmental Science and Technology, 39*(9), 754–781.

Erwin, K. L. (2009). Wetlands and global climate change: The role of wetland restoration in a changing world. *Wetlands Ecology and Management, 17*(1), 71–84.

Gaston, A. J., & Pandey, S. (1987). Sighting of rednecked grebes (*Podiceps grisegena*) on the Pong Dam Lake,Himachal Pradesh. *Journal of the Bombay Natural History Society, 84*(3), 676–677.

Gaur A. Impact of Land Cover Changes on the Ecosystem Services provided by the Renuka Wetland. TERI School of Advanced Studies. 2020.

Gaury, P. K., Meena, N. K., & Mahajan, A. K. (2018). Hydrochemistry and water quality of Rewalsar lake of Lesser Himalaya, Himachal Pradesh, India. *Environmental Monitoring and Assessment, 190*, 1–22.

Ghosh, S. K., & Hegde, V. D. (2013). On a collection of aquatic beetles (Order: Coleoptera: Gyrinidae, Dytiscidae and Hydrophilidae) of Renuka Wildlife Sanctuary, Himachal Pradesh, India. *Records of the Zoological Survey of India, 113*(Part-2), 61–67.

Gopal, B., Shilpakar, R., & Sharma, E. (2010). *Functions and services of wetlands in the Eastern Himalayas: Impacts of climate change*. International Centre for Integrated Mountain Development.

Gupta, G., Khan, J., Upadhyay, A., & Singh, N. K. (2020). Wetland as a sustainable reservoir of ecosystem services: Prospects of threat and conservation. In A. K. Upadhyay, R. Singh, & D. P. Singh (Eds.). *Restoration of Wetland Ecosystem: A Trajectory Towards a Sustainable Environment* (pp. 31–43). Springer.

Harikrishna, K., Appala, R. N., Venkateswara, R. V., Jaisankar, G., & Amminedu, E. (2013). Land use/Land cover patterns in and around Kolleru lake, Andhra Pradesh, India using Remote Sensing and GIS techniques. *International Journal of Remote Sensing and Geoscience, 2*(2), 1–7.

IUCN red list (https://www.iucnredlist.org) assessed on 05 April 2025.

Jhingran, V. G., & Sehgal, K. L. (1978). *Coldwater fishes of India*. Inland Fisheries Society of India.

Jindal, R., & Prajapat, P. (2005). Productivity and status of Renuka wetland (Distt. Sirmour, Himachal Pradesh). *Indian Journal of Ecology, 32*, 180–183.

Jindal, R., Singh, H., & Sharma, C. (2013). Avian fauna of Pong Dam Wetland-a Ramsar site. *International Journal of Environmental Sciences, 3*(6), 2236–2250.

Jindal, R., & Thakur, R. K. (2013). Diurnal variations of plankton diversity and physico-chemical characteristics of Rewalsar Wetland, Himachal Pradesh, India. *Recent Research in Science and Technology, 5*(3), 04–09.

Jindal, R., Thakur, R. K., Singh, U. B., & Ahluwalia, A. S. (2014). Phytoplankton dynamics and water quality of Prashar Lake, Himachal Pradesh, India. *Sustainability of Water Quality and Ecology, 3-4*, 101–113. 10.1016/j.swaqe.2014.12.003.

Jindal, R., & Thakur, R. K. (2014). Hydrobiology and productivity of Kuntbhyog Lake, (District Mandi, Himachal Pradesh), India. *International Journal of Environmental Engineering, 6*(4), 449–459.

Kashyap, R., Verma, K. S., & Brahmi, M. K. (2016). Appraisal of water quality of Rewalsar Lake during Baisakhi fair (harvest festival) in Himachal Pradesh, India. *Journal of Hill Agriculture, 7*(1), 90–95.

Kingsford, R. T., Basset, A., & Jackson, L. (2016). Wetlands: Conservation's poor cousins. *Aquatic Conservation: Marine and Freshwater Ecosystems, 26*(5), 892–916.

Kohli, M. (2015). Wetlands of international importance in Himachal Pradesh. *International Journal of Economic Plants, 2*(1), 023–027.

Kumar, A., & Paliwal, R. (2015). Galloanserae and aquatic neoaves of Pong Dam Wetland, Himachal Pradesh: Status and conservation issues. In M. Rawat (Ed.). *Aquatic Ecosystem: Biodiversity, Ecology and Conservation* (pp. 307–321). Springer.

Kumar, A., Wanganeo, A., & Kumar, K. (2005). Ecological study of Pong Dam: A wetland of Himachal Pradesh. In Bhatia, & K. K. S. SD Khobragade (Eds.). *Urban Lakes in India* (pp. 23–30). Conservation, Management and Rejuvenation.

Kumar, R., Jindal, R., Kulharia, M., Sharma, A. K., Thakur, K., Brar, B., Mahajan, D., Sharma, A., Kumar, S., & Sharma, D. (2023). Knowledge mining for the differentiation pattern with respect to the productivity and trophic status of some selected lakes of Western Himalayas. *Frontiers in Environmental Science, 11*, 1–10.

Kumar, S. (1992). Survey of seasonally ice -covered Suraj Tal lake in Northwest Himalaya. *Records of the Zoological Survey of India, 91*(3-4), 361–366.

Kumari, R., & Kumar, P. (2020). Assessment of aquatic ecological health-a case study of a sacred Lake Rewalsar, Himachal Pradesh: Aquatic ecological health of Lake Rewalsar. *Journal of AgriSearch, 7*(1), 12–17.

Lal, K., & Dua, A. (2018). Check list of ichthyofauna of Gobind Sagar Reservoir: Years 2009-2013. *Journal of Environment and Bio-Sciences, 32*(2), 213–215.

Malik, M., & Rai, S. C. (2019). Drivers of land use/cover change and its impact on Pong Dam wetland. *Environmental Monitoring and Assessment, 191*, 1–14.

Mehta, A., & Thakur, D. R. (2017). Breeding of *Anas strepera* (Linnaeus, 1758) (Anatidae: Anseriformes) from chandertal wetland (an Internationally Known Ramsar Site in Trans Himalayan Ecosystem) - A new record. *European Journal of Zoological Research, 5*(1), 36–39.

Mitsch, W. J., & Gosselink, J. G. (2000). The value of wetlands: Importance of scale and landscape setting. *Ecological Economics, 35*(1), 25–33.

Mitsch, W. J., & Gosselink, J. G (26 Feb 2015). *Wetlands* (5). John Wiley & Sons.

Pandey, S. (1989). The birds of Pong Dam Lake Bird Sanctuary. *Tigerpaper, 16*(2), 20–26.

Pant, H. K., Rechcigl, J. E., & Adjei, M. B. (2003). Carbon sequestration in wetlands: Concept and estimation. *Food, Agriculture and Environment, 1*(2), 308–313.

Parui, P., & Mukherjee, M. (2000). Fauna of Renuka wetland (Diptera: Insecta). Himachal Pradesh, Wetland Ecosystem series. *Zoological Survey*, 135–140.

Pathania, M. S., & Kumar, G. A. (2017). Economic valuation and causes of degradation of Pong Dam wetland of Himachal Pradesh. *International Journal of Social Sciences, 6*(1), 21–29.

Plumpley, D., & Sharma, V. (1990). A report from the University of Newcastle upon Tyne Himalyan Expedition. *World Pheasant Association Newsletter, 55*(28), 28–31.

Prasad, S. N., Ramachandra, T. V., Ahalya, N., Sengupta, T., Kumar, A., Tiwari, A. K., Vijayan, V. S., & Vijayan, L. (2002). Conservation of wetlands of India-a review. *Tropical Ecology, 43*(1), 173–186.

Rana, R. S., Thakur, D. R., Banyal, H. S., & Mehta, A. (2014). Avifauna of chandertal wetland sanctuary of district Lahaul and Spiti, Himachal Pradesh, India. *Asian Journal of Biological Sciences, 7*(4), 151–157.

Rao Rama, T., & Natrajan, A. V. (1979). *Abundance and spatial distribution of commercial fish catches in Govindsagar* (p. 6). Summer Institute, CIFRI, Barrackpore.

Rout, S. P., Vasudevan, S., Selvaganapathi, R., & Singh, K. K. (2020). Lake bathymetry survey and morphometric characteristics of high altitude lake the Chandratal in Western Himalaya. *Journal of Critical Reviews, 7*(13), 4615–4625.

Sarvade, S., Gupta, B., & Singh, M. (2016). Composition, diversity and distribution of tree species in response to changing soil properties with increasing distance from water source- A case study of Gobind Sagar Reservoir in India. *Journal of Mountain Science, 13*, 522–533.

Saunders, D. L., & Kalff, J. (2001). Nitrogen retention in wetlands, lakes and rivers. *Hydrobiologia, 443*(1/3), 205–212. https://doi.org/10.1023/A:1017506914063.

Sengupta, J., Naskar, A., Maity, A., Homechaudhuri, S., & Banerjee, D. (2019). A short taxonomic account of dipteran flies from the Renuka Wetland and adjacent Sanctuary, Himachal Pradesh, India. *Munis Entomology and Zoology, 14*(1), 224–231.

Shan, V., Singh, S. K., & Haritash, A. K. (2021). Present status, conservation, and management of wetlands in India. In R. Al Khaddar, N. D. Kaushika, S. Singh, & R. K. Tomar (Eds.). *Advances in Energy and Environment* (pp. 235–256). Springer.

Sharma, I. (2018). Status of fishery versus exotic fauna in Gobind Sagar Dam Wetland (H.P. & Punjab), India. *International Journal of Fisheries and Aquatic Studies, 6*(2), 396–398.

Sharma, K., Kumari, V., & Lakhanpal, D. (2020). A preliminary study of diversity of butterflies around Govind Sagar Lake in district Una, Himachal Pradesh. *Uttar Pradesh Journal of Zoology, 41*(9), 123–127.

Sharma, K. L., & Kumar, R. (2015). Diversity of butterflies in Renuka Lake and its vicinity of Himachal Pradesh. *International Journal of Science and Research, 4*(9), 1628–1634.

Sharma, L., & Samant, S. S. (2019). Prioritization of habitats and communities for conservation in cold desert biosphere reserve, Trans Himalaya, India. *Ecological Research, 34*(4), 509–523.

Sharma, R. C., & Kumari, R. (2018). Traditional medicinal plants and beliefs system of conservation of sacred lake Prashar, Himachal Pradesh, India. *International Research Journal of Biological Sciences, 7*(3), 46–51.

Sharma, S., Chandan, P., & Thaplyal, V. (2012). Community based participatory conservation of high altitude wetlands: Case of Chandertal and Manimahesh wetlands in Himachal Pradesh of North Western Himalaya, India. *Proceedings of International Wetland Symposium,* 211–219.

Sharma, V., & Walia, Y. K. (2014). Analysis of water quality using physico-chemical parameters of Govind Sagar Lake HP (INDIA). *Asian Journal of Advance Basic Sciences, 2*(3), 83–91.

Sims, A., Zhang, Y., Gajaraj, S., Brown, P. B., & Hu, Z. (2013). Toward the development of microbial indicators for wetland assessment. *Water Research, 47*(5), 1711–1725.

Singh, A. P. (2008). Butterflies of Renuka Wildlife Sanctuary, Sirmour District, Himachal Pradesh, India. *The Indian Forester, 134*, 1326–1338.

Singh, J., Thakur, M. L., & Banyal, H. S. (2014). Avifauna of Prashar Lake and its Surrounding area in Mandi District (Himachal Pradesh), India. *Asian Journal of Biological Sciences, 7*(2), 47–56.

Singh, J., Thakur, M. L., & Banyal, H. S. (2015). Status of Pisces, Amphibia and Reptilia in Prashar Area of Mandi District (Himachal Pradesh), India. *Asian Journal of Biological and Life Sciences, 4*(2), 150–155.

Singh, R., & Thakur, D. R. (2019). Mammalian diversity of Chandertal wild life sanctuary in Lahaul and Spiti district, Himachal Pradesh, India. *Journal of Pharmacognosy and Phytochemistry, 8*(1S), 360–363.

Singh, R., & Thakur, D. R. (2021a). Studies on diversity of Charadriiformes (Chordata: Aves) in Chandertal Wildlife Sanctuary, Himachal Pradesh, India. *International Journal of Fauna and Biological Studies, 8*(4), 48–50.

Singh, R., & Thakur, D. R. (2021b). Studies on Rhopaloceran diversity of high altitude Chandertal Wetland in Lahaul and Spiti district of Himachal Pradesh, India. *Journal of Entomology and Zoology Studies, 9*(4), 448–452.

Singh, S., Dhasmana, M. K., Shrivastava, V., Sharma, V., Pokhriyal, N., Thakur, P. K., Aggarwal, S. P., Nikam, B. R., Garg, V., Chouksey, A., & Dhote, P. R. (2018). Estimation of revised capacity in Gobind Sagar reservoir using Google earth engine and GIS. *The International Archives of the Photogrammetry, Remote Sensing and Spatial Information Sciences, 42*(5), 589–595.

Singh, V., & Banyal, H. S. (2014). Preliminary ecological studies on the Lepidoptera from Khajjiar lake catchment, Himachal Pradesh, India. *Biodiversity Journal, 5*(1), 61–68.

Solanki, V., & Ankanksha (2021). Ramsar wetlands of India-Importance and conservation. *NIU International Journal of Human Rights, 8*, 113–125.

Srivastava, S. K., & Gupta, R. K. (2009). Contribution to the algal flora of Renuka lake wetland Himachal Pradesh. *Indian Journal of Forestry, 32*(1), 107–112.

Srivastava, S. K., & Gupta, R. K. (2004). Studies on the Algal plant diversity of Pong Dam Wetland. *Indian Journal of Forestry, 27*(1), 103–111.

Subramani, S. P., Jishtu, V., Verma, R. K., & Kapoor, K. S. (2007). Floristic composition, life-forms and biological spectrum of Renuka Wildlife Sanctuary, Himachal Pradesh. *Indian Forester, 133*(1), 79–92.

Suman, R., Sharma, P., Gupta, S., Sourirajan, A., & Dev, K. (2015). A novel psychrophilic *Janthinobacteriumlividum*MMPP4 isolated from Manimahesh Lake of Chamba district of Himachal Pradesh, India. *Journal of Biochemical Technology, 6*(1), 846–851.

Tak, P. C., Paliwal, R., & Shrama, R. M. (2001). Occurrence of huge wintering population of Bar-headed Goose,*Anser indicus*, at Pong Dam Wetland, Himachal Pradesh. *Bionotes, 5*(4), 88–89.

Thakur, D. R., & Mehta, A. (2016). Breeding of Ruddy Shelduck *Tadorna ferruginea* (Pallas) (Anatidae: Anseriformes) from Chandertal Wetland in Himachal Pradesh, India. *Journal of the Bombay Natural History Society, 113*, 24–25.

Thakur, D. R., & Mehta, A. (2015). First Record of *Anas querquedula* (Linnaeus, 1758) from Chandertal and marshy meadows of Spiti River near Lossar in Himachal Pradesh, India. *Asian Journal of Scientific Research, 8*(3), 436–441.

Thakur, M., Radha, Kumari, N., Pundir, A., Singh, D., Khosla, P. K., & Sharma, N. (2022). Wild edible plants used by rural people of Pong Dam Wildlife Sanctuary in District Kangra of Himachal Pradesh. *International Journal of Pharmaceutical Research and Applications, 7*(4), 1184–1190.

Thakur, R., Yardi, K., & Ahuja, M. P. (2022). Feeding ecology of the endangered Himalayan gray langur *Semnopithecus ajax* in Chamba, Himachal Pradesh, India. *Journal of Threatened Taxa, 14*(10), 21918–21927.

Yadav, R., Chand, T., & Nautiyal, R. (2021). Valuation of water purification service of Renuka wetland, India: A Ramsar site. *GlasnikSumarskogfakulteta, 123*, 219–226.

Zhang, X., Liu, L., Zhao, T., Wang, J., Liu, W., & Chen, X. (2024). Global annual wetland dataset at 30 m with a fine classification system from 2000 to 2022. *Scientific Data, 11*, 1–12.

Chapter 13

Status of Delhi wetlands: a case study of Yamuna Biodiversity Park wetland, ecosystem services, conservation, and management

Leena Singh
Department of Applied Sciences, Galgotias College of Engineering and Technology, Greater Noida, Uttar Pradesh, India

13.1 Introduction

13.1.1 Wetlands: An Overview

Wetlands are vital ecological systems where land and water meet, serving as the interface between terrestrial and aquatic ecosystems (Xu et al., 2020). These dynamic ecosystems are among the most biologically productive ecosystems on Earth and are vital to the preservation of human life and biodiversity (Ghermandi et al., 2008). The development of unique soil types known as hydric soils, which are defined by the frequent or continuous presence of water, whether through surface runoff or groundwater seepage, is the foundation for the biological peculiarity of wetlands. The interface between the land and water provides a wide range of habitats not observed in land or aquatic ecosystems (Singh et al., 2006).

Wetlands can show a unique characteristic in the interplay among water regimes, biological communities, and soil. The hydrology of a wetland is determined by the quantity, timing, and velocity of water, which in turn shapes its overall ecological functioning. Thus, this impacts both plant and animal life (Mitsch & Gosselink, 2007; Ramsar Convention Secretariat, 2018). Wetlands can support a range of species that are adapted to these overlapping environments due to the "edge effect", which happens where terrestrial and aquatic ecosystems converge. These include a diverse range of fish, amphibians, reptiles, migratory birds, and invertebrates (Ramírez & Fennell, 2014).

In addition to their ecological functions, wetlands play a vital role in human development and welfare. The world's 15% population depends directly on wetlands for their food and livelihoods. Fishing, aquaculture, and rice cultivation play a significant role in the global economy. They support approximately 200 million jobs globally and account for roughly 50% of tourism-related activity. Furthermore, the fishing industry, which provides about 40% of the world's fish, depends on wetlands. Surprisingly, nearly 75% of commercially important marine fish species depend on estuarine wetlands at some stage of their life cycles (Prasad et al., 2002). The aquaculture and pisciculture sectors alone provide a living for more than 650 million people worldwide.

In urban areas, especially megacities like Delhi, wetlands are essential freshwater sources. For example, during the hottest summer months, the Delhi Jal Board is able to supply over 830 million gallons of water per day thanks to the nearby wetland systems, Delhi Water Board (Ramsar Convention on Wetlands, 2021).

But due to massive anthropogenic activities, these ecosystems are under threat. Nearly 66% of the world's wetlands have vanished since 1900, as per the Ramsar Convention Secretariat (Board, 2022). The major threats to these wetlands stem from urbanisation, pollution, and the growing effects of climate change, which have contributed to their degradation. These ecosystems support millions of people, and the biodiversity that relies significantly on their subsistence will be impacted. This has drawn attention to wetland conservation and restoration worldwide.

Wetlands: Biodiversity, Health and their Role in Ecosystem Services. DOI: https://doi.org/10.1016/B978-0-443-34222-6.00018-8

13.1.2 Overview of Yamuna Biodiversity Park

The preservation and revival of the biodiversity of the Yamuna River Basin depend on the 457-acre Yamuna Biodiversity Park (YBP), which is located near Jagatpur Khadar village near Wajirabad on the western bank of the Yamuna River in North Delhi (Fig. 13.1). The park's natural diversity, once shaped along the river's course, is reflected in its wide variety of ecosystems, including wetlands, grasslands, and floodplain forests. The park aims to serve as a repository of biodiversity and a hub for the cultural, ecological, and educational needs of urban residents, thereby restoring the ecological integrity of a landscape once characterized by thriving natural ecosystems (Delhi Biodiversity Foundation).

The park is located on the floodplains of the River Yamuna and is constructed in two phases. The 63.5-ha Phase I and the 121-ha Phase II offer a carefully planned natural landscape that mimics the native ecosystems that the Yamuna has traditionally preserved. These include riparian woodlands, restored wetlands, and native grasslands, and they support rich biodiversity.

One of the park's key features is its 100-acre wetland, which serves several ecological functions. It provides essential habitat for a variety of aquatic life, including resident and migratory bird species, enhances groundwater recharge, and acts as a natural reservoir during floods. The restoration of these wetlands has helped in serving as nurseries for a wide variety of life forms, which has not only improved the water quality but has also helped in the regeneration of aquatic biodiversity (Department of Anthropology) (Fig. 13.2).

The primary objective of YBP is to improve environmental sustainability by restoring harmed habitats. This is achieved by encouraging groundwater replenishment, improving the local microclimate, and providing spaces for ecotourism, research, and environmental education. This has remarkably worked in strengthening the bond between the urban population and nature (Buttke et al., 2014).

The YBP conceptual framework was inspired by the diverse ecosystems that the Yamuna River supports throughout its entire course, from its source to its confluence with the Ganga. As the river traverses different terrains, it sustains a wide variety of forest types and species, some of which are presently undergoing partial restoration within the park.

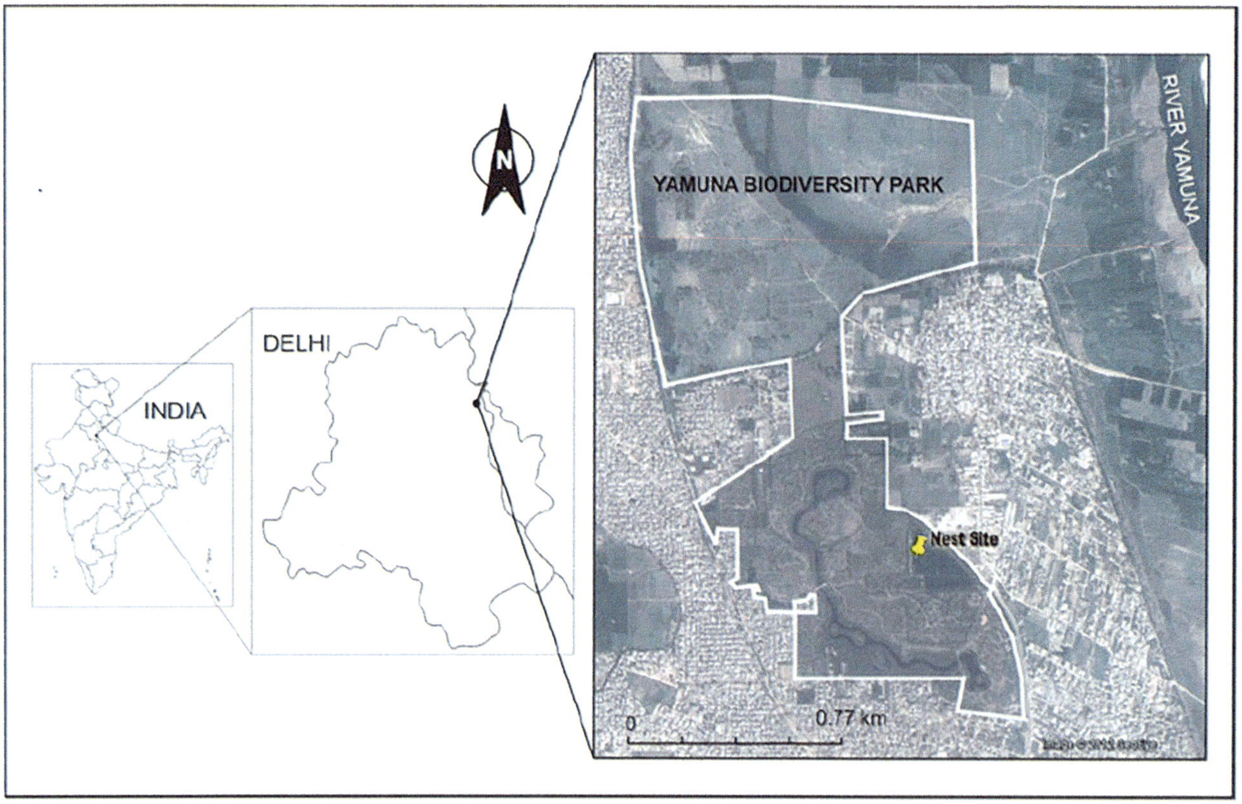

FIGURE 13.1 Satellite imagery of Yamuna Biodiversity Park in Delhi, India. Map lines delineate study areas and do not necessarily depict accepted national boundaries.
Research Gate. Niangthianhoi, S, & Khudsar, F. (2015). The nesting success of the Oriental Darter Anhinga melanogaster in the Yamuna Biodiversity Park, New Delhi, India. *Journal of Threatened Taxa*, *7*, 8148–8153. https://doi.org/10.11609/jott.2430.7.14.8148-8153.

FIGURE 13.2 Wetland of Yamuna Biodiversity Park.

Layered plantation systems, ranging from ground cover to upper canopy plants, are incorporated into YBP's design to mimic the structural and functional components of the natural ecosystems (Sahapedia, 2024).

13.1.3 The history and ecological significance of Yamuna Biodiversity Park

The Yamuna-Aravali ecosystem is the essential foundation for Delhi's existence, which supports life. The main objective of the YBP is to maintain the natural integrity of the Aravali hills and the Yamuna floodplains by protecting and restoring them, which are the most unique landscapes of this region. The main mission of the park includes protecting keystone and endangered species, preserving ecosystems threatened by urbanization, creating field gene banks for rare and traditional plant varieties, and increasing public awareness through outreach and environmental education programmes (Sahapedia, 2024).

One of the main objectives of the park is to restore and conserve native plant and animal species from the Aravalli hills and the Yamuna basin, especially those found in the Delhi region. One of its most notable features is the mosaic of wetlands that were established to support aquatic biodiversity. The park also facilitates ecological monitoring to monitor short-term and long-term environmental changes in the region (Gonçalves et al., 2021).

About 50 endangered populations native to the Aravalli region and the Yamuna basin are found in YBP, a biodiversity heritage site that serves as a living repository. It recharges groundwater, improves local microclimates, offers migratory and resident birds alternative habitats, and serves as a natural sink for pollutants such as carbon dioxide. It also promotes ecotourism and community involvement and provides unique educational opportunities. The park is a great illustration of how urban sustainability and ecological preservation can coexist harmoniously (Hussain).

Delhi's rapid industrialization and extensive human encroachment have, over time, seriously damaged the city's natural ecosystems. The park's development has drawn attention from all over the world, in addition to revitalizing the neighbourhood. Groups of scholars and students from Canada, the United Kingdom, and France have come to study the park's innovative ecological model. The process of restoring this damaged area started under the guidance of Professor C.R. Babu, a renowned environmentalist who is now Professor Emeritus at the University of Delhi. Throughout his distinguished 35-year academic career, Prof. Babu has made significant contributions to the field of ecological restoration and completed several field-based projects. To identify plant species historically native to this region, ecologists were tasked with investigating the Yamuna floodplains upstream and downstream. They then replicated these biological communities in the remote village of Jagatpur, upstream of the Wajirabad reservoir. The area had lost its native vegetation due to the construction of the bund and soil salinity, which allowed only salt-tolerant plants, such as *Suaeda fruticosa*.

To begin the restoration process, scientists reclaimed the area and planted salt-tolerant grasses such as Leptochloa fusca, which gradually reduced the soil pH from 10 to 7. After the soil had been improved with organic additions such as cow dung, a range of native plant saplings was planted. Presently, the park has more than 1200 species of plants, including more than 1000 blooming plant species that were once part of the Yamuna floodplains. These plant species have been classified into 20–25 plant communities, including dry deciduous forests, grasslands, scrublands, and tropical thorn forests.

The restoration is based on a natural ecosystem model. Over the past decade, a self-sustaining ecosystem has emerged, and a diverse range of species have naturally established themselves in the region.

Scientists associated with the research emphasised the expansion of large wetlands, which are expected to cover 60% of the land. The remaining 40% will consist of naturally occurring floodplains that can withstand seasonal floods, forests, and grasslands. These wetlands will store monsoon floodwaters and provide essential habitat for aquatic animals, including birds.

The taxonomist working on the project says that although the restoration is proceeding smoothly, there are still some delays due to unresolved land-delineation issues. But the first step involves creating 100-acre wetlands that have the capacity to hold about 500 million gallons of floodwater. This project is expected to significantly improve groundwater recharge, which will directly benefit the communities along the Yamuna's alluvial plains.

In addition to ecological rehabilitation, the park has evolved into a hub for environmental education. Every year, more than 40,000 tourists visit, including high school and college students from all over India and abroad. To lessen anthropogenic interference, public access is restricted; only escorted groups are permitted entry (Figs. 13.3 and 13.4).

Founded in 2005 on a barren, arid piece of land with salty soil, the YBP has grown to represent ecological success. The conversion of a once-hostile environment into a thriving natural ecosystem demonstrates how scientific intervention and ecological restoration can halt environmental degradation and create sustainable urban biodiversity (Buttke et al., 2014; Delhi Biodiversity Foundation; Department of Anthropology; Sahapedia,2024-a, b).

FIGURE 13.3 Nature reserve zone of Yamuna Biodiversity Park.

FIGURE 13.4 Sacred grove of Yamuna Biodiversity Park.

13.1.4 The vegetation of Yamuna Biodiversity Park

The wide range of plant species of YBP has been carefully conserved and biologically grouped based on interspecies interactions such as competition and symbiosis. The park is home to a variety of ecosystems that have developed naturally as a result of ecological restoration projects. These include subtropical mixed evergreen forests, tropical thorn forests, scrub jungles, moist tropical deciduous forests (dominated by teak), and tropical dry deciduous forests (dominated by both sal and teak). Each of these habitats has a complex structure that includes climbers, herbs, grasses, shrub layers, middle stories, and top canopy. Notable tree species include *Tectona grandis*, *Shorea robusta*, *Dalbergia latifolia*, and *Acacia nilotica*. *Desmostachya bipinnata*, *Vetiveria zizanioides*, and *Bothriochloa pertusa* predominate in the grasslands (Figs. 13.5 and 13.6). More than 20 species of climbers are also protected in the park; many of these have long been used for therapeutic purposes. Examples include *Tinospora cordifolia* (Giloy), which is used as a liver tonic; *Tylophora indica* (Dama-bel), which is used to treat diabetes and asthma; and *Bauhinia vahlii*, whose leaves are used to make ecofriendly plates (Saha, 2006). The YBP now serves as a major conservatory house of about 350 species of herbal plants, taking care of biodiversity and increasing public health awareness. The important genetic resources of tropical and subtropical plants, which include a variety of fruits from the citrus group, like citrus, guava, pomegranate, sapota, and ber. This vegetation of YBP, when developed completely, will lead to an increase in Delhi's forest cover with a variety of floral communities (Delhi Biodiversity Foundation). These not only help maintain ecological balance but also offer a rich food source for local species (Table 13.1). Overall, YBP's floristic diversity is critical to preserving ecosystem services, improving the urban environment, and preserving native biological heritage (Fauzie, 2015).

13.1.5 The Fauna of Yamuna Biodiversity Park

YBP is home to a variety of grassland birds, such as munias, prinias, and francolins, as well as a wide variety of invertebrates, such as mantises, grasshoppers, and beetles. Terrestrial animals like hares, skinks, snakes, and mongooses are also commonly found in the park. Thus, the ecosystem provides complete habitat and food resources to these animals. The wetland area of the park is home to a large population of dragonflies and damselflies, which is very important to the ecosystem. These insects are natural predators of small insects and other predators, which helps in controlling the disease-carrying vector population. Insects, of which more than 35 species are known, improve the overall health of the wetland ecosystem and act as biological control agents. These consist of the Trumpet Tail, Pied Paddy Skimmer, Golden Dartlet, Coromandel Marsh Dart, and Common Clubtail (Niangthianhoi and Khudsar, 2009).

The YBP has established a butterfly-friendly environment for the butterflies, which contribute to pollination and serve as food for birds and other animals and are another significant species of wildlife. Their numbers were declining worldwide, which was of great concern, as they are an indicator of ecological health. This has been achieved by planting plants that are high in nectar and caterpillar host plants. As a result, roughly 80 different species of butterflies, including the common silverline, peacock pansy, common castor, and lime butterfly, thrive here. Visitors can observe butterflies in a

FIGURE 13.5 Butterfly conservatory of Yamuna Biodiversity Park. delhibiodiversityparks.org.

FIGURE 13.6 Faunas of Yamuna Diversity Park.
delhibiodiversityparks.org.

range of stages of life throughout the park. Moths, flies, aphids, bees, wasps, spiders, and other beneficial insects contribute to the region's biodiversity (Koner, 2010).

There is plenty of bird life, which can be exhibited by the presence of about 202 species of both resident and migratory birds. The terrestrial bird species, such as parakeets, bulbuls, peafowls, prinias, munias, and shikras, frequently nest and breed in the park (Niangthianhoi et al.; Khudsar, 2010). The thriving forest zones, which have canopy heights of up to 30–40 ft, create multilayered ecosystems that are home to mammals such as civets, mongooses, hares, nilgai, jungle cats, and wild pigs. The well-organised food web in these areas is composed of three trophic levels, indicating ecological maturity. The nesting of Indian cobras, which helps regulate bird populations like peafowls, further demonstrates the ecological balance and sustainability of these forest communities.

Approximately 70 additional bird species have been observed in Phase II of YBP, including a range of waders and well-known species like the painted stork, Asian openbill, Eurasian spoonbill, ruddy shelduck (Surkhab), sarus crane, pheasant-tailed jacana, and bronze-winged jacana (Khudsar, 2010). The wetlands are home to fish, turtles, invertebrates, plankton, and other aquatic species, in addition to providing these birds with nesting and feeding grounds. By serving as source populations, these bodies of water help repopulate aquatic life in nearby areas that are impacted by seasonal fluctuations in water levels.

TABLE 13.1 Vegetation communities and habitat types developed at Yamuna Biodiversity Park (*YBP*) (*Contributing to Forest Cover Expansion and Ecological Restoration in Delhi).*

S. no.	Habitat/Community type	Dominant vegetation	Ecological significance
1	Mixed deciduous forest with bamboo	Sal, Teak, Bamboo, *Acacia*, etc.	High biodiversity supports herbivores, improves soil stability
2	Mixed deciduous forest without bamboo	Sal, Teak, *Terminalia* spp., etc.	Core forest structure enhances green cover
3	Sal-dominated mixed evergreen forest	*Shorea robusta*, evergreen understorey	Important for shade-tolerant species; high water retention
4	Teak-dominated mixed deciduous forest	*Tectona grandis, Lagerstroemia, Acacia* spp.	Valuable hardwood species foster faunal diversity
5	Sal-dominated mixed deciduous acacia woodlands	*S. robusta, Acacia* spp.	Mixed forest formation improves resilience to drought
6	Acacia woodland	*Acacia nilotica, Acacia catechu*	Supports dryland species; promotes soil nitrogen fixation
7	Grasslands	Native grasses, wild herbs	Key for pollinators, ground-nesting birds, and small mammals
8	Thorn-scrub forests	*Ziziphus, Capparis, Acacia* spp.	Arid-adapted biomes support reptiles and seed-dispersing birds
9	Shallow wetland communities	Emergent and floating vegetation	Breeding ground for amphibians and water birds
10	Deep wetland communities	Submerged aquatic plants, fishes	Nurseries for fish are critical for migratory birds
11	Semi-evergreen low-lying forest	Moist evergreen and semideciduous trees	Water-retentive habitat; controls microclimate
12	Broad-leaved mixed communities	*Ficus, Terminalia, Albizia, Dalbergia* spp.	Supports frugivores, insects, and canopy-dependent fauna
13	Marshy communities	Hydrophytes, sedges, reeds	Water purification; carbon sequestration
14	Riparian communities	*Tamarix, Typha, Barringtonia, Syzygium* spp.	Flood control; habitat for wetland-dependent fauna
15	Peripheral plantations	Mixed native species	Buffer zones prevent edge effects and link fragmented habitats
16	Island communities (acacia–grasses)	*A. nilotica* with native grasses	Roosting and nesting for birds protect the soil from erosion
17	Sacred grove	Culturally important native trees	Integrates cultural heritage with conservation
18	Herbal garden	Medicinal plants like *Tulsi, Ashwagandha*, and *Neem*	Promotes ethnobotany and public health awareness
19	Fruit orchard	*Mangifera indica, Emblica officinalis, Psidium*	Food source for wildlife; community education and nutrition
20	Field gene bank	Rare/endemic trees; seed reserves	Conservation of plant genetic resources for future rewilding

13.1.6 Restoring the ecosystem in Yamuna Biodiversity Park

The restoration of YBP was very challenging to transform, as it was a barren region, which was once a degraded marshland that was unsuitable for preserving natural biodiversity because of rodent infestation. So restoring it into a healthy and self-sufficient ecosystem was very difficult. Unlike traditional restoration methods, YBP focused on nature-based solutions to accelerate the development of a balanced ecological system.

13.1.7 Techniques for restoring soil

One of the first and most significant challenges was the excessive salinity of the soil. To overcome this, a technique known as phytoremediation was used by the researchers. In this method, a specific plant species is used to absorb and neutralise toxins without removing the soil. The grasses, such as Leptochloa, Bothriochloa, Suaeda pluticosa, and Sporobolus, were some of the salt-tolerant species and were among the pioneer flora. These grasses played a crucial role in lowering soil salinity and initiating biological succession.

To increase the soil health, gypsum (calcium sulphate dihydrate, or $CaSO_4 \cdot 2H_2O$) was added to improve soil health. The gypsum has an ability to improve the soil structure due to the presence of calcium and thus increases the water infiltration and keeps plants from absorbing too much sodium. This helped in resulting in favourable conditions for plant growth and seed germination by progressively increasing the stability of organic matter and soil aggregates, resulting in more favourable conditions for plant growth and seed germination (Ravindran et al., 2007; Saha, 2006).

To ensure ongoing soil health, monthly assessments are conducted to monitor critical parameters such as salinity, pH, electrical conductivity, water retention capacity, and concentrations of essential nutrients like phosphates and nitrates. Water bodies in the park are also regularly tested for zinc concentration, total dissolved salts, and other quality indicators to maintain the overall ecosystem balance, forest formation, and natural regrowth (Kaushal et al., 2018).

The wooded areas of the park were first transplanted from indoor nurseries. These same woodlands are now largely self-sufficient and naturally regenerating. Although YBP continues to operate its nursery, its primary focus is now supporting afforestation initiatives in nearby biodiversity parks in the National Capital Region.

13.1.7.1 Polyhouse or controlled propagation unit

Plant propagation at YBP requires a polyhouse, which is constructed of polyethylene and functions similarly to a greenhouse. It helps plants grow under ideal conditions by using solar radiation to heat the interior atmosphere. Modern equipment is used to regulate the interior temperature, humidity, and ventilation, creating a controlled microclimate ideal for year-round growth.

The concept of polyhouses, as an ability to maintain crop growth in the face of outside weather disruptions, has grown in popularity among farmers all over the world. This technique is a vital tool for both agriculture and conservation since it has allowed farmers in India to produce consistent and often higher yields than open-field farming (Rathour & Kumar, 2023).

13.1.7.2 Biodiversity-based ecological restoration for Yamuna Biodiversity Park regeneration

Ecosystems can be as large as lakes and deserts or as small as a tree or a puddle, depending on their size and complexity. A terrarium is an example of a small, artificial ecosystem. Soil, plants, animals, sunlight, water, and temperature are all necessary. If one element is missing or insufficient, for example, insufficient light or poor-nutrient soil, it can lead plants to perish. This can also affect animals that rely on those plants for food, causing a ripple effect throughout the food chain. In a similar vein, natural ecosystems consist of interdependent components that work together to preserve a stable and wholesome environment.

A robust and healthy ecosystem is typically indicated by a high species diversity. These ecosystems are more tolerant of external stresses like human activity, climatic changes, and natural disasters. Every creature has an ecological job, or niche, that contributes to the stability and productivity of the system as a whole (Millennium Ecosystem Assessment, 2005). For instance, in a lake habitat, sunlight promotes the growth of algae. These algae are an important food source for microscopic organisms and produce oxygen for fish and other aquatic life. The carbon dioxide aquatic plants need to grow is released by small fish that consume these bacteria, which then diffuse through the fish's gills. If algae vanished, this delicate balance would be disrupted, affecting all species that depend on them.

Producers, consumers, and decomposers are the three main categories of living organisms that make up ecosystems. Green plants, sometimes referred to as producers, make their own food from sunlight. Consumers get their food from these plants or other organisms. These include:

The herbivores, carnivores, omnivores and ***i*** decomposers, such as specific bacteria and fungi, break down dead organic waste and restore essential nutrients to the soil. Soil is essential to ecosystems because it retains moisture, anchors plants, and provides habitat for a variety of animals. The atmosphere supplies vital gases like oxygen and carbon dioxide and maintains the water cycle. Photosynthesis and environmental warmth, which are both fuelled by solar energy, are essential to life.

Water is another necessary component. It is a major component of living cells and aids in the transfer of nutrients in plants. In fact, humans and other organisms cannot survive for extended periods without water.

Ecosystem processes are influenced by the traits and diversity of the species present. Larger ecological outcomes are influenced by the behaviours of individual organisms, which are defined by their efficacy and functional roles. Therefore, species richness and abundance have a direct impact on an ecosystem's performance. Biodiversity ensures resilience and productivity (Mansi et al.; Rohit et al., 2025).

According to ecological theory, species coexist by occupying ecological niches that are only slightly different from one another. This variation prevents competitive exclusion and promotes diversity. The aforementioned ideas are easily seen in YBP, where species invasions and diversity have naturally led to complex ecological interactions.

The high productivity and abundance of aquatic vegetation in YBP's wetland habitats attract a wide variety of plants and animals (Table 13.2). The annual arrival of migratory birds like the common pochard, northern shoveller, and ferruginous pochard contributes to the park's avian diversity. These wetlands have become breeding grounds for local species like the spot-billed duck, purple heron, oriental darter, and little cormorant, with approximately 35 species already regularly nesting within the park (Niangthianhoi & Khudsar, 2015).

Grasslands and rangelands provide important ecosystem services. Even without intentional introduction, a large number of insects, such as beetles, mantises, and grasshoppers, have established themselves. Similarly, many animals and plants, such as snakes, francolins, and prinias, have made the park their permanent home. The presence of 30 distinct forest groups (Table 13.1), which are home to over 850 plant species, further demonstrates the park's ecological richness. New plant seedlings emerge each year, proving that the environment offers favourable conditions for growth and reproduction, both of which are critical for preserving biodiversity.

The forest and wetland communities of YBP offer an ideal habitat for a wide range of animals. While wetland bird species, such as herons and darters, inhabit riparian zones, terrestrial animals thrive in forest communities. The presence and annual breeding of predators like the Indian Cobra, which help regulate populations of other species such as Peafowl, highlight the park's complex trophic structure and food chain.

Many fruit-bearing trees and shrubs provide additional food sources for wildlife. Winged predators such as damselflies and dragonflies thrive in wetland habitats, controlling mosquito populations and maintaining ecological balance. Pollinators like butterflies not only aid plant reproduction but also provide birds with food. As environmental indicators, they show low resilience in restoring the ecosystem's health.

Rich aquatic ecosystems, home to species like the painted stork, Eurasian spoonbill, and pheasant-tailed jacana, are made possible by the supportive wetland habitats. These ecosystems protect fish, turtles, invertebrates, plankton, and other aquatic genetic resources. Additionally, during times of seasonal water scarcity in nearby areas, these wetlands act as source populations for recolonisation.

Using ecological indicators like phragmites and typha, the park's soil moisture levels are monitored. These indicators help determine habitat quality and inform ecosystem management decisions. Given the interconnectedness of all living things, it is obvious that maintaining species diversity and interactions is essential to ecosystem health.

The relationship between biodiversity and urban ecosystems is often framed in terms of the negative impacts of urbanisation. On the other hand, urban biodiversity itself can be studied and managed using ecological concepts. Given the increasing population of urban areas, it is imperative to restore and preserve biodiversity there. Concepts like species values, habitat fragmentation, and hierarchical design may be applied to the development of sustainable urban ecosystems.

Due to their visibility and sensitivity to habitat changes, birds make excellent indicators in urban settings. Both local and more general landscape-level concerns must be considered for effective management of bird diversity. Planning for urban biodiversity requires consideration of human–wildlife conflicts. Increasing biodiversity in urban areas not only improves ecological health but also enhances the quality of life and education of city dwellers.

The restoration of approximately 30 forest communities that are typical of the Yamuna River Basin has been restored in YBP, offering a range of ecological niches. This has helped in sustaining a wide variety of species and enabling stable populations to flourish. Presently, the park's biodiversity includes 874 plant species, which are aquatic, medicinal, and grass species. 75 different kinds of butterflies (Koner, 2010). There are 35 species of damselflies and dragonflies. Birds come in 196 different species, 18 kinds of reptiles, 18 types of mammals, and 18 species of fish. 400 different species, including invertebrates and amphibians.

The park is a prime example of ecosystem-based management, a comprehensive approach that considers both natural and human interactions. By combining scientific knowledge, policy coordination, and community involvement, YBP ensures long-term species protection and environmental resilience. Instead of focusing on a single goal, it takes a

TABLE 13.2 Flora and fauna of wetlands at Yamuna Biodiversity Park (*YBP*).

Category	Common names/groups	Scientific/Generic names	Remarks
Submerged aquatic plants	Vallisneria, Hydrilla, Pondweed	*Vallisneria* spp., *Hydrilla* spp., *Potamogeton* spp.	Found underwater, crucial for oxygenation, and as primary producers
Floating plants (open water)	Water Ferns	*Salvinia, Azolla*	Found in the limnetic zone; attracts herbivorous ducks
Floating plants (near shore)	Water Lily, Lotus	*Nymphaea* spp., *Nelumbo* spp.	Located in littoral zones, help stabilize temperature and light penetration
Emergent wetland plants	Reeds, Cattails	*Phragmites* spp., *Typha* spp.	Grow above the water surface and support bird nesting and wetland structure
Wetland trees	Tamarisk	*Tamarix* spp.	Thrives in water-saturated soils; part of the Tamarix-Phragmites wetland forest
Indicator plants	Phragmites, Tamarix	*Phragmites* spp., *Tamarix* spp.	Used to locate water-saturated zones during wetland restoration
Introduced fish	Rohu, Catla, Mirgal	*Labeo rohita, Catla catla, Cirrhinus mrigala*	Introduced to support the food web and attract piscivorous birds
Resident water birds	Cormorants, Darters, Herons, Egrets, moorhens, and Spot-Billed Ducks	*Phalacrocorax* spp., *Anhinga* spp., *Ardea* spp., *Egretta* spp.	Year-round residents use the wetland for feeding and nesting
Migratory water birds	Red-Crested Pochard, Tufted Duck, northern shoveller, Pintail, Wigeon, Grebe, Coots, Geese, teal, and Pochard	*Netta rufina, Aythya fuligula, Anas clypeata, Anas acuta*, etc.	Arrive in winter from Europe, Siberia, Ladakh, and China, depending on the wetland habitat
Breeding birds	Purple Swamphen, White-Breasted Waterhen, Herons, egrets, and black-crowned night herons	*Porphyrio porphyrio, Amaurornis phoenicurus, Nycticorax nycticorax*	Nest in dense vegetation (Phragmites-Tamarix); many returned after years of absence
Reptiles and amphibians	Seiboldi's Water Snake	*Fowlea* spp. (likely *Fowlea piscator seiboldii*)	A mildly venomous water snake was observed breeding after ~75 years
Mammals	Wild Boar, Civet, Indian Hare, Porcupine	*Sus scrofa, Viverricula indica, Lepus nigricollis, Hystrix indica*	Regular sightings show ecological health restoration
Other fauna	Cobras, Kraits, Wolf Snakes, Turtles, Frogs	*Naja naja, Bungarus* spp., *Lycodon* spp., *Testudines, Anura*	Return of amphibians and reptiles linked to habitat recovery

multifaceted approach to support ecological evolution and sustainable development (Hussain; Patel, 2019; Saha, 2006; Fauzie, 2015; Niangthianhoi & Khudsar, 2009).

13.1.8 Long-term success and self-healing

The park's ongoing transformation serves as an example of ecological self-rehabilitation. Every discovery of a new plant or the return of a rare animal is celebrated as a victory in the ongoing restoration effort. The field crews closely observe these events and maintain detailed records to track ecological progress (Table 13.3).

With each season, the ecosystem becomes more self-sufficient, and scientists anticipate that in the years to come, even more species will emerge. The success of YBP is not only a triumph for conservation research, but it also offers a model that can be applied to urban ecological restoration in other parts of the country.

TABLE 13.3 Timeline of important occurrences.

Year of development at YBP	Landmark Development in YBP
June 2002	The Biodiversity Park Project was launched
December 2002	The beginning of habitat improvement and shallow wetland development
June 2003	The planting of native vegetation began
June 2004	The landscaping of the visitor area and deep wetland development were completed
June 2005	The first grassing and planting of nitrogen-fixing legumes took place
June 2006	Twenty distinct biotic communities were established along with the creation of a new wetland
June 2007	An additional marshland was created to attract wading birds
In June 2008	The Butterfly conservatory and Amphitheater became operational
June 2009	Wetlands were transformed into functional ecosystems
By June 2010	The tree canopies were fully grown, and several animal and bird species began to nest in the park

YBP, Yamuna Biodiversity Park.

13.2 Ecosystem Services of Yamuna Biodiversity Park

13.2.1 Grasslands (rangelands)

The rangelands of YBP perform a range of essential ecological functions. They serve as primary habitats, supporting numerous plant and animal species and contributing to biodiversity conservation and ecological resilience. These grasslands also help regulate the local microclimate, enhance soil fertility through nutrient cycling, and maintain ecosystem productivity (Gupta & Das, Delhi Biodiversity Foundation; Sahapedia-a, b; Gatti, 2024).

13.2.2 Conservatory of medicinal plants

The medicinal plant conservatory provides vital ecological and educational services. It serves as a habitat for native pollinators such as butterflies and bees, and shelters various bird species. By housing a diverse range of medicinal and aromatic plants—including several rare and endangered taxa—it enriches habitat diversity and promotes conservation of indigenous flora. Additionally, the vegetation contributes to air purification and water filtration, thereby improving the ecological health of the surrounding environment (Gupta & Das, 2025; Delhi Biodiversity Foundation; Sahapedia-a, b).

13.2.3 Climber grove and forest communities

The climber grove and forest communities replicate natural forest ecosystems, providing multilayered habitats that support a diverse assemblage of species. These forest zones play a key role in groundwater recharge, microclimate regulation, and carbon sequestration. Their vegetation filters air and water pollutants while sustaining the ecological processes vital to long-term ecosystem stability (Delhi Biodiversity Foundation; Sahapedia-a, b).

13.2.4 Conservatory of fruit-yelding plants

The conservatory of fruit-yielding species serves as a repository for native fruit plants, including species such as *Manilkara hexandra* (Khirni), which were once locally extinct. It functions as both a food source for wildlife—particularly birds, reptiles, and small mammals—and an educational resource promoting awareness of indigenous horticultural diversity. By conserving genetic resources and sustaining wildlife interactions, this conservatory enhances both biodiversity and ecosystem services (Delhi Biodiversity Foundation; Sahapedia, a, b; Gupta & Das, 2025).

13.2.5 Winged biological control agents

Winged biological control agents, including insectivorous birds and predatory insects, play a critical role in maintaining ecological balance. They naturally suppress pest populations and promote pollination, reducing dependence on chemical pesticides. This biological regulation supports sustainable agriculture and reinforces the park's role as a model of ecological restoration and self-regulating ecosystem dynamics (Delhi Biodiversity Foundation; Sahapedia, a, b; Gupta & Das, 2025).

13.2.6 Butterfly conservatory

The butterfly conservatory represents a key component of the park's pollination network. By hosting a wide variety of butterfly species, it strengthens reproductive success among flowering plants and supports the park's food web dynamics. Beyond its ecological role, the conservatory functions as an important center for conservation education, promoting community engagement in biodiversity preservation (Delhi Biodiversity Foundation; Sahapedia, a, b; Gupta & Das, 2025).

13.2.7 Terrestrial birds

Terrestrial bird species contribute to vital ecosystem functions such as pollination, seed dispersal, and natural insect control. Their presence reflects the success of the park's habitat restoration and the reestablishment of ecological networks within an urban setting. By sustaining multiple trophic interactions, these avian populations enhance biodiversity and demonstrate the park's importance as an urban ecological refuge (Delhi Biodiversity Foundation; Sahapedia, a, b; Gupta & Das, 2025).

13.2.8 Wild animals

Wild fauna at YBP support ecosystem health through multiple ecological processes, including pollination, nutrient cycling, and maintenance of food web dynamics. Their activities aid carbon sequestration and groundwater recharge, while also moderating urban temperatures and improving air quality. The thriving wildlife populations underscore the park's significant contribution to urban environmental sustainability and climate regulation (Delhi Biodiversity Foundation; Sahapedia, a, b; Gupta & Das, 2025).

13.3 Conclusion: a path forward for ecological renewal and sustainable futures

The YBP is situated on the banks of the Yamuna River and is an excellent example of an achievement in ecological restoration. This degraded, salty marshland has flourished into a self-sustaining ecosystem. The native species of flora and fauna of this region have revived, which has brought a significant shift in awakening Delhi's natural legacy.

Thanks to meticulous scientific planning, extensive fieldwork, and the integration of local ecological knowledge, the park now supports species that were previously believed to have been wiped out by urbanisation. The ecosystem has been transformed in harmony, fulfilling its ecological role for all its plant, animal, and microscopic life. The wetlands have developed into vital migratory bird refuges, attracting species from around the globe every year.

The park has naturally shifted from its original focus on ex situ conservation to in situ rehabilitation due to the idea of autorestoration through biodiversity. It has more than 100,000 species, representing 800 plant species and a wide variety of wildlife. The park can function with a minimum of human assistance and is a living example of ecosystem interdependence, indicative of ecological stability and success.

The park is not only a conservation area but also a living school, a genetic reservoir, and a site for scientific research, environmental education, and community engagement. Bringing people closer to nature promotes ecological and social consciousness through recreation, education, and group responsibility.

This initiative has also brought a significant shift in public sentiment. The local people, who were not aware of the biodiversity of their area and its importance, now participate in the plantation drive sponsored by the park, wildlife walks, and various important environmental events. This shows growing interest and involvement of people and reflects the economic and qualitative benefits this ecological endeavour has brought to society.

The mission and objective of Prof. C.R. Babu contend that real progress must be sustainable, meaning it must satisfy present needs while protecting our natural heritage for future generations. This has been reflected in the ecological restoration of this park, which is a living example of the peaceful coexistence of nature and society.

In a world where biodiversity has emerged as a vital intersection of the natural and social sciences, the YBP is a response to global environmental challenges.

The successful restoration of this model has brought research to the forefront of environmental restoration. This asks decision-makers to reconsider ecological goals, challenge conventional development paradigms, and place research at the forefront of environmental restoration.

The success of YBP and the global concern about the environment have inspired other states, Karnataka and Maharashtra, to start similar programs. The residents and local people of Delhi are also promoting the creation of more such biodiversity parks in their communities for aesthetic, ecological, and spiritual advantages.

If this trend continues, the objective of establishing biodiversity parks in every district, safeguarding native species, and promoting ecological balance may not be far off. Such a future promises not only restored ecosystems but also a more sustainable, greener, and healthier Earth for all (Gatti, 2024).

References

Board, D. W. (2022). *Report by the wetland authority of Delhi under the wetland conservation rules, 2017*. https://dpgs.delhi.gov.in/dpgs/wetland-authority-delhi-0.

Buttke, D., Allen, D., & Higgins, C. (2014). Benefits of biodiversity to human health and well-being. *Park Science, 31*, 24–29.

Delhi Biodiversity Foundation, Delhi Development Authority. (n.d.). *Yamuna Biodiversity Park*. https://www.delhibiodiversityparks.org/yamuna-biodiversity-park.html.

Department of Anthropology. University of Delhi. (n.d.): https://anthro.du.ac.in.

Fauzie, A. K., & Sreenivasa. (2015). Study of various plant communities profiles in Yamuna Biodiversity Park, Delhi, India. *International Journal of Science and Research*. ISSN 2319-7064.

Gatti, R. C. (2024). Ecological peace corridors: A new conservation strategy to protect human and biological diversity. *Biological Conservation, 302*, 110947. https://doi.org/10.1016/j.bicon.2024.110947.

Ghermandi, A., et al. (2008). The economic value of wetland conservation and creation: A meta-analysis. *SSRN Electronic Journal*. https://doi.org/10.2139/ssrn.1273002.

Gonçalves, P., Grilo, F., Mendes, R. C., et al. (2021). What's biodiversity got to do with it? Perceptions of biodiversity and restorativeness in urban parks. *Ecology and Society*.

Gupta, D., & Das, S. (2025). Assessing the value of Yamuna biodiversity park in Delhi: A hedonic price approach. *Sustainable Futures, 10*, 01438. https://doi.org/10.1016/j.sftr.2025.101438.

Hussain, A. (n.d.). Impact assessment of biodiversity park in sustainable development parameters—Tughlakabad Biodiversity Park. https://www.scribd.com/document/626937448/Biodiversity-parkAbid.

Kaushal, S. S., Likens, G. E., Pace, M. L., et al. (2018). Freshwater salinization syndrome on a continental scale. *PNAS, 115*, E574–E583. https://doi.org/10.1073/pnas.1711234115.

Khudsar, F. A. (2010). *Kuno wildlife sanctuary: Preparing second home for asiatic lions*. Saarbrücken, Germany: LAP Lambert Academic Publishing, 44–64.

Koner, K. (2010). The butterfly park in Yamuna biodiversity park. *The Economic Times*. https://economictimes.indiatimes.com/the-butterfly-park-in-yamuna-biodiversity-park/articleshow/7056163.cms.

Millennium Ecosystem Assessment. (2005). Ecosystems and Human Well-Being. Vol 5. Washington, DC: Island Press.

Mitsch, W. J., & Gosselink, J. G. (2007). *Wetlands* (4th ed.). John Wiley & Sons Inc.

Niangthianhoi, S., & Khudsar, F. (2015). The nesting success of the oriental darter Anhinga melanogaster in the Yamuna biodiversity park, new Delhi, India. *Journal of Threatened Taxa, 7*, 8148–8153. https://doi.org/10.11609/jott.2430.7.14.8148-8153.

Niangthianhoi, S. L., & Khudsar, F. A. (2009). Assessment of bird diversity in Yamuna biodiversity park and a natural wetland. *Journal of Tropical Forestry, 25*(III & IV), 42–49.

Patel, S. (2019, May 22). From barren land to rich ecosystem: The story behind Yamuna Biodiversity Park. The Indian Express. https://indianexpress.com/article/cities/delhi/from-barren-land-to-rich-ecosystem-story-behind-yamuna-biodiversity-park-5743228/.

Prasad, S. N., Ramachandra, T. V., Ahalya, N., et al. (2002). Conservation of wetlands of India—A review. *Tropical Ecology, 43*(1), 173–186.

Ramírez, F., & Fennell, D. (2014). A comprehensive framework for ecotourism and wetland restoration: The case of Bogotá, Colombia. *Journal of Ecotourism, 13*, 128–151. https://doi.org/10.1080/14724049.2015.1011161.

Ramsar Convention Secretariat. (2018). *Global wetland outlook: State of the world's wetlands and their services to people*. Gland, Switzerland.

Ramsar Convention on Wetlands. (2021). *Global Wetland Outlook: Special Edition 2021*. Secretariat of the Convention on Wetlands, Gland, Switzerland. https://www.ramsar.org/2021.

Rathour, A., & Kumar, R. (2023). Plant propagation techniques in horticulture. In *Recent Trends in Agriculture*. Integrated Publications, 125–134.

Ravindran, K. C., Venkatesan, K., Balakrishnan, V., Chellapan, K. P., & Balasubramanian, T. (2007). Restoration of saline land by halophytes for Indian soils. *Soil Biology & Biochemistry, 39*, 2661–2664.

Rohit, M. L., Pandya, I., & Maitreya, B. (2025). Impact of biodiversity park on the surrounding environment. *International Journal of Scientific Research & Technology, 2*(4), 497–501.

Saha, A. (2006). *Assessment of terrestrial plant diversity in relation to soil parameters at the Yamuna Biodiversity Park, Wazirabad* (MSc dissertation). Department of Environmental Biology, University of Delhi.

Sahapedia, The Wetlands of Yamuna biodiversity park. 2024-a https://www.sahapedia.org/the-wetlands-of-yamuna-biodiversity-park.

Sahapedia, Yamuna Biodiversity Park: An Overview, 2024-b Available at: https://www.sahapedia.org/yamuna-biodiversity-park-overview.

Singh, J. S., Singh, S. P., & Gupta, S. R. (2006). *Ecology, environment, and resource conservation*. New Delhi: Anamaya Publishers.

Xu, X., et al. (2020). Wetland ecosystem services research: A critical review. *Global Ecology and Conservation, 22*, e01027. https://doi.org/10.1016/j.gecco.2020.e01027.

Chapter 14

Ecological and socioeconomic impacts of pollution on Indonesian wetlands

Annisa Utami Rauf[1], Rudi Windra Darisman[2], Sufie Bhaskara[3], and Binsar Oktavianus Togatorop[4]
[1]*Department of Health Behavior, Environment and Social Medicine, Faculty of Medicine, Public Health and Nursing, Universitas Gadjah Mada, Sleman, Indonesia,* [2]*Manggala Agni Kalimantan Daops X/Ketapang, Balai Pengendalian Kebakaran Hutan Wilayah Kalimantan, Pelaihari, South Kalimantan, Indonesia,* [3]*Manggala Agni Kalimantan Daops VI/Tanah Laut, Balai Pengendalian Kebakaran Hutan Wilayah Kalimantan, Pelaihari, South Kalimantan, Indonesia,* [4]*Manggala Agni Kalimantan Daops III/Pangkalan Bun, Balai Pengendalian Kebakaran Hutan Wilayah Kalimantan, Pelaihari, South Kalimantan, Indonesia*

14.1 Introduction

Indonesia has a diverse array of wetland ecosystems, including peatlands, mangroves, riparian zones, swamps, and rice fields. With a total area of 13.4 million hectares, it holds the largest expanse of tropical peatland globally (Yayasan Lahan Basah, 2020). These ecosystems, however, are increasingly threatened by pollution stemming from agricultural runoff, industrial discharge, and untreated sewage. The infiltration of contaminants into wetland soils and water bodies disrupts ecological balance, endangers biodiversity, and poses significant public health risks to communities relying on wetlands for drinking water and agricultural needs (Budi et al., 2024; Fathani et al., 2025; Song et al., 2024).

Beyond its ecological significance, wetland degradation has profound socioeconomic consequences. The decline of these ecosystems disrupts fisheries, agriculture, and tourism, exacerbating economic inequalities among communities reliant on these resources (Reis et al., 2017). Additionally, the loss of wetlands jeopardizes traditional livelihoods and cultural heritage, threatening indigenous knowledge systems and local identities. The urgency to rehabilitate and preserve wetlands has become critical, particularly as Indonesia faces the triple crises of climate change, biodiversity loss, and pollution (Budi et al., 2024; Falahudin et al., 2025).

Recognizing the vital ecological and socioeconomic functions of wetlands, Indonesia has implemented various conservation and restoration initiatives (Sjafrie et al., 2024). The government has committed to large-scale restoration programs, including the rehabilitation of 600,000 ha of mangrove forests and 1.2 million hectares of peatlands, supported by significant financial investments. The adoption of carbon trading mechanisms under Presidential Regulation No. 98 of 2021 aims to regulate carbon emissions while promoting sustainable environmental practices (Yayasan Lahan Basah, 2020). These initiatives align with Indonesia's Nationally Determined Contributions, which emphasize resilience in economic, social, and ecological domains through ecosystem-based adaptation strategies (Naiborhu et al., 2024).

Wetlands are increasingly recognized as "super ecosystems" due to their ability to address multiple environmental challenges simultaneously. Mangrove forests, seagrass meadows, and peatlands function as highly efficient carbon sinks, mitigating climate change while also providing coastal protection and supporting biodiversity (Falahudin et al., 2025; Sjafrie et al., 2024; Suciyono et al., 2024). Unfortunately, these ecosystems face several pressures from unregulated land conversion, excessive resource extraction, and inadequate spatial planning, leading to accelerated degradation. Addressing these threats requires a comprehensive, science-based approach that integrates research, policy, and community engagement. The incorporation of citizen science in biodiversity monitoring and habitat assessment can generate valuable data for policymaking, while strengthening government-led conservation efforts through research-driven strategies and practical implementation can enhance restoration outcomes.

By 2030, Indonesia aims to restore 100,000 ha of critical wetland ecosystems through collaborative efforts involving government agencies, private sector stakeholders, and local communities. Indonesia's wetlands play a crucial role in maintaining ecological stability and supporting economic sustainability (Fathani et al., 2025). However, these ecosystems

Wetlands: Biodiversity, Health and their Role in Ecosystem Services. DOI: https://doi.org/10.1016/B978-0-443-34222-6.00013-9

are increasingly threatened by deforestation, land conversion, and climate change. As one of the most biodiverse countries in the world, Indonesia faces the risk of losing irreplaceable flora and fauna, many of which are integral to ecosystem functioning. The degradation of wetland habitats not only accelerates biodiversity loss but also diminishes their capacity to regulate floods, prevent coastal erosion, and mitigate rising sea levels, thereby heightening disaster risks for vulnerable communities. From an economic perspective, the deterioration of wetlands disrupts key sectors such as fisheries, agriculture, and ecotourism, jeopardizing livelihoods and local economies. Without immediate and coordinated conservation efforts, these interconnected environmental and economic consequences will threaten Indonesia's long-term resilience and sustainable development.

14.2 Degradation of wetland ecosystems from human activities

Human interventions have significantly contributed to the degradation of wetland ecosystems in Indonesia. The primary drivers of this degradation include land conversion, industrial activities, agricultural expansion, and unsustainable land management practices (Taufik et al., 2019; Yahaya et al., 2024). The deterioration of wetlands in Indonesia is a major environmental concern, influenced by both human activities and natural processes. For example, the water quality in the middle Mahakam lakes region has fluctuated, with noticeable improvements occurring between 1998 and 2006. Despite these improvements, the system remains susceptible to pH variations due to its low alkalinity, which can be exacerbated by deforestation, mining, and peatland drainage (De Jong et al., 2015). In another case, extensive agricultural activities, particularly the conversion of peatlands for palm oil plantations, have led to significant degradation. This includes deforestation and drainage of peat swamp forests, which increases carbon emissions and makes these areas more susceptible to fires in Central Kalimantan. Annual CO_2 emissions increase by 79–238 gC m^{-2} for every 0.1 m decline in groundwater level, likely due to the intensified oxidative decomposition of peat. Moreover, CO_2 uptake through vegetation photosynthesis is reduced as dense smoke from peat fires, whether accidental or intentionally ignited for agricultural purposes, limits sunlight penetration (Hirano et al., 2012; Medrilzam et al., 2014).

Rapid industrial development and urbanization have led to the conversion of wetland areas into nonwetland uses. For instance, in Bekasi Regency, extensive development has resulted in a reduction of over 11.5% of wetland area, with nonwetland areas increasing by approximately 3224.40 ha (Novenda et al., 2024). The expansion of urban areas and infrastructure development has also contributed to the loss of coastal wetlands, such as mangroves, which are being converted for other uses. The conversion of wetlands into agricultural land is a significant factor. In South Sumatra, unsustainable practices, including excessive drainage and reclamation for agriculture, have led to extensive wetland degradation. In peatland areas, the conversion to agricultural land, particularly for oil palm and rubber plantations, has been driven by the higher profit margins these crops offer compared to sustainable alternatives (Edi et al., 2018; Hirano et al., 2012; Jalilov et al., 2024; Medrilzam et al., 2014).

Mangrove ecosystems in Indonesia have experienced a significant decline in extent over the last two decades, with projections indicating continued loss if no major interventions are implemented. Nationally, the total mangrove area in 2024 is estimated to be reduced by 21.8% compared to the baseline year 2000. The most severe reductions between 2000 and 2024 occur in Sumatra (53%) and Sulawesi (43%), signaling critical hotspots of degradation. According to Fig. 14.1, the condition of mangroves in Indonesia has declined over the past decades and is projected to continue to decline until 2045 (Kementerian PPN/ Bappenas, 2021).

Logging, land use conversion, and drainage have significantly degraded peatlands, which are a type of wetland. These activities disrupt the hydrological balance, leading to peat subsidence and increased vulnerability to fires, especially during El Niño years (Dohong & Tanika, 2021; Taufik et al., 2019). This shift poses a major threat to the natural environment. In South Sumatra Province, wetland degradation is primarily caused by weak law enforcement and compliance, land reclamation, and excessive drainage. Coastal wetlands, particularly mangrove forests, are increasingly being converted for alternative purposes, leading to the deterioration of Indonesia's mangrove ecosystems (Adi Nugraha et al., 2024; Adriani et al., 2024). Fig. 14.2 illustrates different levels of forest degradation and land cover types used in the classification. Each panel represents a 2 km×2 km area with north oriented upwards. The varying degrees of forest degradation are evident through changes in vegetation, ranging from pristine peat swamp forest to areas dominated by human activities. The different land cover types, such as tall shrubs, industrial plantations, and water bodies, are also depicted.

Wildfires play a significant role in forest destruction, severely impacting ecosystem stability by reducing vegetation, which is vital for sustaining biodiversity and habitat integrity. Studies indicate that biomass levels in unburnt areas are 2.67 times greater than in burnt areas, emphasizing the significant decline in living plant material. This loss not only decreases biodiversity but also disrupts habitat structures, making it harder for various species to thrive. The fires contribute to the accumulation of dead plant material (necromass), which serves as fuel for future fires, perpetuating a

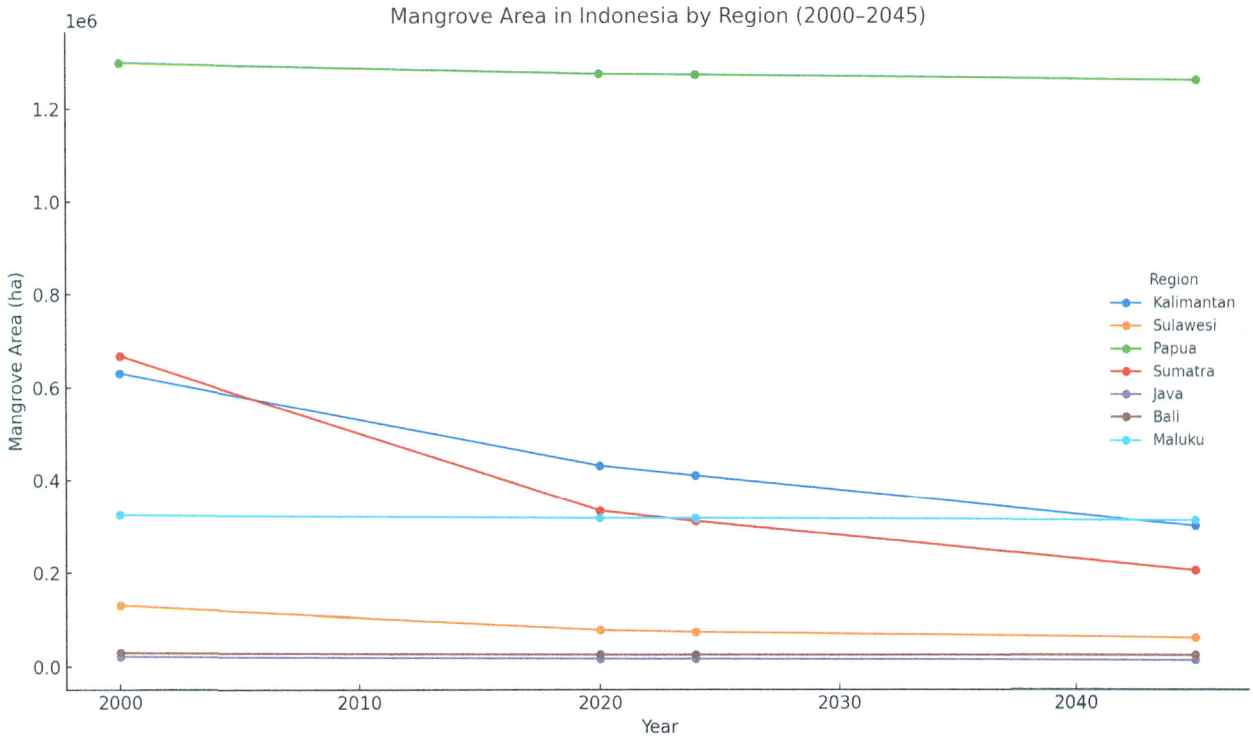

FIGURE 14.1 Observed and projected mangrove decline in Indonesia (2000–45).

destructive cycle (Hammen, 2007). Beyond their ecological consequences, peatland fires are a major contributor to climate change, releasing vast amounts of carbon dioxide (CO_2) into the atmosphere. The 1997/1998 Indonesian fires, for instance, released significant carbon emissions, intensifying greenhouse gas concentrations. Moreover, the burning of both aboveground vegetation and peat soil leads to prolonged carbon depletion, with CO_2 emissions from soil respiration increasing by 46% within the first 9 months after a fire, highlighting the long-term environmental impacts of peatland degradation (Astiani et al., 2018; Silviana et al., 2019).

14.3 Ecosystem disruption

The degradation of wetlands results in a decrease in both species richness and population abundance. For example, bird communities in peat swamp forests experience significant impacts, particularly with a marked decline in nectarivorous and insectivorous species in areas outside intact forest habitats (Posa, 2011). From 1990 to 2010, peat swamp forest cover in Southeast Asia declined from 76.6% to 35.8%, with annual losses of 3.6% in Indonesia and 4.1% in Malaysia (Miettinen et al., 2012). This decline is mainly due to logging, drainage, and conversion to plantations, particularly for oil palm, putting remaining forests at risk. Forest fires have also increased, shifting from dryland to swamp forests between 2002 and 2006, largely due to human disturbances, as intact forests retain moisture and are less likely to burn. In Borneo (Kalimantan Island), only 10% of the original peat swamp forest remains undisturbed, with much of the rest selectively logged. Logging and drainage make these forests highly flammable, and frequent fires hinder regeneration, allowing low vegetation to dominate (Langner & Siegert, 2009; Miettinen et al., 2012).

Intact logged forests in Central Kalimantan support higher species diversity and distinct community structures compared to degraded forest patches and regenerating nonforest areas. Habitat degradation has particularly affected specialized bird species, leading to a decline in nectarivorous and tree foliage-gleaning insectivores that depend on undisturbed environments. Species such as the yellow-vented bulbul (*Pycnonotus goiavier*), an adaptable insectivore-frugivore, have become more dominant in degraded landscapes (Posa, 2011). Although logged peat swamp forests continue to serve as important habitats for threatened and near-threatened bird species, ongoing environmental disturbances disrupt ecological balance by altering seed dispersal, insect populations, and forest regeneration, ultimately affecting overall ecosystem stability (Dohong & Tanika, 2021).

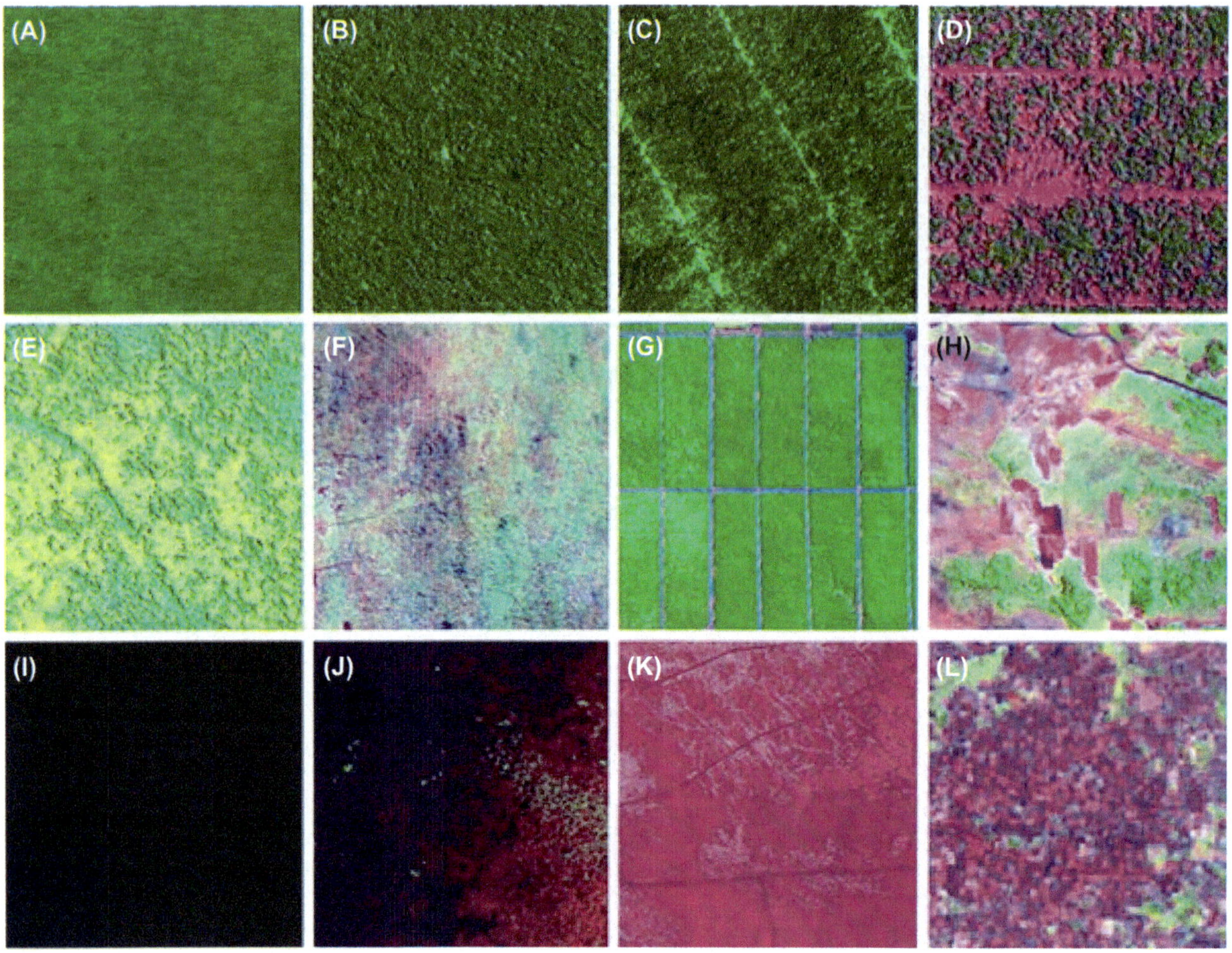

FIGURE 14.2 Classification of forest degradation levels and land cover types: (A) Pristine PSF, (B) slightly degraded PSF, (C) moderately degraded PSF, (D) heavily degraded PSF, (E) tall shrub/secondary forest, (F) ferns/low shrub, (G) industrial plantations, (H) small-holder dominated area, (I) permanent water body, (J) seasonal water, (K) cleared/burnt area, and (L) built-up area. Panels (I) and (L) are derived from SPOT 4 HRVIR imagery (RGB: bands 4, 3, 2), while the rest use SPOT 5 HRG imagery (RGB: bands 4, 3, 2). Slight human activity is noticeable in (B) through faint diagonal pathways (Miettinen & Liew, 2010).

Adapted from Miettinen, J., & Liew, S. C. (2010). Status of peatland degradation and development in Sumatra and Kalimantan. *Ambio*, *39*(5–6), 394–401. https://doi.org/10.1007/s13280-010-0051-2. SPOT images © 2007 CNES.

Dredging activities in Serang, particularly within the Jakarta Bay reclamation project, pose a significant threat to plant biomass, with an estimated 52.96±32.73 Mg C ha^{-1} at risk. Sand extraction further exacerbates the problem, as disturbances to mangrove forests can lead to a decline in soil organic carbon by 2%–50% within eight years of deforestation and up to 56% within a decade of mangrove die-off. This degradation weakens carbon sequestration capacity, disrupts essential ecosystem functions, and contributes to long-term environmental instability (Murdiyarso et al., 2015). The environmental impact assessment for the reclaimed islands indicates that sand mining in Serang could drastically reduce carbon storage, as excavation exposes stored soil carbon and diminishes the ecosystem's ability to regulate climate. Mangrove forests in Serang and Angke contain substantial organic carbon in their deep soil layers, with mining up to 2 m potentially releasing 88.74 Mg C ha^{-1}. However, restricting mining depth to 1 m could significantly mitigate this impact, reducing carbon loss to 25.79 Mg C ha^{-1} and preserving up to 71% of soil carbon reserves (Granek & Ruttenberg, 2008; Otero et al., 2017; Slamet et al., 2020).

In East Indonesia, rapid deforestation and coastal development in the Bird's Head Seascape (BHS) have triggered substantial ecological disturbances. The combination of steep slopes, erodible soils, and high rainfall has heightened the vulnerability of coastal ecosystems, particularly coral reefs, to land-based impacts (Purwanto et al., 2021). These include sediment runoff smothering marine habitats, mangrove loss due to infrastructure expansion, habitat destruction for endangered species, coral reef degradation from land reclamation, altered salinity and temperature at river mouths, and the introduction of invasive species. *Tilapia* spp., a nonnative fish species, has been recorded in disturbed mangrove and riverine habitats, competing with indigenous fish populations. Similarly, the spread of *Chromolaena odorata*, an invasive plant, has been linked to deforested areas, altering soil composition and impeding the natural regeneration of native flora. These disruptions highlight the intricate ecological relationships within the BHS and underscore the urgent need for sustainable coastal management strategies to mitigate further habitat loss and biodiversity decline (Mangubhai et al., 2012).

In the Riau Islands, the wetlands of Natuna support a rich diversity of nonavian species, meeting Ramsar Site criteria, with seagrass playing a vital ecological role. Four documented species—*Enhalus acoroides, Thalassia hemprichii, Halodule pinifolia*, and *Halophila ovalis*—comprise 33% of Indonesia's total seagrass diversity, with *H. ovalis* dominating at 64.73%, likely due to its adaptability to disturbed habitats. However, wetland degradation from coastal development, land reclamation, and pollution threatens biodiversity, reducing habitat for key species like dugongs (*Dugong dugon*), green turtles (*Chelonia mydas*), and commercially valuable fish (Murdiyarso et al., 2015; Warren et al., 2017). The decline of structurally complex seagrasses such as *E. acoroides* suggests increasing environmental stress, weakening ecosystem functions like carbon sequestration and sediment stabilization. As habitat loss accelerates, conservation efforts become critical to preserving seagrass meadows and maintaining biodiversity in Natuna's wetlands (Irawan et al., 2021).

14.4 Public health implications

Pollution from human activities has significantly affected ecosystem services, posing complex challenges for environmental health professionals. Changes to wetlands that interfere with natural water flow can lead to a chain reaction of ecological disturbances, such as biodiversity loss, the proliferation of nonnative species, and disruptions in biogeochemical processes, ultimately diminishing the functionality of ecosystem services (Medrilzam et al., 2014; Novenda et al., 2024). Lower water levels and increased salinity further stress aquatic and plant life, making it difficult for native species to thrive. The consequences extend to human populations, as communities that rely on wetlands for water, food, and livelihoods face displacement. The stagnant or contaminated water in degraded wetlands can promote the spread of waterborne diseases, posing serious public health risks (Guo et al., 2021; Haikerwal & Saxena, 2020; Shridhar et al., 2024). In West Papua, the conversion of primary rainforest to oil palm plantations in the Kais River watershed has significantly altered hydrological processes, posing serious public health risks for downstream communities. Increased surface runoff (21%) and sediment yields (16.9%) contribute to water contamination, while the sharp rise in total nitrogen (78%) and phosphorus (144%) can lead to algal blooms and waterborne diseases. These changes heighten the risk of flooding, which can spread pathogens and pollutants (Asmara & Randhir, 2024).

The degradation of peatland ecosystems in Indonesia has significantly contributed to air pollution, primarily due to recurrent fires that intensify during dry periods, particularly in El Niño years (Hein et al., 2022; Santika et al., 2023). These fires release substantial quantities of fine particulate matter ($PM_{2.5}$), leading to severe deterioration in air quality, notably in highly affected regions such as Sumatra and Kalimantan. The increased concentration of airborne pollutants has been associated with adverse respiratory health outcomes, including a documented 8.5% rise in respiratory illnesses during dry years in Sumatra, driven by changes in land cover and increased fire incidents (Pakhtigian et al., 2024). Children and the elderly are particularly susceptible to the respiratory effects of air pollution caused by peatland fires, with studies indicating increased hospitalization rates among these vulnerable populations during periods of intensified fire activity (Lorenz et al., 2024). Prolonged exposure to $PM_{2.5}$ from peatland fires has been linked to significant morbidity and mortality, contributing to an estimated 33,100 premature adult deaths and 2900 infant deaths annually. Exposure to peat fire emissions is associated with a rise in severe asthma cases and substantial economic losses due to lost workdays and healthcare costs (Hein et al., 2022).

Coastal pollution has severely affected marine ecosystems, particularly coral reefs, which are highly vulnerable to anthropogenic disturbances, as observed in the Cisadane River and Mahakam Lake (De Jong et al., 2015; Purbonegoro et al., 2024). In South Sumatra Province, wetland degradation has resulted in a significant decline in biodiversity, with fish, gymnosperms, and ferns experiencing the most pronounced population reductions and habitat loss (Wulandari et al., 2024). Upper Citarum River Basin in West Java identified significant risks to aquatic ecosystems, despite uncertainties in use estimates, runoff modeling, and effect prediction. While human health risks from oral intake of individual pesticides through river water were predicted to be negligible based on nonmixture toxicity assessments, the potential for cumulative exposure remains a concern. Pesticides such as profenofos, propineb, chlorpyrifos, carbofuran, imidacloprid, methomyl, and chlorantraniliprole were found to pose significant aquatic toxicity risks, which could indirectly affect public health through bioaccumulation in fish and contamination of drinking water sources (Utami et al., 2020).

Effective wetland ecosystem management and pollution control are essential for minimizing the health risks associated with peatland degradation. Approaches such as integrated island management and sustainable peatland restoration play a critical role in reducing pollution-related health burdens by preventing large-scale fires and mitigating air and wetland quality deterioration (Jenkins & Jupiter, 2015; Marlier et al., 2019). Community-based adaptation strategies, including early warning systems, reforestation efforts, and improved healthcare infrastructure, are vital for strengthening resilience and reducing the long-term health impacts of wetland pollution disasters (Kurniawan et al., 2024).

14.5 Social and cultural impacts

The Mega Rice Project, initiated in the mid-1990s under President Suharto, aimed to convert 1 million ha of peatland in Central Kalimantan into rice fields to achieve food self-sufficiency. However, poor planning, limited understanding of peatland ecosystems, and inadequate infrastructure led to its failure, resulting in severe environmental and socioeconomic consequences. Extensive deforestation and peatland drainage not only increased carbon emissions and heightened fire risks but also displaced indigenous Dayak communities, depriving them of their traditional livelihoods in shifting cultivation and agroforestry (Akbar, 2011; Medrilzam et al., 2014). Additionally, the nutrient-poor, acidic peat soil proved unsuitable for rice farming, leading to abandoned fields and long-term ecological degradation, with ongoing land conflicts and frequent forest fires persisting as lasting impacts of the project.

The peat swamp forests in Aceh Province are experiencing widespread degradation due to extensive land conversion, which has contributed to poverty and underdevelopment among local communities. Although government interventions have been introduced to enhance both ecosystem sustainability and community well-being, such as community empowerment initiatives, these efforts have largely failed to achieve their intended social objectives, raising concerns about their overall effectiveness (Agussabti et al., 2022).

Indigenous communities, particularly those reliant on forest resources, face growing threats from climate change (Habibullah et al., 2024). In Bengkulu, communities surrounding Kerinci Seblat National Park depend heavily on these resources for their livelihoods. Their vulnerability is shaped by extreme weather events, shifting climate patterns, livelihood strategies, and access to essential resources such as food and water (Widiono et al., 2024). Climate change exacerbates their challenges, increasing the risk of poverty and food insecurity due to crop failures, which in turn forces them to seek alternative income sources, often leading to resource conflicts. The vulnerability of indigenous communities to climate change is determined by multiple socioeconomic factors, including educational attainment, access to financial resources, availability of critical irrigation infrastructure, and the diversification of household income sources. Limited access to these essential resources constrains adaptive capacity, increasing susceptibility to climate-induced disruptions and economic instability.

The degradation of critical ecosystems, driven by uncontrolled fires and large-scale developments, has resulted in declining incomes and diminishing livelihood opportunities for local communities. The loss of mangrove forests, for example, has been directly associated with reduced annual income among fishery households, underscoring the significant economic importance of wetland conservation in sustaining local economies and livelihoods (Yamamoto, 2023). Balancing economic growth with environmental sustainability necessitates integrated management strategies that incorporate both socioeconomic and ecological considerations (Angi et al., 2022). The social forestry (SF) program represents one such initiative, aiming to resolve land conflicts and encourage sustainable resource use. However, challenges such as institutional inefficiencies and inadequate support often hinder its success. Simultaneously, the restoration of degraded peatlands and the promotion of traditional agroforestry systems play a vital role in preserving wetland ecosystems while supporting local livelihoods. These conservation efforts not only safeguard ecological functions but also enhance long-term socioeconomic resilience by mitigating the adverse effects of environmental degradation (Rebelo et al., 2010; Samsudin et al., 2020).

14.6 Current conservation and restoration efforts

According to Alexander (2018), effective wetland management should be an adaptive, iterative, and dynamic process that involves all stakeholders. This includes recognizing essential characteristics, setting well-defined objectives, and outlining the necessary actions and resources for sustainable management (Alexander, 2018). A strong legal framework for Integrated Coastal Zone Management in Indonesia is vital for sustainable coastal development. To optimize natural resource use and support growth in coastal and island regions, the government has established a zoning plan, as stated in East Nusa Tenggara Provincial Regulation No. 04/2017. This policy plays a key role in driving economic development by facilitating regulated investment in the tourism sector. One notable example is the Savu Marine National Park, which offers rich opportunities for both nature-based and cultural tourism, emphasizing the need for sustainable management to protect ecosystems while benefiting local communities (Tussadiah et al., 2021). Collaborative efforts, such as reforestation along rivers, fishing restrictions, and buffer zones, could help address these challenges (Thahira et al., 2023). Cross-border collaboration through joint monitoring and legally binding agreements is crucial for safeguarding these ecosystems.

TABLE 14.1 Conservation and management efforts for wetland ecosystems in Indonesia.

No	Locations	Method	Results	Source
1	Sumatra Selatan and Jambi Provinces	InVEST (Integrated Valuation of Ecosystem Services and Tradeoffs) model	The water supply per hectare is estimated to be 10,248 $m^3\ ha^{-1}$, with a total supply of 16.46 billion m^3 per year, which is not significantly different from the current land use. In the FOR *scenario, the water supply per hectare slightly increases* to 10,526 $m^3\ ha^{-1}$, about 3% higher than the current condition. The total water supplied reaches approximately 16.91 billion m^3 per year, an increase of about 500 million m^3 compared to the current land use	Song et al. (2024)
2	Kampar Regency, Riau Province	Household knowledge	Training programs on nursery establishment, fire-free land preparation, postharvest processing, and honeybee cultivation enhanced participants' knowledge and skills in peatland protection. Proximity to village centers with better communication infrastructure, including internet access, also supported improved awareness and practices	Yunus et al. (2025)
3	Sriwulan, Bedono, Timbulsloko, and Surodadi, Central Java Province	Mangrove conservation	Mangrove ecosystems, which are crucial for coastal protection and biodiversity, can be restored through community-based management and conservation strategies. Combining effective local management with external scientific and technological assistance can improve outcomes	Damastuti and De Groot (2017)
4	Kayan Sembakung Delta, North Kalimantan Province	SF	Implementing SF programs can help resolve land conflicts and promote the sustainable use of wetland resources. These programs provide a structured approach to resource management, fostering cooperation among stakeholders and reducing disputes over land ownership and usage	Angi et al. (2022)
5	South Sumatra Province	Policy and regulatory measures	Effective enforcement of existing policies and regulations is crucial. Weak enforcement and inadequate policies are major causes of wetland degradation. Enhancing policy support and improving coordination among stakeholders can help protect wetlands	Edi et al. (2018)
6	Makassar, South Sulawesi	Constructed wetlands and biological monitoring	The study on Hasanuddin University Lake aimed to improve water quality by controlling contamination sources and implementing biological monitoring systems	Ibrahim et al. (2024)
7	Mempawah and Kubu Raya Regencies of West Kalimantan	Rewetting peatlands and paludiculture	Replanting native peat swamp forest plants in degraded areas can help restore soil properties and microbial activity, leading to improved ecosystem functions and reduced emissions. It not only helps in reducing GHG emissions but also provides biomass for bioenergy, which can replace fossil fuels	Novita et al. (2024)

(Continued)

TABLE 14.1 (Continued)

No	Locations	Method	Results	Source
8	Citarum watershed, West Java	Ecohydrological approaches	Implementing ecohydrological methods can help regulate water quality and quantity, reduce soil erosion, and improve overall ecosystem health. This approach has been applied in the Saguling reservoir catchment area with positive results	Sutapa et al. (2021)
9	Sebangau, Central Kalimantan	Fish population conservation	Efforts to conserve fish populations in wetlands and rivers are necessary to maintain biodiversity and support local livelihoods	Thornton et al. (2018)
10	Riau Province	Hybrid governance models and management	New governance strategies that manage peatlands based on their hydrological characteristics are emerging. These models aim to address the socio-ecological crises associated with peatland conversion and involve multiple stakeholders to ensure effective management	Astuti (2020)
11	Raja Ampat, West Papua Province	Bird's Head Seascape	Raja Ampat has implemented a tourism licensing system and environmental regulations, aiming to balance economic growth with conservation efforts	Mangubhai et al. (2012)
12	Citarum River, West Java Province	Runoff modeling and effect prediction	Profenofos, propineb, chlorpyrifos, carbofuran, imidacloprid, methomyl, and chlorantraniliprole have been identified as posing considerable risks to aquatic ecosystems. To safeguard water quality and biodiversity, it is recommended that water managers implement measures to minimize pesticide application	Utami et al. (2020)
13	Central and East Kalimantan Province	Fire-free village	Indonesia's plan to expand the "fire-free village" model to over 700 villages aims to reduce peatland fires through targeted incentives, yet its scalability remains a challenge. Effective peatland fire management will depend on balancing decentralized innovation	Sloan et al. (2017)

SF, Social forestry.

The cumulative impacts of wetland degradation underscore the urgent need for conservation and restoration efforts to ensure their long-term sustainability (W. Utami et al., 2024).

Between 2004 and 2015, Indonesia experienced six significant peatland fire events, resulting in an estimated economic loss of $93.9 billion USD and thousands of fatalities. In 2015 alone, fire-related damage accounted for approximately 3.3% of the country's gross domestic product (Page et al., 2011). Beyond the economic consequences, peatland degradation has led to the erosion of traditional livelihoods, threatening indigenous knowledge systems and weakening cultural connections to these ecosystems. Furthermore, the loss of biodiversity disrupts ecological functions that are essential for sustaining local communities. In response, ecosystem-based disaster risk reduction interventions in Sumatra have been implemented to address peatland degradation through community-led restoration efforts (Vicarelli et al., 2024; Warren et al., 2017). These initiatives encompass fire prevention training, peatland monitoring, and the promotion of sustainable livelihood practices such as paludiculture and aquaculture (Agus et al., 2019). By integrating ecological restoration with socioeconomic strategies, these interventions not only enhance environmental resilience but also contribute to the preservation of cultural heritage and community stability in affected regions.

Indonesia has set ambitious goals for managing land cover in peatland and mangrove ecosystems as part of its sustainable development agenda and low-carbon vision for 2045. In the case of peatlands, the government plans to restore 1.05 million hectares during 2022–24, with the cumulative area under restoration projected to reach 3.44 million hectares by 2030–34, a target that will be sustained until 2045. These restoration programs are concentrated in nonforest areas already classified as moderately to severely degraded. Alongside restoration, efforts are also directed at slowing vegetation cover degradation in peatlands, with reduction targets ranging from 0.32 to 1.26 million hectares in 2022 to 2024, narrowing to 0.47 to 0.93 million hectares by 2040 to 2045. The focus is on landscapes with sparse vegetation or recurring fire damage, such as shrublands, swamps, grasslands, and open lands.

For mangroves, the priority is on rehabilitating degraded and critical ecosystems. The initial target is 23,320 ha between 2022 and 2024, expanding to 245,100 ha in the following decades. After 2029, rehabilitation will also include potential mangrove habitats within aquaculture ponds through methods such as silvofishery practices and Associated Mangrove Aquaculture, while ensuring compliance with green belt regulations in protected coastal areas. Rehabilitation sites are selected based on the Coastal Vulnerability Index, which highlights the areas most exposed to risks. In parallel, Indonesia has committed to steadily reducing mangrove deforestation. From a baseline average of 26,121 ha annually in 2015–20, the country aims to cut losses by as much as 19,132 ha by 2045 (Kementerian PPN/ Bappenas, 2021). This will be achieved progressively, starting with annual reductions of 1572–4020 ha in 2022–24, and scaling up to 15,832–19,132 ha per year by 2040–45.

Table 14.1 presents various efforts implemented in different regions of Indonesia to protect and restore wetland ecosystems. A range of approaches, including ecosystem valuation models, mangrove conservation, social-based management policies, and ecohydrological strategies, have been utilized to address diverse environmental challenges. These studies highlight how a combination of science-based strategies, policies, and community involvement can enhance the effectiveness of conservation efforts and promote sustainable natural resource management.

14.7 Conclusion

Protecting Indonesian wetlands is essential for preserving biodiversity, sustaining local communities, and safeguarding public health. However, rising pollution from agriculture, industry, and household waste poses serious threats to these ecosystems, leading to environmental degradation and socioeconomic challenges. A holistic approach is needed to tackle these issues, combining stronger policy enforcement, active community participation, and sustainable resource management. Through effective mitigation measures, wetland health can be restored, pollution minimized, and vital ecosystem services maintained. Ensuring the long-term stability of Indonesian wetlands is not only crucial for the environment but also for sustainable development and societal well-being.

References

Adi Nugraha, P., Hayati, N., & Taufik, M. (2024). The use of sentinel-1 imagery for degraded peatland analysis in South Kalimantan area with SSM algorithm and Wosten model. *IOP Conference Series: Earth and Environmental Science, 1418*(1), 012002. https://doi.org/10.1088/1755-1315/1418/1/012002.

Adriani, D., Yazid, M., Riswani, Damayanthy, D., Choi, E., & Yang, H. (2024). Livelihood alternatives in restored peatland areas in South Sumatra province, Indonesia. *Land, 13*(5), 643. https://doi.org/10.3390/land13050643.

Agus, C., Azmi, F. F., Widiyatno, Ilfana, Z. R., Wulandari, D., Rachmanadi, D., Harun, M. K., & Yuwati, T. W. (2019). The impact of forest fire on the biodiversity and the soil characteristics of tropical peatland. In W. Leal Filho, J. Barbir, & R. Preziosi (Eds.). *Handbook of climate change and biodiversity* (pp. 287–303). Springer International Publishing. https://doi.org/10.1007/978-3-319-98681-4_18.

Agussabti, A., Zikri, I., Rahmaddiansyah, R., Humam Hamid, A., Baihaqi, A., & Takahashi, M. (2022). Exploring the social science of tropical peatland restoration: Towards more effective community empowerment initiatives for the Aceh peatland ecosystem. *Mires and Peat, 28*(06), 1–16. https://doi.org/10.19189/MaP.2022.OMB.StA.2376.

Akbar, A. (2011). Studi Kearifan Lokal Penggunaan Api Persiapan Lahan: Studi Kasus di Hutan Mawas, Kalimantan Tengah. *Neliti*. https://www.neliti.com/publications/29049/studi-kearifan-lokal-penggunaan-api-persiapan-lahan-studi-kasus-di-hutan-mawas-k#cite.

Alexander, M. (2018). Wetland Management Planning: Overview. In C. M. Finlayson, M. Everard, K. Irvine, R. J. McInnes, B. A. Middleton, A. A. Van Dam, & N. C. Davidson (Eds.). *The wetland book* (pp. 1877–1891). Springer Netherlands. https://doi.org/10.1007/978-90-481-9659-3_269.

Angi, E. M., Kartika, K., & Wiati, C. B. (2022). The potential, wetlands utilization through the social forestry program in Kayan Sembakung Delta, North Kalimantan, Indonesia. *IOP Conference Series: Earth and Environmental Science, 976*(1), 012023. https://doi.org/10.1088/1755-1315/976/1/012023.

Asmara, B., & Randhir, T. O. (2024). Modeling the impacts of oil palm plantations on water quantity and quality in the Kais River Watershed of Indonesia. *Science of the Total Environment, 928*, 172456. https://doi.org/10.1016/j.scitotenv.2024.172456.

Astiani, D., Curran, L., Burhanuddin, Taherzadeh, Mujiman, M., Hatta, M., Pamungkas, W., & Gusmayanti, E. (2018). Fire-driven biomass and peat carbon losses and post-fire soil CO_2 emission in a west kalimantan peatland forest. *Journal of Tropical Forest Science, 30*(4), 570–575. https://doi.org/10.26525/jtfs2018.30.4.570575.

Astuti, R. (2020). Fixing flammable Forest: The scalar politics of peatland governance and restoration in Indonesia. *Asia Pacific Viewpoint, 61*(2), 283–300. https://doi.org/10.1111/apv.12267.

Budi, D. S., Priyadi, A., Permana, A., Herjayanto, M., Slembrouck, J., Mubarak, A. S., & Mustofa, I. (2024). Sustainable captive breeding practices for native Indonesian freshwater fish. *Animal Reproduction Science, 271*, 107623. https://doi.org/10.1016/j.anireprosci.2024.107623.

Damastuti, E., & De Groot, R. (2017). Effectiveness of community-based mangrove management for sustainable resource use and livelihood support: A case study of four villages in Central Java, Indonesia. *Journal of Environmental Management, 203*, 510–521. https://doi.org/10.1016/j.jenvman.2017.07.025.

de Jong, E. B. P., Ragas, A. M. J., Nooteboom, G., & Mursidi, M. (2015). Changing water quality in the Middle Mahakam Lakes: Water quality trends in a context of rapid deforestation, mining and palm oil plantation development in Indonesia's Middle Mahakam wetlands. Wetlands, 35(4), 733–744. https://doi.org/10.1007/s13157-015-0665-z

Dohong, A., & Tanika, L. (2021). Hydrological management practices. In M. Osaki, N. Tsuji, N. Foead, & J. Rieley (Eds.). *Tropical peatland eco-management* (pp. 567–593). Springer Singapore. https://doi.org/10.1007/978-981-33-4654-3_20.

Edi, A. M., Elisa, W., & Bella, S. (2018). Dynamics, degradation and future challenges of wetlands in South Sumatra province, Indonesia. *E3S Web of Conferences, 68*, 04001. https://doi.org/10.1051/e3sconf/20186804001.

Fathani, M. H., Mulyani, Y. A., Mardiastuti, Ani, & Djatmiko, W. A. (2025). Factors affecting bird diversity in the wetlands of muara gembong, Indonesia. *Jurnal Manajemen Hutan Tropika (Journal of Tropical Forest Management), 31*(1), 61–71. https://doi.org/10.7226/jtfm.31.1.61.

Falahudin, D., Arifin, Z., Yogaswara, D., Edward, E., Wulandari, I., Dharmawan, I. W. E., Sudaryanto, A., Hoang, A. Q., & Takahashi, S. (2025). Halogenated organic compounds in mangrove sediments from Bintan Island, Indonesia: Occurrence, profiles, sources, and potential ecological risk. *Environmental Pollution, 367*, 125667. https://doi.org/10.1016/j.envpol.2025.125667.

Granek, E., & Ruttenberg, B. I. (2008). Changes in biotic and abiotic processes following mangrove clearing. *Estuarine, Coastal and Shelf Science, 80*(4), 555–562. https://doi.org/10.1016/j.ecss.2008.09.012.

Guo, S.-Y., Li, L., Zhang, L.-J., Li, Y.-L., Li, S.-Z., & Xu, J. (2021). From the one health perspective: Schistosomiasis Japonica and Flooding. *Pathogens, 10*(12), 1538. https://doi.org/10.3390/pathogens10121538.

Habibullah, Weningtyastuti, Risyanto, R., & Tampung, M. (2024). Gathering palm fruit, alternative job for orang rimba during climate crisis and poverty in Indonesia. In U. Mukhopadhyay, S. Bhattacharya, P. Chouhan, S. Paul, I. R. Chowdhury, & U. Chatterjee (Eds.). *Climate crisis, social responses and sustainability* (pp. 521–540). Springer Nature Switzerland. https://doi.org/10.1007/978-3-031-58261-5_22.

Haikerwal, A., & Saxena, S. K. (2020). Impact of climate change on water-associated infectious diseases. In S. K. Saxena (Ed.). *Water-associated infectious diseases* (pp. 53–62). Springer Singapore. https://doi.org/10.1007/978-981-13-9197-2_6.

Hammen, V. C. (2007). Long-time risk of groundwater/drinking water pollution with sulphuric compounds beneath burned peatlands in Indonesia. *Water Science and Technology, 56*(1), 253–258. https://doi.org/10.2166/wst.2007.459.

Hein, L., Spadaro, J. V., Ostro, B., Hammer, M., Sumarga, E., Salmayenti, R., Boer, R., Tata, H., Atmoko, D., & Castañeda, J.-P. (2022). The health impacts of Indonesian peatland fires. *Environmental Health, 21*(1), 62. https://doi.org/10.1186/s12940-022-00872-w.

Hirano, T., Segah, H., Kusin, K., Limin, S., Takahashi, H., & Osaki, M. (2012). Effects of disturbances on the carbon balance of tropical peat swamp forests. *Global Change Biology, 18*(11), 3410–3422. https://doi.org/10.1111/j.1365-2486.2012.02793.x.

Ibrahim, R., Akil, A., Zubair, A., Samudro, G., Samudro, H., & Mangkoedihardjo, S. (2024). Improving Hasanuddin University Lake water quality by controlling contamination sources and biological monitoring systems. *Journal of Advanced Research in Fluid Mechanics and Thermal Sciences, 114*(1), 166–177. https://doi.org/10.37934/arfmts.114.1.166177.

Irawan, B., Putra, R. D., Suryanti, A., Razai, T. S., & Putra, I. P. (2021). Biodiversity of natuna as a potential Ramsar site of wetlands with international importance in Indonesia. *E3S Web of Conferences, 324*, 03015. https://doi.org/10.1051/e3sconf/202132403015.

Jalilov, S.-M., Lestari, S., Winarno, B., Yuwati, T. W., Sakuntaladewi, N., & Mendham, D. S. (2024). Why is tropical peatland conservation so challenging? Findings from a livelihood assessment in Sumatra, Indonesia. *Mires and Peat, 30*(03), 1–20. https://doi.org/10.19189/MaP.2022.OMB.Sc.1985391.

Jenkins, A. P., & Jupiter, S. (2015). Natural disasters, health and wetlands: A pacific small island developing state perspective. In C. M. Finlayson, P. Horwitz, & P. Weinstein (Vol. Eds.), *Wetlands and human health: 5*, (pp. 169–191). Springer Netherlands. https://doi.org/10.1007/978-94-017-9609-5_9.

Kementerian PPN/ Bappenas. (2021). *Strategi Nasional Pengelolaan Lahan Basah*. https://eos.co.id/id/wp-content/uploads/2021/07/20210707-1245-P-KKSDA-Bappenas-Stranas-Lahan-Basah-part-1.pdf.

Kurniawan, T. A., Meidiana, C., Goh, H. H., Zhang, D., Jiang, M., Othman, M. H. D., Anouzla, A., Aziz, F., Mahmoud, M., Khan, M. I., Ali, I., Khan, M. M. H., & Goh, K. C. (2024). Social dimensions of climate-induced flooding in Jakarta (Indonesia): The role of non-point source pollution. *Water Environment Research, 96*(9), e11129. https://doi.org/10.1002/wer.11129.

Langner, A., & Siegert, F. (2009). Spatiotemporal fire occurrence in Borneo over a period of 10 years. *Global Change Biology, 15*(1), 48–62. https://doi.org/10.1111/j.1365-2486.2008.01828.x.

Lorenz, C., Libonati, R., Coelho Belém, L. B., Oliveira, A., Chiaravalloti, R. M., Nunes, A. V., Luciano Batista, E. K., Fernandes, G. W., Chiaravalloti-Neto, F., Damasceno-Junior, G. A., Berlinck, C. N., & De Oliveira Roque, F. (2024). Historical association between respiratory diseases hospitalizations and fire occurrence in the Pantanal wetland, Brazil. *Atmospheric Pollution Research, 15*(8), 102182. https://doi.org/10.1016/j.apr.2024.102182.

Mangubhai, S., Erdmann, M. V., Wilson, J. R., Huffard, C. L., Ballamu, F., Hidayat, N. I., Hitipeuw, C., Lazuardi, M. E., Muhajir, Pada, D., Purba, G., Rotinsulu, C., Rumetna, L., Sumolang, K., & Wen, W. (2012). Papuan bird's head seascape: Emerging threats and challenges in the global center of marine biodiversity. *Marine Pollution Bulletin, 64*(11), 2279–2295. https://doi.org/10.1016/j.marpolbul.2012.07.024.

Marlier, M. E., Liu, T., Yu, K., Buonocore, J. J., Koplitz, S. N., DeFries, R. S., Mickley, L. J., Jacob, D. J., Schwartz, J., Wardhana, B. S., & Myers, S. S. (2019). Fires, smoke exposure, and public health: An integrative framework to maximize health benefits from peatland restoration. *GeoHealth, 3*(7), 178–189. https://doi.org/10.1029/2019GH000191.

Medrilzam, M., Dargusch, P., Herbohn, J., & Smith, C. (2014). The socio-ecological drivers of forest degradation in part of the tropical peatlands of Central Kalimantan, Indonesia. *Forestry: An International Journal of Forest Research, 87*(2), 335–345. https://doi.org/10.1093/forestry/cpt033.

Miettinen, J., & Liew, S. C. (2010). Status of peatland degradation and development in Sumatra and Kalimantan. *Ambio, 39*(5–6), 394–401. https://doi.org/10.1007/s13280-010-0051-2.

Miettinen, J., Shi, C., & Liew, S. C. (2012). Two decades of destruction in Southeast Asia's peat swamp forests. *Frontiers in Ecology and the Environment, 10*(3), 124–128. https://doi.org/10.1890/100236.

Murdiyarso, D., Purbopuspito, J., Kauffman, J. B., Warren, M. W., Sasmito, S. D., Donato, D. C., Manuri, S., Krisnawati, H., Taberima, S., & Kurnianto, S. (2015). The potential of Indonesian mangrove forests for global climate change mitigation. *Nature Climate Change, 5*(12), 1089–1092. https://doi.org/10.1038/nclimate2734.

Naiborhu, N. S. R., Mulya, J. H., Rini, N. S., & Vico, N. (2024). Under the shadows of business as usual: Nationally determined contribution (NDC) for climate action. *Cogent Social Sciences, 10*(1), 2419984. https://doi.org/10.1080/23311886.2024.2419984.

Novenda, S. B. A., Farda, N. M., & Yudhanto, F. I. (2024). Dynamics of multi-temporal analysis of coastal wetland area change using cloud computing. In C. M. Roelfsema, A. Besse Rimba, S. Arjasakusuma, & A. Blanco (Eds.). *Eighth geoinformation science symposium 2023: Geoinformation science for sustainable planet* (pp. 77). SPIE. https://doi.org/10.1117/12.3009761.

Novita, N., Asyhari, A., Ritonga, R. P., Gangga, A., Anshari, G. Z., Jupesta, J., Bowen, J. C., Lestari, N. S., Kauffman, J. B., Hoyt, A. M., Perryman, C. R., Albar, I., Putra, C. A. S., Adinugroho, W. C., Winarno, B., Castro, M., Yeo, S., Budiarna, T., Yuono, E., & Sianipar, V. C. (2024). Strong climate mitigation potential of rewetting oil palm plantations on tropical peatlands. *Science of The Total Environment, 952*, 175829. https://doi.org/10.1016/j.scitotenv.2024.175829.

Otero, X. L., Méndez, A., Nóbrega, G. N., Ferreira, T. O., Santiso-Taboada, M. J., Meléndez, W., & Macías, F. (2017). High fragility of the soil organic C pools in mangrove forests. *Marine Pollution Bulletin, 119*(1), 460–464. https://doi.org/10.1016/j.marpolbul.2017.03.074.

Page, S. E., Rieley, J. O., & Banks, C. J. (2011). Global and regional importance of the tropical peatland carbon pool. *Global Change Biology, 17*(2), 798–818. https://doi.org/10.1111/j.1365-2486.2010.02279.x.

Pakhtigian, E. L., Pattanayak, S. K., & Tan-Soo, J.-S. (2024). Forest fires, smoky kitchens, and human health in Indonesia. *Environmental and Resource Economics, 87*(8), 2115–2141. https://doi.org/10.1007/s10640-024-00865-y.

Posa, M. R. C. (2011). Peat swamp forest avifauna of Central Kalimantan, Indonesia: Effects of habitat loss and degradation. *Biological Conservation, 144*(10), 2548–2556. https://doi.org/10.1016/j.biocon.2011.07.015.

Purbonegoro, T., Damar, A., Riani, E., Butet, N. A., & Cordova, M. R. (2024). Accumulation of Cd and Pb in sediments and Asian swamp eels (Monopterus albus) from downstream area of Cisadane River, Indonesia. *Environmental Monitoring and Assessment, 196*(6), 496. https://doi.org/10.1007/s10661-024-12635-w.

Purwanto, Andradi-Brown, Matualage, D. A., Rumengan, D., Awaludinnoer, I., Pada, D., Hidayat, N. I., Amkieltiela, Fox, Fox, H. E., Mangubhai, M., Hamid, S., Lazuardi, L., Mambrasar, M. E., Maulana, R., Mulyadi, N., Tuharea, S., Pakiding, F., & Ahmadia, G. N. (2021). The bird's head seascape marine protected area network—Preventing biodiversity and ecosystem service loss amidst rapid change in Papua, Indonesia. *Conservation Science and Practice, 3*(6), e393. https://doi.org/10.1111/csp2.393.

Rebelo, L.-M., McCartney, M. P., & Finlayson, C. M. (2010). Wetlands of Sub-Saharan Africa: Distribution and contribution of agriculture to livelihoods. *Wetlands Ecology and Management, 18*(5), 557–572. https://doi.org/10.1007/s11273-009-9142-x.

Reis, V., Hermoso, V., Hamilton, S. K., Ward, D., Fluet-Chouinard, E., Lehner, B., & Linke, S. (2017). A global assessment of inland wetland conservation status. *Bioscience, 67*(6), 523–533. https://doi.org/10.1093/biosci/bix045.

Samsudin, Y. B., Puspitaloka, D., Rahman, S. A., Chandran, A., & Baral, H. (2020). Community-based peat swamp restoration through agroforestry in Indonesia. In J. C. Dagar, S. R. Gupta, & D. Teketay (Eds.). *Agroforestry for Degraded Landscapes* (pp. 349–365). Springer Singapore. https://doi.org/10.1007/978-981-15-4136-0_12.

Santika, T., Muhidin, S., Budiharta, S., Haryanto, B., Agus, F., Wilson, K. A., Struebig, M. J., & Po, J. Y. T. (2023). Deterioration of respiratory health following changes to land cover and climate in Indonesia. *One Earth, 6*(3), 290–302. https://doi.org/10.1016/j.oneear.2023.02.012.

Shridhar, K., Aggarwal, A., & Mani, S. K. (2024). Public health approaches for prevention and control of water-related diseases (SDG 3). In *Water Matters*. Elsevier, 159–170. https://doi.org/10.1016/B978-0-443-15537-6.00013-6.

Silviana, S., Saharjo, B. H., & Sutikno, S. (2019). Effect of wildfires on tropical peatland vegetation in Meranti Islands District, Riau Province, Indonesia. *Biodiversitas Journal of Biological Diversity, 20*(10). https://doi.org/10.13057/biodiv/d201039.

Sjafrie, N. D. M., Rahmadi, P., Triyono, T., Kurniawan, F., Supriyadi, I. H., Zulpikar, F., Adrianto, L., Rahmawati, S., & Hernawan, U. E. (2024). Monetary value of ecosystem services in unhealthy seagrass meadows in Indonesia. *Ecosystem Services, 70*, 101668. https://doi.org/10.1016/j.ecoser.2024.101668.

Slamet, N. S., Dargusch, P., Aziz, A. A., & Wadley, D. (2020). Mangrove vulnerability and potential carbon stock loss from land reclamation in Jakarta Bay, Indonesia. *Ocean & Coastal Management, 195*, 105283. https://doi.org/10.1016/j.ocecoaman.2020.105283.

Sloan, S., Locatelli, B., Wooster, M. J., & Gaveau, D. L. A. (2017). Fire activity in Borneo driven by industrial land conversion and drought during El Niño periods, 1982–2010. *Global Environmental Change, 47*, 95–109. https://doi.org/10.1016/j.gloenvcha.2017.10.001.

Song, C., Choi, H.-A., Choi, E., Yang, A.-R., Lee, W.-K., & Lim, C.-H. (2024). Setting the direction of sustainable restoration projects in peatlands considering ecosystem services: Case of Jambi and Sumatra Selatan, Indonesia. *Ecological Indicators, 160*, 111784. https://doi.org/10.1016/j.ecolind.2024.111784.

Suciyono, Kenconojati, H., Ulkhaq, M. F., Anggreani, S. F., Santanumurti, M. B., Kadim, M. K., Arbi, U. Y., Amran, R. H., & Imlani, A. H. (2024). Profile of Pangpang Bay (Banyuwangi, Indonesia) based on water, sediment type, and macrobenthic diversity. *Egyptian Journal of Aquatic Research, 50*(3), 414–423. https://doi.org/10.1016/j.ejar.2024.08.001.

Sutapa, I. D. A., Apip, Fakhrudin, M., & Yogaswara, H. (2021). Implementation of ecohydrology to support sustainable water resources management in tropical region, Indonesia. *Ecohydrology & Hydrobiology, 21*(3), 501–515. https://doi.org/10.1016/j.ecohyd.2021.08.010.

Taufik, M., Veldhuizen, A. A., Wösten, J. H. M., & Van Lanen, H. A. J. (2019). Exploration of the importance of physical properties of Indonesian peatlands to assess critical groundwater table depths, associated drought and fire hazard. *Geoderma, 347*, 160–169. https://doi.org/10.1016/j.geoderma.2019.04.001.

Thahira, A., Adhayanto, O., Irman, & Arjuna, H. (2023). Addressing mangrove degradation in Littoral Bintan, Indonesia by applying the common consent principle: An education for Littoral states. *BIO Web of Conferences, 79*, 04002. https://doi.org/10.1051/bioconf/20237904002.

Thornton, S. A., Dudin, Page, S. E., Upton, C., & Harrison, M. E. (2018). Peatland fish of Sebangau, Borneo: Diversity, monitoring and conservation. *Mires and Peat, 22*, 1–25. https://doi.org/10.19189/MaP.2017.OMB.313.

Tussadiah, A., Sujiwo, A. S., Andesta, I., & Daeli, W. (2021). Assessment of coastal ecosystem services and its condition for policy management plan in East Nusa Tenggara, Indonesia. *Regional Studies in Marine Science, 47*, 101941. https://doi.org/10.1016/j.rsma.2021.101941.

Utami, R. R., Geerling, G. W., Salami, I. R. S., Notodarmojo, S., & Ragas, A. M. J. (2020). Environmental prioritization of pesticide in the Upper Citarum River Basin, Indonesia, using predicted and measured concentrations. *Science of The Total Environment, 738*, 140130. https://doi.org/10.1016/j.scitotenv.2020.140130.

Utami, W., Sugiyanto, C., & Rahardjo, N. (2024). Mangrove area degradation and management strategies in Indonesia: A review. *Journal of Degraded and Mining Lands Management, 11*(3), 6037–6047. https://doi.org/10.15243/jdmlm.2024.113.6037.

Vicarelli, M., Georgescu, A., & Sudmeier-Rieux, K. (2024). Cost-benefit and equity analysis of nature-based solutions in Haiti, India, Indonesia and Uganda. *Nature-Based Solutions, 6*, 100196. https://doi.org/10.1016/j.nbsj.2024.100196.

Warren, M., Hergoualc'h, K., Kauffman, J. B., Murdiyarso, D., & Kolka, R. (2017). An appraisal of Indonesia's immense peat carbon stock using national peatland maps: Uncertainties and potential losses from conversion. *Carbon Balance and Management, 12*(1), 12. https://doi.org/10.1186/s13021-017-0080-2.

Widiono, S., Wahyuni, E. S., Kolopaking, L. M., & Satria, A. (2024). Livelihood vulnerability of indigenous people to climate change around the Kerinci Seblat National Park in Bengkulu, Indonesia. *Regional Sustainability, 5*(4), 100181. https://doi.org/10.1016/j.regsus.2024.100181.

Wulandari, R., Iswara, A. P., Qadafi, M., Prayogo, W., Astuti, R. D. P., Utami, R. R., Jayanti, M., Awfa, D., Suryawan, I. W. K., Fitria, L., & Andhikaputra, G. (2024). Water pollution and sanitation in Indonesia: A review on water quality, health and environmental impacts, management, and future challenges. *Environmental Science and Pollution Research, 31*(58), 65967–65992. https://doi.org/10.1007/s11356-024-35567-x.

Yahaya, N. S., Pereira, J. J., Taha, M. R., & Yaacob, W. Z. W. (2024). Natech potential due to emerging climate hazards and awareness of the local stakeholders in the coastal industrial site of Kuala Selangor, Malaysia. *Chemosphere, 367*, 143551. https://doi.org/10.1016/j.chemosphere.2024.143551.

Yamamoto, Y. (2023). Living under ecosystem degradation: Evidence from the mangrove–fishery linkage in Indonesia. *Journal of Environmental Economics and Management, 118*, 102788. https://doi.org/10.1016/j.jeem.2023.102788.

Yayasan Lahan Basah. (2020). Lahan Basah untuk Manusia dan Alam Arahan Strategis 2020 – 2030. Wetlands International Indonesia. https://indonesia.wetlands.org/id/arahan-strategis-2020-2030/.

Yunus, M., Pagdee, A., & Baral, H. (2025). Local livelihoods and determinants of peatland protection in Indonesia. *Trees, Forests and People*100800. https://doi.org/10.1016/j.tfp.2025.100800.

Chapter 15

Linkage between wetlands and wellbeing to enhance ecosystem services: a sustainable approach

Krishma Kumari, Richa Kothari, and Deepak Pathania
Department of Environmental Sciences, Central University of Jammu, Jammu, Jammu and Kashmir, India

15.1 Introduction

Wetlands are one of the most dynamic and intricate ecosystems where a clear trade-off between environmental preservation and development exists. Wetlands are among the most productive ecosystems on Earth (Balwan & Kour, 2021). It protects livelihoods, biodiversity, and environmental quality because it generates food, stores carbon, replenishes groundwater, and manages flooding. Furthermore, the vegetation in wetlands serves various functions such as housing, mat construction, food, and species shelter (Bhowmik, 2022). These are the land regions that are nearly entirely water-saturated or covered with water either year-round or seasonally.

Wetlands have various advantages, including fish and fiber, water supply, water purification, flood management, climate regulation, coastal protection, recreational activities, and increasingly, tourism. People living near wetlands frequently wholly or partially depend on the services provided by wetland ecosystems. They are immediately harmed by loss or deterioration (Chakraborty et al., 2023). In Malaysia, medicinal plants collected from wetlands contribute about US$80 per month to rural households. Inland wetlands, such as lakes, rivers, and swamps, provide the majority of freshwater for humans worldwide. Five million people in Florida, United States, rely on the Everglades wetlands for their water supply. During the dry season, 10,000 residents in El Salvador receive water from Laguna El Jocotal, a shallow floodplain lake (McCartney & Acreman, 2009). It also helps maintain the quality of the environment by absorbing and processing waste. The organic cycling of carbon dioxide, methane, and hydrogen sulfide occurs in wetlands (Abou-Elela, 2019). Wetland peat deposits hold 14%–16% of the global soil carbon pool but cover only 3% of the land surface. A wide variety of flora and animals are found in wetlands. Wetland functions like water storage, sediment trapping, and nutrient cycling support species and their genetic diversity.

Wetlands have intangible advantages that enhance people's mental and spiritual health. For instance, the social cohesion of the Lozi people in western Zambia was strengthened by rituals associated with their migrations to and from the floodplain as water levels rose and fell. Additionally, many visitors travel to the beauty of wetlands like Trinidad's Caroni Swamp and Botswana's Okavango Delta, which have contributed to the economic growth of the regions. Beneficiaries are frequently located far from wetlands (Keitumetse et al., 2023). As a result, when decisions are made regarding the use of wetlands, the relationship between wetland services and human well-being may be undervalued.

Wetland forms frequently merge into one another and broader landscapes due to their hydrological and ecological connections. Consequently, wetlands should be considered a part of the river basin or coastal zone.

15.2 Classification and distribution of wetlands

Wetlands may be found on all continents except Antarctica, from the Arctic to the tropics. Wetlands may differ considerably due to local and regional alterations in water chemistry, hydrology, temperature, flora, soils, and other factors, including human disturbance. There are several types of wetlands and classifications for them. Based on the proximity to the coast or shore, wetlands are classified into two main categories.

2.1 Inland wetlands
2.2 Coastal wetlands

Wetlands: Biodiversity, Health and their Role in Ecosystem Services. DOI: https://doi.org/10.1016/B978-0-443-34222-6.00005-X

15.2.1 Inland wetlands

Inland wetlands are water bodies and other areas that are continually or intermittently covered with water due to poorly drained soils. Inland wetlands include marshy areas, ponds, lakes, and swamps. Based on the nature of formation are divided into two main types, such as natural wetlands and man-made wetlands in Fig. 15.1

15.2.2 Coastal wetlands

Coastal wetlands are areas of land that are periodically or permanently submerged in fresh, brackish, or saltwater. These include lagoons, creeks, coral reefs, tidal flats, mangroves, salt marshes, salt pans, and aquaculture. It provides invaluable biological benefits and is recognized as one of the most productive ecosystems. On the other hand, coastal wetlands are one of the most endangered ecosystems due to human activity (Kolli, 2020). Based on their classification, wetlands are categorized into two main types: natural wetlands and man-made wetlands, and their distribution in India by type, area, and number is also presented in Fig. 15.1 (https://indianwetlands.in/wetlands-overview/national-wetlands-statistics/).

15.2.2.1 Natural wetland

Wetlands that are either seasonally or continuously saturated with water are known as natural wetlands. This natural environment supports the growth of wetland soils and provides habitat for aquatic plants. These are found in wetlands in many forms, including bogs, marshes, swamps, wooded wetlands, wet meadows, and coastal wetlands such as mangroves. The natural wetlands area is roughly twice the size of the man-made wetlands area. Natural wetlands are of the following types.

15.2.2.1.1 Lakes

These wetlands are typically fed by streams and rivers and found in natural depressions. Lakes are extremely valuable to people. It facilitates the control of a river's flow. It helps maintain equal water flow throughout the dry season and minimizes flooding during periods of heavy rainfall. Hydroelectricity was also produced from the lakes (Dhar, 2002). In addition to supporting the aquatic ecosystem, they improve the area's natural beauty, contribute to tourism growth, and

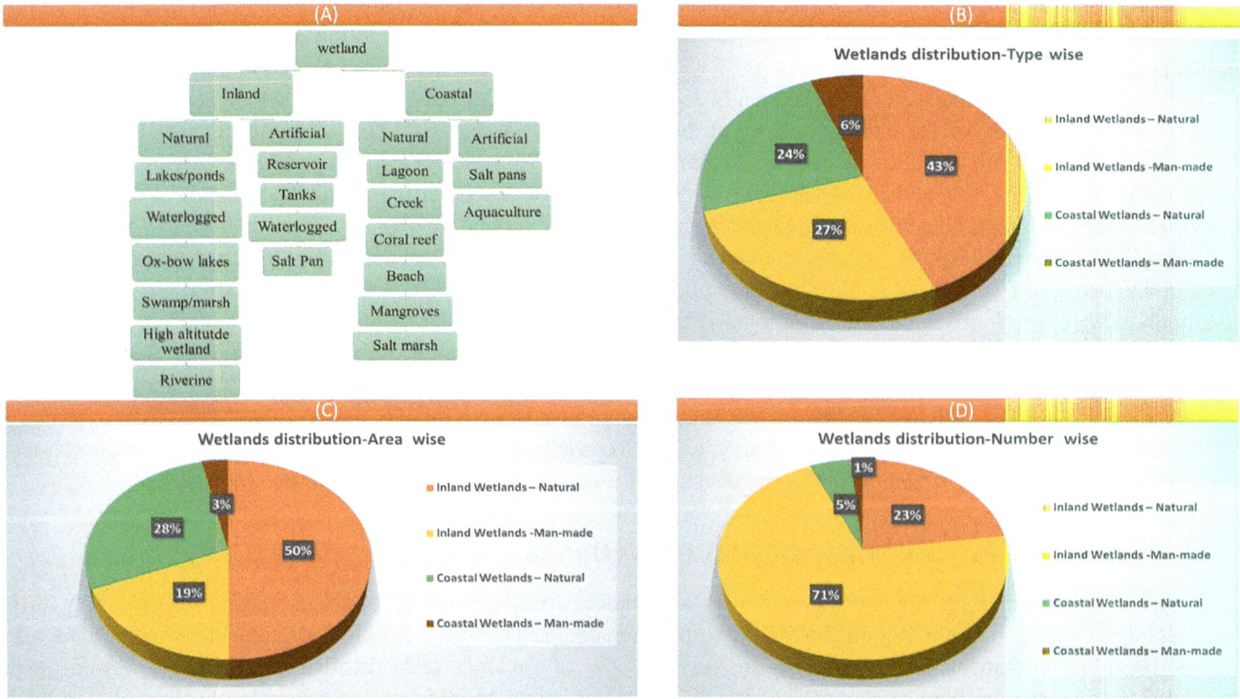

FIGURE 15.1 (A) Classification of wetlands; (B) distribution of wetlands in India (type-wise); (C) distribution of wetlands in India (area-wise); (D) distribution of wetlands in India (number-wise).

offer recreational opportunities. Lakes may offer excellent opportunities for leisure, travel, and cottage or home ownership. Additionally, it also helps in groundwater replenishment.

15.2.2.1.2 Ponds

A pond is a mud wall built over a small stream in the valley to keep the monsoon for arid regions. Ponds are commonly considered small bodies of standing water that are shallow enough to allow rooted plants to grow from one shore to the other (Sharma & Naik, 2024).

15.2.2.1.3 Oxbow lakes

An oxbow lake is a shallow, crescent-shaped body of water created when a meandering river or stream erodes the outer banks of a bend, resulting in a curved lake (Ayeni et al., 2019). Oxbow lakes are visible in the satellite image in flat areas near rivers. During the premonsoon season, some areas of the lake typically have aquatic flora in a reddish-pink color (Mitsch et al., 2015).

15.2.2.1.4 High-altitude wetlands

High-altitude wetlands (HAWs) are defined as "areas of swamp, marsh, meadow, fen, peat land, or water located at an altitude above 3000 m, permanent or temporary, with water that is static or flowing, fresh, brackish, or saline, and are generally located at an altitude between the continuous natural forest border and the permanent snow." HAWs encompass a variety of water bodies, including rivers, lakes, ponds, glaciers, and glacial lakes. In contrast to other wetland ecosystems, they are distinguished by a distinctive diversity of water sources, habitats, species, and populations, and they have not seen widespread human disturbance (O'Neill, 2019). High-altitude desert wetlands are intricate ecosystems impacted by permafrost, glacial, fluvial, and cryogenic phenomena. They occupy mountain ranges with cold or dry climates (Mętrak et al., 2023).

15.2.2.1.5 Riverine wetlands

Water accumulates along the main rivers, especially in the plains, leading to the growth of marshes and swamps. Herbaceous plants predominate in marshes maintained by water sources such as surface runoff and direct rainfall (Perez-Ruzafa et al., 2019). They have an adequate water supply and poor drainage, which keeps the ground wet and mineral levels high enough to encourage organic decomposition.

15.2.2.1.6 Waterlogged

In a region where water is present near the land surface, submerging the roots of all plants except hydrophytes leads to the demise of other vegetation. Canals and other man-made features led to seepage, causing waterlogging in nearby areas. Waterlogging is caused by floods, seepage from unlined canals, and other irrigation networks. From a spectral perspective, waterlogged landscapes resemble lakes or ponds when surface water is present.

15.2.2.1.7 Lagoons

Lagoons are shallow bodies of water in coastal areas that are connected to the sea but are isolated from the ocean by a barrier. The barrier might be a shingly or sandy wave-built structure. Lagoons are classified into two main categories: (1) long or irregular channels that are situated between the seashore and coastal barrier islands, and (2) circular or irregular channels that are encircled by coral atoll reefs or shielded from direct wave action by barrier coral reefs.

15.2.2.1.8 Creeks

A small body of water on land is called a creek. These often serve as a sea entry into the land. These streams form similarly to rivers, with small imperfections that eventually redirect the water into several channels (Zhao et al., 2016).

15.2.2.1.9 Coral reefs

Microscopic organisms live in consolidated living colonies in warm tropical seas. The rock-like reefs are referred to as coral reefs or organic reefs. The clusters of calcareous coral and coral-algae deposits in sand composed of foraminiferal shells connect the interstitial area. Reefs are living communities that are developing on top of this historical accumulation (Ye & Sun, 2021). The world's coral reefs are threatened by overfishing, harmful fishing methods, water pollution, and coastal development. These stresses become worse due to climate change, which may eventually reduce the healthy and mature reef ecosystems worldwide, and more frequent catastrophic coral bleaching events (Bayraktarov et al., 2019).

15.2.2.1.10 Beach

Unvegetated parts along the shoreline consist of loose sand. A rocky beach is composed of rocky material. With the majority of natural environments, beaches provide a variety of recreational activities and certain beaches could be more appropriate for particular purposes than others. One beach could be more suited for swimming, while another might be better for surfing. Beaches typically include a variety of people engaged in a variety of activities (such as swimming, running, and surfing), and the importance of these pursuits will vary depending on the individual (Pascoe, 2019).

15.2.2.1.11 Mangroves

This is defined as a halophytic tree and bushes thriving along tropical and subtropical coastlines in brackish to saltwater tidal waters. These are the coastal marsh areas around the main deltas of nations. Mangroves appear red on satellite imagery when they form a continuous patch. Brick red may be seen in place of red when mangrove associations are dispersed or deteriorated.

15.2.2.1.12 Salt marshes

These are halophytic grasslands that are either natural or seminatural, consisting of dwarf brushwood on alluvial sediments bordering salty bodies of water, with tidally or nontidally variable water levels, and are known as salt marshes. Salt marshes are gray-blue in color (Chen et al., 2022). Salt marshes are distinct from sand-dominated beach ecosystems because they are made up of a surface unit of organic-rich, low-permeability peat, mud, and clay that sits on top of sandy estuary sediments (Guimond & Tamborski, 2021).

15.2.2.2 Man-made wetlands

These wetlands were created to meet specific needs, including storing water for drinking and irrigation, fish farming, or leisure. Reservoirs, barrages, salt pans, and aquaculture ponds are all examples of man-made wetlands (Mitsch & Gosselink, 2000).

15.2.2.2.1 Reservoirs

A lake/pond created to store water, typically by building a dam across a river. Reservoirs have irregular boundaries, with a noticeable dyke visible in remote-sensing images. In the case of a reservoir, the wetland boundary includes aquatic vegetation, water, and the water's footprint. The rich pink tone of the reservoir's aquatic flora is visible in the subsequent photos. Deep reservoirs are dark blue, while reservoirs with a heavy silt load are ink blue in color toned (Mitsch et al., 2014).

15.2.2.2.2 Barrages

In plain areas over rivers, dykes are built to create water and irrigation facilities. These water storage areas become wetlands, for example, Okhla and Harike. Water has a smooth texture and a dark blue appearance. Depending on its density, the pink-colored aquatic vegetation might be contiguous or dispersed. The designation "reservoir/barrage" will be applied to reservoirs created by barrages (Seelan, 2015).

15.2.2.2.3 Salt pan(inland)

These are man-made, shallow, rectangular depressions where saltwater is collected and left to evaporate in the sun to produce salt.

15.2.2.2.4 Salt pans (coastal)

A man-made, undrained, rectangular, shallow depression or hollow where saltwater collects and evaporates to leave behind salt. The shape of salt pans is either square or rectangular. The look is blue when water is present, and white when salt is formed (Orkhonselenge et al., 2022).

15.2.2.2.5 Aquaculture ponds

These are the bodies of water used for the aforementioned purposes. Aquaculture ponds are often square or rectangular in design. It's blue in color.

15.3 Importance of wetland

Wetlands are sometimes called "nature's kidneys" due to their capacity to filter contaminants out of water through sediment and vegetation storage as well as atmospheric losses. Wetland is treated with wastewater from a variety of sources, including mining, pulp and paper mills, landfill leachates, agricultural production, rainwater runoff, and petroleum refineries. These pollutants include biological contaminants such as viruses and pathogens, physical contaminants like suspended sediment, and chemical contaminants that are both organic (e.g., petroleum hydrocarbons, pesticides, and medicines) and inorganic (e.g., heavy metals and excessive nutrient concentrations). Wetlands have shown long-term removal capacity of 80%–95% for chemical oxygen demand (COD) and biological oxygen demand (BOD), 99% for bacterial pollution, 35% for Nitrogen, and 25% for Phosphorus in domestic wastewater (Verhoeven & Meuleman, 1999). Wetlands are among the most productive ecosystems on Earth and provide some benefits to human society. However, they are adaptable and ecologically sensitive systems. Wetlands are enormously diverse ecosystems depending on their genesis, geographical location, water regime, chemistry, dominating species, and soil and sediment structures. Globally, the area of wetland ecosystems ranges between 917 million hectares (m ha) to more than 1275 m ha, with an estimated annual economic worth of around US$15 trillion (Bassi et al., 2014).

In India, different types of wetlands are found, including freshwater, saltwater, and marine wetlands. Inland wetlands are primarily artificial or temporary. Historically, people have depended on wetlands for millennia, and have also impacted human lifestyles. Human interference has affected even the most natural wetlands. Wetlands and humans continue to interact in a variety of ways, causing their traits to change constantly and frequently in unpredictable ways.

With its varied flora and fauna, the ecoclimatic regimes have produced a range of wetland systems, from hot and humid wetlands in coastal zones to frigid desert wetlands at high elevations. India's various terrain and climatic regimes, along with 115 cm of annual rainfall, support and sustain a variety of distinctive wetland ecosystems (Barksdale et al., 2013). Mangrove swamps, coral reefs, marine wetlands, saline and temporary wetlands of arid and semiarid regions, HAW lakes, wetlands in the flood plains of the major river systems, and coastal wetlands like lagoons, backwaters, and estuaries are among India's natural wetlands. Indian wetlands encompass the entire spectrum of ecological types. Numerous artificial wetlands have increased plant and animal diversity, in addition to the many kinds of natural wetlands. These artificial wetlands were created to meet the demands of most inland wetlands, and are dependent on rivers like the Ganga, Narmada, Krishna, Godavari, Kaveri, etc., and other important rivers are either directly or indirectly responsible for the survival of inland wetlands (Ayeni et al., 2019). They can be found in the central Indian highlands, the deltaic areas of the east and west coasts, the hot, desert regions, the wet and humid zones of the Indian south peninsula, as well as the Andaman-Nicobar Islands and Lakshadweep Island (Mitsch et al., 2015). India is home to numerous wetland ecosystems that sustain a wide variety of habitats. Numerous ecological products and services have been provided by these wetlands, but their hydrological, economic, and ecological functions are declining as a result of growing urbanization, industrialization, and agricultural intensification (Chadwick et al., 2006).

Wetlands play an important role in all terrestrial ecosystems. In addition to storing carbon dioxide and mitigating climate change, the vast diversity of wetland ecosystems is essential for regulating hydrological processes, water quality, carbon fixation, flood control and management, and nutrient cycling. In addition to providing food and fisheries, these wetlands help in the replenishment of groundwater. They serve as a filter, storing, absorbing, and changing the pollutants that have been transported away from the land before they reach natural river systems. Wetlands provide millions of people with both direct and indirect utility benefits, including food, fiber, and raw materials; storm and flood control; clean water supply; scenic beauty; and educational, aesthetic, and recreational advantages (Greene & Blossey, 2011). The ecosystem services provide the benefits shown in Fig. 15.2.

15.3.1 Provisioning services

Wetlands provide a wide variety of ecological services that benefit humans both directly and indirectly. Ecosystem services are the products and services that are necessary for human well-being (Ali et al., 2020). People can get fish, animals, trees, fruits, and vegetables directly from the ecosystem. A provisioning service is any advantage that nature offers to humanity. Ecosystem products, which are used to manufacture textiles, apparel, and medicinal benefits, are part of provisioning services.

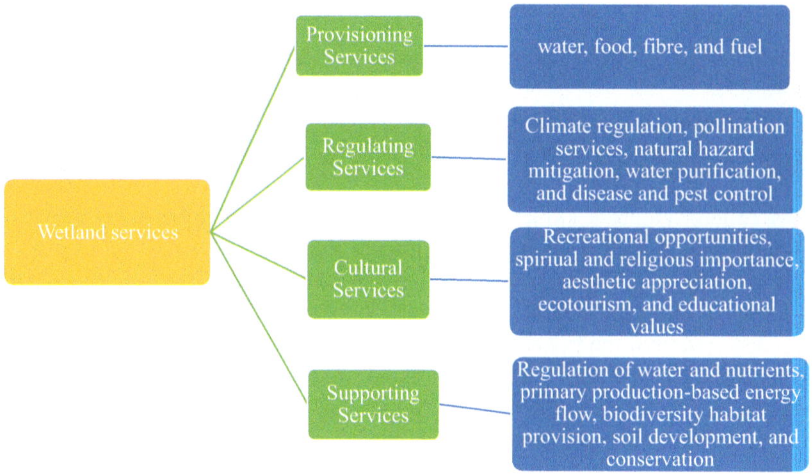

FIGURE 15.2 Ecosystem services provided by wetlands and their benefits to human well-being.

15.3.2 Regulating services

Human existence on Earth is possible due to the ecosystem's support. Microbes help decompose waste, plants help clean the air and water, pollinators help pollinate flowers, and tree roots assist in stabilizing the soil to avoid soil erosion. These natural processes interact to keep ecosystems productive, sustainable, adaptable, and healthy. Regulating services benefit ecological processes, including carbon storage, climate regulation, flood control, erosion control, pollination, decomposition, and water purification (Shen et al., 2022).

15.3.3 Cultural services

It consistently supports our social, intellectual, and cultural development. Ecosystems have been significant to human thought since ancient civilizations etched pictures of flora, animals, and weather patterns on cave walls. A nonmaterial benefit that contributes to the growth and progress of a culture, including formal and informal education, ecotourism, inspiration, aesthetics, spiritual enrichment, recreation, and cultural heritage (Butt et al., 2021). The significance of ecosystems in local, national, and global cultures; the production and sharing of knowledge; the creativity that results from engagement with nature (music, art, and architecture); and leisure (Davidson et al., 2018).

15.3.4 Supporting services

Wetlands provide essential ecosystem services that support life and enable other ecosystem services. Ecosystems could not survive without the stability of underlying natural processes such as soil formation, nitrogen cycling, photosynthesis, and the water cycle. These mechanisms enable basic life forms, much less large ecosystems, and human populations. Without supporting services, the provisional, regulating, and cultural services would not be feasible (Cahoon et al., 2020).

15.4 Threats to the wetlands

Wetlands have suffered significant harm in the last few centuries. The negative consequences of increased human activity include pollution and degradation of wetlands. Wetland problems are caused by unsustainable, environmentally unfriendly human activities. Human activities such as agricultural development and drainage have resulted in considerable wetland degradation and loss. Since 1900, an estimated 54%–57% of natural wetlands have disappeared globally (Kulik et al., 2024). Another study clearly shows that the globe has lost 50% of its wetlands (or 50% since 1900 AD). According to 189 accounts of wetland area change, the observed long-term loss of natural wetlands ranges from 54% to 57%, but since 1700 AD, the loss may have reached 87%. In the 20th and early 21st centuries, the pace of wetland loss has increased significantly (i.e., 3.7 times), with 64%–71% of wetlands lost since 1900 AD (Davidson, 2014).

The reason for this loss of wetland is the opening of land for agriculture, industry, and habitation. If the wetlands are destroyed, all living organisms will be displaced or extinct. This habitat loss will pose a major wildlife threat, especially to the species that are in an endangered situation (Duraiappah, 2011). Due to degradation of habitat, pollution, aquatic resources overuse, tourism, as well as foreign viruses and parasites, wetlands and their biodiversity are lost globally (Barbier, 2018). The main reasons for both acute and chronic wetland losses are shown in Fig. 15.3, which presents a diagrammatic sketch categorizing the causes of wetland loss into acute (sudden) and chronic (gradual) types, further divided into natural and anthropogenic (human-induced) origins:

- **Acute natural losses:** Unexpected natural catastrophes that instantly damage wetland habitats include floods, cyclones, and droughts.
- **Acute anthropogenic losses:** Infrastructure development (like dam construction), rapid urbanization, and industrial accidents can result in sudden loss or deterioration.
- **Chronic natural losses:** Gradual consequences of climate change, including changes in rainfall patterns and increasing sea levels, change the hydrology and diminish wetland functions over time.
- **Chronic anthropogenic losses:** Long-term human effects include land encroachment, excessive water consumption, invasive species introduction, and pollution from agrochemicals and untreated sewage.

The Government of India, in association with the State Governments, started a wetland conservation initiative in 1985–86 after realizing the significance of preserving wetlands. The encroachment, siltation, weed infestation, catchment erosion, surface runoff carrying fertilizers and pesticides from agricultural fields, and discharge of domestic sewage and effluents caused deterioration of water quality, weed proliferation, loss of biodiversity, and other related issues. Necessary steps have been taken to stop further degradation and shrinkage of water bodies (Chaikumbung et al., 2016).

Industrialization, agriculture, and urbanization also prevent wetlands from carrying out their vital ecological functions. Because of the waste they produce, these activities degrade wetland water quality, deplete food supplies, and endanger all life in this ecosystem. Agricultural and industrial activities interact to produce a variety of problems, including altered river and stream flow patterns, disruptions to the water cycle in wetlands, increased salinity of the water, and the accumulation of pollutants such as industrial waste and pesticides. Consequently, it harms other living things that interact with the wetland and poisons aquatic species (Ekumah et al., 2020).

India has ratified the Ramsar Convention on Wetland Management, which expands biodiversity conservation. It includes a wide range of habitats and many man-made wetlands, such as lakes, coastal lagoons, and rivers. Under the National Wetlands Protection Programme, the government has designated certain wetlands for management and protection, carrying out the Management Action Plan.

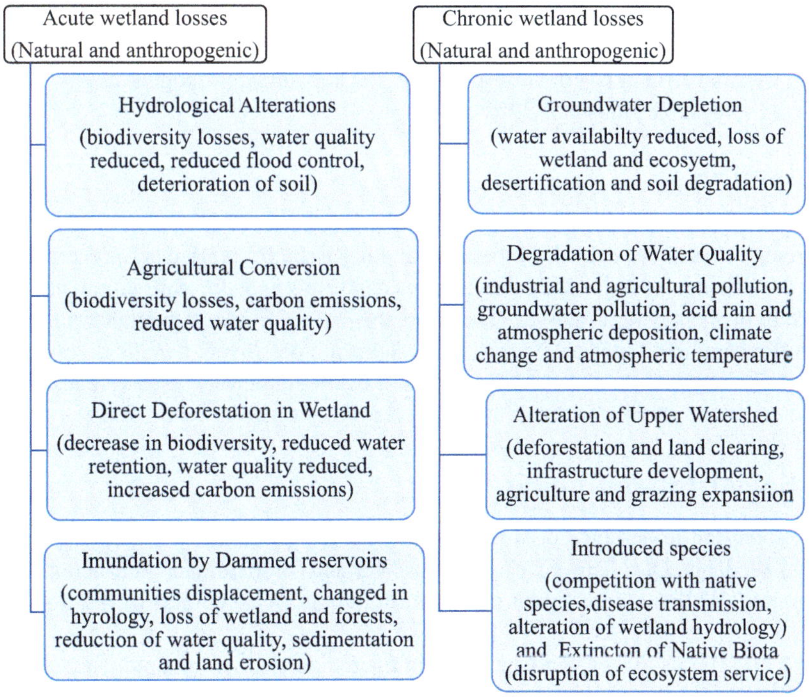

FIGURE 15.3 Diagrammatic sketch of acute (natural and anthropogenic) and chronic (natural and anthropogenic) losses of wetland.

15.4.1 Urbanization

There is growing development pressure on wetlands close to metropolitan centers for commercial, industrial, and residential uses. Urban wetlands are necessary to preserve the public water supply (Chaikumbung et al., 2018).

15.4.2 Agriculture

Paddy fields have replaced large areas of wetlands. The hydrology of the surrounding marshes was drastically changed by the construction of numerous reservoirs, canals, and dams for irrigation. Some of the most significant effects induced by humans include agricultural irrigation from rivers and streams that supply wetlands, as well as dams used to produce energy and provide drinking water. When the water regime's normal flow is not followed, the wetlands either lose their wetland characteristics by using more water than they can handle or dry out because they don't receive enough water. As a result, wetland breeding birds depart, fish migration and breeding cycles are disrupted, and plant diversity is lost.

15.4.3 Pollution

Another significant issue that wetlands face is pollution. Overall, 80% of wastewater worldwide is dumped untreated into wetlands. Wetlands can remove pollutants from water because they act as natural filters. They can trap dangerous gases and solid sediments up to 90% of the time, and they can remove 60% of the metals from water. Naturally, when the wetland ecosystem functions well, all of this is feasible (McCartney et al., 2015). The wetland environment is severely harmed by pesticides from agricultural practices and chemical contamination from industrial sites. Wetlands filter water naturally. They can remove mercury from industrial sources or other forms of pollution; they can only remove pesticides and fertilizers from agricultural runoff (Sharma & Naik, 2024). Concern over how industrial pollution affects drinking water sources and wetland biological diversity is growing (Pedersen et al., 2018).

15.4.4 Change of climate

Climate change is a severe concern for the world, including erratic precipitation patterns and abrupt temperature fluctuations that harm the fundamental features of the wetland ecosystem and the organisms. Wetlands serve as sinks for carbon emissions and other greenhouse gases that contribute to climate change, thus protecting and removing threats. Wetlands are also affected by rising sea levels, humidity, storm frequency, precipitation change, floods, and a rise of CO_2 in the atmosphere.

15.4.5 Sand mining

This includes removing material from a riverbed or marsh. When streams are dredged, nearby wetlands are dried up and the water table is lowered (Nayak & Bhushan, 2022).

15.4.6 Introduced species

Exotic imported plant species like salvinia and water hyacinth pose a threat to Indian wetlands. They compete with native vegetation and choke rivers. Invasive species have become another issue for wetlands in recent years. The global wildlife trade, the increase in the number of animals maintained in captivity, and the cultivation of various plant species for landscape reasons outside of their natural habitat. These organisms are found in wetlands in a variety of forms, which harm biodiversity by suffocating local species. Natural ecosystems are destroyed, genetic diversity is disrupted, and native species are eventually excluded from the food chain as a result of the numerous diseases caused by invasive species and the pressure they place on native species for land and food resources (Reeves et al., 2019).

15.5 Wetland linked to well-being

To increase awareness about the importance of wetlands for both the environment and people, the 2nd of February has been observed as World Wetlands Day. This year's theme is Wetlands and Human Well-being, which follows last year's focus on wetland restoration. Wetland protection is essential for human well-being as well as for ecological reasons (Zhao et al., 2016).

Offering a variety of ecosystem services, wetlands help in maintaining environmental and ecological balance. Just like natural sponges, powerhouses absorb and store surplus rainfall, important for flood control since it slows and buffers water flow, thus lowering the extent of inundation. Wetlands release stored water during dry spells, reducing water scarcity and preserving natural hydrological cycles. They have an equally important role in purifying water. By removing pollutants from the water, wetlands guarantee better water quality, which is necessary for agriculture, wildlife, and human health (Maund et al., 2019).

Wetlands are essential for livelihoods by supporting agriculture and fisheries and also have significant cultural and recreational value in addition to their environmental roles. Because they control the microclimates in their immediate surroundings and play an important role as carbon sinks, helping to stabilize the global climate (Zhou et al., 2020).

Wide-ranging effects would result from the loss of wetlands, including increased flood risk, water-quality problems, and economic and cultural losses. According to the most recent estimate from the Government of India's Space Application Center in Ahmedabad, approximately 15.98 million ha, or 4.86% of India's total land area, are covered by wetlands. India has played a crucial role in wetland conservation by implementing laws and policies like the Wildlife Act 1972, the Water Act 1974, and the Wetland Rules, 2017, addressing the need for wetland ecosystems and management (Let & Pal, 2022)

15.6 Conservation of wetland

In the last 50 years, considerable efforts have been made to preserve and restore wetlands. Even though these initiatives are primarily driven by the advantages and services that wetlands provide to humans, their significance to the environment and living things is beginning to influence conservation efforts. A multitude of parties are addressing the risks to wetlands. Wetlands are being protected by a wide range of actors, from local organizations and worldwide collaborations to institutional initiatives and individual assistance (Birch et al., 2022). Timeline of wetland conservation is shown in Fig. 15.4.

From a broad standpoint, efforts to eradicate threats to wetlands and restore them to support healthy ecosystems have been ongoing for many years through worldwide conservation agreements, national policies, action plans, initiatives, and campaigns run by nongovernmental organizations. Individual efforts, however, are crucial for the preservation of the wetlands. Financial donations, the preservation and development of flora to prevent the spread of alien species through voluntary assistance to local organizations' conservation initiatives, and compliance with rules when visiting wetlands. In addition to these, a few adjustments to our everyday routines help to preserve wetlands. Significant adjustments were made, for instance, by using recyclable materials, reducing pesticide use in agricultural operations, properly disposing of oil and chemical cleaning products that pollute water, and adopting reusable items to reduce waste (Gómez-Baggethun et al., 2019).

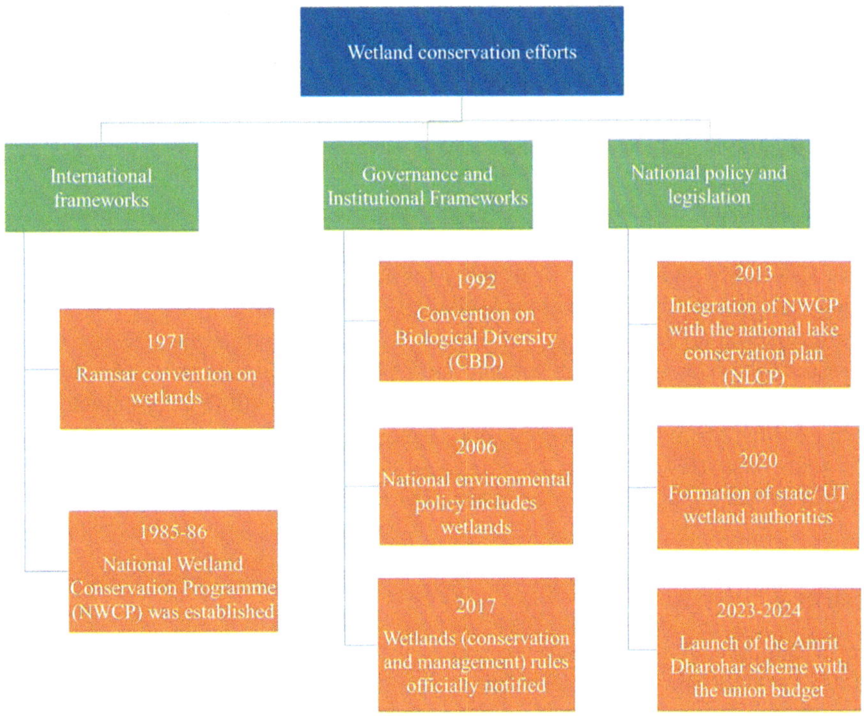

FIGURE 15.4 Timeline of wetland conservation.

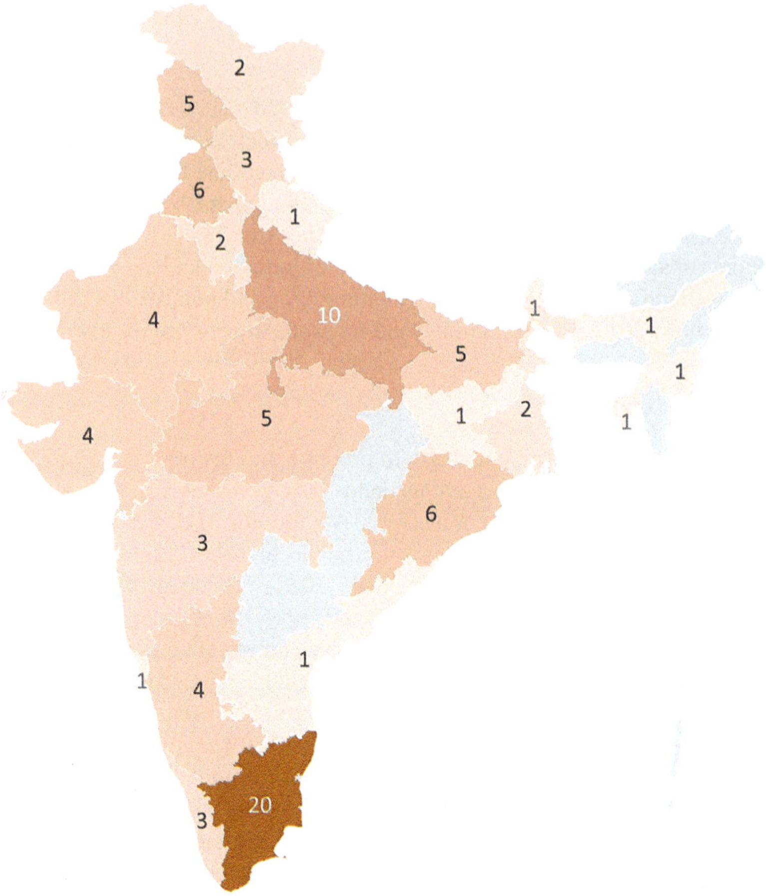

FIGURE 15.5 Ramsar site distribution in India. Map lines delineate study areas and do not necessarily depict accepted national boundaries.

Protecting and restoring wetlands were substantially aided by active participation in the policymaking process, with civic consciousness and responsibility. Working together is necessary to combat threats to wetlands. A major international agreement known as the Ramsar Convention came into force in 1971 to encourage conservation efforts and the wise use of wetlands to protect them. Over 2300 wetlands in 171 countries, with a total area of 2.5 million km^2, are designated as significant wetlands globally under this convention. Wetlands are protected by various national laws, even though some countries, such as the United States and Canada, have very few Ramsar sites (Thorslund et al., 2017). As per the Ramsar Convention, India has also shown significant responsibility towards wetland conservation. India has 91 Ramsar wetland sites distributed throughout the country, as shown in Fig. 15.5.

International nongovernmental organizations play a significant role in conservation efforts, even though the nations that have ratified this treaty are the ones using their national laws to protect the wetlands. The Ramsar Convention is thus coordinated by six organizations, namely: Wildfowl and Wetlands Trust, Birdlife International, the International Water Management Institute (IWMI), the World Wide Fund for Nature (WWF), Wetlands International, and the International Union for Conservation of Nature (IUCN). These organizations support the Ramsar Convention by offering financial help, technical support for field research, and experience sharing (Xu et al., 2020).

15.7 Conclusion

Wetlands are places where people seek to survive and sustain homes, and many cultures and communities have socioeconomic aspects of health. This is the relationship of wetlands between the cultural, social, economic, and political aspects of human well-being. The fundamental idea of the study, which is to identify cobenefits and the causes of the decline in human health linked to wetlands, should be addressed by preserving or improving ecosystem services. For example, the preservation of wetland ecosystem services should not be compromised by any required disease-eradication efforts in or around wetlands.

Wetland managers need to inform national organizations responsible for water supply, sanitation, and health about the well-established health and well-being benefits of naturally existing wetland ecosystems. For wetland managers, this will require greater collaboration and the search for new partnerships within the field of wetland conservation, as well as among the government, nongovernmental organizations, and the commercial sector. Furthermore, countries having development sectors like mining, industrial activities, sanitation, agriculture, transportation, etc, ought to take all precautions to avoid operations that have a direct or indirect negative impact on wetlands and degrade ecosystem services promoting human health and well-being. It is recommended that governments prioritize the connection between human health and wetland ecosystems worldwide through policies and strategies, especially regarding domestic and agricultural water supplies and sanitation.

The current understanding of how climate change increases the risk of disease should be considered when making decisions about co-managing wetlands and human health. Given that a changing climate is predicted to continue increasing the risk to human health, it should aim to preserve wetlands' ability to adapt to climate change. Members of the human health and wetland management sectors will need to work together more effectively. The majority of these health issues will resonate with the field of public health, which encompasses the broader areas of health and medicine, according to wetland managers. Professionals in health promotion will be equipped to look upstream at the social and environmental factors that contribute to poor health outcomes; preventive interventions focus on these factors.

References

Abou-Elela, S. I. (2019). Constructed wetlands: The green technology for municipal wastewater treatment and reuse in agriculture. *Unconventional Water Resources and Agriculture in Egypt,* 189–239.

Ali, S., Abbas, Z., Rizwan, M., Zaheer, I. E., Yavaş, İ., Ünay, A., & Kalderis, D. (2020). Application of floating aquatic plants in phytoremediation of heavy metals polluted water: A review. *Sustainability, 12*(5), 1927.

Ayeni, A. O., Ogunsesan, A. A., & Adekola, O. A. (2019). Provisioning ecosystem services provided by the Hadejia Nguru Wetlands, Nigeria–Current status and future priorities. *Scientific African, 5*, e00124.

Balwan, W. K., & Kour, S. (2021). Wetland-an ecological boon for the environment. *East African Scholars Journal of Agriculture and Life Sciences, 4*(3), 38–48.

Barbier, E. B. (2018). *The value of Coastal Wetland ecosystem services. Elsevier eBooks.*, 947–964. 10.1016/b978-0-444-63893-9.00027-7.

Barksdale, W. F., Anderson, C. J., & Kalin, L. (2013). The influence of watershed run-off on the hydrology, forest floor litter and soil carbon of headwater wetlands. *Ecohydrology, 7*(2), 803–814. https://doi.org/10.1002/eco.1404.

Bassi, N., Kumar, M. D., Sharma, A., & Pardha-Saradhi, P. (2014). Status of wetlands in India: A review of extent, ecosystem benefits, threats and management strategies. *Journal of Hydrology: Regional Studies, 2*, 1–19.

Bayraktarov, E., Stewart-Sinclair, P. J., Brisbane, S., Boström-Einarsson, L., Saunders, M. I., Lovelock, C. E., & Wilson, K. A. (2019). Motivations, success, and cost of coral reef restoration. *Restoration Ecology, 27*(5), 981–991.

Bhowmik, S. (2022). Ecological and economic importance of wetlands and their vulnerability: A review. *Research Anthology on Ecosystem Conservation and Preserving Biodiversity,* 11–27.

Birch, W. S., Drescher, M., Pittman, J., & Rooney, R. C. (2022). Trends and predictors of wetland conversion in urbanizing environments. *Journal of Environmental Management, 310*, 114723. https://doi.org/10.1016/j.jenvman.2022.114723.

Butt, M. A., Zafar, M., Ahmed, M., Shaheen, S., Sultana, S., Butt, M. A., & Sultana, S. (2021). Types of wetland and wetland plants. *Wetland Plants: A Source of Nutrition and Ethno-medicines,* 35–54.

Cahoon, D. R., McKee, K. L., & Morris, J. T. (2020). How plants influence resilience of salt marsh and mangrove wetlands to Sea-Level rise. *Estuaries and Coasts, 44*(4), 883–898. https://doi.org/10.1007/s12237-020-00834-w.

Chadwick, M. A., Dobberfuhl, D. R., Benke, A. C., Huryn, A. D., Suberkropp, K., & Thiele, J. E. (2006). Urbanization affects stream ecosystem function by altering hydrology, chemistry, and biotic richness. *Ecological Applications, 16*(5), 1796–1807. https://doi.org/10.1890/1051-0761(2006)016.

Chaikumbung, M., Doucouliagos, H., & Scarborough, H. (2016). The economic value of wetlands in developing countries: A meta-regression analysis. *Ecological Economics, 124*, 164–174. https://doi.org/10.1016/j.ecolecon.2016.01.022.

Chaikumbung, M., Doucouliagos, H., & Scarborough, H. (2018). Institutions, culture, and Wetland values. *Ecological Economics, 157*, 195–204. https://doi.org/10.1016/j.ecolecon.2018.11.014.

Chakraborty, S. K., Sanyal, P., & Ray, R. (2023). Introductory and basic eco-biological aspects of wetlands. In *Wetlands Ecology: Eco-biological uniqueness of a Ramsar site (East Kolkata Wetlands, India).* Springer International Publishing, 1–38.

Chen, X., Zhang, M., & Jiang, H. (2022). Morphological characteristics and hydrological connectivity evaluation of tidal creeks in coastal wetlands. *Land, 11*(10), 1707.

Davidson, N. C. (2014). How much wetland has the world lost? Long-term and recent trends in global wetland area. *Marine and Freshwater Research, 65*(10), 934–941.

Davidson, N. C., Fluet-Chouinard, E., & Finlayson, C. M. (2018). Global extent and distribution of wetlands: Trends and issues. *Marine and Freshwater Research, 69*(4), 620. https://doi.org/10.1071/mf17019.

Dhar, U. (2002). National biodiversity strategy and action plan wild plant biodiversity.

Duraiappah, A. K. (2011). Ecosystem services and human well-being: Do global findings make any sense? *Bioscience, 61*(1), 7–8. https://doi.org/10.1525/bio.2011.61.1.2.

Ekumah, B., Armah, F. A., Afrifa, E. K. A., Aheto, D. W., Odoi, J. O., & Afitiri, A. R. (2020). Assessing land use and land cover change in coastal urban wetlands of international importance in Ghana using Intensity Analysis. *Wetlands Ecology and Management, 28*(2), 271–284. https://doi.org/10.1007/s11273-020-09712-5.

Greene, B. T., & Blossey, B. (2011). Lost in the weeds: Ligustrum sinense reduces native plant growth and survival. *Biological Invasions, 14*(1), 139–150. https://doi.org/10.1007/s10530-011-9990-1.

Guimond, J., & Tamborski, J. (2021). Salt marsh hydrogeology: A review. *Water, 13*(4), 543.

Gómez-Baggethun, E., Tudor, M., Doroftei, M., et al. (2019). Changes in ecosystem services from wetland loss and restoration: An ecosystem assessment of the Danube Delta (1960–2010). *Ecosystem Services, 39*, 100965. https://doi.org/10.1016/j.ecoser.2019.100965.

Keitumetse, S. O., Mwale, K. P., Satau, G., Velempini, K., Baitsiseng, V. O., Ntema, O. P. B. M., & Mogotsi, S. T. (2023). Exploring cultural values of African wetlands for sustainable conservation: Okavango Delta World heritage site, Botswana. *Journal of Cultural Heritage Management and Sustainable Development, 13*(3), 501–516.

Kolli, M. K. (2020). Lake catchment interaction analysis by using remote sensing and GIS techniques–the case study of Kolleru Lake, South India.

Kulik, M., Urban, D., Grzywaczewski, G., Bochniak, A., Grzywna, A., & Sender, J. (2024). Half a century of wetland degradation: The present state and trends of changes in Western Polesie-Long-term wetland degradation. *Global Ecology and Conservation, 56*, e03324.

Let, M., & Pal, S. (2022). Socio-ecological well-being perspectives of wetland loss scenario: A review. *Journal of Environmental Management, 326*, 116692. https://doi.org/10.1016/j.jenvman.2022.116692.

Maund, P. R., Irvine, K. N., Reeves, J., et al. (2019). Wetlands for wellbeing: Piloting a nature-based health intervention for the management of anxiety and depression. *International Journal of Environmental Research and Public Health, 16*(22), 4413. https://doi.org/10.3390/ijerph16224413.

McCartney, M. P., & Acreman, M. C. (2009). 17 Wetlands and water resources. *The Wetlands Handbook*, 357.

McCartney, M. P., Rebelo, L. M., & Sellamuttu, S. S. (2015). Wetlands, livelihoods and human health. *In: Wetlands: Ecology, Conservation and Management*, 123–148. https://doi.org/10.1007/978-94-017-9609-5_7.

Mitsch, W. J., Bernal, B., & Hernandez, M. E. (2015). Ecosystem services of wetlands. *International Journal of Biodiversity Science, Ecosystem Services & Management, 11*(1), 1–4.

Mitsch, W. J., & Gosselink, J. G. (2000). The value of wetlands: Importance of scale and landscape setting. *Ecological Economics, 35*(1), 25–33. https://doi.org/10.1016/s0921-8009(00)00165-8.

Mitsch, W. J., Zhang, L., Waletzko, E., & Bernal, B. (2014). Validation of the ecosystem services of created wetlands: Two decades of plant succession, nutrient retention, and carbon sequestration in experimental riverine marshes. *Ecological Engineering, 72*, 11–24. https://doi.org/10.1016/j.ecoleng.2014.09.108.

Mętrak, M., Chachulski, Ł., Pawlikowski, P., Rojan, E., Sulwiński, M., & Suska-Malawska, M. (2023). Potential role of high-altitude wetlands in preservation of plant biodiversity under changing climatic conditions in the arid Eastern Pamir. *Catena, 220*, 106704.

Nayak, A., & Bhushan, B. (2022). Wetland ecosystems and their relevance to the environment: importance of wetlands. In *Handbook of research on monitoring and evaluating the ecological health of wetlands*. IGI Global, 1–16.

O'Neill, A. R. (2019). Evaluating high-altitude Ramsar wetlands in the Eastern Himalayas. *Global Ecology and Conservation, 20*, e00715.

Orkhonselenge, A., Uuganzaya, M., & Davaagatan, T. (2022). Genesis of lakes in Mongolia. In *Lakes of Mongolia: Geomorphology, Geochemistry and Paleoclimatology*. Springer International Publishing, 51–62.

Pascoe, S. (2019). Recreational beach use values with multiple activities. *Ecological Economics, 160*, 137–144.

Pedersen, E., Weisner, S. E. B., & Johansson, M. (2018). Wetland areas' direct contributions to residents' well-being entitle them to high cultural ecosystem values. *The Science of the Total Environment, 646*, 1315–1326. https://doi.org/10.1016/j.scitotenv.2018.07.236.

Perez-Ruzafa, A., Pérez-Ruzafa, I. M., Newton, A., & Marcos, C. (2019). Coastal lagoons: Environmental variability, ecosystem complexity, and goods and services uniformity. In *Coasts and estuaries*. Elsevier, 253–276.

Reeves, J. P., Knight, A. T., Strong, E. A., et al. (2019). The application of wearable technology to quantify health and wellbeing co-benefits from urban wetlands. *Frontiers in Psychology, 10*. https://doi.org/10.3389/fpsyg.2019.01840.

Seelan, S. K. (2015). Groundwater targeting using remote sensing. In *Remote Sensing Handbook, Volume V*, CRC Press, 65–109.

Sharma, L. K., & Naik, R. (2024). Wetland Ecosystems. In *Conservation of saline wetland ecosystems: An initiative towards un decade of ecological restoration*. Springer Nature Singapore, 3–32.

Shen, S., Pu, J., Xu, C., Wang, Y., Luo, W., & Wen, B. (2022). Effects of human disturbance on riparian wetland landscape pattern in a coastal region. *Remote Sensing, 14*(20), 5160. https://doi.org/10.3390/rs14205160.

Thorslund, J., Jarsjo, J., Jaramillo, F., et al. (2017). Wetlands as large-scale nature-based solutions: Status and challenges for research, engineering and management. *Ecological Engineering, 108*, 489–497. https://doi.org/10.1016/j.ecoleng.2017.07.012.

Verhoeven, J. T., & Meuleman, A. F. (1999). Wetlands for wastewater treatment: Opportunities and limitations. *Ecological Engineering, 12*(1-2), 5–12.

Xu, X., Chen, M., Yang, G., Jiang, B., & Zhang, J. (2020). Wetland ecosystem services research: A critical review. *Global Ecology and Conservation, 22*, e01027. https://doi.org/10.1016/j.gecco.2020.e01027.

Ye, C., & Sun, F. (2021). Development of a social value evaluation model for coastal wetlands. *Ecological Informatics, 65*, 101417. https://doi.org/10.1016/j.ecoinf.2021.101417.

Zhao, Q., Bai, J., Huang, L., Gu, B., Lu, Q., & Gao, Z. (2016). A review of methodologies and success indicators for coastal wetland restoration. *Ecological indicators, 60*, 442–452.

Zhou, L., Guan, D., Huang, X., Yuan, X., & Zhang, M. (2020). Evaluation of the cultural ecosystem services of wetland park. *Ecological Indicators, 114*, 106286. https://doi.org/10.1016/j.ecolind.2020.106286.

Further reading

Tulu, F. D., & Desta, M. A. (2015). Human development and wetland conservation policy. *International Journal of Environmental Sciences, 4*(3), 126–138.

Zhao, Q., Bai, J., Huang, L., Gu, B., Lu, Q., & Gao, Z. (2015). A review of methodologies and success indicators for coastal wetland restoration. *Ecological Indicators, 60*, 442–452. https://doi.org/10.1016/j.ecolind.2015.07.003.

Chapter 16

Protecting wetlands from invasive species through effective communication strategies—a mini review

Kavyanjali Shukla[1], Kantharajan Ganesan[2], and Uttam Kumar Sarkar[2]
[1]*Indian Science Communication Society, Lucknow, Uttar Pradesh, India,* [2]*ICAR-National Bureau of Fish Genetic Resources, Lucknow, Uttar Pradesh, India*

16.1 Introduction

Invasive species are those nonnative organisms whose introduction causes or is likely to cause environmental, economic or (human/animal) health-related harm (Colautti & Richardson, 2009; USDA, 2023). The Convention on Biological Diversity defines invasive species as those "Non-Native Species whose introduction by human activity outside their natural past or present distribution threatens biodiversity" (Convention on Biological Diversity, 2010). Invasive nonnative species (INNS) are a major threat to biodiversity and ecological balance. INNS have high colonization rates and continue to spread at an increasing pace (Ricciardi, 2013; Seebens et al., 2020), with introductions now affecting entire ecological communities rather than individual species (Skillings, 2016). Invasive species establishment often involves microbial symbionts, including viruses, bacteria, fungi and other parasites, which may promote their survival by offering advantages like enhanced nutrient absorption or immune system modulation, thereby increasing invasive nonnative species populations exponentially, outcompeting, outnumbering and disrupting the native species populations in the new zone where they have been introduced. This process is called biological invasion, which causes ecosystem imbalances, altering competition dynamics, reducing biodiversity and facilitating the spread of new diseases (Foster et al., 2021). Throughout this paper, we refer to nonnative species as NNS and invasive nonnative species as INNS.

Wetlands, with their ecological, educational and recreational importance, are particularly vulnerable to biological invasions due to their natural connectivity to rivers, lakes and coastal zones (Sandilyan, 2022). These pathways enable invasive species to spread rapidly through water flow, nutrient exchange and biodiversity interactions (Wisse, 2025). Wetlands are often connected to other waterbodies, facilitating the cotransport of symbionts and pathogens (e.g., microorganisms) along with invasive nonnative species (INNS), enabling INNS spread through water movement. Many aquatic species have larval stages that enable long-distance dispersal (Foster et al., 2021). Some nonnative species (NNS) have been known to be cultured illegally in aquaculture ponds near wetlands and rivers in many parts of India; as such, when flooding occurs, these nonnative fish species may enter the aquatic systems or wetlands. Climate change has further compounded the challenge of dealing with invasive species (Raj et al., 2021). It is seen that warmer temperatures can enhance the reproductive rates of invasive species, extend their growing seasons and increase the survival rates of their microbial symbionts, further accelerating ecosystem disruptions (Foster et al., 2021), often making wetlands as suitable ground for invasive species to settle and thrive (Lázaro-Lobo & Ervin, 2021; Raj et al., 2021).

Given the complexity of these interactions, effective detection and management strategies are necessary. Advanced techniques, such as modern histology, environmental DNA (eDNA) and molecular diagnostics, present us with novel tools for monitoring introduced invasive species and enable early identification (Foster et al., 2021). eDNA is a rapid, sensitive, efficient and noninvasive technique that is increasingly employed to monitor aquatic ecosystems and detect invasive species (Wang et al., 2023). The eDNA findings can be geotagged which may be disseminated to the public through mobile applications and other social media platforms.

Wetlands: Biodiversity, Health and their Role in Ecosystem Services. DOI: https://doi.org/10.1016/B978-0-443-34222-6.00010-3

Additionally, interdisciplinary research, including contributions from ecological science and social science can help formulate better policies and public awareness initiatives, strengthen conservation efforts and mitigate the consequences of biological invasions in wetlands (Beaury et al., 2020). This paper systematically reviews and highlights the role of communication, recent research conducted worldwide and recommendations about effective communication strategies for invasive species management (ISM), with special reference to wetlands.

16.2 Review methodology

A systematic literature search was conducted using Google Scholar in three iterative rounds. **Round 1:** Focused on the keywords "invasive species," "India," and "wetland" to gather a foundational understanding of the issue within the Indian context. **Round 2:** Refined the search to "invasive species" and "wetland" to broaden the geographic scope and ecological relevance.

Round 3: Targeted "invasive species communication strategies" to explore public awareness, stakeholder engagement and policy responses. Research papers were screened for relevance based on titles and abstracts, with full-text review conducted for these studies.

16.3 Invasive species and wetland biodiversity

The negative impact of invasive species on the biodiversity and ecological stability of the wetland is reported globally. These species alter ecosystems, disrupt food chains and pose challenges for conservation efforts (Lázaro-Lobo & Ervin, 2021). *Python bivittatus* (Burmese python), a large invasive snake in Florida's Everglades, has significantly reduced native wildlife populations like raccoons, opossums and bobcats while some rabbit and fox populations have vanished in the wetland habitats (Wisse, 2025). *Myocastor coypus* (nutria or coypu), are invasive semiaquatic rodents introduced to North America in the late 1800s, causes damage to wetlands by burrowing into banks and overfeeding on native plants. Their year-round feeding reduces biomass, depleting food sources for other species, causes habitat destruction and increase in wetland vulnerability to flooding. The introduction of *Sus scrofa* (feral pigs) as livestock by early settlers has become a highly destructive invasive species in the United States and Australia. They threaten wetlands by spreading weeds and disease, trampling vegetation, degrading soil, water and preying on native species. Understanding their environmental impact is essential for developing effective control measures. *Rhinella marina* (cane toad), an invasive species in Australia, was introduced in 1935 to control cane beetles but has since rapidly spread across the northern regions of Australia. Lacking natural predators, it poses a significant threat to wetlands by outcompeting native species and poisoning animals that consume it. Its presence has contributed to population declines in northern quolls, large goannas and it also poses a risk to wetland-dwelling snakes (Wisse, 2025).

The negative effects of invasive species on the structure and functioning of Indian wetland ecosystems have been reported by many researchers. The aquatic weeds, including *Eichhornia crassipes* (water hyacinth) and *Salvinia molesta* (water fern), which are fast-growing, free-floating plant cover the wetland surface (Garg & Singh, 2018; Ingole et al., 2018), causing eutrophication-like conditions leading to oxygen depletion and threats to the survival of fish species (Howard & Harley, 1997). Some invasive terrestrial plant species, such as *Lantana camara,* have also recently been recorded from the Nawabganj bird sanctuary (a wetland) in Uttar Pradesh, India (Garg & Singh, 2018). *L. camara* is known to disrupt the natural regeneration of native vegetation (Sharma & Raghubanshi, 2010) and alter local plant communities. It has invaded several Indian biodiversity hotspots (Bhagwat et al., 2012), threatening native ecosystems (Sandilyan, 2019). *Mikania micrantha* is another invasive terrestrial plant species, threatening some of the Indian wetlands (Banerjee & Dewanji, 2012). It forms dense thickets over the native plants and smothers them by blocking sunlight and nutrients (Sandilyan, 2019). Further, the spread of invasive alien fish species such as *Aristichthys nobilis, Ctenopharyngodon idella, Cyprinus carpio, Hypophthalmichthys molitrix, Oreochromis mossambicus, O. niloticus, Clarias gariepinus, Mylopharyngodon piceus, Gambusia affinis, Poecilia reticulata, Pterygoplichthys disjunctivus, P. multiradiatus,* P. pardalis and *P. anisitsi* impacts the native fish genetic resources and economic contribution of the wetlands in India (Sarkar et al., 2020; Sandilyan, 2022). Exotic fishes negatively impact native fish populations in the wetlands by means of predation, competition for food and space, genetic introgression and significant alterations to community structure and ecosystem processes (Singh & Lakra, 2011; Singh et al., 2010).

In Hong Kong Wetland Park, several invasive plant species such as *Mikania micrantha*, *Salvinia molesta* and *Sonneratia sp.* have been recorded. Invasive animal species, including *O. mossambicus*, *O. niloticus*, *Solenopsis invicta* (red imported fire ant) and *Pomacea canaliculata* (golden apple snail), have also been documented. Control measures are implemented periodically to manage their spread and minimize ecological impacts (Agriculture Fisheries & Conservation Department & Hong Kong Wetland Park, 2021). Surveying/tracking of invasive species occurrences strengthens long-term protection of native biodiversity. Continuous and consistent monitoring of the occurrence of such invasive species should be recorded and communicated to the public globally.

16.4 Importance of science communication in invasive species management

Hamilton et al. (2007) highlight how risk perception differs between scientists and the public. Science assesses risks objectively, while people perceive risks subjectively. This discrepancy adds uncertainty when conveying risk-related information to stakeholders and that is why proper communication is needed (Copp et al., 2021). In the past, biologists and community groups have disagreed when it comes to tackling ISM on many occasions. In a situation, where researchers have proposed and the government has implemented methods of control, based on scientific research (top-down approach) but, the methods do not comply with the understanding of the public due to reasons like cultural issues or opinions. As such, a conflict situation has arisen and the methods have been put to stop (Shine & Doody, 2011). For instance, scientists have been largely unsuccessful in communicating their research findings and potential control methods for invasive *Rhinella marina* (cane toads) to community groups, sparking debate. In such conditions, implementing better communication strategies and methods involving how effectively and convincingly information has been communicated, without hurting the sentiments or challenging the opinions of the general public/community groups, helps to succeed the process (Shine & Doody, 2011). Thematic diagram of the technical areas to be involved for appropriate communication among stakeholders for invasive species management (ISM) is depicted in (Fig 16.1).

16.4.1 Role of languages in communication

NNS have been transported for ages and now the rate has accelerated due to global trade and travel (Chapman et al., 2017). Amid India's present economic growth (which involves increased global trade and tourism), raising awareness about the causes and impacts of biological invasions remains crucial (Shukla, 2022). As such, it is necessary that scientists communicate their findings to the general public and other stakeholders in easy and understandable language, risk assesors are able to do risk assessment about introductions, invasive potential in their mother tongue(s) while, awareness about NNS, among different stakeholders may be done in local/regional languages. However, NNS risk analysis faces uncertainties due to limited research, insufficient data and communication (Copp et al., 2021).

16.4.1.1 Language evolution concerning climate change and ISM

Scientific terminology evolves across languages. As new terms and phrases are added among different native languages, they clarify environment-related concepts and enhance the understanding of natural and artificial environments among the public. Recent language evolution among various languages reflects growing awareness of climate change impacts (Copp et al., 2021).

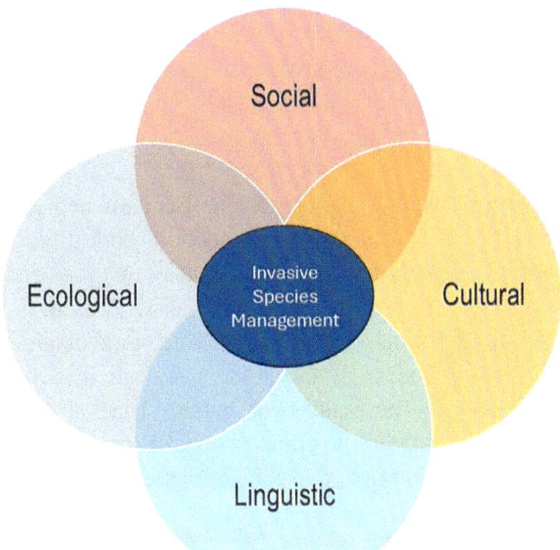

FIGURE 16.1 Thematic diagram of the technical areas to be involved for appropriate communication among stakeholders for invasive species management (*ISM*).

This happens to be true in the Indian scenario as well. There has been an increase in Hindi-language literature on climate change issues in India in the past decade (Toppo, 2025). Studies have shown that using native languages in NNS risk assessments improves understanding, accuracy and communication, reducing linguistic uncertainty and strengthening ISM, especially when tools like Aquatic Species Invasiveness Screening Kit (AS-ISK)[1] (a toolkit) offer multilingual support (Copp et al., 2021).

16.4.1.2 Understanding uncertainties in communication

Language uncertainties on climate change topics have been reported due to varied interpretations among people (Copp et al., 2021). During the invasive species risk analysis, linguistic uncertainty has been diagnosed in how the questions are framed in a questionnaire in risk assessment toolkits, the assessors own understanding, interpretation and use of invasive biology terms (Carey & Burgman, 2008; Copp et al., 2021; Ibabe & Sporer, 2004). Vilà et al. (2019) categorized various types of "uncertainties in communication" among NNS risk analysis as, linguistic—uncertainty in how risk analysis guidelines are communicated; stochastic—uncertainty from unknown variations in the invasion process and epistemic—uncertainty from gaps in knowledge about NNS or the invaded ecosystem (Copp et al., 2021).

Communication gaps among technical experts and stakeholders can hinder effective ISM. According to Li et al., (2020) in China, regional field scientists/ managers understand local invasiveness well due to increased field work exposure but find difficulties in understanding and interpreting "English-language" risk analysis protocols. This slows down the adoption of certain English language risk toolkits/ schemes among Chinese to combat NNS and its management strategies (Li et al., 2020). Therefore, from this example, scientists and policymakers must recognize how awareness of linguistic barriers is important as it impacts NNS research, risk assessment and ISM (Verbrugge et al., 2016).

Terminological uncertainties in NNS risk assessment increase when risk-assessors work/evaluate in a non-native language (Matthews et al., 2017). Even with precise language, risk assessors may interpret and judge information differently (Verbrugge et al., 2016). Discrepancies may occur among native speakers due to personal understanding of terms (Doupnik & Richter, 2003), interpersonal variation (Regan et al., 2002) and geographical/ cultural context differences (Matthews et al., 2017). Translating risk assessment outcomes into easy, understandable, relatable language is helpful for stakeholders, the public and policymakers in drafting legislation and integrating policy into management strategies (Copp et al., 2021; Wei, 2018).

16.4.1.3 Language role in developing risk assessment and screening toolkits

Copp et al. (2021) emphasized the importance of language inclusivity to enhance stakeholder engagement in ISM. A multilingual NNS screening toolkit improves clarity about assessment procedures/ methods (questions, responses, justification) and communication of the invasive potential of NNS among various stakeholders. At present, most NNS risk analysis toolkits are in English, with AS-ISK available in more than thirty other languages (Copp et al., 2021).

16.4.2 Combining socio-ecological approaches in invasive species management

Conservation is no longer just about protecting ecosystems; it now recognizes the deep connections between environmental health and social issues. Human activities, policies and community dynamics all shape ecological outcomes (Crain et al., 2014; Omondiagbe et al., 2017). Combining social and ecological approaches requires an understanding of interdependent individual/ community interactions within a given ecological setting. This involves examining social networks, which shape relationships, provide resources and influence individual/collective impact (Cooper et al., 2007). Omondiagbe et al. (2017) utilized social network analysis (SNA) to map stakeholders' relationships among individuals and groups involved in ISM. They identified central actors (Government agencies and NGOs that had more connections and influence) and peripheral actors (Community groups and private landowners with lesser connections) in a network. Key metrics used were degree centrality (Betweenness Centrality: Identified stakeholders who acted as bridges between others; Closeness Centrality: Assessed how quickly a stakeholder could reach others in the network). Their paper includes network diagrams showing stakeholder positions and connections, highlighting clusters and isolated nodes. If stakeholders are not involved in different steps/ stages of ISM, a conflict may be generated, and the process may be delayed or halted (Blackie et al., 2014 & Ogden & Gilbert, 2011). Therefore, viewing collaboration as the ultimate solution to complex/ challenging issues such as ISM, it may be more useful to analyze how groups handle conflict, allocate resources and establish cooperative efforts (Schoon et al., 2016 & Omondiagbe

[1] The AS-ISK is a decision-support tool designed to assess the potential invasiveness of nonnative aquatic species. Developed by CEFAS (Centre for Environment, Fisheries and Aquatic Sciences, an executive body of the UK Department for Environment, Food and Rural Affairs) and its international partners, AS-ISK replaces previous taxon-specific toolkits and can be applied to all aquatic plants and animals across marine, brackish and freshwater systems. (CEFAS, 2025).

et al., 2017). Additionally, there have been various models of study to see what type and how communication among and within the communities works for sustainable ISM, what stakeholders and communities can be/ may be involved towards contributing to solutions on ISM (Omondiagbe et al., 2017). The analysis revealed opportunities to strengthen ties between community groups and central actors (Government agencies and NGOs). By increasing community involvement and connectivity, the network could become more resilient and effective. SNA helped pinpoint which stakeholders should be engaged to improve coordination and resource sharing (Strategic stakeholders' engagement). SNA provided critical insights into stakeholder dynamics, revealing both strengths and weaknesses in the collaboration landscape. It also emphasized the need for inclusive/ strategic partnerships to enhance sustainable ISM.

16.4.3 Communication and conservation involving the public

16.4.3.1 *Understanding invasive species management communication among children and youth*

The studies conducted by Matos (2014) on Portuguese participants showed higher concern and knowledge linked to more outdoor exposure and strong conservation communication strategies. The study highlights the need for direct engagement and shows that nationality can influence environmental awareness. It was also found that children and youth categories are less aware of invasive species' long-term impacts on the global economy, climate and health. Therefore, communication strategies must be better addressed in future to close the specific knowledge gaps and improve outreach for better ISM (Matos, 2014). The concept of human–nature relationships (HNR) links people's values, beliefs and worldviews with environmental engagement (Flint et al., 2013 & Kemp et al., 2017).

"Human behaviour on INNS is shaped by knowledge, concern and environmental attitudes" (Connelly et al., 2016). However, studies connecting HNR to invasive species behavior are limited (Verbrugge et al., 2013) and their application across regions for policy and INNS management is also not well understood (De Groot & de Groot, 2009; De Groot et al., 2011).

16.4.3.2 *Communication via outreach programs*

Effective communication among stakeholders in outreach programs is crucial for ISM (Golebie & van Riper, 2023). It facilitates information exchange between scientists, policymakers, conservationists and local communities, ensuring that efforts are well-coordinated and responsive to emerging threats (Kemp et al., 2017). Clear communication in outreach programs helps raise awareness about invasive species risks, encourages proactive detection, monitoring and fosters collaboration for policy development along with enforcement (Davis et al., 2018). Engaging diverse perspectives, including social scientists and educators, can improve outreach strategies/programs, making the implementation of mitigation measures easier. Strong stakeholder communication ultimately leads to better decision-making, resource allocation and public participation, enhancing overall success on ISM (Davis et al., 2018; Omondiagbe et al., 2017; Palmér et al., 2023).

16.4.3.3 *Devising campaigns and logos in outreach programs*

Recreation happens to be one of the ways of NNS introduction and establishment in wetlands (Golebie & van Riper, 2023). Recreationists happen to transport NNS via boats and water recreation equipment (Rothlisberger et al., 2010), mostly unintentionally. As such, campaigns and logos may play an important role in displaying messages to impact the behavioral psychology of these recreationists (Golebie & van Riper, 2023). Golebie and van Riper (2023) tested values-framed messages among Illinois recreational water users, finding that biospheric and altruistic campaign messages connected deeply among these stakeholders. They concluded that applying values framing in INNS outreach campaigns may contribute towards fewer nonnative/ invasive species introductions by significantly lowering the propagule pressure of invasive species (Drake & Mandrak, 2014) and reducing their spread risk (Golebie & van Riper, 2023). Since removing INNS is often unfeasible, prevention remains the most effective strategy (Vander Zanden & Olden, 2008). Educating water-recreationists is a key priority (Pradhananga et al., 2015; Seekamp et al., 2016 a, b) and understanding how they interpret outreach campaigns can improve the outreach strategies (Cole et al., 2019). To design persuasive messages, psychological frameworks like the value-belief-norm (VBN) theory,[2] which highlights values as long-term drivers of

[2] VBN theory of environmentalism, developed by Paul Stern and colleagues, suggests that such behavior stems from a chain of psychological processes: Values- People/individuals who hold strong biospheric or altruistic values are more likely to care about the environment; Beliefs-These values influence beliefs, such as awareness of environmental consequences and a sense of personal responsibility & Norms-When individuals believe their actions can make a difference, they feel a moral obligation (personal norm) to act in environmentally friendly ways. It's been widely used to understand behaviors like recycling, energy conservation and support for environmental policies. The theory proposes that people support environmental causes not just because of external pressures, but because they internalize a sense of duty based on their values and beliefs (Stern, 1999).

environmental behavior (Stern et al., 1999) and Protection Motivation Theory (PMT)[3] (Rogers, 1975) can be applied. These help tailor messages that align with water recreationists' motivations and beliefs. The effectiveness of such campaigns also depends on how messages are presented. Key components of INNS awareness campaigns include message processing, elaboration (Petty & Cacioppo, 1986) and reactance (Brehm & Brehm, 1981). To bridge the "knowledge-action gap" (Kollmuss & Agyeman, 2002), future research should explore how biospheric values, social norms and beliefs enhance values-based messaging effectiveness. These elements can increase/ amplify stakeholders' motivation, capturing attention and communicating prevention strategies, making the ISM campaign more effective (Golebie et al., 2021; Peloza & White, 2007; Shin et al., 2022). Findings also suggest that self-transcendent messages[4] foster deeper reflection and stronger action against INNS spread and help in ISM (Golebie & van Riper, 2023).

According to Kemp et al.(2017) logos should be aesthetically appealing, complement educational materials (Francis, 2014), especially for audiences with limited prior experience. Management agencies should tailor/customize, outreach/ communication strategies to diverse stakeholder subgroups (Kemp et al., 2017).

16.4.3.4 Science—society studies

Science and society studies on invasion science examine how scientists and institutions shape public understanding of nature (Milton, 2000). Warner and Kinslow (2013) highlighted how limited public engagement and one-way communication (Top-down approach) during efforts to control invasive *Epiphyas postvittana* (light brown apple moth) in the United States led to public protests, legally blocking the control actions taken by the government and scientists. This case emphasize the need for a two-way dialog to build trust and support in conservation efforts. The three levels at which environmental conservation communication is generally undertaken include (1) top-down (regulations by government agencies); (2) bottom-up (grassroots, self-governing) and (3) middle-out (civic environmentalism) (Epanchin-Niell, et al., 2010; John et al., 2006). When communication is unidirectional (from government/scientists towards public, top-down approach), a gap emerges between scientific reasoning and public concerns, potentially leading to debate over invasive species programs. If management strategies do not align with cultural beliefs, values or societal systems, they may face resistance/ controversy/ backlash from the general public (Warner & Kinslow, 2013). Therefore incorporating a bottom-up approach (about public/ stakeholders' views) and participatory/ middle-out (civic environmentalism) approach on ISM is recommended (Epanchin-Niell et al., 2010).

16.4.3.5 Science communication and public engagement

Public engagement facilitates two-way learning between citizens and scientists. It helps government agencies/scientists/ policymakers understand local values, anticipate public concerns, improve support and efforts towards better management of invasive species (Kleinman et al., 2011; Warner & Kinslow, 2013).

Scientific outreach is essential, but often a formal communication training among scientists has been lacking (Woitowich et al., 2022). Empowering scientists with a formal science communication training develops deeper connections with public, bridging gaps, sparking curiosity and nurturing long-term public engagement through shared learning and trust (Swords et al., 2023).

16.4.4 Evaluation of invasive species prevention campaign effectiveness

Haley et al. (2023) found that public awareness campaigns on invasive species often lack clear methods/ procedures, had inconsistencies, making their effectiveness hard to assess. In a study done on the effectiveness of public awareness campaigns for the invasive species management (ISM), it was found that out of the 24 public awareness campaigns, four were successful campaigns and success could not be determined in the remaining 20 campaigns (Haley et al., 2023).

Poorly designed campaigns may backfire or trigger public resistance, may be counterproductive, waste resources or even worsen the issue (Hiom, 2006), where people may act against the planned recommendations (Kahan, 2013). INNS awareness linked to behavioral change remains unclear and understudied (Seekamp et al., 2016b & Wallen & Kyle, 2018).

[3] PMT is a psychological framework that describes how people/individuals analyse/assess threats and then choose to engage in protective behavior. It was developed by R.W. Rogers in 1975 to interpret responses to "fear appeals" and was later expanded in 1983 into a broader theory of persuasive communication (Rogers, 1975).

[4] Self-transcendent framing is a term used in psychology, communication and behavioral studies to inspire and motivate individuals to act for the greater good. It refers to presenting information in a way that encourages individuals to look beyond their personal interests and connect with a greater purpose such as humanity, nature or a shared societal goal. This approach promotes empathy, cooperation and develops a sense of belonging, making people more receptive to messages about social or environmental issues.

To improve impact, campaigns should use standardized evaluations like (Before-After Control-Impact) designs that measure changes in knowledge, behavior and biological/ behavioral outcomes (Haley et al., 2023).

While invasive species research is often quantitative, qualitative studies are needed to address social, cultural and political contexts (Kapitza et al., 2019). Integrating social science assessments into public engagement strategies has supported INNS management actions (Glen & Hoshino, 2020; Liang, 2023) as seen in successful invasive species eradication programs like those in New Zealand (Oppel et al., 2011; Russel et al., 2018; Haley et al., 2023).

it is seen that studies with negative/null results often go unpublished (Rosenthal, 1979). This omission is concerning and needs attention in environmental research, as failed campaigns still offer valuable learning opportunities, leading to more effective future outreach strategies (Cooke et al., 2020; Haley et al., 2023). Incorporating gray literature like NGO reports/ surveys/interviews can fill knowledge gaps, improve insights and broaden perspectives beyond English-language sources (Haley et al., 2023; Lawrence, 2017).

16.4.5 Role of technology-based tools

Although there has been much criticism about the young being so invested in mobile technology and social media apps, leading to a disconnect with nature (Matos, 2014), we still support the idea that technology, when used sensibly, contributes to the betterment of humankind and environment (Hopper, 2020 & Santos et al., 2020). Tools like mobile technology and games enhance public engagement and environmental awareness effectively and hold great promise for conservation, where general public may be involved in invasive species mapping projects. Though their full potential is yet to be realized. Leveraging collaboration between scientists, developers and users is key to maximizing impact on ISM. Use of emerging technologies enables faster, smarter and more proactive ISM in a globally connected world (Fricke & Olden, 2023).

Santos et al. (2020) developed a mobile game using the "AEIOU"[5] framework (Awareness, Enjoyment, Interest, Opinion, Understanding) to raise awareness about invasive plants in Portugal. The game focused on *Eichhornia* sp. (water hyacinth) and taught players how to identify and manage it, while demonstrating proper removal and disposal methods to prevent further spread. User feedback about this game and app was positive, showing strong potential for broader outreach and public engagement (Santos et al., 2020).

16.4.6 Media and communication

The rise of diverse media channels has expanded communication possibilities (Rania, 2017 & Davis et al., 2018). Each communication medium has its advantages and limitations but leveraging broadcast and social media to complement targeted outreach, such as citizen science, public events and workshops, has the potential to help build sustained momentum around INNS awareness. According to Davis et al. (2018), broadcast and social media offer a wide reach but low interaction/ engagement (0% of 593,000 interactions in broadcast) while face-to-face public events (33% of 820 incidences of reach in two public events)/ workshops (58% of 118 incidences of reach in four workshops), a citizen science event (100% of 152 incidences of reach in two events) develops deeper engagement despite engaging fewer people. An effective INNS communication plan should have researcher and project credibility. The plan should also include correct, clear, locally relevant and understandable information to avoid miscommunication and impactful outreach using minimal resources (Davis et al., 2018). Among online and social media outreach, various metrics can be analyzed to gain valuable insights into effectiveness and resource allocation (Wolf, 2017 & López-Bosch et al., 2025). Careful assessment helps optimize communication strategies for maximum engagement (Davis et al., 2018).

Communication plans often include asking outreach events/workshop attendees how they learned about the event, a common practice in many countries to assess media impact. In a study by Davis et al. (2018), social media proved valid and effective in reaching policymakers and international groups. Thus, it may be concluded that strategic use of citizen science, events and workshops can amplify outreach and create lasting community impact (Davis et al., 2018; López-Bosch et al., 2025).

[5] The **AEIOU principle**—*Awareness, Enjoyment, Interest, Opinion formation and Understanding* is a powerful framework used in science communication and education to evaluate how people engage with scientific content. This model is often used in evaluating science festivals, museum exhibits, public lectures or educational media. It helps educators and communicators design experiences that don't just inform, but also delight, provoke thought and empower action.

16.4.7 Communication challenges for invasive species management in wetlands

The transport of INNS sometimes facilitates other symbionts to get introduced to a new place. e.g., if a nonnative fish species is introduced to another country, it might carry some microorganisms (ecto/ endo parasites) which might be pathogenic/potentially pathogenic in a new environment. Such introductions may pose a threat to animal, human and ecological health in an aquatic environment (Foster et al., 2021). Gaps in research, knowledge, policy and communication can lead to insufficient/inadequate risk assessment of invasive species and associated pathogen movement into new regions, increasing the likelihood of disease spread (Dunn & Hatcher, 2015; Hess-Erga et al., 2019; Hulme, 2014).

In addition, wetland management is hindered by poor accessibility and navigation (due to swampy, waterlogged terrain), dense vegetation, low visibility, unreliable global positioning system, labor-intensive tracking and delay in effective detection, documentation and communication about invasive species (Wisse, 2025).

16.5 Role of climate change in increasing challenges

Climate change poses significant challenges to ISM globally, affecting the distribution, spread and impact of invasive species in complex ways (Beaury et al., 2020). Addressing these challenges requires stronger collaborative and interdisciplinary efforts, including the development and applications of tools and approaches. For instance, range shift maps and climate-informed risk assessments can support scientists and managers in tracking the movement and expansion of invasive species under changing climatic conditions. When used appropriately, these tools can provide information and contribute to strategic and adaptive management plans for effective ISM. Based on these principles, this working methodology was formally known as "*Translational ecology*"[6] among scientists. It is defined as "the process by which ecologists, stakeholders and decision-makers work collaboratively to develop scientific research that results in improved decision-making" (Enquist et al., 2017).

Climate change exacerbates/compounds ISM challenges by rising water temperatures, allowing NNS and their symbionts (mostly microorganisms) to survive and establish. Warmer conditions may accelerate pathogenic/disease-causing microbial life cycles, leading to their amplification in population densities and virulence (Dutta & Dutta, 2016). Additionally, shifting hydrodynamics and recurring extreme weather events contribute to the spread of invasive aquatic species. These effects are particularly relevant to wetland ecosystems, where climate-driven changes may further impact species interactions and ecological stability (Foster et al., 2021; Rahel & Olden, 2008). Understanding and assessing INNS introduction risks, pathogenic threats to native species for trade/consumption and changes to the invasive species symbiome, potentially altering species ecological or behavioral traits in a new environment, is essential to interpret in a climate change scenario (Foster et al., 2021).

Climate change is more likely to cause an increased spread of invasive species (Foster et al., 2021 & Palmér et al., 2023). Over time, some NNS that may be initially noninvasive may gradually become invasive, due to delayed detection (INNS). The time lag means that there is a window, the period between their initial introduction and their spread. If this window is missed, the introduced species initially considered harmless could later pose ecological challenges by becoming invasive once they spread (Palmér et al., 2023). A conceptual framework illustrating the influence of climate change on the alteration of invasive species traits and their cascading impacts on wetland ecosystems is depicted in Fig. 16.2.

In a study conducted by Beaury et al. (2020), assessing insights from managers incorporating climate change into ISM, based on a questionnaire survey response from 211 invasive species managers revealed that 65% of them, incorporate climate change knowledge in strategic planning, control and preventive management. However often they face challenges due to lack of clear, understandable, user-friendly and applicable research (Beaury et al., 2020), limiting their ability to implement effective management strategies (Peters et al., 2018) that hinders effective action leading to a "knowing-doing" gap (Pfeffer & Sutton, 1999; Schwartz et al., 2021). Bridging this gap by "*Translational ecology*" that focuses on collaboration between scientists and decision-makers to create useful, science-driven, real-world solutions for emerging ecological challenges and pressing environmental issues like climate change and ISM will be helpful (Barney et al., 2019). A knowledge coproduction (identifying barriers to effective management, especially gaps between research and practical perspectives) from both experts' and stakeholders' points of view from the start of a research project will be useful. Strong, trust-based relationships and continuous communication ensure that scientific findings lead to effective decision-making and resource management (Beaury et al., 2020).

[6] Translational ecology term was coined by William H. Schlesinger (2010) where, ecological findings by scientists can be promptly translated/interpreted and communicated (to all end users/ stakeholders) for useful practical solutions/ actions on ecological problems (Schlesinger 2010).

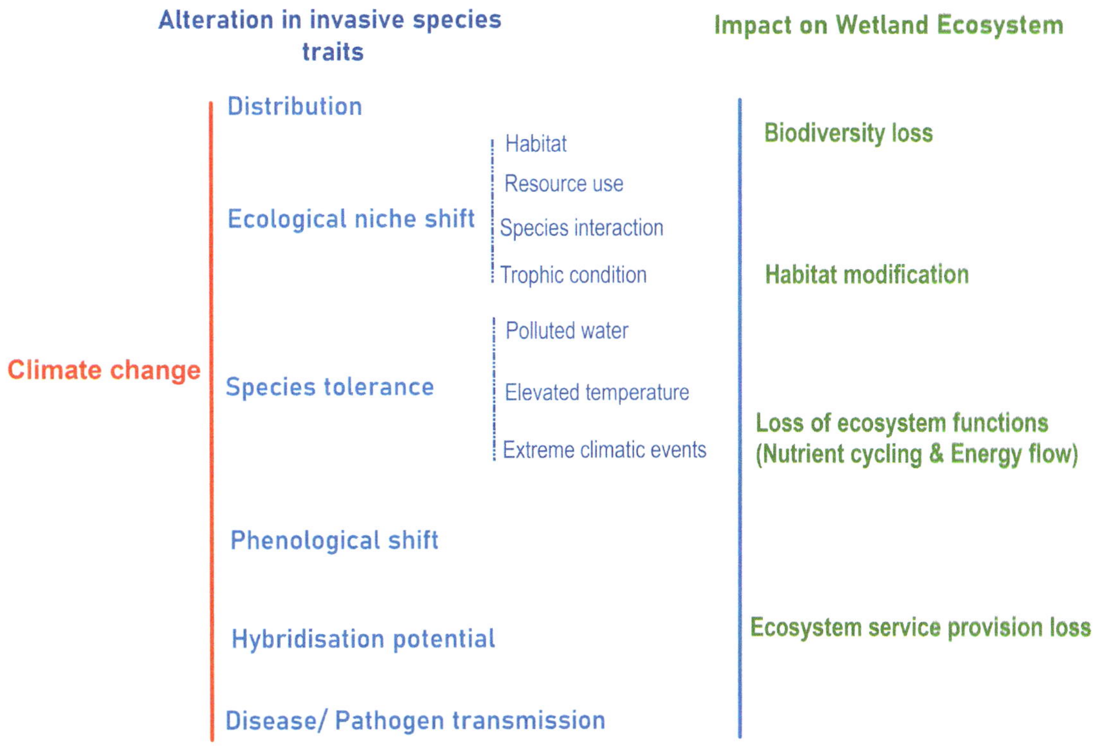

FIGURE 16.2 Conceptual framework illustrating the influence of climate change on the alteration of invasive species traits and their cascading impacts on wetland ecosystems.

According to Rahel and Olden (2008), climate change threatens to blur the boundaries of what we call "invasive species." Invasive species were once defined as outsiders crossing continents only; the term may soon include native species that, driven by shifting climates, spread aggressively, overwhelm ecosystems and destabilize the very communities they once belonged to. For example, in Japan, climate change may let native *Salvelinus leucomaenis* (white-spotted charr) outcompete and displace *Salvelinus malma* (Dolly Varden), showing that even indigenous species can spread, dominate and harm others, showing the traits typical of invasive species in a climate change scenario (Nakano et al., 1996).

16.6 Recommendations

To effectively disseminate information about invasive nonnative species (INNS), especially in wetland ecosystems, it's crucial to link outreach strategies and recommendations with global biodiversity frameworks like the Ramsar Convention, Aichi Biodiversity Targets and IPBES (Intergovernmental Science-Policy Platform on Biodiversity and Ecosystem Services) priorities. The strategic relevance of the Ramsar Convention involves the wise use of wetlands through education, public awareness and international cooperation. Creative communication about INNS supports Ramsar's Strategic Plan by enhancing public engagement and knowledge sharing. The Aichi Biodiversity Targets, especially target nine, focus specifically on invasive alien species, aiming to identify, prioritize and control them to reduce their impact on biodiversity (Convention on Biological Diversity, 2020). Keeping these in view, we summarize/suggest the following recommendations:

1. The development of interactive mobile apps/games for the management of invasive species shall be encouraged. Presently, a mobile app game is being developed by the Portuguese to manage invasive *Eichhornia* sp. . This not only helped users with its management and avoidance aspects but also led to the citizen science approach for collecting the data on invasive species (Santos et al., 2020). More such citizen science apps may be launched to report invasive species sightings, also spread in Indian wetlands.
2. A standardized, accessible and easily understandable watch list and checklist system should be developed to enhance invasive species monitoring (Reaser et al., 2020). A checklist records species observed and identified in a specific area, showing known occurrences. A watch list includes species that may or may not have been detected in the area; we have to watch for their presence or absence. To maximize their utility, it is essential to improve the standardization, accessibility and communication strategies of watch lists (Reaser et al., 2020).

3. Establishing an institutional framework for the management of invasive species across various domains, including marine, freshwater, terrestrial and wetland domains (Sandilyan, 2022). In addition to biological research, encouraging interdisciplinary collaboration through the inclusion of social science, education and psychology departments may foster a more comprehensive approach to addressing the spread of invasive species. A stronger social science approach can improve outreach and communication strategies to prevent biological invasions in freshwater ecosystems (Kemp et al., 2017).
4. Direct engagement of resource managers in ISM is crucial to integrating climate change-related information into the management. As such, a collaborative, two-way research incorporating the research needs from the manager's point of view and scientists working on ISM will be of great value (Beaury et al., 2020).
5. *Translational ecology* offers a promising approach to overcoming barriers in ISM. Breaury et al. (2020) support the ISM strategies from managers who integrate climate adaptation into their decisions, demonstrating the practicality of adaptive management in a changing environment with respect to climate change.
6. Effective communication efforts are required to equip the local people to identify common NNS. Strengthening the belief that NNS (nonnative species)/INNS (invasive nonnative species) both affect local biodiversity. This may encourage control efforts among the public. Communication efforts should also incorporate how climate change influences NNS and their invasive potential (Beaury et al., 2020).
7. Designing campaigns for targeted stakeholders, for early detection of invasive species, rapid response and communication will help better management of potential INNS (Golebie et al., 2021; Golebie & van Riper, 2023).
8. Devising creative approaches to disseminate information about invasive species (INNS) spread and management among the public will be helpful (Lázaro-Lobo & Ervin, 2021). For example, developing augmented reality experiences, community art projects, storytelling campaigns (puppetry, folk songs, theatre) supports the Ramsar Convention's education goals and Aichi Targets 1 and 9 by making biodiversity action fun, memorable and child-led. Using local narratives and folklore to frame invasive species as ecological villains, encourage youth participation in INNS awareness.
9. Using organizational network analysis tools, such as social network analysis (SNA) (Omondiagbe et al., 2017). It is found that integrating social and ecological factors can strengthen regional control of NNS/INNS and promote more effective research towards better ISM. More SNA among communities is recommended to understand the position of Govt agencies/NGOs and community groups within a conservation network, so that better stakeholders' collaborations are developed and utilized for sustainable ISM.
10. Urban wetlands may have greater access to digital outreach tools, while rural communities may rely on interpersonal communication or traditional/folk media. Addressing this socio-geographic variability in communication effectiveness is essential (Sinthumule & Netshisaulu, 2022). Ramsar Convention's Policy Brief on Small Wetlands also emphasizes the importance of inclusive communication strategies in wetland conservation across diverse geographic settings (Ramsar, n.d.).
11. We identified an urgent need to establish a dedicated day or week for invasive species awareness among all stakeholders in India and neighboring countries as the "National Biological Invasion Awareness Day/Week" (Shukla, 2025). This should not be limited to a conference-like celebration (only among science community) but involve larger communities (rural/urban/peri-urban public, women and youth participation, tribal/indegenous groups for diverse perspectives on ISM), incorporating diverse local languages/dialects in the awareness programs to maximize the outreach on ISM. In our opinion, this initiative would facilitate greater understanding of biodiversity loss, highlight the risks associated with the introduction of NNS and promote proactive efforts in their identification, monitoring and prevention. By engaging scientists, policymakers, community members and educators, such a campaign can build momentum toward collaborative action and long-term ecological resilience, also promoting stakeholders' inclusivity. The 2023 IPBES thematic assessment report on invasive alien species and their control supports this point of view, by emphasizing the importance of cross-sectoral knowledge sharing, to address the growing threat of INNS through collaborative policy and public engagement. The report also highlights that INNS contribute to 60% of global plant and animal extinctions, leading to major driver of biodiversity loss (IPBES, 2023).
12. We advocate for a greater number of expert-written popular science articles that demystify the concept of invasive species for the public, especially in the Indian context. These should be crafted in simple English as well as in Hindi and other regional languages to ensure inclusivity and broad reach. By making scientific knowledge more accessible and culturally relevant, such articles can spark public interest, improve awareness of biodiversity threats and empower communities to participate actively in monitoring and mitigation efforts (based on our own perspective).
13. We recommend that toolkits presently available (e.g., AS-ISK, developed by CEFAS) and developed in the future, shall integrate Hindi and other regional languages, as it would significantly enhance and strengthen the toolkit's role as a preventive measure against the spread of invasive species within India as well as globally.

14. Soundscape studies on invasive species show that vocalizations of nonnative animals can disrupt the communication of native species. Disrupted species communication can cascade into changes in predator–prey dynamics, reproductive success and long-term biodiversity (Hopkins et al., 2022). Further research on acoustic monitoring could become a tool for early detection of invasive species, aiding conservation and management strategies.

16.7 Conclusion

Communication strategies should encourage positive environmental behavior in society (Kemp et al., 2017). Science communication in ISM plays a crucial role in shaping public perception, engagement and conservation efforts. Effective strategies help bridge the gap between scientific expertise and public awareness, promoting informed decision-making and encouraging stronger environmental stewardship (Warner & Kinslow, 2013). Effective INNS management requires international, national and transboundary coordination, cooperation and collaboration to enhance assessment, resource allocation, guidance and risk assesment tools for prevention and control (Foster et al., 2021). Strengthening awareness across key sectors and stakeholders is essential, alongside globally standardized prevention and mitigation measures. Effective actions include INNS pathway management, international border checks (including molecular screening for symbionts such as microrganisms on NNS/INNS), stringent quarantine for intentionally introduced species, routine monitoring and rapid response for unintended/ accidental introductions (Foster et al., 2021).

While these measures remain important for prevention, we strongly believe that communicating these points to the general public is crucial for effective management of invasive species. Ineffective communication hinders conservation efforts, emphasizing the need for meaningful outreach programs/ strategies for effective ISM. Simply delivering messages is insufficient, public engagement is vital to ensure that information resonates, which can be brought about by removing language uncertainties in communication, resolving conflicts among the policymakers and general public, use of technology, interdisciplinary approaches, targeted communication using standardized assessment tools, campaigns and using diverse media forms for its proper outreach as affirmed through the findings and recommendations of other researchers working on ISM.

Declaration

The authors have used generative AI and AI-assisted technologies to improve the readability and language of the write-up.

References

Agriculture Fisheries & Conservation Department & Hong Kong Wetland Park. (2021). Habitat management plan (pp. 1-5). https://www.wetlandpark.gov.hk/filemanager/files/public/wcms/HKWP_Habitat_Management_Plan_on_Website_ver_May_2021.pdf.

Banerjee, A., & Dewanji, A. (2012). *Mikania micrantha* H.B.K.—A potential and economical threat to global biodiversity with special emphasis on Indian context. In *Proceedings of the 18th Australasian Weeds Conference. Weed Society of Victoria Inc*, 17–20 6Eldershaw V.

Barney, J. N., Schenk, T., Haak, D. C., Salom, S., Brown, B., & Hotchkiss, E. R. (2019). Building partnerships and bridging science and policy to address the biological invasions crisis. *Invasive Plant Science Management*. https://doi.org/10.1017/inp.2018.33.

Beaury, E. M., Fusco, E. J., Jackson, M. R., Laginhas, B. B., Morelli, T. L., Allen, J. M., Pasquarella, V. J., & Bradley, B. A. (2020). Incorporating climate change into invasive species management: Insights from managers. *Biological Invasions, 22*, 233–252. https://doi.org/10.1007/s10530-019-02087-6.

Bhagwat, S. A., Breman, E., Thekaekara, T., Thornton, T. F., & Willis, K. J. (2012). A battle lost? Report on two centuries of invasion and management of *Lantana camara* L. in Australia, India and South Africa. *PLoS ONE, 7*(3), e32407. https://doi.org/10.1371/journal.pone.0032407.

Blackie, H. M., MacKay, J. W. B., Allen, W. J., Smith, D. H. V., Barrett, B., Whyte, B. I., Murphy, E. C., Ross, J., Shapiro, L., Ogilvie, S., Sam, S., MacMorran, D., Inder, S., & Eason, C. T. (2014). Innovative developments for long-term mammalian pest control. *Pest Management Science, 70*(3), 345–351. https://doi.org/10.1002/ps.3627.

Brehm, J. W., & Brehm, S. S. (1981). *Psychological reactance: A theory of freedom and control. Academic Press.*

CEFAS, (2025). Decision support tools for the identification and management of invasive non-native aquatic species. Centre for environment fisheries and aquaculture science. https://www.cefas.co.uk/expertise/research-advice-and-consultancy/non-native-species/decision-support-tools-for-the-identification-and-management-of-invasive-non-native-aquatic-species/.

Carey, J. M., & Burgman, M. A. (2008). Linguistic uncertainty in qualitative risk analysis and how to minimize it. *Annals of the New York Academy of Sciences, 1128*, 13–17. https://doi.org/10.1196/annals.1399.003.

Chapman, D., Purse, B. V., Roy, H. E., & Bullock, J. M. (2017). Global trade networks determine the distribution of invasive non-native species. *Global Ecology and Biogeography, 26*, 907–917. https://doi.org/10.1111/geb.12599.

Colautti, R. I., & Richardson, D. M. (2009). Subjectivity and flexibility in invasion terminology: Too much of a good thing? *Biological Invasions, 11*, 1225–1229. https://doi.org/10.1007/s10530-008-9387-7.

Cole, E., Keller, R. P., & Garbach, K. (2019). Risk of invasive species spread by recreational boaters remains high despite widespread adoption of conservation behaviours. *Journal of Environmental Management, 229*, 112–119. https://doi.org/10.1016/j.jenvman.2018.06.078.

Connelly, N. A., Lauber, T. B., Stedman, R. C., & Knuth, B. A. (2016). The role of anglers in preventing the spread of aquatic invasive species in the Great Lakes region. *Journal of Great Lakes Research. 42*(3), 703–707.

Convention on Biological Diversity. (2010). Guiding principles for the prevention, introduction and mitigation of impacts of alien species that threaten ecosystems, habitats or species. Secretariat of the Convention on Biological Diversity. https://www.cbd.int/invasive.

Convention on Biological Diversity. (2020). Aichi biodiversity targets: Strategic plan for biodiversity 2011–2020. Secretariat of the Convention on Biological Diversity. https://www.cbd.int/sp/targets.

Cooke, S. J., Rytwinski, T., Taylor, J. J., ... Smol, J. P. (2020). On 'success' in applied environmental research – What is it, how can it be achieved, and how does one know when it has been achieved? *Environmental Reviews, 28*, 357–372.

Cooper, C. B., Dickinson, J., Phillips, T., & Bonney, R. (2007). Citizen science as a tool for conservation in residential ecosystems. *Ecology Society, 12*, 11. http://www.ecologyandsociety.org/vol12/iss2/art11/.

Copp, G. H., Vilizzi, L., Wei, H., Li, S., ... Mendoza, R. (2021). Speaking their language–Development of a multilingual decision-support tool for communicating invasive species risks to decision makers and stakeholders. *Environmental Modelling & Software, 135*, 104900. https://doi.org/10.1016/j.envsoft.2020.104900.

Crain, R., Cooper, C., & Dickinson, J. L. (2014). Citizen science: A tool for integrating studies of human and natural systems. *Annu Rev Environ Resour, 39*, 641–645.

Davis, E., Caffrey, J. M., Coughlan, N. E., Dick, J. T., & Lucy, F. E. (2018). Communications, outreach and citizen science: Spreading the word about invasive alien species. *Management of Biological Invasions, 9*(4).

De Groot, M., & de Groot, W. T. (2009). Room for river'' measures and public visions in the Netherlands: A survey on river perceptions among riverside residents. *Water Resources Research, 45*, 1–11.

De Groot, M., Drenthen, M., & De Groot, W. T. (2011). Public visions of the human/nature relationship and their implications for environmental ethics. *Environmental Ethics, 33*, 44.

Doupnik, T. S., & Richter, M. (2003). Interpretation of uncertainty expressions: A cross-national study. *Accounting, Organizations and Society, 28*, 15–35. https://doi.org/10.1016/S0361-3682(02)00010-7.

Drake, D. A. R., & Mandrak, N. E. (2014). Bycatch, bait, anglers, and roads: Quantifying vector activity and propagule introduction risk across lake ecosystems. *Ecological Applications, 24*(4), 877–894. https://doi.org/10.1890/13-0545.1.

Dunn, A. M., & Hatcher, M. J. (2015). Parasites and biological invasions: Parallels, interactions, and control. *Special Issue: Wildlife Parasitology. Trends in Parasitology, 31*, e5. https://doi.org/10.1016/j.pt.2014.12.003.

Dutta, H., & Dutta, A. (2016). The microbial aspect of climate change. *Energy Ecology & Environment, 1*, 209–232. https://doi.org/10.1007/s40974-016-0034-7.

Enquist, C. A. F., Jackson, S. T., Garfin, G. M., ... Shaw, M. R. (2017). Foundations of translational ecology. *Frontiers in Ecology and the Environment, 15*(10), 541–550. https://doi.org/10.1002/fee.1733.

Epanchin-Niell, R. S., Hufford, M. B., Aslan, C. E., Sexton, J. P., Port, J. D., & Waring, T. M. (2010). Controlling invasive species in complex social landscapes. *Frontiers in Ecology and the Environment, 8*(4), 210–216. https://doi.org/10.1890/090029.

Flint, C. G., Kunze, I., Muhar, A., Yoshida, Y., & Penker, M. (2013). Exploring empirical typologies of human–nature relationships and linkages to the ecosystem services concept. *Landscape and Urban Planning, 120*, 208–217.

Foster, R., Peeler, E., Bojko, J., Clark, P. F., Morritt, D., Roy, H. E., Stebbing, P., Tidbury, H. J., Wood, L. E., & Bass, D. (2021). Pathogens co-transported with invasive non-native aquatic species: Implications for risk analysis and legislation. *NeoBiota, 69*, 79–102. https://doi.org/10.3897/neobiota.69.71358.

Francis S. (2014) Perceptions of symmetry and color in environmental logos. Master's thesis, Virginia Polytechnic Institute and State University, Blacksburg VTechWorks. https://vtechworks.lib.vt.edu/handle/10919/48433.

Fricke, R. M., & Olden, J. D. (2023). Technological innovations enhance invasive species management in the anthropocene. *Bioscience, 73*(4), 261–279. https://doi.org/10.1093/biosci/biad018.

Garg, A., & Singh, P. (2018). Floristic diversity of nawabganj bird sanctuary–A wetland in upper gangetic plains of Uttar Pradesh, India. *Indian Journal of Forestry, 41*(3), 273–284.

Glen, A. S., & Hoshino, K. (2020). Social and logistical challenges in managing invasive predators: Insights from islands in Japan and New Zealand. *Pacific Conservation Biology, 26*, 344–352.

Golebie, E., van Riper, C. J., Suski, C., & Stedman, R. (2021). Reducing invasive species transport among recreational anglers: The importance of values and risk perceptions. *North American Journal of Fisheries Management, 41*(6), 1812–1825. https://doi.org/10.1002/nafm.10696.

Golebie, E. J., & van Riper, C. J. (2023). Enhancing aquatic invasive species outreach through values-framed messages. *Environmental Communication, 17*(1), 67–86.

Haley, A. L., Lemieux, T. A., Piczak, M. L., ... Cooke, S. J. (2023). On the effectiveness of public awareness campaigns for the management of invasive species. *Environmental Conservation, 50*(4), 202–211. https://doi.org/10.1017/S037689292300019X.

Hamilton, C., Adolphs, S., & Nerlich, B. (2007). The meanings of risk: A view from corpus linguistics. *Discourse & Society, 18*, 163–181. https://doi.org/10.1177/0957926507073374.

Hess-Erga, O. K., Moreno-Andres, J., Enger, O., & Vadstein, O. (2019). Microorganisms in ballast water: Disinfection, community dynamics, and implications for management. *The Science of the Total Environment, 657*, 704–716. https://doi.org/10.1016/j.scitotenv.2018.12.004.

Hiom, S. (2006). Public awareness regarding UV risks and vitamin D – The challenges for UK skin cancer prevention campaigns. *Progress in Biophysics and Molecular Biology, 92*, 161–166.

Hopkins, J. M., Edwards, W., & Schwarzkopf, L. (2022). Invading the soundscape: Exploring the effects of invasive species' calls on acoustic signals of native wildlife. *Biological Invasions, 24*(11), 3381–3393. https://link.springer.com/article/10.1007/s10530-022-02856-w.

Hopper, A. (2020). How technology can be harnessed for the good of the planet. *The Royal Society Blog (news and views)*. https://royalsociety.org/blog/2020/03/how-technology-can-be-harnessed-for-the-good-of-the-planet/.

Howard, G. W., & Harley, K. L. S. (1997). How do floating aquatic weeds affect wetland conservation and development? *Wetlands Ecology and Management, 5*, 215–225. https://doi.org/10.1023/A:1008209207736.

Hulme, P. E. (2014). Invasive species challenge the global response to emerging diseases. *Trends in parasitology, 30*, 267–270. https://doi.org/10.1016/j.pt.2014.03.005.

IPBES (2023). Thematic assessment report on invasive alien species and their control. *Intergovernmental Science-Policy Platform on Biodiversity and Ecosystem Services*. (https://www.ipbes.net/ias).

Ibabe, I., & Sporer, S. L. (2004). How you ask is what you get: On the influence of question form on accuracy and confidence. *Applied Cognitive Psychology, 18*, 711–726. https://doi.org/10.1002/acp.1025.

Ingole, N. A., Nain, A. S., Kumar, P., & Chalal, R. (2018). Monitoring and mapping invasive aquatic weed salvinia molesta using multispectral remote sensing technique in Tumaria Wetland of Uttarakhand, India. *Journal of the Indian Society of Remote Sensing, 46*, 863–871. https://doi.org/10.1007/s12524-018-0764-4.

John, D., Wu, C., Horta, K., et al. (2006). Top-down, grassroots, and civic environmentalism: Three ways to protect ecosystems. *Frontiers in Ecology and the Environment, 4*(2), 45–51.

Kahan, D. M. (2013). A risky science communication environment for vaccines. *Science, 342*, 53–54.

Kapitza, K., Zimmermann, H., Martin-Lopez, B., & von Wehrden, H. (2019). Factors driving social perceptions of invasive species: A systematic literature review. *NeoBiota, 43*, 47–63.

Kemp, C., van Riper, C. J., BouFajreldin, L., P Stewart, W., Scheunemann, J., & van den Born, R. J. (2017). Connecting human–nature relationships to environmental behaviors that minimize the spread of aquatic invasive species. *Biological Invasions, 19*, 2059–2074.

Kleinman, D. L., Delborne, J. A., & Anderson, A. A. (2011). Engaging citizens: The high cost of citizen participation in high technology. *Public Understanding of Science, 20*(2), 221–240.

Kollmuss, A., & Agyeman, J. (2002). Mind the Gap: Why do people act environmentally and what are the barriers to pro-environmental behavior? *Environmental Education Research, 8*(3), 239–260. https://doi.org/10.1080/13504620220145401.

Lawrence, A. (2017). Grey literature publishing in public policy: Production and management, costs and benefits. In L. Chan, & F. Loizides (Eds.). *Expanding perspectives on open science: Communities, cultures and diversity in concepts and practices* (pp. 85–99). IOS Press.

Li, S., Wei, H., Vilizzi, L., Zhan, A., ... Copp, G. H. (2020). The future of legislation, policy, risk analysis, and management of non-native freshwater fishes in China. *Reviews in Fisheries Science & Aquaculture*. https://doi.org/10.1080/23308249.2020.1782830.

Liang, J. (2023). Igniting public engagement with biodiversity conservation: Exploring the communication of predator free 2050 in Aotearoa New Zealand. *Communication Research and Practice, 9*, 1–16.

Lázaro-Lobo, A., & Ervin, G. N. (2021). Wetland invasion: A multi-faceted challenge during a time of rapid global change. *Wetlands, 41*, 64.

López-Bosch, D., Fernández-Llamazares, Á., Torrent, L., & López-Baucells, A. (2025). How do social media posts enhance public interest in a citizen science project? Insights from the bat monitoring programme. *European Journal of Wildlife Research, 71*(2025), 65. https://doi.org/10.1007/s10344-025-01930-7.

Matos, M. (2014). On controlling invasive species: How to tackle conservation, ethics and communication to the general public. Alfred toepfer natural heritage scholarship study visit report. *Europarc Federation*. https://www.europarc.org/wp-content/uploads/2015/02/ATS-2014_On-controlling-invasive-species-how-to-tackle-conservation-ethics-and-communication-to-the-general-public_Milene-MATOS.pdf.

Matthews, J., Van der Velde, G., Collas, F. P., de Hoop, L., Koopman, K. R., Hendriks, A. J., & Leuven, R. S. E. W. (2017). Inconsistencies in the risk classification of alien species and implications for risk assessment in the European Union. *Ecosphere, 8*, e01832. https://doi.org/10.1002/ecs2.1832.

Milton, K. (2000). Ducks out of water: Nature conservation as boundary maintenance. In J. Knight (Ed.). *Natural enemies: People–Wildlife conflicts in anthropological perspective* (pp. 229–248). *London*: Routledge.

Nakano, S., Miyasaka, H., & Kuhara, N. (1996). Effects of an introduced predatory fish on native stream fishes in Japan. *Conservation Biology, 10*(1), 1–10.

Ogden, J., & Gilbert, J. (2011). Running the gauntlet: Advocating rat and feral cat eradication on an inhabited island—Great Barrier Island, New Zealand. In C. R. Veitch, M. N. Clout, & D. R. Towns (Eds.). *Island invasives: Eradication and management (IUCN)* (pp. 467–471). *Gland*. http://www.issg.org/pdf/publications/Island_Invasives/pdfHQprint/4Ogden.pdf.

Omondiagbe, H. A., Towns, D. R., Wood, J. K., & Bollard-Breen, B. (2017). Stakeholders and social networks identify potential roles of communities in sustainable management of invasive species. *Biological Invasions, 19*, 3037–3049.

Oppel, S., Beaven, B. M., Bolton, M., Vickery, J., & Bodey, T. W. (2011). Eradication of invasive mammals on islands inhabited by humans and domestic animals. *Conservation Biology, 25*, 344–352.

Palmér, C., Wallin, A., Persson, J., Aronsson, M., & Blennow, K. (2023). Effective communications on invasive alien species: Identifying communication needs of Swedish domestic garden owners. *Journal of Environmental Management, 340*, 117995.

Peloza, J., & White, K. (2007). Hey, what gives? The effects of altruistic versus egoistic charity appeals on donation intentions. In G. Fitzsimons, & V. Morwitz (Eds.). Advances in Consumer Research *(Vol. 34* (pp. 347–350). Association for Consumer Research.

Peters, C. B., Schwartz, M. W., & Lubell, M. N. (2018). Identifying climate risk perceptions, information needs, and barriers to information exchange among public land managers. *Science of the Total Environment 616–617,* 245–254. https://doi.org/10.1016/j.scitotenv.2017.11.015.

Petty, R. E., & Cacioppo, J. T. (1986). The elaboration likelihood model of persuasion. In *Communication and persuasion: Central and peripheral routes to attitude change. Springer*, 1–24.

Pfeffer, J., & Sutton, R. I. (1999). Knowing ''What'' to Do Is Not Enough: Turning knowledge into action. *California Management Review, 42*(1), 83–108. https://doi.org/10.2307/41166020.

Pradhananga, A., Davenport, M. A., Seekamp, E., & Bundy, D. (2015). Preventing the spread of aquatic invasive species: Boater concerns, habits, and future behaviors. *Human Dimensions of Wildlife, 20*(5), 381–393. https://doi.org/10.1080/10871209.2015.1030479.

Rahel, F. J., & Olden, J. D. (2008). Assessing the effects of climate change on aquatic invasive species. *Conservation Biology, 22*(3), 521–533. http://doi/epdf/10.1111/j.1523-1739.2008.00950.x.

Raj, S., Kumar, A. B., Tharian, J., & Raghavan, R. (2021). Illegal and unmanaged aquaculture, unregulated fisheries and extreme climatic events combine to trigger invasions in a global biodiversity hotspot. *Biology Invasions, 23*(2021), 2373–2380. https://doi.org/10.1007/s10530-021-02525-4.

Ramsar., n.d. Small Wetlands – Why is it essential to preserve them? Available: https://www.ramsar.org/sites/default/files/2025-05/en-final.pdf (accessed 29 October 2025).

Rania, Q. (2017). Sciencedirect using social hub media to expand public participation in municipal urban plans. *Procedia Engineering, 198*, 34–42. https://doi.org/10.1016/j.proeng.2017.07.072.

Reaser, J. K., Frey, M., & Meyers, N. M. (2020). Invasive species watch lists: Guidance for development, communication, and application. *Biological Invasions, 22*(1), 47–51.

Regan, H. M., Colyvan, M., & Burgman, M. A. (2002). A taxonomy and treatment of uncertainty for ecology and conservation biology. *Ecological Applications, 12*, 618–628. 10.1890/1051-0761.

Ricciardi, A. (2013). Invasive species. In R. Leemans (Ed.). *Ecological systems: Selected entries from the encyclopedia of sustainability science and technology* (pp. 161–178). Springer. 10.1007/978-1-4614-5755-8.

Rogers, R. W. (1975). A protection motivation theory of fear appeals and attitude change. *The Journal of Psychology, 91*(1), 93–114. 10.1080/00223980.1975.9915803.

Rosenthal, R. (1979). The 'file drawer problem' and tolerance for null results. *Psychological Bulletin, 86*, 638–641.

Rothlisberger, J. D., Chadderton, W. L., McNulty, J., & Lodge, D. M. (2010). Aquatic invasive species transport via trailered boats: What is being moved, who is moving it, and what can be done. *Fisheries, 35*(3), 121–132. https://doi.org/10.1577/1548-8446-35.3.121s10539-016-9544-0.

Russel, J. C., Taylor, C. N., & Aley, J. P. (2018). Social assessment of inhabited islands for wildlife management and eradication. *Australasian Journal of Environmental Management, 25*, 24–42.

Sandilyan, S. (2019). Strategies for control and management of some selective Invasive Alien Plant Species Endangering Indian Biodiversity. In *National biodiversity authority*. Centre for Biodiversity Policy and Law (CEBPOL) National Biodiversity Authority ISBN:978-81-940589-8-4.

Sandilyan, S. (2022). Alien fish species in Indian inland wetlands: current status and future challenges. *Wetlands Ecology Management, 30*, 423–437. https://doi.org/10.1007/s11273-022-09870-8.

Santos, L., Reis, P., Costa, F., Esteves, M., Teixeira, R., & Coelho, A. (2020). "Invasive plants"–A serious game to bring awareness about invasive species. *In INTED2020 Proceedings,* 6158–6164. https://doi.org/10.21125/inted.2020.1672 IATED.

Sarkar, U. K., Mishal, P., Borah, S., Karnatak, G., Chandra, G., Kumari, S., Meena, D. K., Debnath, D., Yengkokpam, S., Das, P., & DebRoy, P. (2020). Status, potential, prospects, and issues of floodplain wetland fisheries in India: Synthesis and review for sustainable management. *Reviews in Fisheries Science & Aquaculture, 29*(1), 1–32.

Schoon, M., York, A., Sullivan, A., & Baggio, J. (2016). The emergence of an environmental governance network: The case of the Arizona borderlands. *Regional Environmental Change*. https://doi.org/10.1007/s10113-016-1060-x.

Schwartz, M. W., Fleishman, E., Williamson, M. A., Williams, J. N., & Morelli, T. L. (2021). The use of boundary-spanning organizations to bridge the knowledge-action gap in North America. In C. C. Ferreira, & C. F. C. Klütsch (Eds.). *Closing the knowledge-implementation gap in conservation science: Evidence transfer across spatiotemporal scales and different stakeholders* (pp. 229–254). Springer. https://doi.org/10.1007/978-3-030-81085-6_9.

Seebens, H., Bacher, S., Blackburn, T. M., Capinha, C., Dawson, W., Dullinger, S., Genovesi, P., Hulme, P. E., van Kleunen, M., Kuhn, I., Jeschke, J. M., Lenzner, B., Liebhold, A. M., Pattison, Z., Pyšek, P., Winter, M., & Essl, F. (2020). Projecting the continental accumulation of alien species J, through to 2050. *Global Change Biology, 27*, 970–982. https://doi.org/10.1111/gcb.15333.

Seekamp, E., Mayer, J. E., Charlebois, P., & Hitzroth, G. (2016a). Effects of outreach on the prevention of aquatic invasive species spread among organism-in-trade hobbyists. *Environmental Management, 58*(5), 797–809. https://doi.org/10.1007/s00267-016-0748-5.

Seekamp, E., McCreary, A., Mayer, J., Zack, S., Charlebois, P., & Pasternak, L. (2016b). Exploring the efficacy of an aquatic invasive species prevention campaign among water recreationists. *Biological Invasions, 18*(6), 1745–1758. https://doi.org/10.1007/s10530-016-1117-2.

Schlesinger, W. H. (2010). Translational ecology. Science, 329(5992), 609. https://doi.org/10.1126/science.1195624.

Sharma, G. P., & Raghubanshi, A. S. (2010). How *Lantana* invades dry deciduous forest: A case study from Vindhyan highlands, India. *Tropical Ecology, 51*(2S), 305–316.

Shin, S., van Riper, C. J., Stedman, R. C., & Suski, C. D. (2022). The value of eudaimonia for understanding relationships among values and pro-environmental behavior. *Journal of Environmental Psychology, 80*, 101778. https://doi.org/10.1016/j.jenvp.2022.101778.

Shine, R., & Doody, J. S. (2011). Invasive species control: Understanding conflicts between researchers and the general community. *Frontiers in Ecology and the Environment, 9*(7), 400–406.

Shukla, K. (2022). Understanding biological invasions: A quick read. In T. Ahmad, D. Verma, A. K. Shukla, V. K. Dixit, S. Bajpai, & R. Prasad (Eds.). *Recent frontiers in environment, ecology and biodiversity* (pp. 196–207). Akhand publishing house. ISBN:978-93-90870-77-6 https://www.researchgate.net/publication/371490676_Understanding_Biological_invasions_A_Quick_Read.

Shukla, K. (2025). Aao manayein aakramak prajati jagrukta diwas. In S. Mishra, K. Shukla, & V. P. Singh (Eds.). *Vanyajeevon ki Pukaar* (pp. 50–51). Indian Science Communication Society ISBN:978-81-956248-7-4.

Singh, A. K., & Lakra, W. S. (2011). Risk and benefit assessment of alien fish species of the aquaculture and aquarium trade into India. *Reviews in Aquaculture, 3*(1), 3–18. https://doi.org/10.1111/j.17535131.2010.01039.x.

Singh, A. K., Pathak, A. K., & Lakra, W. S. (2010). Invasion of an exotic common carp, *Cyprinus carpio* L. (Actinopterygii: Cypriniformes: Cyprinidae) in the Ganga River, India and its impacts. *Acta Icth et Piscat, 40*(1), 11–19. https://doi.org/10.3750/AIP2010.40.1.02.

Sinthumule, N. I., & Netshisaulu, K. H. (2022). Wetland resource use and conservation attitudes of rural vs. urban dwellers: A comparative analysis in Thohoyandou, Limpopo Province, South Africa. *Water, 14*(8), 1290. https://doi.org/10.3390/w14081290.

Skillings, D. (2016). Holobionts and the ecology of organisms: Multi-species communities or integrated individuals? *Biology & Philosophy, 31*, 875–892. https://doi.org/10.1007/s10539-016-9544-0.

Stern, P. C., Dietz, T., Abel, T., Guagnano, G. A., & Kalof, L. (1999). A value-belief-norm theory of support for social movements: The case of environmentalism. *Human Ecology Review, 6*(2), 81–97.

Swords, C. M., Porter, J. S., Hawkins, A. J., Li, E., Rowland-Goldsmith, M., Koci, M. D., Tansey, J. T., & Woitowich, N. C. (2023). Science communication training imparts confidence and influences public engagement activity. *Journal of Microbiology & Biology Education, 24*, e00037-23. https://doi.org/10.1128/jmbe.00037-23.

Toppo, M., (2025). Down to Earth. https://www.downtoearth.org.in/environment/nature-writes-back-the-hindi-literary-world-has-seen-a-sea-change-in-environmental-awareness.

USDA. (2023). U.S. Department of Agriculture. National Invasive species information center https://www.invasivespeciesinfo.gov/what-are-invasive-species.

Vander Zanden, M. J., & Olden, J. D. (2008). A management framework for preventing the secondary spread of aquatic invasive species. *Canadian Journal of Fisheries and Aquatic Sciences, 65*(7), 1512–1522. https://doi.org/10.1139/F08-099.

Verbrugge, L. N. H., van den Born, R. J. G., & Lenders, H. J. R. (2013). Exploring public perception of non-native species from a vision of nature perspective. *Environmental Management, 52*, 1562–1573.

Verbrugge, L. N. H., Leuven, R. S. E. W., & Zwart, H. A. E. (2016). Metaphors in invasion biology: Implications for risk assessment and management of non-native species. *Ethics Policy Environmental, 19*, 273–284. https://doi.org/10.1080/21550085.2016.1226234.

Vilà, M., Gallardo, B., Preda, C., García-Berthou, E., Essl, F., Kenis, M., Roy, H. E., & González-Moreno, P. (2019). A review of impact assessment protocols of non-native plants. *Biology Invasions, 21*, 709–723. https://doi.org/10.1007/s10530-018-1872-3.

Wallen, K. E., & Kyle, G. T. (2018). The efficacy of message frames on recreational boaters' aquatic invasive species mitigation behavioural intentions. *Human Dimensions of Wildlife, 23*, 297–312.

Wang, L., Wan, F., & Qian, W. (2023). Research and prospects of environmental DNA (eDNA) for detection of invasive aquatic species in East Asia. *Frontiers in Marine Science, 10*, 1284953.

Warner, K. D., & Kinslow, F. (2013). Manipulating risk communication: Value predispositions shape public understandings of invasive species science in Hawaii. *Public Understanding of Science, 22*(2), 203–218.

Wei, L. (2018). Translanguaging as a practical theory of language. *Applied Linguistics, 39*, 9–30. https://doi.org/10.1093/applin/amx039.

Wisse, A. (2025). Insights: Wetlands and invasive species. *Wildlife Drones*. https://wildlifedrones.net/wetlands-and-invasive-species/.

Woitowich, N. C., Hunt, G. C., Muhammad, L. N., & Garbarino, J. (2022). Assessing motivations and barriers to science outreach within academic science research settings: A mixed-methods survey. *Front Communication, 7*. https://doi.org/10.3389/fcomm.2022.907762.

Wolf, J. M. (2017). The multipurpose tool of social media: Applications for scientists, science communicators, and educators. *Clinical Microbiology Newsletter, 39*, 75–79. https://doi.org/10.1016/j.clinmicnews.2017.04.003.

Chapter 17

Fisheries and aquaculture in wetland ecosystems: assessing benefits, risks, and future prospects

Shumaila Batool[1], Urwah Sheikh[1], Khadija Batool[1], Anam Shehzadi[1], Neha Anees[1], and Sadia BiBi[2]
[1]*Department of Zoology, Wildlife and Fisheries, University of Agriculture, Faisalabad, Pakistan,* [2]*College of Life Sciences and Technology, Tarim University, Xinjiang, P.R. China*

17.1 Introduction

Wetlands, the transitional areas between aquatic and terrestrial environments, cover 6% of Earth's surface (917 to over 1275 million hectares) and water saturation predominantly influences soil formation and the composition of animal and plant groups inhabiting the region (Balwan & Kour, 2021). This chapter considers all major wetland types, including inland freshwater wetlands (e.g., swamps, marshes, peatlands) and coastal systems (e.g., mangroves, tidal flats), recognizing their diverse yet interconnected ecological functions.

Wetlands are among the most prolific ecosystems in the biosphere and significantly contribute to the ecological sustainability of an area. They play a crucial role in preserving biodiversity, mitigating flooding, safeguarding coastal regions from erosion, serving as nurseries for terrestrial and aquatic-terrestrial organisms, and enhancing and maintaining water quality. These ecosystem services support human livelihoods and food security (Gumbricht et al., 2017). The primary factors contributing to the degradation of wetlands include rapid urbanization and infrastructure expansion, overexploitation of land, conversion to agricultural use, pollution from agricultural and industrial sources, variability, and climate change. According to WWF's Living Planet Index, populations of wetland-dependent species have declined by 76% since 1970 (Almond et al., 2020), threatening both ecological integrity and economic stability.

Fisheries and aquaculture are intrinsically linked to wetland health. Wetlands provide essential spawning, feeding, and nursery habitats for fish, crustaceans, and mollusks, supporting capture fisheries that supply high-quality protein, essential omega-3 fatty acids, and micronutrients such as selenium, iodine, and phosphorus (Farmery et al., 2022). Aquaculture—ranging from onshore ponds to inshore pens and offshore cages benefits from wetland-associated waters due to their nutrient richness and ecological support (Asche & Smith, 2018). Sustainable aquaculture can complement capture fisheries, reduce pressure on wild stocks, and even aid ecosystem restoration through integrated multitrophic systems and mangrove–aquaculture integration.

Yet, overfishing, habitat conversion, invasive species, and aquaculture disease outbreaks compound the challenges posed by wetland degradation. Effective management must therefore integrate wetland conservation with sustainable fisheries and aquaculture practices, ensuring ecological resilience, equitable livelihoods, and long-term food system sustainability. This approach aligns with global blue economy strategies, recognizing wetlands as critical socioecological assets in the face of climate change.

17.2 Wetland ecosystems

17.2.1 Definition and types of wetlands

Wetlands are complex ecosystems characterized by soil flooding, resulting in low-oxygen environments that support a specific assemblage of plants, animals, and microorganisms that have evolved to tolerate periods of slow-moving or

Wetlands: Biodiversity, Health and their Role in Ecosystem Services. DOI: https://doi.org/10.1016/B978-0-443-34222-6.00002-4

stationary water. Bogs, swamps, marshes, and other comparable ecosystems are typical classifications for wetlands (as shown in Fig. 17.1) based on their soil and plant life.

Wetlands are present globally across all biomes. Certain wetlands, like tidal marshes, exemplify a transitional zone, as they exist at the interface between open water and land. Other formations, including the prairie potholes of central North America and the Carolina bays (elliptical depressions) of the Atlantic Coastal Plain, primarily receive water from groundwater or precipitation and are not linked to a specific body of water (De Steven & Harrison, 2022).

Wetlands are predominantly found in boreal and tropical climates. However, diverse coastal and inland wetlands also exist in temperate areas. This distribution typically results from conditions that lead to a surplus of water. The Peruvian Amazon and Bornean peatlands are part of the tropical rainforest biome (Ribeiro et al., 2021). In the treeless tundra of Canada and Alaska, waterlogged and inundated marshes rest upon permafrost. Numerous sections of the terrain consist of peatlands formed by *Picea glauca* (white spruce) and *Picea mariana* (black spruce), which are nourished by precipitation or the thawing of snow (Anderson, 2016). Wetlands also exist in the arid desert biome, such as the Mesopotamian marshlands located at the junction of the Euphrates River and Tigris (AlMaarofi, 2015). In temperate regions, wetlands are generally located adjacent to beaches, lakes, rivers, or other areas where local water input surpasses output.

17.2.1.1 Classification of wetland

Wetlands have been categorized in various ways for different reasons. Marine, riverine, estuarine, palustrine, and lacustrine ecosystems formed the basis of the USFWS's hierarchical approach in 1979's Classification of Wetlands and Deepwater Habitats of the United States (Chakraborty et al., 2023a). Similarly, the USFWS model served as the basis for the Ramsar Convention's categorization system, which added a cultural wetland type that humans created (McKee et al., 2018). Other classification systems are more uniformly based on structure, function, or management goals when it comes to wetlands. Structure refers to the physical appearance of the wetlands, while function describes the hydrologic regimes and the wetlands' position within them.

17.2.1.1.1 Coastal systems

17.2.1.1.1.1 Mangroves Mangroves are located between 32°N and 38°S latitude in subtropical and tropical coastal regions. They are sensitive to low temperatures and are typically found in areas that do not experience severe frosts. The average temperature of the air during the coldest months remains above 20°C. Particular mangroves, however, may endure temperature declines to 7°C or below (Ward et al., 2016). Mangrove populations located farthest from the Equator are in areas heated by equatorial currents or in somewhat fresher estuarine waters, where diminished salt stress may allow the trees to endure lower temperatures.

Water currents carry mangrove seeds, and environments conducive to mangrove expansion are somewhat shielded from wave action, where reduced energy conditions facilitate the establishment of mangrove seedlings. The most extensive mangrove areas are found in humid deltaic regions, including the Sundarbans in the Ganges delta of Bangladesh and India, the Niger Delta in Cameroon and Nigeria, and the Gulf of Paria and the Orinoco deltas along Venezuela's eastern coast. Additional beaches featuring practically uninterrupted mangrove stands encompass the West African coast, the deltaic coast of southern Papua New Guinea, and the northern Brazilian coast (Rieley & Page, 2016).

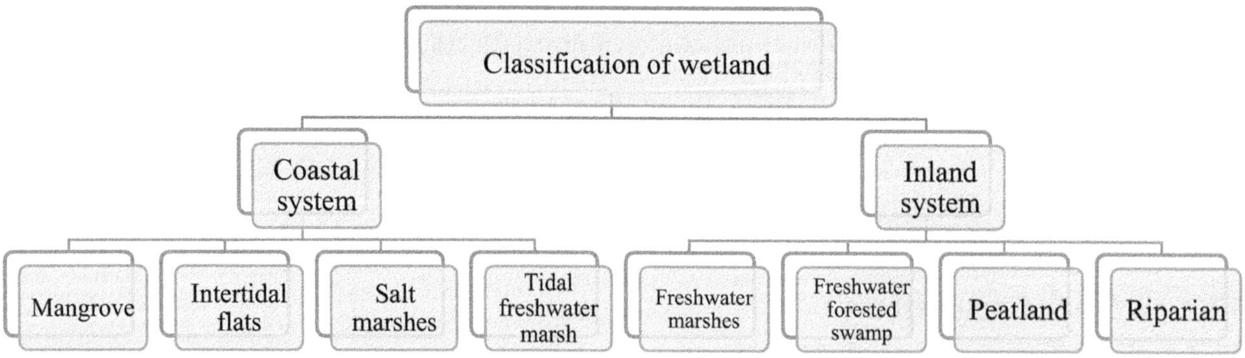

FIGURE 17.1 Diagrammatic representation of wetland types.

17.2.1.1.1.2 Salt marshes Salt marshes are most common along boreal and temperate coastlines where silt deposits and mangroves do not predominate; nevertheless, they can also be found in the Arctic and tropical climates (Alongi & Alongi, 2018). Salt marshes thrive in areas where sediment accumulation equals or exceeds the rate of land subsidence and there is appropriate protection from disruptive waves and storms. Some inland wetland systems become saline when evapotranspiration's rate (the combined loss of water from soil and plant) increases the precipitation rate or when they come into contact with saline groundwater. In locations with saline groundwater, such as eastern Nebraska, flora resembles that of coastal salt marshes (Finkl & Makowski, 2021).

17.2.1.1.1.3 Tidal freshwater marshes These wetlands occur in extensive river systems worldwide, ranging from subarctic regions to the Equator. The most extensive areas of these wetlands are found at temperate latitudes. Tidal freshwater marshes occur in sections of river systems sufficiently near the shore to undergo considerable tidal influence while remaining free from salinity. The mouths of these rivers downstream undergo tidal fluctuations. In contrast, the constriction of upstream river channels limits the volume of the incoming tide, resulting in a more significant alteration in water depth upriver compared to the coastal area (Hoitink & Jay, 2016).

This kind of marsh is common in the main river basins on the west coast of North America and along the eastern and Gulf coasts. The Thames and Rhine rivers in Europe, the Yellow River (Huang He) in Asia, and the St. Lawrence River between the United States and Canada are other significant river systems that contain enormous tidal freshwater marshes (Chen et al., 2024).

17.2.1.1.1.4 Intertidal flats Intertidal flats are predominantly flat, unvegetated regions between deeper aquatic zones and tidal marshes. They inhabit somewhat sheltered regions with minimal wave activity; however, they are affected by wind, currents, and waves. Their sediments are too unstable, and the intensity and length of flooding are insufficient to establish communities of big plants (Paterson et al., 2019).

17.2.1.1.2 Inland systems

17.2.1.1.2.1 Freshwater marshes The primary characteristic that unites these varied wetland ecosystems is that they are all freshwater systems not bordered by forests and are characterized by the dominance of aquatic plants such as grasses and sedges. Freshwater wetlands do not accumulate peat. They vary in geological origins and hydrological factors, ranging in size from small pothole marshes under one hectare to the vast sawgrass plains of the Florida Everglades (Batzer & Sharitz, 2014).

Freshwater marshes are found globally in low-lying regions where water accumulates, or where an impervious soil or geological layer causes water to pool. This assemblage of wetlands encompasses the vernal pools, prairie potholes and playas of North America, in addition to the vegetated margins of small lakes, the coastal lagoons situated behind barrier island beaches, and the delta marshes of tributary rivers feeding substantial lakes, such as the Great Lakes of North America (McKee & Baldwin, 1999). The group encompasses marsh systems that provide significant seasonal influxes of freshwater to more saline regions, exemplified by the Everglades in South Florida, where a sheet of freshwater flows from Lake Okeechobee to the ocean during the wet season (Batzer & Sharitz, 2014).

17.2.1.1.2.2 Peatlands Peatlands, also known as moors in Europe and mires when they are actively creating peat, form in regions where peat, the organic matter, accumulates each growing season due to factors that slow down plant decomposition. Two forms of peatlands, fens and bogs, have been researched extensively in high latitudes. Bogs form in low-nutrient depressions that are primarily fed by rainfall, whereas fens form on flats, depressions, or slopes due to prolonged flows of mineral-rich groundwater into the root zone. Bogs or fen areas may mix to form a blanket across a vast area.

Tropical peatlands are located in the lowland regions of Southeast Asia, East Asia, the Caribbean, Central America, Africa, and South America. The most extensive areas of these peatlands are located in Indonesia, encompassing 10%–12% of the nation. Certain peatlands are tropical bogs found in Southeast Asia and tropical regions of South America and Africa; nevertheless, they have not been as extensively researched as in higher latitudes (Rieley & Page, 2016).

17.2.1.1.2.3 Freshwater forested swamps The prevalence of woody vegetation, including trees, characterizes freshwater forested wetlands. These wetlands are globally \distributed. In the tropics, extensive swamps adjacent to significant rivers frequently flood them for extended periods. The entire Atlantic coastline of the temperate United States is notably abundant in wetlands. Freshwater forested marshes dominated by *Acer rubrum* (red maple) are in the previously glaciated northeastern United States (Rheinhardt, 2007). Coastal swamps, extending from the coast of Maine to the Gulf Coast, are predominantly characterized by *Chamaecyparis thyoides* (Atlantic white cedar). These swamps, located in areas drier than deepwater swamps, experience flooding in winter and for an extended duration in spring. Red maple wetlands, conversely, endure reduced flooding.

Deepwater swamps dominated by *Cupressus* (cypress) and *Nyssa* (tupelo) trees are located along the Mississippi River, from Delaware to Texas, and extend north to Illinois. Nonetheless, they predominantly arise along the expansive, winding rivers of the Atlantic Coastal Plain. The flat geography of this plain, which impedes the drainage of rivers into the ocean, along with the rising sea levels subsequent to glacial retreat, has resulted in advantageous wet conditions. Deepwater cypress-tupelo swamps are typically submerged 90%–100% of the year, experiencing low to moderate current strength (Mitsch & Gosselink, 2015).

17.2.1.1.2.4 Riparian wetlands Riparian wetlands are sometimes referred to as riverine or floodplain wetlands. They interact laterally with nearby aquatic and terrestrial systems, as well as with upstream and downstream channels. Alongside riparian wetlands are rivers and streams that occasionally overflow their banks, causing floods to which the wetland flora has adapted. Additionally, they are situated where a narrow channel provides fresh opportunities for plants to establish themselves and thrive.

Riparian wetlands exist globally and exhibit various morphologies across locations due to climatic and geographical influences. Large, nearly flat alluvial valleys, such as those in Bangladesh, the Amazon Basin of South America, and the floodplains of major rivers like the Mississippi in the United States, are examples of riparian ecosystems. In arid locations, riparian ecosystems may consist of narrow strips of vegetation along streambanks that are susceptible to instability and flooding. In mountainous areas like the Pacific Northwest, riparian systems may be limited along steep headwater streams but can expand into extensive floodplains in the lowland areas (Larsen & Alp, 2015).

17.2.2 Ecological functions and services

Wetlands are essential for the integrity of the ecosystem. They are known as "The Kidneys of the Landscape" and play a crucial role in the hydrological cycle, preserving drinking water quality, enhancing groundwater recharge, and acting as organic filters. They also enhance food production by facilitating the movement and delivery of sediment and nutrients into estuaries or river floodplains. Wetlands' ability to act as organic filters has reduced the amount of wastewater from fisheries and aquaculture, contributing to pollution management, as shown in Fig. 17.2.

Coastal wetlands serve as important carbon reservoirs. They sequester significant quantities of carbon in their soils and flora. Carbon sequestration directly influences atmospheric CO_2 concentrations and global climate dynamics, including

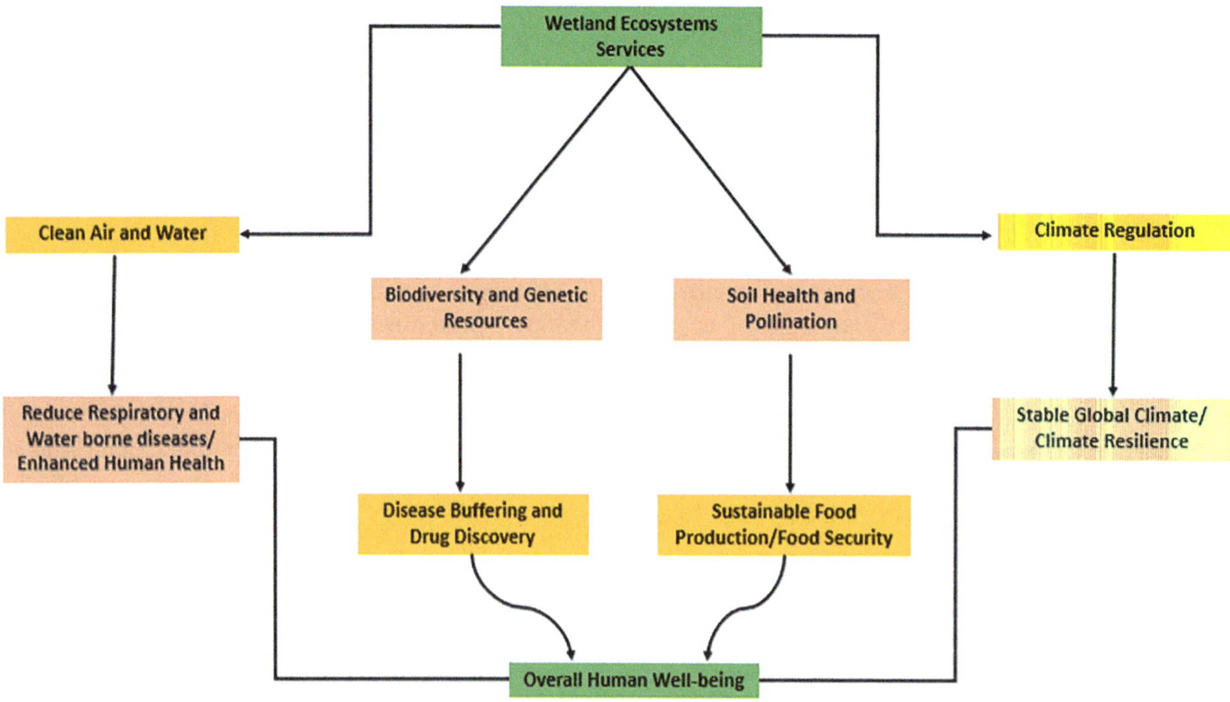

FIGURE 17.2 Diagrmmatic representation of wetland ecosystem services linkage to climate outcomes and human well-being.

sea-level elevation and ocean temperatures. The function of wetlands in carbon sequestration is important concerning marine climate change. Wetlands are also significant for mass tourism and ecotourism, as they enhance urban air quality, provide humid air, and mitigate heat waves, allowing for recreational activities such as boating, fishing, hunting, and tourism as shown in (Fig. 17.3). Esthetic preferences and coastal wetlands indirectly improve coastal resilience by providing financial incentives for wetland protection and restoration (Fig. 17.3).

17.2.3 Biodiversity in wetlands

Wetlands are typically rich in biodiversity, often much more so than many other environments. Extended waterlogging of substrates results in oxygen deficit (hypoxia) or complete absence (anoxia) in the soil, leading to various chemical alterations in soil properties. This makes the environment inappropriate for terrestrial plants, yet numerous herbaceous and woody species are well adapted to wetland conditions. Hydrological factors and water quality directly or indirectly govern various biological processes affecting biodiversity. The morphology of wetland basins, coupled with significant seasonal water-level fluctuations, leads to considerable niche diversification, resulting in a more dynamic and diverse array of flora and fauna (Semeraro et al., 2015).

Wetland flora comprises representatives from all taxonomic categories, ranging from unicellular algae to bryophytes, mosses, ferns, and woody angiosperms. A superficial examination of a wetland can be deceptive since it may reveal the dominance of only one or two plant species, such as *Typha*, *Phragmites*, *Cyperus*, or *Scirpus* (Zhang et al., 2020). A thorough investigation is necessary to ascertain the complete diversity of flora and fauna. In addition to many submerged, free-floating, rooted, floating-leaved, and emergent plants that occupy distinct niches, countless planktonic and filamentous algae also inhabit diverse wetlands. A substantial array of fungi and bacteria significantly contributes to wetland activities.

The faunal diversity includes representatives from nearly all taxonomic groupings, ranging from protozoa to mammals. Wetland fauna consists of two categories of animals: (1) those that rely exclusively on wetlands for their whole existence, and (2) those that utilize wetlands for a portion of their life cycle or specific requirements. There are six types of wetland animals: (1) native to the area, (2) that migrate there regularly from deepwater habitats, (3) that migrate there from terrestrial uplands, (4) that migrate there from other wetlands, (5) that visit occasionally, and (6) that depend indirectly on wetland biota, such as canopy insects. Various fish species, amphibians, reptiles, birds, and insects rely on environments with distinct hydrological properties for feeding, breeding, nesting, or during different life cycle stages. These animals may inhabit wetlands permanently or migrate seasonally from other habitats to wetlands. Numerous animals frequently

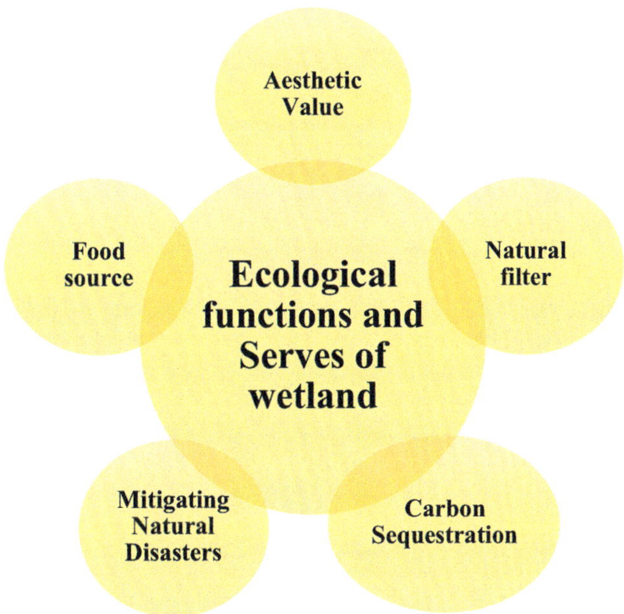

FIGURE 17.3 Schematic representation of ecological functions and services of wetlands.

utilize wetlands directly (e.g., grazing) yet do not inhabit them. They remain to rely on other wetland organisms (Knapp et al., 2019).

Waterfowl are widely recognized and extensively studied elements of wetland biodiversity. Their diversity correlates directly with habitat diversity, which offers a range of roosting, nesting, feeding, and breeding places, along with various food sources, including planktonic algae, seeds, insects, and fish.

Among invertebrates, arthropods, including crustaceans and 11 orders of insects, and mollusks are the predominant elements of wetland fauna. Oligochaetes are prevalent as well. Among vertebrates, amphibians, fish, and birds are predominant. Numerous species, including beavers, otters, and rhinoceroses, inhabit wetlands exclusively, while others, such as swamp deer and water buffalo, primarily utilize wetlands for grazing (Hu et al., 2017). Most terrestrial insects in their adult form undergo larval stages in wetlands; numerous terrestrial birds consume aquatic organisms; and the existence of certain marine species, including turtles, crocodiles, and countless fish, is contingent upon wetlands. Long-distance migratory birds, which frequently traverse continents, include waterfowl, turtles, and fish.

17.3 Fisheries in wetland ecosystems

Wetlands provide a feeding and breeding environment for most fish and crustacean species. Wetlands offer abundant nourishment, attracting numerous fish species and serving as nurseries and breeding grounds for several commercially significant fish species. This, in turn, safeguards biodiversity by shielding the young from predators. The roots and trunks of mangroves provide a substrate for organisms such as oysters to proliferate. Their roots also capture small particles, resulting in soft soils optimal for burrowing mollusks and crustaceans. Some fish ingest plant matter, while others feed on crustaceans and small insects that inhabit or feed on plants or the nutrient-rich organic material produced from the decomposition of wetland plant matter by bacteria, worms, and aquatic organisms. Numerous essential species inhabit unspoiled and distinctive environments. This habitat supports both capture fisheries and aquaculture activities. They are regarded as "biological supermarkets," hosting significant food webs and a diverse array of animals (Ghosh et al., 2024).

The intricate combination of fresh and saltwater coastal wetlands yields a rich assortment of microbes, invertebrates, and algae, establishing the foundation of the food web. Thus these wetlands underpin essential recreational and commercial fisheries, bolstering economies and food security. The safeguarding and conserving of coastal wetlands are crucial for maintaining the ongoing fertility of these vital breeding habitats for aquatic organisms. Robust coastal wetland ecosystems sustain a variety of fauna and flora, including species adapted to saline environments (Sarkar et al., 2020). This diversification augments environmental resilience, as various species may respond differently to fluctuating conditions, helping preserve ecosystem stability.

17.3.1 Traditional fishing practices

Fishing is a primordial activity integral to human existence for more than 40,000 years (Raju et al., 2016). In many regions of the world, local communities had to be proficient at fishing because it was their main source of sustenance. Throughout the centuries, various cultures devised unique fishing strategies to thrive in their specific settings (Aarif et al., 2017).

17.3.2 Modern fisheries techniques

These are the methods employed in today's world. They comprise the utilization of intricate nets such as gillnets, cast nets, bay nets, drift nets, and encircling nets. Advanced equipment is also being utilized. This includes motorized equipment now used for commercial fishing. Diesel marine engines, ships, and trawlers are specifically designed and utilized for fishing expeditions lasting 2–3 days to preserve catches. The following (Tables 17.1 and 17.2) shows the most common fishing nets being used nowadays.

17.3.3 Economic importance of wetland fisheries

The primary wetland functions essential for fish survival include refuge, spawning and nursery habitats, food production, and mitigating hazardous contaminants in water (Ringler & Cai, 2006).

17.3.3.1 Food production

Wetlands offer an abundant food source due to their rapid plant growth. Certain fish consume plant matter, while others feed on microscopic crustaceans and insects that inhabit or feed on plants. Specific fish favor decomposing wetland plant matter, transformed by bacteria, aquatic insects, and worms into a nutrient-dense organic substance called detritus. Packed with

TABLE 17.1 Overview of some common traditional fishing practices.

Name	Description	References
Spearfishing	Catching fish with spears or similar sharp tools	Sbragaglia et al. (2023)
Angling	Traditional fishing method using a hook and line; line fishing and rod angling	Beverly et al. (2003)
Ice fishing	Fishing by spearing or angling through a hole in the ice on a frozen aquatic surface	Lawrence et al. (2022)
Trapping	Stone or wooden weirs trap fish for capture by hand or with a spear	Raju et al. (2016)
Hand gathering	Manual fish collection methods, including catfish noodling and trout tickling	Getu et al. (2015)
Netting	Netting is used to increase the fishing yield, which involves capturing fish with nets, such as casting nets and gillnets	Olaniyan (2015)
Kite fishing	An activity of fishing using a kite to suspend bait over water for fish capture, usually in inaccessible waters	Aarif et al. (2017)

TABLE 17.2 Summarize the most commonly used fishing nets.

Fishing nets	Descriptions	References
Scine net	This is a type of net in which one end is attached to a base and the other can be moved to cover a particular area. After that, the net is pulled back to keep the fish inside the marked area	He et al. (2021)
Trap net	This net includes a series of wooden stakes embedded in the seabed and the trap positioned at the terminal end	Lucchetti et al. (2023)
Gill net	This rectangular netting is fastened at the top and secured at the bottom with a foot rope. Floats buoy the headline, while the foot-pole is burdened with lead and stones	Northridge et al. (2017)
Bag net	A bag net features an extensive vertical wall of netting, typically several hundred meters long, positioned perpendicular to the coastline	Okwuosa et al. (2018)
Drift net	This net comprises a set of lightweight rectangular nets connected end-to-end, forming an extensive vertical curtain that suspends freely in the water	Ichii et al. (2017)
Cast net	This is a conical-shaped net with small weights and stones hung around its edges	Kim et al. (2020)
Hooks and lines	This is a most straightforward net consisting of a line and multiple hooks. Bait is placed on fixed hooks	Cooke and Danylchuk (2020)
Surrounding net	It consists of large walls of netting that surround the fish from the sides and bottom	He et al. (2021)

essential nutrients for fish growth, detritus circulates throughout the aquatic system. Wooded floodplains' cyclical inundation and desiccation enhance vegetative growth, detrital output, and nutrient accessibility. These products are subsequently washed downstream, benefiting distant fish species, such as menhaden, which consume one-third of the detritus in their diet.

Moreover, fish need not consume detritus directly to benefit from this substantial wetland food supply. Striped bass eat smaller fish and shellfish that feed on trash. Humans depend on wetlands for sustenance. While a platter of steaming cattails may not be attractive to most, many would relish the taste of grilled largemouth bass. Largemouth bass do not directly ingest marsh vegetation or debris; they prey on red swamp crawfish. Red swamp crawfish ingest detritus. The interplay between wetland vegetation producing detritus, red swamp crawfish that feed on detritus, largemouth bass that prey on red swamp crawfish, and people that consume largemouth bass illustrates a food chain (Mwakubo et al., 2007). Wetland vegetation supports fish and many eaters, such as birds, reptiles, bears, otters, sea lions, and people, via interconnected feeding chains known as food webs.

17.3.3.2 Spawning and nursery grounds

Fish eggs and juvenile fish possess distinct requirements compared to adult fish, prompting many adult fish to inhabit alternate environments before spawning in wetlands. Vulnerable and immobile eggs can be hidden from predators in marsh margins and submerged flora. Wetland flora, decomposed organic matter, and debris provide substrates for fish to attach their eggs (Graff & Middleton, 2001). Upon hatching, the vegetation functions as both a protective shield and a source of nourishment. Young fish rapidly seek refuge in wetland vegetation for camouflage. In contrast, the juvenile stages of bay scallops, hard clams, and some other shellfish attach to salt marsh vegetation and seagrasses for several weeks before settling on the substrate. The predominant shrimp harvested in the Gulf of Mexico depend on salt marshes for nursery habitats. The vegetation protects young shellfish until they reach a size large enough for life on the estuary floor. Shallow waters in wetland ecosystems shelter juvenile fish from adult predators, as many adults cannot thrive in such habitats. Wetlands serve as prolific spawning grounds for fish. Northern pike migrates from the profundities of Lake Erie to reproduce in freshwater marshes adjacent to the seashore (Chen et al., 2023). Adults vacate the marshes soon after spawning, whereas larvae and juveniles remain under the protective vegetation until midsummer, when the shallow waters become too warm. Anadromous fish, such as river herring and salmon, migrate significant distances from saltwater to freshwater streams and nearby wetlands for reproduction. Blueback herring spawn when inundation grants access to abundant food supplies in riparian hardwood forests. Many salmon species lay their eggs in the gently flowing waters of streambeds, while the larvae and juveniles use the plentiful food supplies available in streamside wetlands during flooding. Specific populations of chum salmon spawn directly in coastal wetlands (Table 17.3). Striped bass migrate to freshwater areas for spawning, enabling juveniles to exploit tidal freshwater wetlands as nurseries and feeding habitats.

17.3.3.3 Shelter from predators

Both adult and adolescent fish utilize wetlands as a refuge from predators. Thick foliage can hinder predators' sight and hide little fish. Northern pike, Juvenile muskellunge, and other small fish with green speckled coloring can blend into marsh grasses, successfully avoiding identification by predators (Davis et al., 2012). Dense vegetation and shallow waterways impede many predators from reaching coastal marshes and other wetlands. Gobies, juvenile snook, anchovies, and immature spotted seatrouts utilize the complex root systems of mangrove wetlands for protection against larger predators. The root systems of trees and shrubs in floodplain wetlands allow streambanks to overhang the water, providing a safeguarded habitat for cutthroat trout, chinook salmon, and many fish species. Fish employ wetlands not just to avoid predators but also to find shelter from variations in environmental conditions, such as changes in water level, flow rate, or adverse weather. Coho salmon rely on the serene waters of forested wetlands adjacent to streams to escape the rapid currents of Pacific Northwest rivers during winter floods (Whitfield, 2017).

17.3.3.4 Purified water

Clean water is essential for fish, just as clean air is for humans. Wetlands cleanse water by removing potentially harmful pollutants. Specific contaminants may be sequestered by wetland flora and then retained within sedimentary layers. Wetland plants and microorganisms convert other pollutants into less dangerous forms. Wetlands offer natural filtration that is significantly more cost-effective and efficient than technological alternatives. Wetlands mitigate the impacts of natural fluctuations in water quality on fish populations (Haarstad et al., 2012). The stems, roots, and leaves of wetland vegetation impede water movement, reducing erosion along streambanks and coastlines. Mitigating erosion decreases the silt load in the water, which can obstruct fish gills or suffocate fish eggs. Lake Erie marshes capture silt, benefiting species that require clear water and sandy or gravel substrates for spawning. Forested wetlands control the influx of freshwater into downstream estuarine and brackish nursery zones. Without this essential function, salinity levels would vary considerably with each downpour, placing stress on fish and other aquatic organisms that rely on a fragile balance between freshwater and saltwater (Scholz, 2023).

TABLE 17.3 Some important fish species found in various wetlands.

Wetland	Fish species	References
Mangrove	Dwarf seahorse (*Hippocampus zosterae*) Gray snapper (*Lutjanus griseus*) Mangrove rivulus (*Rivulus marmoratus*) Ladyfish (*Elops saurus*) Spotfin mojarra (*Eucinostomus argenteus*) Tarpon (*Megalops* spp.)	Graff and Middleton (2001)
Intertidal flats	Northern clingfish (*Gobiesox maeandricus*) Monkeyface eel/prickleback (*Cebidichthyes violaceus*) Epaulette shark (*Hemiscyllium ocellatum*)	Koyama (2022)
Salt marshes	Atlantic silverside (*Menidia menidia*) Black drum (*Pogonias cromis*) Bay anchovy (*Anchoa mitchilli*)	Teixeira et al. (2014)
Tidal freshwater marshes	American shad (*Alosa sapidissima*) Chain pickerel (*Esox niger*) Pacific lamprey (*Lampetra tridentata*) Common snook (*Centropomus undecimalis*) Channel catfish (*Ictalurus punctatus*)	Koyama (2022)
Freshwater marshes	Bluegill (*Lepomis macrochirus*) Muskellunge (*Esox masquinongy*) White bass (*Morone chrysops*) Northern pike (*Esox lucius*) Yellow bullhead (*Ameiurus natalis*) Minnows (Cyprinidae)	Veitch and Sawynok (2005)
Freshwater forested swamp	Alewife (*Alosa pseudoharengus*) Atlantic needlefish (*Strongylura marina*) Bowfin (*Amia calva*) Crawfish (*Procambarus* spp.) Darters (*Etheostoma* spp.) Paddlefish (*Polyodon spathula*)	Ho et al. (2016)
Peatland	Snakehead (*Channa striata*) Kissing gourami (*Helostoma temminckii*) Catfish (*Clarias batrachus*) Nile tilapia (*Oreochromis niloticus*)	Thornton et al. (2016)
Riparian	Taillight shiner (*Notropis maculatus*) Quillback (*Carpiodes cyprinus*) Spotted gar (*Lepisosteus oculatus*) Pugnose minnow (*Opsopoeodus emiliae*)	Dala-Corte et al. (2016)

17.4 Benefits of fisheries and aquaculture in wetlands

17.4.1 Economic benefits

17.4.1.1 Employment and economic opportunities

Wetland fisheries and aquaculture are vital for employment and sustenance, particularly in rural and underdeveloped regions. These activities generate direct employment in aquaculture, propagation, and collection, providing secure incomes to millions of laborers. Indirect job opportunities may emerge in related sectors such as fish processing, packing, transportation, and equipment production. Women are essential, especially in postharvest tasks such as marketing, processing, and cleaning, which give them economic influence and increase household income. Fisheries and aquaculture mitigate rural poverty, enhance living standards, and bolster economic stability in wetland-dependent communities by generating different employment options (Sarkar et al., 2020).

17.4.1.2 Efficient resource management

Sustainable fisheries and aquaculture in wetlands enable the enduring vitality of these ecosystems while optimizing their productivity. Wetlands possess high biodiversity, offering habitats for numerous aquatic species, and their careful management helps preserve this natural resource. Methods like rice-fish farming and aquaponics amalgamate aquaculture with agriculture, enhancing resource utilization and recycling nutrients to minimize waste (Saini et al., 2024). Moreover, aquaculture mitigates the strain on wild fish populations by providing a regulated alternative, facilitating the recovery of natural ecosystems. By-products from aquaculture activities can be utilized as fertilizer, fostering a circular economy. Regulatory measures, including fishing quotas and seasonal prohibitions, are crucial to avert overexploitation and guarantee the sustainable utilization of wetland resources for future generations.

17.4.1.3 Recreation and tourism

Communities receive significant revenue from the tourism and recreation industries, which rely on wetlands for fisheries and aquaculture. Recreational fishermen, birdwatchers, and ecotourists interested in water habitats frequent these regions. Wetland fisheries are already popular destinations for travelers looking for authentic cultural experiences, but the traditional fishing methods and cultural festivities that surround them just add to the allure. Recreational fishing, sometimes known as sport fishing, is a form of low-impact tourism that generates revenue through guided tours, equipment rentals, and licenses (Aazami & Shanazi, 2020). Investments in tourist infrastructure, including lodging and entertainment venues, boost local economies and generate more employment opportunities. Also, by reinvesting tourist spending in wetland conservation, we can strengthen fisheries and aquaculture and raise environmental awareness among tourists.

17.4.2 Nutritional and food security

Fisheries significantly contribute to the global food economy. Fisheries employ over 200 million individuals who rely directly on ocean fishing for their sustenance. Fish serves as the principal protein source for over 950 million individuals globally and constitutes a significant component of the diet for many others. Numerous researchers have demonstrated the correlation between population expansion and overall fish consumption. In under 50 years, the global average per capita fish intake has nearly doubled (Obiero et al., 2019). Worldwide, fish constitutes around 16% of the animal protein consumed by humans and is a significant source of minerals and essential fatty acids. Fish is the principal source of omega-3 fatty acids in the human diet. Omega-3 fatty acids are essential nutrients for the proper development of newborns' brains and eyes, and they play a protective function in various human diseases, including lupus, cardiovascular disease, depression, and other mental disorders. Asia leads in capture aquaculture and fisheries production. Most expansion in the fisheries sector is anticipated to occur in emerging nations, representing 79% of global food fish output. The link between fish yields and food safety suggests that traditional large-scale fish farming usually produces positive results (Ghosh et al., 2024). Fish production significantly contributes to food security by enhancing overall food supplies for the population. A significant consequence to consider is aquaculture's role in improving food security for underprivileged populations, especially those vulnerable to malnutrition.

17.4.2.1 Fisheries' role in food security

Fisheries are essential for food security, supplying animal protein to over 1 billion individuals worldwide. In low-income nations, fish constitute 22% of animal protein intake. Small-scale fisheries are particularly significant for nutrition as they are consumed in their entirety, supplying calcium and many vitamins. Aquaculture can enhance food security and nutrition through multiple avenues, including increased local food availability, job creation, and higher fish consumption (Boyd et al., 2022).

Aquaculture within small-scale integrated farming systems can yield high-quality animal protein and other nutrients at economical costs, generating "own enterprise" job opportunities for women and children and facilitating income through the sale of high-value goods. It also aids the landless by employing shared resources such as finfish cage culture, mollusk cultivation, and fisheries enhancement in communal aquatic environments.

Aquaculture enhances farm productivity and sustainability by utilizing agricultural by-products as fertilizers and feed. Aquaculture in rice paddies aids in integrated pest management and the control of human medical vectors. Ponds serve as crucial on-farm water reserves for irrigation and animals in regions experiencing seasonal water shortages.

TABLE 17.4 Represents some factors and their impact on wetlands

Factors affecting the wetlands	Negative impacts	References
Pollution	Water quality degradation, loss of biodiversity, eutrophication, habitat destruction, altered hydrology, and bioaccumulation	Ballut-Dajud et al. (2022)
Climate change	Increased flooding, drought, altered water quality, seawater intrusion, loss of carbon sequestration	Taillardat et al. (2020)
Agriculture and urbanization	Water consumption, soil erosion, sedimentation, wetland loss by land filling, and altered hydrology	Sarkar and Maji (2022)
Invasive species	Competition for resources, displacement of native species, change in hydrology, soil degradation, disruption of the food chain, spread of disease and pathogens	Ren et al. (2021)

Technological interventions have increased production and reduced costs in aquaculture, leading to lower prices and greater availability of fishery products, thereby enhancing food security. Current trends suggest price volatility favors enhancing local production, supporting rural livelihoods and food security (Pradeepkiran, 2019).

17.4.3 Biodiversity conservation and enhancement

Wetlands serve as a breeding and feeding habitat for most crustaceans and fish. Wetlands offer abundant sustenance, attracting numerous fish species. Thus this safeguards biodiversity by shielding juveniles from predators.

The roots and trunks of mangroves provide a substrate for organisms such as oysters to colonize. Their roots also capture small particles, resulting in soft soils optimal for burrowing mollusks and crustaceans. Some fish ingest plant debris, while others feed on small insects and crustaceans that inhabit or feed on plants or the nutrient-rich organic material produced when wetland vegetation decomposes through the action of bacteria, worms, and aquatic organisms. Numerous essential species inhabit unspoiled and distinctive environments. This habitat supports both capture fisheries and aquaculture activities. They are regarded as "biological supermarkets," supporting significant food webs and a diverse array of animals (Gopal, 2009). Coastal wetlands serve as critical breeding habitats for numerous fish and crustacean species, including marine varieties, and are integral to marine ecosystems. These biologically diverse habitats provide refuge, abundant food resources, and protection from predators, creating optimal conditions for the early life stages of several aquatic species. The complex interplay of saltwater and freshwater in coastal wetlands produces a diverse array of bacteria, algae, and invertebrates, forming the basis of the food web. Consequently, these wetlands sustain vital commercial and recreational fishing, enhancing coastal economies and food security. The safeguarding and conservation of coastal wetlands are crucial for maintaining the sustained productivity of these key breeding grounds for aquatic species (Barbier, 2011). Resilient coastal wetland habitats will support diverse flora and fauna, including species suited to brackish or saline conditions. This diversification enhances environmental resilience, as different species may respond differently to changing conditions, helping maintain ecosystem stability.

17.5 Risks and challenges

Wetlands are a vital part of the biosphere that support various life forms. However, human activities and some other factors risk the wetlands' stability. Some challenges are being discussed here (Table 17.4).

17.5.1 Environmental impact

Several environmental factors are deteriorating the wetland's health. These are climate change, pollution, deforestation, and urbanization.

17.5.1.1 Climate change

Climate change can impact wetlands in several ways, including direct and indirect consequences, higher temperatures, shifts in rainfall intensity and duration, and other factors such as drought, flooding, and cyclones. The biogeochemistry and function of the wetland can be significantly altered by changes in hydrology and rising temperatures, potentially resulting in the functional loss of wetlands. This implies that they will cease to offer a water purification service, which could result in the release of nutrients into surface waters and the deleterious effects of decomposition, including eutrophication and acidification in water bodies. As an outcome of climate change, a wetland that decomposes faster than it produces may transition from a carbon sink to a carbon source, releasing CO_2 and methane into the atmosphere. Milder conditions may lead to greater nitrous oxide emissions from wetlands due to increased microbial activity and faster rates of nitrification and denitrification (Taillardat et al., 2020).

17.5.1.2 Agriculture and the urbanization

Wetlands play a crucial role in agricultural units by supplying water for livestock, farming, and aquaculture, and by contributing to environmental regulation. Agriculture significantly affects wetlands through land conversion, excessive use of nutrients and pesticides, unsustainable water extraction or diversion, and biodiversity overexploitation. Humans have significantly increased their use of the Earth's land surface over the last century. The percentage of natural land declined from 70.1% to 46.5%. Crop and pasture fields rose from 27.2% to 46.5% of the total land area. Numerous studies have found that agricultural development is frequently the primary cause of wetland loss (Sarkar & Maji, 2022). Farming practices and systems must change to combat wetlands' ongoing loss and degradation. This transformation must also ensure food security for the world's growing population while maintaining sufficient food production amid rapid environmental change.

In urbanization, changes in wetland morphology are influenced by the combined action of man and nature, encompassing both natural and human factors. Natural force is eternal and always present in the process. Human dynamics, which refers to the transformation and utilization of wetlands by humans, shifts from weak to strong, resulting in more widespread and profound alterations to wetland morphology. This transformation makes wetlands the primary force of the entire ecosystem for a specific period. Wetlands have been significantly damaged, resulting in the collapse of entire ecosystems, particularly due to a combination of natural and human dynamics. The natural environment and biological characteristics of urban and nonurban environments exhibit substantial differences. The hydrological characteristics of urban wetlands have been altered by ongoing municipal development, which directly affects their structure and function. For instance, the hydrological properties of rivers will be influenced by the construction of municipal infrastructure, including roads, drainage systems, and river surges. This, in turn, will result in a cascade effect on the wetland and terrestrial ecology of the surrounding areas. Wetlands' biological and chemical properties have been altered by urbanization, leading to changes in species assemblages, dispersal capabilities, and interactions. The region contains a few uncommon, specific, or unique forests. The primary determinants of plant and animal growth and reproduction ecology are the size, morphology, and hydrological conditions of regions.

17.5.1.3 Pollution

Pollution is particularly detrimental to wetlands. Our global effluent is discharged into wetlands without treatment, accounting for 80% of the total. Pollution from factories, fertilizers, herbicides, or significant spills poses serious threats to wetlands. They are frequently located in the lower reaches of watercourses or are supplied by various streams and waterways, which means they can "receive and accumulate" pollution from a wide area. The accumulation of this pollution can significantly impact the safety of a wetland. Plastics are also a significant source of pollution for wetlands, particularly in coastal regions. Each year, 8–12 million kilograms of plastic are introduced into our oceans. However, it is anticipated that the production of virgin plastics will have increased by 40% by 2030 (Stegmann et al., 2022). One of the greatest threats to wetlands is the use of pesticides and fertilizers in contemporary intensive cultivation. These chemicals, as well as poisons like mercury from polluted factories, can substantially influence plant and animal health and reproduction, reducing biodiversity. Pollution from a single source, such as a factory or an overflow conduit, is becoming more manageable in the United Kingdom (Mansaray et al., 2018). However, due to a lack of legislation, this type of "point-source" pollution remains a significant concern in developing countries.

17.5.2 Social conflicts

Many people rely on wetlands for a living. This resource delivers vital services and products, bringing hundreds of millions of dollars into national economies each year. These ecosystems experience social and economic conflict due to the differing interests of stakeholders and the local community.

The locals carry out a variety of activities in the marshes. Some of them were farming and other agricultural activities. A number of individuals grow rice, yams, cabbage, and various other wetland plants. Some crops are grown year-round, while others are grown during the dry season (Lynch et al., 2016). Farmers both sell and grow their produce. Papyrus cultivation is also a popular activity in the marshes. People frequently manufacture mats and other crafts, which are marketed in town, mainly to tourists. While these activities are economically beneficial, there are other, more personal uses for the wetlands. Some community members collect water for drinking, washing, and bathing. Participants stated this is especially true for individuals who cannot buy tap water. Other activities described were fishing, small-scale fish farms, livestock grazing, brickmaking, piggery, and collecting building materials.

Everyone is focused on expanding the economy and infrastructure while paying little attention to the environment and conservation. Wetlands serve as the foundation for factory construction. Wetlands provide factories with inexpensive land. While wetlands may not deliver direct benefits to companies, they do have significant impacts on them. People in the communities that use the wetlands are often poorer. They live in humble dwellings and survive on very little. For many community members, the wetlands provide a source of income; for others, they are their primary means of sustenance. Many of these issues stem from communities' reliance on wetlands. Wetlands are under significant strain due to both industrial and community-based degradation. However, stakeholders believe that while local encroachment is widespread, industrial encroachment is dense and will have longer-term repercussions for the wetlands. The various stakeholders recognized the importance of conserving wetlands while supporting development but noted that this is difficult due to inadequate implementation of environmental policies. One reason communities are frustrated with the industries is that they are so close to the wetlands people rely on, whereas the manufacturers do not often rely on them (Scholte et al., 2016).

According to some, wetlands are causing social and economic problems. Conversion for industrial expansion, settlements, agriculture, and clay mining, as well as overuse by the local population, threaten these. Most of these degrading activities are considered more critical than wetland conservation. A reasonable estimate is that 50% of wetlands in the world have disappeared (Hu et al., 2017).

17.5.3 Biosecurity concerns

Invasive species are a significant biosecurity hazard. Although most non-native species do not pose a threat to indigenous species, some can be very harmful. These are categorized as invasive nonnative species. Invasive nonnative species contribute to the degradation of native biodiversity by spreading diseases or competing with indigenous species for resources like food and habitat. Invasive mosquito fish devour zooplankton and certain amphibian larvae, resulting in higher nutrient concentrations that can stimulate phytoplankton blooms and boost populations of some amphibian larvae, but not all. Invasion meltdown transpires when the biotic interactions of one invasive species facilitate the establishment and success of another species, occurring through competitive and facilitative interactions. Alligator weed, an invasive macrophyte, indirectly enhances the abundance and persistence of invasive mosquito fish in the Yangtze River by offering overwintering thermal refuges. The red swamp crayfish enhances the propagule pressure of the invasive large-flower primrose willow by severing the stalks during foraging, thereby releasing fragments that facilitate spread (Aldridge et al., 2015).

The complexity of impact arises from variations in behavior or species traits. Variations in individual behavior may lead to distinct ecological consequences. Cane toads at the invasion front exhibit greater endurance than those in the invasive population's core, facilitating increased dispersal and foraging capabilities. Cane toads exhibit population-level effects on native predators, influenced by the phylogenetic, behavioral, and morphological characteristics of the native species. The toxicity of cane toads upon consumption, their relatively large size, and their presence in terrestrial habitats significantly impact the predators adversely affected. Predators with head sizes and foraging behaviors adapted to consuming cane toads in terrestrial habitats experienced population declines, whereas those foraging in arboreal habitats showed population increases.

Wetlands boast a wide variety of flora and fauna. These may serve as sites for disease vectors. In the case of malaria and dengue, wetlands are the breeding ground for mosquitoes.

17.6 Sustainable practices and management

The wetland system performs many functions, including protecting water sources, controlling floods, providing habitats for wildlife, and maintaining shoreline stability (Jisha & Puthur, 2021). Nonetheless, the rapid economic boom and population growth have harmed the wetland habitat. Wetlands are sensitive systems that are vulnerable to even little disturbances. Implementing a comprehensive and methodical management approach is critical to ensuring their long-term viability.

17.6.1 Integrated wetland management

A systematic approach that involves the protection of not only wetlands but also considers ecological, economic, and social factors. This system ensures the conservation of wetlands and their wise use by man.

The basic principles of IWM state that it should be developed while considering a variety of other system dynamics, such as environmental economics, social development, and ecological conservation. Wetlands are inextricably linked to people's lives; hence, it is critical to consider humans as part of the ecosystem while managing them. Most wetlands cannot be managed effectively using traditional measures, such as designating them as protected areas and excluding human activity (Chen et al., 2014). The government, stakeholders, nongovernmental organizations, and researchers should all be involved in planning the management approach. As previously stated, natural conditions and human activities are destroying wetlands, necessitating additional measures, such as climate adaptation.

Wetland evaluation and management are dynamically complicated and require an integrated approach. Reductionism is inefficient for understanding the interactions in wetland environmental systems. The GMB approach can be used to understand the wetland's holistic system, which includes all regional characteristics. The wetland system's behavior may then be stimulated over time using a GMB-based SD model, and the effectiveness of the authorized management measures can be evaluated (Chen et al., 2014).

Integrated wetland management is complex due to conflicts of interest among stakeholders, limited awareness, regulatory gaps, and financial constraints. However, these issues can be addressed by collaborating with the local community and engaging stakeholders.

17.6.2 Sustainable fishing practices

Environmentally friendly fishing practices seek to sustain fish stocks while protecting aquatic ecosystems. These involve deploying more selective fishing gear, restricting fishing zones, and boosting aquaculture output.

Adopting sustainable fishing methods is critical for protecting marine ecosystems and ensuring long-term profitability in fishing operations. Some of the main concerns regarding sustainable development are that elevated fishing pressures, such as excessive harvesting, development or recruitment overfishing, will result in a reduction of significant potential yield; that stock framework will erode to the point where it loses diversity and endurance to climate shifts; that biological systems and their economic infrastructures will cycle between collapse and recovery, with each cycle being less productive than the previous one; and that change (Kenny et al., 2018). By employing ethical fishing methods, fishermen can reduce their impact on non-target species such as seabirds, turtles, and dolphins. Climate change, which affects ocean temperatures and ecosystems, is another issue that sustainable fisheries management addresses.

Sustainable fishing methods can be efficiently adopted by collaborating with stakeholders such as fishermen, scientists, and policymakers, guaranteeing the health of wetland ecosystems and the livelihoods of coastal people for future generations.

17.7 Policy and governance

Effective policy and governance are required to conserve and sustainably use wetlands, as they are vital ecosystems.

17.7.1 National and international policies

17.7.1.1 Ramsar convention

International organizations such as the Ramsar Convention establish criteria for wetland conservation and resource management. The Convention was ratified in the Iranian city of Ramsar in 1971 and put into effect in 1975. Since then, almost 90% of UN member states across all geographic regions have decided to participate as "Contracting Parties." The

Ramsar Convention was founded in 1975 to conserve and wisely use all wetland regions via individual and national efforts, as well as international cooperation, to contribute to global sustainable development (Rattan et al., 2021).

The Ramsar Convention is based on three pillars that define its mission: to promote the appropriate use of all wetland habitats, to designate wetlands of international importance on the Ramsar List for effective management, and to foster international cooperation on global wetlands, shared wetland systems, and species.

17.7.1.2 National policies

A national wetland policy is key to implementing the Ramsar Convention's sustainable use concept (McInnes et al., 2017). However, many Contracting Parties to the Convention on Wetlands struggle to define, establish, and implement national policies that promote wetland protection and management. Several countries, particularly those with a commonwealth or government structure, such as Australia, Belgium, Canada, Pakistan, and the United States, have reported adopting or exploring wetland policies and plans at the subnational level. This reveals shared constitutional authority for wetlands preservation at both the national and subnational levels throughout these countries. The national strategy is to establish comprehensive wetland conservation regulations, strengthen collaboration with national and international governments, manage wetland data, promote practical wetland science, achieve local wetland conservation, and educate people about the value of wetlands (Moomaw et al., 2018).

17.8 Case studies

Emerging case-based evidence shows that fisheries and aquaculture activities within wetland ecosystems are closely linked to human health, both positively and negatively. A case in point is the Volta Estuary in Ghana, where the rehabilitation of degraded mangrove forests has yielded numerous co-benefits for local populations (Boafo et al., 2025). In addition to enhancing fish nursery habitats and boosting catches, mangrove vegetation restoration has also led to a quantifiable decrease in the self-reported incidence of malaria. Field surveys and qualitative comparative analysis indicated that local residents had fewer incidences of mosquito-borne diseases after mangrove habitats were restored. The restoration work included the removal of stagnant pools and the restoration of tidal flow, which indirectly reduced the number of suitable breeding sites for Anopheles mosquitoes (Lah, 2021). This case study shows that wetland restoration and fisheries improvement initiatives can be used to provide both ecosystem and human health outcomes.

Conversely, the Acre region of the Brazilian Amazon shows evidence that certain types of aquaculture can pose greater health risks when wetland systems are poorly managed. Small-scale farmers in this area usually build artificial ponds to raise fish, thus changing the local hydrology. According to a mathematical modeling study, in cases where pond vegetation is left to grow freely, the ensuing habitat may prove to be extremely productive for mosquito larvae. Malaria transmission rates were significantly lower in communities where clearing of vegetation around ponds was done regularly (Pacheco et al., 2025). The case demonstrates the value of ecosystem-based management practices in aquaculture systems for reducing the threat of vector-borne diseases.

The other pertinent example is the East Kolkata Wetlands in India, a UNESCO World Heritage site where wastewater from the city of Kolkata is directed into a multifaceted system of wetland-based agriculture and fish farming (Chakraborty et al., 2023b). Fish are reared in ponds fed by sewage, and local communities grow vegetables using the nutrient-rich water of the wetland system. Besides providing food and income to over 50,000 inhabitants, the system also functions as a natural wastewater treatment system, enhancing sanitation and reducing exposure to untreated effluent. By so doing, the wetland acts as a nature-based health infrastructure, providing indirect health outcomes in the form of better environmental quality and food security (Tella et al., 2025).

Similar processes can be seen in Chilika Lake in eastern India, the largest brackish-water lagoon in the world. A significant restoration program—desiltation and opening up of a blocked sea mouth—managed to restore water circulation and salinity balance across the lake. The result of this revitalization was the restoration of commercially valuable fish and shrimp species, a rise in fish catch and incomes of traditional fishing communities. Better livelihoods were reflected in better access to food and health services, with local households reporting less nutritional stress and seasonal out-migration (Patra et al., 2024). This case shows the importance of healthy wetland functions to sustain local economies and community well-being.

Lastly, an interesting case of how integrated ecosystem management can enhance both ecological and social resilience is found in the Bangweulu Wetlands in Zambia. The program alleviated pressure on wetland fish stocks and dealt with the underlying causes of health vulnerability through a combination of fisheries comanagement, livelihood diversification (e.g., beekeeping, bird tourism), and public health initiatives (including reproductive health education). Communities stated that they had better nutrition in the household and less exploitative fishing methods (Lyoba, 2024). All these case studies show that aquaculture and fisheries in wetland ecosystems have significant implications for human health; therefore, integrative strategies should be adopted that explicitly consider health issues in planning and management.

17.9 Future prospects for fisheries and aquaculture in wetland ecosystems

Fisheries and aquaculture in wetland ecosystems present a complex set of technological and environmental challenges and changing market demands for the future, as much for the developing world as for the developed. Wetlands shall play an essential role in sustainable aquaculture production as global fish consumption increases. Implementations in aquaculture practices have high potential to increase productivity and reduce the ecological footprint, including recirculating aquaculture systems (RAS) and integrated multi-trophic aquaculture (Ruiz-Vanoye et al., 2025).

Although there is a growing body of literature documenting interactions among fisheries, aquaculture, and wetland ecosystems, several research gaps remain. To begin with, there have been relatively few studies that have been conducted to determine the long-term health effects of wetland-based fisheries and aquaculture projects. The majority of current research focuses on short-term outcomes (e.g., reductions in the incidence of vector-borne diseases or changes in nutritional status), and longitudinal studies that determine whether these benefits are sustained over time or vary with the dynamics of the socio-ecological environment are lacking.

Second, the wetland microbiome and its potential effects on the health of aquatic species and humans have not been studied empirically much. Knowledge of the way in which microbial communities respond to various aquaculture activities and environmental stressors may assist in the optimization of ecosystem management in terms of cobenefits in biodiversity, food production, and disease control.

Third, the effects of climate change are not adequately incorporated into existing analyzes; adaptation models should be developed in the future to understand how climate-induced alterations in wetland hydrology, salinity, and temperature could influence fish species distribution, emergence of pathogens, and dynamics of vectors. Lastly, additional transdisciplinary research is necessary between epidemiology, ecology, and aquaculture sciences. These methods are especially relevant in determining trade-offs and synergies of food production and health. Filling these gaps will give stronger evidence to guide sustainable and health-sensitive wetland management and policy formulation.

Despite climate change still being a critical challenge, rising temperatures, variable rainfall, and saltwater intrusion are causing the loss of wetland ecosystems, one of its significant impacts. Future strategies should focus on climate-resilient aquaculture practices and habitat restoration efforts to address these risks. Breeding stress-resistant fish species may also provide a promising solution.

Digital technologies such as remote sensing, artificial intelligence, and blockchain can revolutionize monitoring, traceability, decision-making, wetland fisheries, and aquaculture. One of these tools would not only increase operational efficiency. Still, they will also add transparency to supply chain services in response to growing consumer demand for sustainably sourced products.

In addition, policy frameworks will have to shift to accommodate conservation priorities with local communities' socio-economic needs. Community-led aquaculture initiative and comanagement will foster stakeholder participation and equitable resource distribution.

In the future, fisheries and aquaculture in wetlands will rely on a synergistic strategy that integrates technological innovation, climate adaptation, and community participation. Wetland ecosystems can sustain their ecological integrity for future generations if they are managed more sustainably and with forward-looking policies.

17.10 Conclusion

Wetlands are vital ecosystems that provide various services to the ecosystems, such as carbon storage, flood control, water purification, and habitats for various species. Significant threats faced by wetlands are also discussed in this chapter, specifically those from human activities such as habitat destruction, climate change, and pollution. Moreover, the negative impacts posed by agriculture and urban development are also mentioned.

Wetlands also significantly affect coastal and rural economies, providing livelihoods for numerous people. Everyone wants to get their profit; therefore, these sites often face conflict. One of the key concerns in this chapter is that aquaculture, related to wetlands, plays an essential role in global food sustainability. As the human population grows daily, the demand for aquatic food also increases. So, it is time to adopt sustainable aquaculture techniques to meet global dietary requirements and support ecological stability. Sustainable fisheries management is discussed as an essential factor in this chapter. Various solutions, such as improved techniques and stricter policies, can help address these challenges.

Furthermore, ecofriendly innovative practices such as integrated wetland management, RAS, and IMTA can enhance productivity.

Moving forward, conservation management of wetlands depends on collaboration among local communities, the government, and industry. Ensuring the sustainability of wetlands will be beneficial for both current and future generations.

References

Aarif, K., Nefla, A., Muzaffar, S., Musammilu, K., & Prasadan, P. (2017). Traditional fishing activities enhance the abundance of selected waterbird species in a wetland in India. *Avian Research, 8*, 1–10.

Aazami, M., & Shanazi, K. (2020). Tourism wetlands and rural sustainable livelihood: The case from Iran. *Journal of Outdoor Recreation and Tourism, 30*, 100284.

AlMaarofi, S. (2015). *Ecological assessment of re-flooded mesopotamian marshes (Iraq)*. University of Waterloo.

Aldridge, D., Aldridge, S., Mead, A., Scales, H., Smith, R., Zieritz, A., & Sutherland, W. (2015). *Control of freshwater invasive species: global evidence for the effects of selected interventions*.

Almond, R. E., Grooten, M., & Peterson, T. (2020). *Living planet report 2020-Bending the curve of biodiversity loss*. World Wildlife Fund.

Alongi, D. M., & Alongi, D. M. (2018). *Salt marshes. Blue carbon: Coastal sequestration for climate change mitigation*. Springer, 9–22.

Anderson, O. R. (2016). The Alaskan tundra: plant and terrestrial microbial communities in a changing climate.

Asche, F., & Smith, M. D. (2018). Induced innovation in fisheries and aquaculture. *Food Policy, 76*, 1–7.

Ballut-Dajud, G. A., Sandoval Herazo, L. C., Fernández-Lambert, G., Marín-Muñiz, J. L., López Méndez, M. C., & Betanzo-Torres, E. A. (2022). Factors affecting wetland loss: A review. *Land, 11*(3), 434.

Balwan, W. K., & Kour, S. (2021). Wetland-an ecological boon for the environment. *East African Scholars Journal of Agriculture and Life Sciences, 4*(3), 38–48.

Barbier, E. B. (2011). Wetlands as natural assets. *Hydrological Sciences Journal, 56*(8), 1360–1373.

Batzer, D. P., & Sharitz, R. R. (2014). *Ecology of freshwater and estuarine wetlands*. Univ of California Press.

Beverly, S., Chapman, L., & Sokimi, W. (2003). Horizontal longline fishing methods and techniques: A manual for fishermen. In *Secretariat of the Pacific Community Nouméa, New Caledonia*.

Boafo, Y. A., Lah, G. S., & Manteaw, B. O. (2025). Wetlands as nature-based solutions to climate-induced coastal hazards: Institutional perspectives from the Keta Lagoon Complex, Ghana. *Wetlands ecology and management, 33*(4), 60.

Boyd, C. E., McNevin, A. A., & Davis, R. P. (2022). The contribution of fisheries and aquaculture to the global protein supply. *Food Security, 14*(3), 805–827.

Chakraborty, S. K., Sanyal, P., & Ray, R. (2023a). Diversity and classification of wetlands in international and national perspectives. In *Wetlands ecology: Eco-biological uniqueness of a Ramsar site (East Kolkata Wetlands, India*. Springer, 167–226.

Chakraborty, S. K., Sanyal, P., & Ray, R. (2023b). Eco-biological Uniqueness of East Kolkata Wetland, a Ramsar Site in India. In *Wetlands ecology: Eco-biological uniqueness of a Ramsar site (East Kolkata Wetlands, India*. Springer, 257–301.

Chen, H., Chang, Y.-C., & Chen, K.-C. (2014). Integrated wetland management: An analysis with group model building based on system dynamics model. *Journal of Environmental Management, 146*, 309–319.

Chen, M., Xian, Y., Huang, Y., Sun, Z., & Wu, C. (2024). Geographical features and development models of estuarine cities. *Journal of Geographical Sciences, 34*(1), 25–40.

Chen, Q., Li, Q., Lin, Y., Zhang, J., Xia, J., Ni, J., Cooke, S. J., Best, J., He, S., & Feng, T. (2023). River damming impacts on fish habitat and associated conservation measures. *Reviews of Geophysics, 61*(4), e2023RG000819.

Cooke, S. J., & Danylchuk, A. J. (2020). Hook disgorgers remove deep hooks but kill fish: A plea for cutting the line. *Fisheries Management and Ecology, 27*(6), 622–627.

Dala-Corte, R. B., Becker, F. G., & Melo, A. S. (2016). Riparian integrity affects diet and intestinal length of a generalist fish species. *Marine and Freshwater Research, 68*(7), 1272–1281.

Davis, B., Johnston, R., Baker, R., & Sheaves, M. (2012). Fish utilisation of wetland nurseries with complex hydrological connectivity. *PLoS ONE, 7*(11), e49107.

De Steven, D., & Harrison, C. A. (2022). A hydrogeologic-landscapes framework for depressional-wetland vegetation in the southeastern coastal plain, USA. *Wetlands, 42*(7), 72.

Farmery, A. K., Alexander, K., Anderson, K., Blanchard, J. L., Carter, C. G., Evans, K., Fischer, M., Fleming, A., Frusher, S., & Fulton, E. A. (2022). Food for all: Designing sustainable and secure future seafood systems. *Reviews in Fish Biology And Fisheries, 32*(1), 101–121.

Finkl, C. W., & Makowski, C. (2021). Coastal ecosystems of the world. *Journal of Coastal Research*, 1–230.

Getu, A., Misganaw, K., & Bazezew, M. (2015). Post-harvesting and major related problems of fish production. *Fisheries and Aquaculture Journal, 6*(4), 1000154.

Ghosh, S., Anju, P., Pattanayak, R., & Sahu, N. C. (2024). Fisheries and aquaculture in wetland ecosystems: A review of benefits, risks, and future prospects in India. *Journal of Coastal Research, 40*(3), 598–612.

Gopal, B. (2009). Biodiversity in wetlands. *The Wetlands Handbook, 2*, 65–95.

Graff, L., & Middleton, J. (2001). *Wetlands and fish: Catch the link*. US Department of Commerce, National Oceanic and Atmospheric Administration.

Gumbricht, T., Roman-Cuesta, R. M., Verchot, L., Herold, M., Wittmann, F., Householder, E., Herold, N., & Murdiyarso, D. (2017). An expert system model for mapping tropical wetlands and peatlands reveals South America as the largest contributor. *Global Change Biology, 23*(9), 3581–3599.

Haarstad, K., Bavor, H., & Mæhlum, T. (2012). Organic and metallic pollutants in water treatment and natural wetlands: A review. *Water Science and Technology, 65*(1), 76–99.

He, P., Chopin, F., Suuronen, P., Ferro, R. S., & Lansley, J. (2021). Classification and illustrated definition of fishing gears. *FAO Fisheries and Aquaculture Technical Paper, 672*, 94 I-.

Ho, J. K., Ramchunder, S. J., Memory, A., Theng, M., Li, T., Clews, E., Cai, Y., Tan, H. H., & Yeo, D. C. (2016). A guide to the freshwater fauna of Nee Soon Swamp Forest. *Lee Kong Chian Natural History Museum & Tropical Marine Science Institute*.

Hoitink, A. F., & Jay, D. A. (2016). Tidal river dynamics: Implications for deltas. *Reviews of Geophysics, 54*(1), 240–272.

Hu, S., Niu, Z., Chen, Y., Li, L., & Zhang, H. (2017). Global wetlands: Potential distribution, wetland loss, and status. *Science of the Total Environment, 586*, 319–327.

Ichii, T., Nishikawa, H., Igarashi, H., Okamura, H., Mahapatra, K., Sakai, M., Wakabayashi, T., Inagake, D., & Okada, Y. (2017). Impacts of extensive driftnet fishery and late 1990s climate regime shift on dominant epipelagic nekton in the transition region and subtropical frontal zone: Implications for fishery management. *Progress in Oceanography, 150*, 35–47.

Jisha, K., & Puthur, J. T. (2021). *Ecological importance of wetland systems. Wetlands conservation: Current challenges and future strategies*, 40–54.

Kenny, A. J., Campbell, N., Koen-Alonso, M., Pepin, P., & Diz, D. (2018). Delivering sustainable fisheries through adoption of a risk-based framework as part of an ecosystem approach to fisheries management. *Marine Policy, 93*, 232–240.

Kim, J.-H., Park, S.-H., Baek, S.-H., Jang, M.-H., Lee, H.-J., & Yoon, J.-D. (2020). Does different performance of sampling gears (Cast Net versus Gill Net) bring the inappropriate estimation of freshwater fish in a large river? *Korean Journal of Ecology and Environment, 53*(2), 156–164.

Knapp, S., Schmauck, S., & Zehnsdorf, A. (2019). Biodiversity impact of green roofs and constructed wetlands as progressive eco-technologies in urban areas. *Sustainability, 11*(20), 5846.

Koyama, A. (2022). Coastal fishes in tidal flats and salt marshes. In *Fish diversity of Japan: Evolution, zoogeography, and conservation*. Springer, 379–394.

Lah, G. S. (2021). Exploring the trends and impacts of climate change-induced coastal hazards in Keta, Ghana: Existing adaptation strategies and possibilities of nature-based solutions University Of Ghana].

Larsen, S., & Alp, M. (2015). Ecological thresholds and riparian wetlands: An overview for environmental managers. *Limnology, 16*(1), 1–9.

Lawrence, M. J., Jeffries, K. M., Cooke, S. J., Enders, E. C., Hasler, C. T., Somers, C. M., Suski, C. D., & Louison, M. J. (2022). Catch-and-release ice fishing: Status, issues, and research needs. *Transactions of the American Fisheries Society, 151*(3), 322–332.

Lucchetti, A., Petetta, A., Bdioui, M., Gökçe, G., Saber, M., Sacchi, J., Ozbilgin, H., Carlson, A., & Carpentieri, P. (2023). Catalogue of fishing gear in the Mediterranean and Black Sea region *(Vol. 695)*. Food & Agriculture Org.

Lynch, A. J. J., Kalumanga, E., & Ospina, G. A. (2016). Socio-ecological aspects of sustaining Ramsar wetlands in three biodiverse developing countries. *Marine and Freshwater Research, 67*(6), 850–868.

Lyoba, C. (2024). *Land use land cover change and perception of drivers in the Bangweulu wetland and surrounding areas, Zambia* The University of Zambia].

Mansaray, A. S., Aamodt, J., & Koroma, B. M. (2018). Water pollution laws in Sierra Leone—A review with examples from the UK and USA. *Natural Resources, 9*(11), 361.

McInnes, R. J., Simpson, M., Lopez, B., Hawkins, R., & Shore, R. (2017). Wetland ecosystem services and the Ramsar Convention: An assessment of needs. *Wetlands, 37*(1), 123–134.

McKee, K. L., & Baldwin, A. H. (1999). Disturbance regimes in North American wetlands. *Ecosystems of the World*, 331–364.

McKee, P., Batterson, T., Dahl, T., Glooschenko, V., Jaworski, E., Pearce, J., Raphael, C., Whillans, T., & LaRoe, E. (2018). Great Lakes aquatic habitat classification based on wetland classification systems. In *The development of an aquatic habitat classification system for lakes*. CRC Press, 59–72.

Mitsch, W. J., & Gosselink, J. G. (2015). *Wetlands*. John wiley & sons.

Moomaw, W. R., Chmura, G., Davies, G. T., Finlayson, C., Middleton, B. A., Natali, S. M., Perry, J., Roulet, N., & Sutton-Grier, A. E. (2018). Wetlands in a changing climate: Science, policy and management. *Wetlands, 38*(2), 183–205.

Mwakubo, S. M., Ikiara, M. M., & Abila, R. (2007). Socio-economic and ecological determinants in wetland fisheries in the Yala Swamp. *Wetlands Ecology and Management, 15*, 521–528.

Northridge, S., Coram, A., Kingston, A., & Crawford, R. (2017). Disentangling the causes of protected-species bycatch in gillnet fisheries. *Conservation Biology, 31*(3), 686–695.

Obiero, K., Meulenbroek, P., Drexler, S., Dagne, A., Akoll, P., Odong, R., Kaunda-Arara, B., & Waidbacher, H. (2019). The contribution of fish to food and nutrition security in Eastern Africa: Emerging trends and future outlooks. *Sustainability, 11*(6), 1636.

Okwuosa, O. B., Eyo, J. E., Omovwohwovie, E. E., & Amadiibiam, C. O. (2018). A review on fishing gear technology of the world and its application. *Iconic Research and Engineering Journals, 2*, 177–196.

Olaniyan, R. (2015). Fishing methods and their implications for a sustainable environment. *Fisheries and Aquaculture Journal, 6*(3), 1.

Pacheco, F. S., Heilpern, S. A., DiLeo, C., Almeida, R. M., Sethi, S. A., Miranda, M., Ray, N., Barros, N. O., Cavali, J., & Costa, C. (2025). Towards sustainable aquaculture in the Amazon. *Nature Sustainability, 8*(3), 234–244.

Paterson, D. M., Fortune, I., Aspden, R. J., & Black, K. S. (2019). Intertidal flats: Form and function. In *Coastal wetlands*. Elsevier, 383–406.

Patra, S., Basuri, C. K., Muduli, P. R., Kanuri, V. V., RS, R., Dipnarayan, G., KR, A., Avvari, L., Panda, U. S., & SK, D. (2024). Determination of biogeochemical rate constants for Chilika Lake, a tropical brackish water lagoon on the east coast of India. *Frontiers in Marine Science, 11*, 1323229.

Pradeepkiran, J. A. (2019). Aquaculture role in global food security with nutritional value: A review. *Translational Animal Science, 3*(2), 903–910.

Raju, C. S., Rao, J. C. S., Rao, K. G., & Simhachalam, G. (2016). Fishing methods, use of indigenous knowledge and traditional practices in fisheries management of Lake Kolleru. *Journal of Entomology and Zoology Studies, 4*(5), 37–44.

Rattan, R., Sharma, B., Kumar, R., Saigal, V., & Shukla, S. (2021). *Ramsar convention: History, structure, operations, and relevance. Wetlands conservation: Current challenges and future strategies,* 17–39.

Ren, J., Chen, J., Xu, C., van de Koppel, J., Thomsen, M. S., Qiu, S., Cheng, F., Song, W., Liu, Q.-X., & Xu, C. (2021). An invasive species erodes the performance of coastal wetland protected areas. *Science Advances, 7*(42), eabi8943.

Rheinhardt, R. D. (2007). Hydrogeomorphic and compositional variation among red maple (Acer rubrum) wetlands in southeastern *Massachusetts. Northeastern Naturalist, 14*(4), 589–604.

Ribeiro, K., Pacheco, F. S., Ferreira, J. W., de Sousa-Neto, E. R., Hastie, A., Krieger Filho, G. C., Alvalá, P. C., Forti, M. C., & Ometto, J. P. (2021). Tropical peatlands and their contribution to the global carbon cycle and climate change. *Global Change Biology, 27*(3), 489–505.

Rieley, J., & Page, S. (2016). Tropical peatland of the world. *Tropical Peatland Ecosystems,* 3–32.

Ringler, C., & Cai, X. (2006). Valuing fisheries and wetlands using integrated economic-hydrologic modeling—Mekong River Basin. *Journal of Water Resources Planning and Management, 132*(6), 480–487.

Ruiz-Vanoye, J. A., Diaz-Parra, O., Márquez Vera, M. A., Fuentes-Penna, A., Barrera-Cámara, R. A., Ruiz-Jaimes, M. A., Toledo-Navarro, Y., Bernábe-Loranca, M. B., Simancas-Acevedo, E., & Trejo-Macotela, F. R. (2025). A Comprehensive review of quality of aquaculture services in integrated multi-trophic systems. *Fishes, 10*(2), 54.

Saini, V., Singh, A. K., Paul, T., Thakur, A., & Singh, M. (2024). Advances in environment management for sustainable fisheries & livestock production.

Sarkar, D., & Maji, N. (2022). Status and threats of wetland change in land use pattern and planning: Impact of land use patterns and urbanization. In *Handbook of research on monitoring and evaluating the ecological health of wetlands.* IGI Global Scientific Publishing, 106–127.

Sarkar, U. K., Mishal, P., Borah, S., Karnatak, G., Chandra, G., Kumari, S., Meena, D., Debnath, D., Yengkokpam, S., & Das, P. (2020). Status, potential, prospects, and issues of floodplain wetland fisheries in India: Synthesis and review for sustainable management. *Reviews in Fisheries Science & Aquaculture, 29*(1), 1–32.

Sbragaglia, V., Arlinghaus, R., Blumstein, D. T., Diogo, H., Giglio, V. J., Gordoa, A., Januchowski-Hartley, F. A., Laporta, M., Lindfield, S. J., & Lloret, J. (2023). A global review of marine recreational spearfishing. *Reviews in Fish Biology and Fisheries, 33*(4), 1199–1222.

Scholte, S. S., Todorova, M., Van Teeffelen, A. J., & Verburg, P. H. (2016). Public support for wetland restoration: What is the link with ecosystem service values? *Wetlands, 36*, 467–481.

Scholz, M. (2023). *Wetlands for water pollution control.* Elsevier.

Semeraro, T., Giannuzzi, C., Beccarisi, L., Aretano, R., De Marco, A., Pasimeni, M. R., Zurlini, G., & Petrosillo, I. (2015). A constructed treatment wetland as an opportunity to enhance biodiversity and ecosystem services. *Ecological Engineering, 82*, 517–526.

Stegmann, P., Daioglou, V., Londo, M., van Vuuren, D. P., & Junginger, M. (2022). Plastic futures and their CO_2 emissions. *Nature, 612*(7939), 272–276.

Taillardat, P., Thompson, B. S., Garneau, M., Trottier, K., & Friess, D. A. (2020). Climate change mitigation potential of wetlands and the cost-effectiveness of their restoration. *Interface Focus, 10*(5), 20190129.

Teixeira, A., Duarte, B., & Caçador, I. (2014). *Salt marshes and biodiversity. Sabkha ecosystems: Volume IV: Cash crop halophyte and biodiversity conservation,* 283–298.

Tella, T. A., Festus, B., Olaoluwa, T. D., & Oladapo, A. S. (2025). Water and wastewater treatment in developed and developing countries: Present experience and future plans. In *Smart Nanomaterials for Environmental Applications*. Elsevier, 351–385.

Thornton, S. A., Dudin, L. S., Limin, S., Page, C. U., & Harrison, M. E. (2016). Peatland fish of Sabangau, Borneo: ecology and implications for future monitoring and conservation. In *Proceedings of the 15th International Peat Congress, Kuching, Malaysia.*

Veitch, V., & Sawynok, B. (2005). Freshwater wetlands and fish. In *Great Barrier Reef Marine Park Authority and Sunfish Queensland Inc Report No. SQ200401.*

Ward, R. D., Friess, D. A., Day, R. H., & Mackenzie, R. A. (2016). Impacts of climate change on mangrove ecosystems: A region by region overview. *Ecosystem Health and Sustainability, 2*(4), e01211.

Whitfield, A. K. (2017). The role of seagrass meadows, mangrove forests, salt marshes and reed beds as nursery areas and food sources for fishes in estuaries. *Reviews in Fish Biology and Fisheries, 27*(1), 75–110.

Zhang, C., Wen, L., Wang, Y., Liu, C., Zhou, Y., & Lei, G. (2020). Can constructed wetlands be wildlife refuges? A review of their potential biodiversity conservation value. *Sustainability, 12*(4), 1442.

Chapter 18

Role of coastal wetlands and mangrove ecosystem in different ecosystem services and human health

Biswajit Patra[1], Rohit Goutam[2], Mihir Tanay Das[3], Soumya Ranjan Dash[4], Arun Kumar Mohanta[5], and Surya Narayan Pradhan[6]

[1]*PG Department of Botany, Fakir Mohan University, Balasore, Odisha, India,* [2]*School of Environmental Sciences, Jawaharlal Nehru University, New Delhi, India,* [3]*School of Environmental Sciences, Fakir Mohan University, Balasore, Odisha, India,* [4]*School of Biological Sciences, AIPH University, Bhubaneswar, Odisha, India,* [5]*Indira Gandhi Institute of Technology, Sarang, Utkal University, Bhubaneswar, Odisha, India,* [6]*School of Life Sciences, Sambalpur University, Sambalpur, Odisha, India*

18.1 Introduction

Mangrove ecosystems and coastal wetlands are some of the most important and productive places on Earth, offering priceless services that support both the well-being of human populations worldwide and the health of the environment (Kemarau et al., 2025). A dynamic interface between land and sea, where freshwater and saltwater combine to create distinctive habitats that sustain a diverse range of plant and animal species, is what defines these ecosystems. Mangroves and coastal wetlands have a wide range of roles in providing ecosystem services that affect human health, water quality, biodiversity, and climate control (Takal et al., 2025). Beyond their immediate effects on the environment, these ecosystems are important because they have an impact on the social, cultural, and economic facets of human existence. Understanding these ecosystems' roles in maintaining ecosystem services and advancing human health has become crucial for communities, scientists, and policymakers as a result of the growing dangers these ecosystems face from human activities like urbanization, industrialization, and climate change (Feng et al., 2025). Seagrass beds, mangrove forests, and saltmarshes are examples of coastal wetlands that serve as essential natural barriers against environmental deterioration. They are essential to both natural systems and human cultures because they provide a range of ecosystem services that enhance coastal resilience. Particularly well-suited to the challenges of coastal habitats, mangroves flourish in saline conditions and perform a number of vital ecological roles (Omoregie et al., 2025). Carbon sequestration, storm protection, water purification, habitat provision, and nutrient cycling are among these roles. Through these methods, they promote local economies via sustainable tourism, fishing, and natural resource management while also mitigating the effects of climate change, such as rising sea levels and extreme weather. Carbon sequestration, a vital function in light of global climate change, is one of the main functions of mangrove ecosystems and coastal wetlands (Wijgerde et al., 2020). Some of the planet's most efficient carbon sinks are coastal wetlands, particularly mangroves, which absorb significant amounts of carbon dioxide from the atmosphere and store it in the soil as organic matter. Because their moist soils inhibit the breakdown of organic matter and, consequently, the return of carbon to the atmosphere, mangrove ecosystems are very effective at sequestering carbon at rates far greater than those of terrestrial forests (Provencher et al., 2020). As a result, they play a vital role in combating climate change by reducing greenhouse gas emissions and preserving atmospheric equilibrium. Protecting coastal populations from the destructive effects of storms, erosion, and floods is another essential ecological function that coastal wetlands and mangroves offer (Yu et al., 2022). During hurricanes, typhoons, and tropical storms, mangrove forests' complex root systems serve as natural barriers against storm surges, lessening the effects of strong winds and waves. Additionally, by stabilizing silt through their intricate root systems, they stop coastal erosion and protect important infrastructure, such as houses, roads, and agricultural areas. Communities are often more vulnerable to

Wetlands: Biodiversity, Health and their Role in Ecosystem Services. DOI: https://doi.org/10.1016/B978-0-443-34222-6.00004-8

the devastating effects of natural disasters in areas where mangroves and coastal wetlands have been destroyed or degraded (Simha et al., 2020). This emphasizes how crucial it is to preserve and rebuild these ecosystems as part of plans for catastrophe risk reduction and climate adaptation. Another essential function that mangroves and coastal wetlands offer is water filtering. By capturing and filtering contaminants from rivers, runoff, and agricultural practices before they reach the open ocean, these ecosystems contribute to the preservation of water quality. As natural filtering systems, mangrove forests and coastal wetlands clear the water column of excess nutrients, silt, and dangerous substances (Pateiro et al., 2020). This procedure lessens the possibility of hazardous algal blooms, prevents water pollution, and safeguards delicate coral reef ecosystems that depend on clear, nutrient-balanced waters to survive. Additionally, mangroves help control salt levels, which keeps coastal waters acceptable for a variety of aquatic species (Das et al., 2020). Mangrove ecosystems and coastal wetlands are among the planet's most ecologically diverse environments in terms of biodiversity. For many kinds of fish, crabs, mollusks, and other marine creatures, these ecosystems serve as vital breeding, feeding, and rearing grounds. Juvenile fish use the extensive root systems of mangrove trees as vital habitats, seeking refuge from potential predators. Mangrove habitats are essential for both sustainable fisheries and global food security because many economically significant fish species, including snapper, grouper, and shrimp, depend on them for their early life stages (Morgenstern et al., 2024). Mangroves are home to a wide variety of bird species, including wading birds and migratory waterfowl, which depend on them for breeding and food in addition to marine life. Communities who rely on fishing and tourism for their livelihoods would be at risk since biodiversity would dwindle as a result of the loss of these habitats (Benedetti-Cecchi et al., 2024). Mangroves, coastal wetlands, and human health have a complex and intertwined relationship. Human well-being is impacted by these ecosystems both directly and indirectly, especially in coastal areas where millions of people live and work. Particularly in underdeveloped nations where these ecosystems serve as the foundation of subsistence economies, coastal wetlands sustain local residents' lives by supplying resources, including fish, freshwater, and lumber (Zhou et al., 2024). Additionally, mangrove forests are essential for cultural and recreational pursuits, providing areas for ecotourism, fishing, and environmental conservation education. Mangroves and coastal wetlands also benefit human health by enhancing the quality of the air and water. These ecosystems aid in controlling atmospheric carbon dioxide levels as natural carbon sinks, which helps the world's efforts to slow down climate change (Simpson et al., 2011). Additionally, they lower the risk of waterborne illnesses by filtering out contaminants and extra nutrients. Since polluted water may cause cholera, dysentery, and other waterborne infections that disproportionately impact disadvantaged communities in coastal areas, clean water is crucial for human health. Coastal wetlands and mangroves provide human beings with psychological and recreational advantages in addition to environmental and health benefits (Afvari & Elston, 2024). It has been demonstrated that having access to nature improves mental health by lowering stress, anxiety, and depression. People may participate in ecotourism activities like birding, kayaking, and nature walks thanks to the preservation of mangroves and coastal wetlands. In addition to fostering mental health, these activities increase public awareness of the value of conservation and the necessity of preserving these priceless ecosystems (Curren et al., 2021). Even though mangroves and coastal wetlands are extremely important, human activities, including pollution, land reclamation, deforestation, and climate change, pose serious risks to these ecosystems. These habitats have been widely lost as a result of the conversion of mangrove forests into infrastructure projects, urbanization, and agricultural land, especially in tropical and subtropical areas. Because the loss of mangroves reduces carbon sequestration, increases vulnerability to storms, and reduces biodiversity, this degradation has major ramifications for ecosystem services (Middleton & David, 2022). Additionally, sea-level rise, ocean acidification, and more frequent severe weather events are consequences of climate change that exacerbate these risks and jeopardize the well-being of coastal ecosystems. Mangrove and coastal wetland restoration and protection have emerged as top concerns for local populations, environmental groups, and governments. Techniques like protected area creation, sustainable management techniques, and ecosystem-based adaptation can help guarantee that ecosystem services will continue to be provided and safeguard the well-being of human and natural systems (Laspidou & Ziliaskopoulos, 2022). Along with safeguarding current ecosystems, initiatives are underway to restore damaged coastal wetlands and mangrove forests through pollution control, sustainable agriculture, and replanting programs. One cannot stress how crucial coastal wetlands and mangrove ecosystems are to preserving biodiversity, controlling the climate, safeguarding coastlines, and promoting human health. Both the sustainability of the environment and the welfare of millions of people worldwide depend on the vital services that these ecosystems provide (Honnay & Jacquemyn, 2007). We must appreciate these ecosystems and give conservation and sustainable management top priority as the problems caused by climate change, population increase, and environmental degradation worsen (Smith et al., 2006). By doing this, we can guarantee that mangroves and coastal wetlands will continue to play a crucial role in promoting human and ecological health for many years to come.

18.2 Methods

Literature search was conducted by using academic databases like Scopus, PubMed, Google Scholar, and Web of Science. The search was also performed using a combination of keywords, refined with the language English, conference abstracts, and article types. Reference lists are also included to identify additional related research.

18.3 Biodiversity support and habitat provision

Mangroves and coastal wetlands provide vital habitats for a variety of species, many of which are endangered or of commercial importance, making them hotspots for biodiversity. Many fish, crabs, mollusks, and other marine creatures use these habitats as breeding, feeding, and rearing grounds (Dunn et al., 2009). For instance, juvenile fish species frequently seek cover in the thick root systems of mangroves, which provide them with optimal growth and development circumstances as well as protection from predators. Numerous bird species, such as wading birds, shorebirds, and migrating waterfowl, are also supported by the mangrove habitat. A varied and well-balanced food web is supported by the habitats that coastal wetlands and mangroves provide for terrestrial animals, amphibians, and reptiles, as well as marine species (Wang et al., 2012). Since many species would lose their vital habitats, the loss of these ecosystems would have a substantial negative impact on local fisheries and entire marine and coastal ecosystems.

18.4 Carbon sequestration and climate regulation

The ability of mangrove forests and coastal wetlands to sequester carbon makes them crucial for reducing the effects of climate change. Large volumes of carbon are stored in the biomass and soil of these ecosystems, with mangroves being especially effective at absorbing and retaining carbon from the atmosphere. Long-term carbon storage is made possible by the wet nature of mangrove soils, which slows down the breakdown of organic matter (Li et al., 2018). Given that their capacity to sequester carbon is far greater than that of terrestrial forests, mangroves are among the most efficient carbon sinks on the planet. Coastal wetlands and mangroves help mitigate the impacts of global warming by lowering the amount of greenhouse gases in the atmosphere through carbon sequestration (Honnay & Jacquemyn, 2007). Maintaining and restoring these ecosystems is essential to achieving carbon neutrality and mitigating the effects of climate-related hazards such as sea-level rise and extreme weather events, as the world grapples with the growing problem of climate change.

18.5 Coastal protection and erosion control

Coastal wetlands, especially mangrove forests, are essential for shielding coastal towns from erosion, storm damage, and tsunamis. Mangroves' intricate root systems serve as organic barriers that disperse wave energy and lessen the intensity of storm surges. Mangroves protect infrastructure, agricultural fields, and human populations from floods and coastal erosion during extreme weather events like hurricanes and cyclones (Lannoo et al., 2009). Mangrove roots aid in stabilizing coastal sediments, halting soil erosion, and preserving the integrity of the shoreline in addition to providing storm protection (Smith et al., 2006). Mangrove destruction caused by human activities such as deforestation and land reclamation makes coastal towns more vulnerable to erosion and flooding during extreme weather events.

18.6 Water filtration and quality maintenance

Mangroves and coastal wetlands are powerful natural filters that assist in purifying water by capturing and eliminating sediments, pollutants, and excess nutrients. By absorbing nutrients from wastewater or agricultural runoff, such as phosphorus and nitrogen, through their thick vegetation, they keep coastal waterways from becoming eutrophic. Coral reef deterioration, toxic algal blooms, and hypoxia (depletion of oxygen) are all consequences of eutrophication that negatively impact marine life (Kato & Miura, 1996). Additionally, contaminants that might damage aquatic life and disturb marine ecosystems, such as pesticides and heavy metals, are filtered by mangroves. Coastal wetlands and mangroves contribute significantly to the preservation of the health of marine ecosystems, such as coral reefs and seagrass beds, which are vital for biodiversity, by protecting water quality and lowering the concentrations of toxic compounds (Li et al., 2022).

18.7 Flood regulation and hydrological functioning

Coastal wetlands and mangrove ecosystems function as organic sponges, absorbing and holding onto extra water during times of intense precipitation or storm surges. By preventing low-lying coastal regions from flooding, this flood management function reduces the risk of property damage and fatalities (Ahmad et al., 2022). By affecting groundwater recharge and preserving water availability for home and agricultural use, wetlands' and mangroves' capacity to hold water also contributes to the hydrological balance of coastal areas (Wang et al., 2022). The flood-regulating role of these ecosystems is especially crucial for protecting infrastructure, towns, and agricultural fields during monsoons or tropical storms. Coastal wetlands and mangroves help coastal towns become more resilient to climate change and extreme weather events by reducing the effects of floods.

18.8 Human health and well-being

Mangrove habitats and coastal wetlands play a direct and indirect role in promoting human health. These ecosystems offer resources that support human lives and enhance mental and physical health in a number of ways:

18.8.1 Food security and livelihoods

Local fisheries are supported by mangroves and coastal wetlands, giving coastal populations access to a plentiful supply of seafood. Mangrove forests provide breeding and rearing sites for a variety of fish species, including shrimp, crabs, and finfish (Sarath et al., 2022). These resources are essential to the subsistence, nutrition, and economic well-being of coastal populations. Coastal wetlands are essential to the livelihoods of millions of people who work in aquaculture, fishing, and other associated sectors in many tropical and subtropical countries.

18.8.2 Medicinal and traditional uses

Mangroves and wetlands have long provided medicinal plants and materials for traditional healing methods to many coastal people. In indigenous medicine, mangrove plants have been used to treat conditions such as infections, digestive problems, and skin illnesses (Ma et al., 2022). Local customs are intricately linked to the cultural value of these ecosystems, and protecting them is crucial to maintaining biodiversity and cultural heritage (Smith et al., 2006).

18.8.3 Mental and physical health through ecotourism

Mangrove habitats and coastal wetlands provide ecotourism possibilities that enhance mental and physical health. Activities that offer both recreational and therapeutic advantages include kayaking, birding, and nature hikes in mangrove forests (Grenyer et al., 2006). Natural settings have been demonstrated to improve mental health by lowering stress, anxiety, and sadness. Additionally, ecotourism brings in revenue for nearby communities, creating jobs and promoting healthy human-environment interaction.

18.8.4 Waterborne disease regulation

Coastal wetlands and mangroves help stop the spread of waterborne illnesses by filtering out pollutants and excess nutrients from water. Polluted stagnant water can serve as a haven for mosquitoes and other disease-carrying organisms (Liu et al., 2021). These ecosystems lower the danger of illnesses like cholera, dengue, and malaria—which are especially common in coastal areas—by preserving clean water.

18.9 Economic benefits and livelihood support

Mangroves and coastal wetlands not only directly improve human health but also have a substantial economic impact by supporting tourism, agriculture, fisheries, and carbon trading. Long-term economic stability in coastal regions is facilitated by the preservation and sustainable utilization of natural ecosystems (Pratiwi et al., 2021). For instance, mangrove forests provide lumber, fuel, honey, and medicinal plants that help millions of people make a living. Furthermore, by lowering the expenses related to pollution control and natural catastrophes, the function of coastal wetlands and mangroves in sustaining ecosystem services, including water filtering, flood control, and storm protection, indirectly promotes economic resilience (Samal et al., 2022).

18.10 Mental health benefits of nature exposure

Mental health is one of the most important areas where exposure to the environment affects human well-being. Spending time in natural settings has been shown in several studies to help lessen the symptoms of sadness, anxiety, and stress.

People can recuperate from the mental and emotional stresses of city life by being in the serene, tranquil presence of natural surroundings (Shi et al., 2022). Exposure to nature is thought to trigger what psychologists refer to as the "restorative effect," which is nature's capacity to aid the mind in recuperating from mental exhaustion. It is believed that being in a natural setting helps people focus and pay attention again, feel happier, and have reduced levels of the stress hormone cortisol. People who live close to green spaces or have easy access to natural places also report feeling less stressed and having better mental health overall, according to studies. For this reason, activities such as taking a stroll in a park, going on a hike in a forest, or even just looking out a window at nature have been associated with increased emotional resilience, mental clarity, and relaxation (Lal et al., 2020). Exposure to nature also helps lessen the symptoms of more serious mental health issues like depression, anxiety disorders, and posttraumatic stress disorder (PTSD). For instance, spending time in natural environments or engaging in nature-based therapies like ecotherapy has been demonstrated to significantly reduce the symptoms of PTSD in patients (Iqbal et al., 2021). The heightened physiological arousal that often accompanies these illnesses may be mitigated by the calming effects of nature.

18.11 Physical health benefits

Exposure to nature has a substantial positive impact on both mental and physical health. Hiking, biking, walking, and gardening are some of the ways people who spend time in nature become more physically active. Spending time in nature raises physical activity levels, which can reduce the risk of chronic illnesses, including obesity, diabetes, and cardiovascular disease. The cardiovascular system is strengthened, physical endurance is increased, and good weight management is encouraged when engaging in outdoor activities (Alyousef et al., 2022). Additionally, there is mounting evidence that the immune system may be strengthened by spending time in natural settings. According to studies, being in nature, especially in green areas, can boost the body's immune cells, which aids in the body's defense against inflammation and illnesses (Pratiwi et al., 2021). In the modern world, when illnesses linked to a poor lifestyle are becoming more prevalent, this is especially crucial. Exposure to nature is also linked to better air quality. High levels of air pollution are common in urban areas, and they can negatively impact cardiovascular and respiratory health. Conversely, natural areas are usually linked to cleaner surroundings, lower levels of pollutants, and improved air quality (Alyousef et al., 2022). The ability of forests, wetlands, and coastal regions to filter pollutants and provide clean air is often superior, benefiting lung health overall.

18.12 Cognitive benefits and improved focus

Exposure to nature has been associated with enhanced cognitive function, including enhanced memory, problem-solving skills, and focus. According to research, those who frequently spend time in nature or reside near green areas typically have longer attention spans, greater mental clarity, and better cognitive function. Indeed, research indicates that even a short period of time spent in nature can enhance cognitive function and focus, particularly in kids and people with attention issues like ADHD. The Attention Restoration Theory, a well-known theory that explains this effect, contends that natural settings have a restorative effect on attention by offering a mild and nondemanding stimulus that enables the brain to recuperate from the mental exhaustion that results from extended periods of directed, focused attention (Mallik et al., 2012). People can restore concentration and clarity by connecting with nature in a way that is both engaging and healing. The advantages of nature for mental well-being are also evident as people age. Regularly spending time in natural environments or interacting with nature helps older adults avoid diseases such as dementia and Alzheimer's and experience less cognitive decline (Woldeyohannis et al., 2022). Opportunities for interaction and stimulation are provided by natural settings, which are crucial for preserving mental abilities.

18.13 Nature exposure and social well-being

Social well-being is also positively impacted by nature. Having access to green areas and natural settings can improve social bonds and promote community engagement. Urban parks, gardens, and leisure spaces frequently turn into meeting spots where people socialize and build relationships. In addition to providing emotional support, this social engagement can help fight loneliness and isolation, two emotions that are frequently connected to negative mental health outcomes (Sadeghi et al., 2022). Social connectivity may also be improved by engaging in nature-based activities like outdoor sports, hiking with others, or volunteering for environmental conservation initiatives. According to studies, group nature-based activities provide social support, a sense of belonging, and a sense of purpose—all of which are protective factors for mental health (Lal et al., 2020). Additionally, neighborhood green-space restoration and protection initiatives can foster a sense of pride and shared responsibility. Collaboratively creating or maintaining natural areas can improve social cohesiveness and provide a stronger feeling of communal well-being.

18.14 Children's health and development

For youngsters, the advantages of being in nature are especially noticeable. Children who have access to nature are more likely to grow up with stronger immune systems, improved motor skills, and greater creativity and curiosity, according to studies. Because it offers a stimulating environment that promotes exploration, problem-solving, and social connection, nature is essential for cognitive and emotional development. Playing outside in natural settings has been linked to increased attention spans, greater mood control, and stress management skills (Chaturvedi et al., 2021). Additionally, nature gives kids the chance to engage in unstructured play, which is critical for building resilience and creativity. Because they have a stronger bond with nature and a sense of duty to preserve it, youngsters who spend time in nature are also more likely to engage in proenvironmental actions, according to research (Ricolfi et al., 2020).

18.15 Nature-based interventions for health and healing

The importance of nature in the healing process is becoming more widely acknowledged by medical practitioners. Ecotherapy and forest therapy are two examples of nature-based interventions that are increasingly being used as supplemental therapies for mental health issues. For example, a Japanese technique known as "forest bathing" or Shinrin-yoku is submerging oneself in a forest setting to enhance mood, ease tension, and encourage relaxation (Guo et al., 2012). It has been demonstrated that the practice improves general well-being, lowers blood pressure, and lowers stress hormones. Depression, anxiety, and PTSD are among the disorders that have been treated using ecotherapy, a therapeutic approach that entails supervised encounters with nature (Atangana & Oberholster, 2021). People may enhance their physical and mental well-being and experience the healing benefits of nature by participating in outdoor activities such as hiking, gardening, or nature walks (Fig. 18.1).

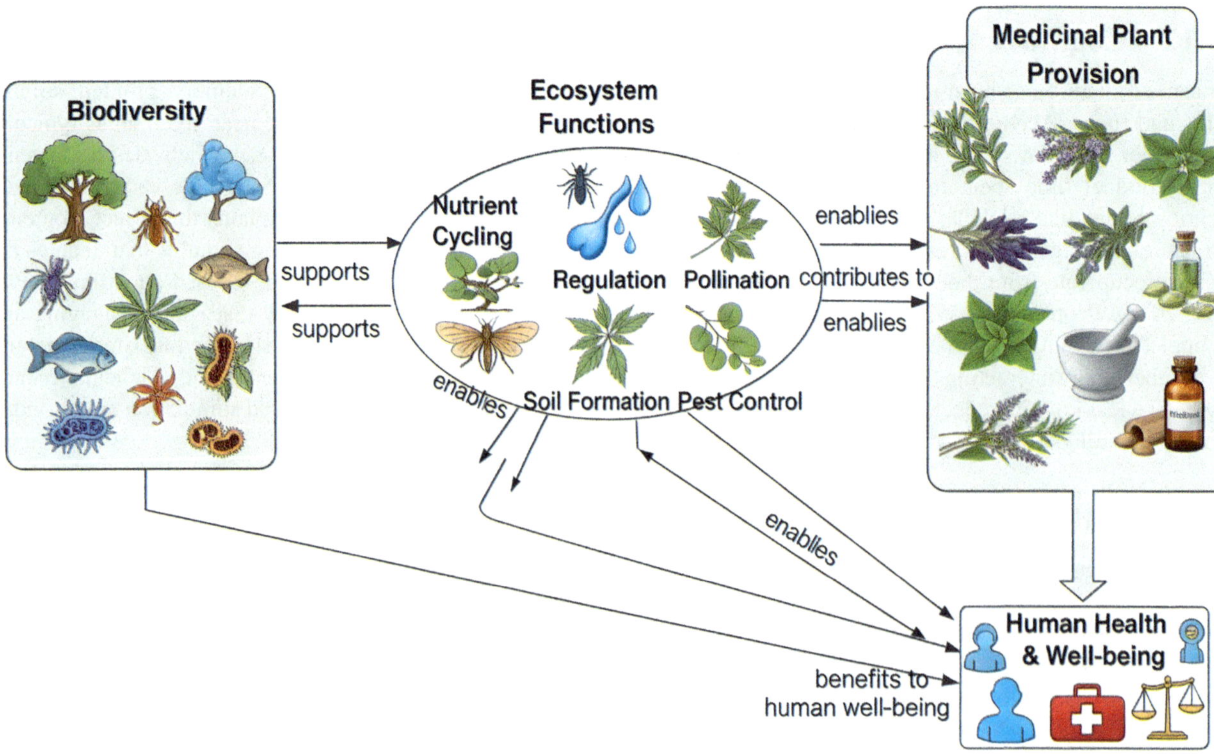

FIGURE 18.1 Mangrove ecosystem and its benefits to human well-being.

18.16 Major threats to mangrove ecosystems

Mangrove habitats, also known as "the lungs of the coast," are essential to the well-being of coastal environments and offer necessary ecosystem services such as water filtering, carbon sequestration, coastal protection, and biodiversity conservation. But there are growing risks to these special ecosystems, threatening both their existence and the important functions they provide. Deforestation is one of the biggest dangers to mangrove ecosystems, and it is caused by human activities, including industrialization, urbanization, and agricultural growth (Galunder & Kolar, 2010). Mangrove forests are being removed for infrastructural development, land reclamation, and agricultural pursuits like rice or shrimp aquaculture in many coastal places. In addition to destroying the mangrove habitat, this conversion lowers biodiversity and interferes with the ecological services that mangroves offer (Pradhan et al., 2012). For instance, in nations like Thailand, Indonesia, and India, extensive mangrove damage has resulted from large-scale shrimp cultivation. Local ecosystems are drastically changed when mangroves are converted to aquaculture ponds, which substitute monoculture farming methods—which are frequently polluted and unsustainable—for the intricate mangrove environment. Another significant danger to mangrove ecosystems is pollution, especially from sewage, agricultural runoff, and industrial waste (Pathak et al., 2017). Because of their extreme sensitivity, mangroves can be adversely affected by pollution, which can reduce oxygen availability, degrade water quality, and disrupt the fragile ecosystem's balance. Coastal waterways experience nutrient overload as a result of excessive fertilizer and pesticide usage in agriculture. This may result in eutrophication, a condition in which an overabundance of nutrients promotes the development of toxic algal blooms that choke off marine life, including mangrove seedlings and aquatic creatures that depend on mangroves, and lower oxygen levels (Rather et al., 2024). Oil spills can suffocate mangrove roots and damage the vegetation in coastal locations. Numerous marine species are poisoned by oil, which also interferes with the mangrove ecosystem's natural processes and causes long-term harm. Bottles, bags, and other plastic waste can become lodged in mangrove roots, preventing the plants from receiving light and oxygen (Qi et al., 2022). Marine life is harmed by the buildup of plastic in mangrove regions because animals frequently confuse plastics for food, which can result in entanglement and ingestion. Sea-level fluctuations significantly impact mangroves (Oktariani et al., 2023). Mangrove forests may be submerged by rising waters, especially in low-lying coastal regions. Mangroves can occasionally move inland to decrease sea-level rise, but this potential is limited by urbanization and the conversion of coastal lands for human use. Sea-level rise can also change tide patterns and make freshwater sources more salinized, which can impact mangrove sustainability. Mangrove forests may sustain physical harm from storms that are more frequent and intense due to climate change. Mangroves' capacity to shield coasts from erosion and floods can be disrupted by storm surges and strong winds (Zhang et al., 2021). The capacity of mangrove forests to recover is also negatively impacted over time when they are destroyed by storms. Certain temperature ranges are ideal for mangroves. Global warming can cause ocean temperatures to rise, which can impact nutrition availability and cause coral bleaching (Blasco et al., 1996). Mangrove species, particularly those nearing the limits of their environmental tolerance, can also be stressed by abrupt changes in temperature (Table 18.1).

18.17 Overharvesting of mangrove resources

Timber, firewood, honey, and medicinal plants are just a few of the many resources that mangroves offer to nearby populations. Degradation of mangrove ecosystems may result from overharvesting of these resources (Mazda et al., 2002). For instance, illegal logging combined with the removal of mangrove trees for the manufacture of charcoal or wood might result in habitat fragmentation and loss (Kimani et al., 2020). Unsustainable harvesting of mangrove products, such as honey and medicinal plants, can strain these ecosystems and reduce mangroves' ability to recover and regenerate. Overexploitation of mangroves reduces their capacity to sustain biodiversity and deliver essential ecosystem services like water filtering and coastal protection (Charrua et al., 2020).

18.18 Invasive species

Nonnative plants, animals, or microbes that damage the ecosystems they infiltrate are known as invasive species. Invasive species have the ability to change habitat structure, upend food webs, and outcompete native plants in mangrove ecosystems. For instance, some nonnative mangrove species, like *Rhizophora mangle*, have the ability to displace native mangrove species and lower the ecosystem's total diversity (Rodella et al., 2020). In a similar vein, invasive species, such as certain mollusks or crabs, may harm the forest by disturbing the environment or feeding on mangrove roots. Additionally, by increasing mangroves' susceptibility to disease and extreme weather events, invasive species might exacerbate the consequences of climate change.

TABLE 18.1 Traditional medicinal effects of mangrove forests (Bao, 2011; Charrua et al., 2020).

Mangrove plants	Medicinal properties	Uses
Avicennia marina (gray mangrove)	Antiinflammatory, antimicrobial, antioxidant	Used for treating wounds, inflammation, skin infections, and as an antimicrobial agent
Rhizophora mangle (red mangrove)	Antiinflammatory, anticancer, antibacterial	Extracts are used for treating wounds, ulcers, and as a remedy for diarrhea
Bruguiera gymnorrhiza (large-leafed mangrove)	Antidiabetic, antioxidant, antiinflammatory	Antidiabetic, antioxidant, antiinflammatory
Ceriops tagal (yellow mangrove)	Antimicrobial, antiinflammatory, antioxidant	Used for treating cough, fever, and inflammation
Lumnitzera racemosa	Antidiabetic, antioxidant, antibacterial	Used to treat infections, diabetes, and improve digestive health
Xylocarpus granatum	Anticancer, antiinflammatory, antioxidant	The bark and leaves are used for treating cancer, inflammation, and as a general tonic
Sonneratia caseolaris (mangrove apple)	Antibacterial, antiviral, antioxidant	Used for treating respiratory diseases, skin infections, and as a general health tonic
Kandelia candel (mangrove tree)	Antimicrobial, antioxidant, antiinflammatory	Used for treating wounds, skin infections, and inflammatory diseases
Aegiceras corniculatum	Antiinflammatory, antimicrobial, antioxidant	Used for managing inflammation, infections, and fever
Avicennia officinalis	Antioxidant, hepatoprotective, antimicrobial	Known for liver protection and for treating various infections and wounds
Excoecaria agallocha (blind-your-eye mangrove)	Anticancer, antibacterial, antiinflammatory	Used in traditional medicine for cancer treatment, skin infections, and inflammatory disorders
Nypa fruticans (nipa palm)	Antidiabetic, antiinflammatory, antioxidant	Used to lower blood sugar levels, treat wounds, and improve gastrointestinal health
Heritiera littoralis	Antioxidant, antimicrobial, antiinflammatory	Used in the treatment of skin diseases, infections, and as an antiinflammatory
Lumnitzera littorea	Antioxidant, antiinflammatory, antimicrobial	Used to treat skin infections, fever, and reduce inflammation

18.19 Coastal development and infrastructure

Mangrove ecosystems are being destroyed as coastal populations increase due to the need for land for infrastructure, industry, housing, and tourism. One of the main reasons why mangroves are disappearing from coastal areas is urbanization (Table 18.2). This involves building ports, resorts, and other projects that need filling tidal marshes and reclaiming land, frequently at the price of mangrove forests (Rodríguez Guitián et al., 2020). Many nations drain their mangroves and replace them with urban areas or industrial areas such as shipping ports or oil refineries. In addition to destroying mangrove habitats, these actions lessen the mangroves' capacity to offer crucial coastal protection functions, including halting coastal erosion and lessening the effects of storm surges (Macusi et al., 2020).

18.20 Overfishing and destructive fishing practices

Many fish and crustacean species rely on mangrove ecosystems as breeding, nesting, and feeding grounds, making them vital habitats. Mangrove environments and the animals that inhabit them are harmed by destructive fishing methods, such as using poison or dynamite to capture fish (Bandara et al., 2024). Furthermore, overfishing causes fish populations to decline, which lowers biodiversity in the environment and upsets food webs. Degradation of habitats and increased

TABLE 18.2 Case study support for ecosystem services in Odisha's mangrove forest.

Case study	Location	Ecological monitoring
Badakot village plantation	Kendrapara	Local community-led mangrove plantation offering protection, livelihoods
InVEST ecosystem services study	Bhitarkanika region	Carbon sequestration/loss and sediment–nutrient export quantified under scenarios
Economic valuation	Odisha coast	Storm protection per hectare immensely outweighs land opportunity costs
Ethnopharmacology review	Bhitarkanika NW region	Nine mangrove species with therapeutic properties; Sonneratia caseolaris phytochemistry
Ecological monitoring and nursery	Bhitarkanika	Monitoring for adaptive management; nursery preserves high species diversity
Faunal diversity study	Bichitrapur	Rich estuarine animal diversity linked to mangrove plant structure

environmental stress result from the removal of fish and other marine life from mangrove regions, thereby upending the natural balance.

18.21 Lack of effective management and conservation policies

Mangrove habitats are also seriously threatened by ineffective governance and conservation measures. Mangrove habitats are not sufficiently protected by laws or regulations in many nations, and illicit activities such as fishing, logging, and land reclamation can go unchecked. Even in cases where mangroves are legally protected, laws may not be well implemented, and enforcement is frequently lax (Kimani et al., 2020). These crucial ecosystems frequently continue to deteriorate when mangrove protection is not incorporated into municipal or national planning (Table 18.3).

18.22 Discussion

Human health and well-being have long been found to be significantly influenced by exposure to nature. As the globe grows more urbanized, many people are losing touch with the natural settings that used to be an integral part of their everyday lives. Spending time in natural environments, such as parks, woods, beaches, or wetlands, has been linked to significant improvements in both physical and mental health, according to new research (Charrua et al., 2020). The concept that being in nature is not only a luxury or a leisure activity, but also an essential part of human health and well-being is supported by this expanding body of research. Bioactive compounds are abundant in mangrove environments, and numerous species show substantial therapeutic promise. Scientific studies have demonstrated the mangrove plants' many therapeutic qualities over time, emphasizing both their potential for contemporary therapeutic uses and their value in traditional medicine (Rodella et al., 2020). *Avicennia marina, R. mangle, Bruguiera gymnorrhiza*, and other species have bioactive qualities, such as antibacterial, antiinflammatory, antidiabetic, antioxidant, and anticancer activity, as shown in the above table. These characteristics result from the distinct chemical makeup of mangrove plants, which produce specific chemicals to defend themselves against environmental stresses as they adapt to salty, wet conditions. Mangrove plants' antibacterial and antiinflammatory properties have garnered considerable interest due to their potential to address global health issues such as inflammation, infections, and antibiotic resistance (Rodríguez Guitián et al., 2020). The antioxidant properties of mangrove plants, especially those of species such as *Sonneratia caseolaris* and *Xylocarpus granatum*, have also shown promise in preventing illnesses like cancer and cardiovascular diseases linked to oxidative stress. Furthermore, the medicinal potential of these plants in the treatment of chronic illnesses is highlighted by their anticancer and antidiabetic qualities, which are demonstrated by species like *B. gymnorrhiza* and *Lumnitzera racemosa* (Marin-Monroy et al., 2020). The existence of many of these plant species is threatened by the damaging effects of human activities on mangrove ecosystems, including pollution, coastal development, and deforestation. Priceless medical resources may be lost along with the ongoing destruction of mangroves. Thus the future of mangrove-based medical

TABLE 18.3 Variability in mangrove ecosystem service in different geographic regions.

Dimension	Region	Ecosystem services	Threats	Management
Geographic region	Odisha (Eastern India)	Coastal protection, carbon sequestration, ethnomedicinal value, subsistence fishing	Cyclones, overharvesting, and habitat fragmentation	Community participation, mangrove plantations supported by NGOs and local governments
	Gujarat (Western India)	Shoreline stabilization, salt-tolerant biodiversity, and fisheries	Industrial pollution, salinity intrusion, and land reclamation	Top-down conservation, industrial regulation, less enforced
	Sundarbans (India/ Bangladesh)	Biodiversity hotspot, livelihood provisioning, storm protection	Sea-level rise, embankment failure, human-wildlife conflict	Transboundary governance challenges; integrated coastal zone management strategies underway
Socio-political setting	Community-managed in Odisha	Sustainable harvesting, local ecological knowledge integration	Limited funding, lack of formal recognition	Joint forest management, local by laws, NGO collaboration
	State-controlled (e.g., Andhra Pradesh)	Legal protection of mangroves, biodiversity conservation	Weak enforcement, encroachment for aquaculture	Forest department-driven conservation; limited local involvement
	Conflict-affected or marginalized areas	Emergency shelter, water filtration, subsistence use	Governance vacuum, illegal logging, and displacement	Weak institutional presence, need for humanitarian-environmental integration
Development level	Low-income coastal villages	Basic provisioning (fuelwood, fish), traditional medicine, and flood protection	Dependency pressure, lack of alternatives	High need for livelihood-linked conservation incentives
	Urban mangrove zones	Recreation, carbon sinks, flood mitigation	Urban sprawl, infrastructure projects, sewage	Competing land use priorities; scope for ecorestoration and green infrastructure integration
	Commercial coastal regions	Blue carbon trade potential, tourism	Aquaculture expansion, tourism pressure, and pollution	Scope for PES (payment for ecosystem services), stricter land use zoning

research depends on protecting mangrove environments and making sure that harvesting methods are sustainable (Yu et al., 2020). Even if a number of mangrove species have demonstrated encouraging therapeutic qualities, thorough clinical studies and scientific validation are still lacking to verify their effectiveness. Furthermore, it is challenging to draw firm conclusions about the research's potential medical benefits because many of them are based on small-scale laboratory tests. Standardization in the production of plant extracts is also necessary to ensure reliability and consistency in therapeutic applications. Numerous bioactive chemicals are produced by mangrove trees, and these molecules frequently differ depending on the plant's age, location, and level of environmental stress (Mafi-Gholami et al., 2020). Because of this intricacy, it is challenging to identify certain bioactive substances and comprehend how they work. The active ingredients that give mangrove plants their therapeutic qualities must be found and isolated, and their possible synergistic effects must be investigated. Indigenous societies have long utilized several mangrove species for medicinal purposes (Hisamatsu et al., 2020). This traditional wisdom, however, is frequently transmitted informally and might not be extensively available or well-documented. A more comprehensive approach to the medical use of mangrove plants may result from a determined effort to record and incorporate this traditional knowledge with contemporary scientific study. There are a number of intriguing directions for further study and advancement, and the therapeutic potential of mangrove plants remains mainly unexplored (Miguel, 2018). The development of sustainable harvesting practices that safeguard the long-term health of mangrove forests should be prioritized due to the ecological relevance of mangrove ecosystems. Restoration projects should concentrate on repairing damaged mangrove regions, and conservation plans should be created to prevent the overexploitation of medicinal plants. Supporting sustainable resource use and community-based mangrove management can help protect these ecosystems' therapeutic value and biodiversity (Onofri & Maynou, 2020). Comprehensive chemical and pharmacological studies are necessary due to the diverse phytochemical composition of mangrove species. To understand their modes of action and evaluate their efficacy through clinical trials, future research should concentrate on finding and isolating the bioactive chemicals that are responsible for the observed therapeutic characteristics. Bioactive chemicals will be more accurately and efficiently isolated from mangroves thanks to developments in biotechnology and analytical methods (Moschetto et al., 2021). Drug research should also concentrate on developing formulations that optimize these substances' bioavailability. Thorough clinical investigations are required to verify therapeutic claims regarding mangrove plants. The safety, effectiveness, and dose guidelines for several medications made from mangroves will be established in part by these experiments. To close the gap between traditional knowledge and contemporary medical research, ethnobotanists, pharmacologists, and clinicians must work together (Onofri & Maynou, 2020). To ensure safety and quality control, regulatory bodies should also establish precise rules governing the use of mangrove plant extracts in pharmaceuticals. Enhancing the therapeutic potential of mangrove plants may be greatly aided by technological developments such as genetic engineering and synthetic biology. For instance, in controlled settings like bioreactors, high-value bioactive chemicals might be produced from mangrove trees using synthetic biology (Rangel-Buitrago et al., 2020). More respect and preservation of these habitats may result from promoting ecotourism and increasing public knowledge of the value of mangrove ecosystems. Community engagement in mangrove conservation and sustainable usage should be encouraged by educational initiatives that highlight the medical benefits of mangrove plants. In addition to highlighting the potential of mangrove habitats as sources of novel medications, ecotourism may also offer financial incentives for their preservation. Integrating traditional knowledge with current scientific research is one of the main future directions (Mafi-Gholami et al., 2020). Over the course of millennia, many indigenous cultures have developed a profound awareness of the therapeutic benefits of mangrove plants. Scientists and indigenous knowledge holders working together may hasten the identification of novel bioactive substances and enhance the creation of long-lasting and efficient therapies (Islam & Haque, 2004). Restoring damaged mangrove ecosystems should be a top priority because of the vital ecological function that mangrove forests play in reducing coastal erosion, preserving biodiversity, and sequestering carbon. Restoring mangrove ecosystems will present an opportunity to revitalize both the ecological services and the therapeutic benefits they provide. Restoration initiatives can support the sustainable collection of mangrove species' medicinal qualities while also preserving their genetic variety (Wang et al., 2021).

18.23 Conclusion

Coastal wetlands and mangrove ecosystems are indispensable to the health of both the environment and human societies. Their role in providing ecosystem services—ranging from biodiversity support to carbon sequestration, water purification, and coastal protection—cannot be overstated (Rodríguez Guitián et al., 2020). These ecosystems also contribute to human health by providing resources for food, water, medicine, and recreation, while reducing the risks of waterborne diseases and climate-related disasters. As coastal ecosystems face increasing pressures from urbanization, pollution, and climate change, it is crucial to prioritize their conservation and restoration to ensure that these essential services continue to benefit both nature and humanity for generations to come. Mangrove ecosystems offer a wealth of medicinal properties that have long been utilized in traditional medicine and present significant potential for modern therapeutic applications (Díaz-Cuevas et al., 2020). The diverse bioactive compounds found in mangrove plants, such as antiinflammatory, antimicrobial, antioxidant, and anticancer properties, make them a valuable resource for treating various health conditions. However, the sustainability of these resources is threatened by habitat destruction, overharvesting, and environmental degradation. Future research should focus on preserving mangrove ecosystems, integrating traditional knowledge with modern science, and conducting rigorous clinical trials to validate the medicinal claims (Macusi et al., 2020). By advancing sustainable practices, protecting mangrove habitats, and exploring their pharmacological potential, we can ensure that mangrove plants continue to contribute to both ecological health and human well-being for generations to come. Traditional medicine has long made use of the many therapeutic benefits that mangrove habitats provide, and there is a great deal of promise for contemporary therapeutic uses (Mafi-Gholami et al., 2020). Mangrove plants have a variety of bioactive chemicals that have antiinflammatory, antibacterial, antioxidant, and anticancer effects, making them an excellent tool for treating a range of illnesses. However, habitat destruction, overharvesting, and environmental degradation pose a danger to these resources' sustainability (Moschetto et al., 2021). To verify the medical claims, future studies should concentrate on maintaining mangrove habitats, fusing traditional wisdom with contemporary technology, and carrying out thorough clinical studies. We can guarantee that mangrove plants will continue to support ecological health and human well-being for many generations to come by promoting sustainable practices, preserving mangrove ecosystems, and investigating their medicinal potential.

References

Afvari, S., & Elston, D. M. (2024). Aquatic antagonists: Scorpionfish envenomation. *Cutis; Cutaneous Medicine for the Practitioner, 113*, 133–135. https://doi.org/10.12788/cutis.0973.

Ahmad, M. N., Zia, A., van den Berg, L., Ahmad, Y., Mahmood, R., Dawar, K. M., Alam, S. S., Riaz, M., & Ashmore, M. (2022). Effects of soil fluoride pollution on wheat growth and biomass production, leaf injury index, powdery mildew infestation and trace metal uptake. *Environmental Pollution (Barking, Essex: 1987), 298*, 118820. https://doi.org/10.1016/j.envpol.2022.118820.

Alyousef, R. A., Alfaifi, H. J., Zaidi, F. K., & Al, M. (2022). Geostatistical and pollution index - based approach for assessing heavy metal pollution in the Cambro - Ordovician Saq Aquifer in Central Saudi Arabia. *Environmental Earth Sciences*. https://doi.org/10.1007/s12665-022-10498-3.

Atangana, E., & Oberholster, P. J. (2021). Using heavy metal pollution indices to assess water quality of surface and groundwater on catchment levels in South Africa. *Journals of African Earth Sciences, 182*, 104254. https://doi.org/10.1016/j.jafrearsci.2021.104254.

Bandara, R. M. W. J., Curchitser, E., & Pinsky, M. L. (2024). The importance of oxygen for explaining rapid shifts in a marine fish. *Global Change Biology, 30*, 1–12. https://doi.org/10.1111/gcb.17008.

Bao, T. Q. (2011). Effect of mangrove forest structures on wave attenuation in coastal Vietnam. *Oceanologia, 53*, 807–818. https://doi.org/10.5697/oc.53-3.807.

Benedetti-Cecchi, L., Bates, A. E., Strona, G., Bulleri, F., Horta e Costa, B., Edgar, G. J., Hereu, B., Reed, D. C., Stuart-Smith, R. D., Barrett, N. S., Kushner, D. J., Emslie, M. J., García-Charton, J. A., Gonçalves, E. J., & Aspillaga, E. (2024). Marine protected areas promote stability of reef fish communities under climate warming. *Nature Communication, 15*. https://doi.org/10.1038/s41467-024-44976-y.

Blasco, F., Saenger, P., & Janodet, E. (1996). Mangroves as indicators of coastal change. *Catena, 27*, 167–178. https://doi.org/10.1016/0341-8162(96)00013-6.

Charrua, A. B., Bandeira, S. O., Catarino, S., Cabral, P., & Romeiras, M. M. (2020). Assessment of the vulnerability of coastal mangrove ecosystems in Mozambique. *Ocean & Coastal Management, 189*. https://doi.org/10.1016/j.ocecoaman.2020.105145.

Chaturvedi, A., Bhattacharjee, S., Kumar, V., & Singh, A. K. (2021). A study on the interdependence of heavy metals while contributing to groundwater pollution index. *Environmental Science and Pollution Research, 28*, 25798–25807. https://doi.org/10.1007/s11356-021-12352-8.

Curren, E., Kuwahara, V. S., Yoshida, T., & Leong, S. C. Y. (2021). Marine microplastics in the ASEAN region: A review of the current state of knowledge. *Environmental Pollution (Barking, Essex: 1987), 288*, 117776. https://doi.org/10.1016/j.envpol.2021.117776.

Das, I., Lauria, V., Kay, S., Cazcarro, I., Arto, I., Fernandes, J. A., & Hazra, S. (2020). Effects of climate change and management policies on marine fisheries productivity in the north-east coast of India. *The Science of the Total Environment, 724*. https://doi.org/10.1016/j.scitotenv.2020.138082.

Dunn, R. R., Harris, N. C., Colwell, R. K., Koh, L. P., & Sodhi, N. S. (2009). The sixth mass coextinction: Are most endangered species parasites and mutualists? In: *Proceedings of the Royal Society B: Biological Sciences, 276*, 3037–3045. https://doi.org/10.1098/rspb.2009.0413.

Díaz-Cuevas, P., Prieto-Campos, A., & Ojeda-Zújar, J. (2020). Developing a beach erosion sensitivity indicator using relational spatial databases and Analytic Hierarchy Process. *Ocean & Coastal Management, 189*. https://doi.org/10.1016/j.ocecoaman.2020.105146.

Feng, J., Yu, H., Wu, L., Yuan, C., Zhao, X., Sun, H., Cheng, C., Li, Y., Sun, J., Li, Y., Wang, X., Shang, Y., Xu, J., & Zhang, T. (2025). Evaluating ecosystem characteristics and ecological carrying capacity for marine fauna stock enhancement within a marine ranching system. *Animals, 15*, 1–48. https://doi.org/10.3390/ani15020165.

Galunder, B. K., & Kolar, M. (2010). *Chemical analysis and the river mura water quality, 4*, 4–17.

Grenyer, R., Orme, C. D. L., Jackson, S. F., Thomas, G. H., Davies, R. G., Davies, T. J., Jones, K. E., Olson, V. A., Ridgely, R. S., Rasmussen, P. C., Ding, T. S., Bennett, P. M., Blackburn, T. M., Gaston, K. J., Gittleman, J. L., & Owens, I. P. F. (2006). Global distribution and conservation of rare and threatened vertebrates. *Nature, 444*, 93–96. https://doi.org/10.1038/nature05237.

Guo, G., Wu, F., Xie, F., & Zhang, R. (2012). Spatial distribution and pollution assessment of heavy metals in urban soils from southwest China. *Journal of Environmental Sciences, 24*, 410–418. https://doi.org/10.1016/S1001-0742(11)60762-6.

Hisamatsu, R., Tabeta, S., Kim, S., & Mizuno, K. (2020). Storm surge risk assessment for the insurance system: A case study in Tokyo Bay, Japan. *Ocean & Coastal Management, 189*, 105147. https://doi.org/10.1016/j.ocecoaman.2020.105147.

Honnay, O., & Jacquemyn, H. (2007). Susceptibility of common and rare plant species to the genetic consequences of habitat fragmentation. *Conservation Biology: The Journal of the Society for Conservation Biology, 21*, 823–831. https://doi.org/10.1111/j.1523-1739.2006.00646.x.

Iqbal, Z., Abbas, F., Mahmood, A., Ibrahim, M., Gul, M., Yamin, M., Aslam, B., Imtiaz, M., Elahi, N. N., Qureshi, T. I., & Sial, G. Z. H. (2021). Human health risk of heavy metal contamination in groundwater and source apportionment. *International Journal of Environmental Science and Technology, 19*, 7251–7260. https://doi.org/10.1007/s13762-021-03611-9.

Islam, M. S., & Haque, M. (2004). The mangrove-based coastal and nearshore fisheries of Bangladesh: Ecology, exploitation and management. *Reviews in Fish Biology and Fisheries, 14*, 153–180. https://doi.org/10.1007/s11160-004-3769-8.

Kato, M., & Miura, R. (1996). Flowering phenology and anthophilous insect community at a threatened natural lowland marsh at Nakaikemi in Tsuruga, Japan. *Contributions from the Biological Laboratory, Kyoto University, 29*, 1–48.

Kemarau, R. A., Suab, S. A., Eboy, O. V., Sa, Z., Echoh, D. U., & Sakawi, Z. (2025). Integrative approaches in remote sensing and GIS for assessing climate change impacts across Malaysian ecosystems and societies.

Kimani, P., Wamukota, A., Manyala, J. O., & Mlewa, C. M. (2020). Analysis of constraints and opportunities in marine small-scale fisheries value chain: A multi-criteria decision approach. *Ocean & Coastal Management, 189*, 105151. https://doi.org/10.1016/j.ocecoaman.2020.105151.

Lal, K., Sehgal, M., Gupta, V., & Sharma, A. (2020). Assessment of groundwater quality of CKDu affected uddanam region in Srikakulam district and across Andhra Pradesh, India. *Groundwater for Sustainable Development, 11*, 100432. https://doi.org/10.1016/j.gsd.2020.100432.

Lannoo, M. J., Kinney, V. C., Heemeyer, J. L., Engbrecht, N. J., Gallant, A. L., & Klaver, R. W. (2009). Mine spoil prairies expand critical habitat for endangered and threatened amphibian and reptile species. *Diversity, 1*, 118–132. https://doi.org/10.3390/d1020118.

Laspidou, C., & Ziliaskopoulos, K. (2022). Using system dynamics modelling to visualize the effects of resource management and policy interventions on biodiversity at a regional scale. *Ecological Indicators, 145*, 109630. https://doi.org/10.1016/j.ecolind.2022.109630.

Li, J., Yang, J., Liu, M., Ma, Z., Fang, W., & Bi, J. (2022). Quality matters: Pollution exacerbates water scarcity and sectoral output risks in China. *Water Research, 224*, 119059. https://doi.org/10.1016/j.watres.2022.119059.

Li, X., Li, M., Hou, L., Zhang, Z., Pang, X., & Li, Y. (2018). De novo transcriptome assembly and population genetic analyses for an endangered Chinese endemic *Acer miaotaiense* (Aceraceae). *Genes (Basel), 9*. https://doi.org/10.3390/genes9080378.

Liu, H. Y., Jay, M., & Chen, X. (2021). The role of nature-based solutions for improving environmental quality, health and well-being. *Sustain, 13*, 1–56. https://doi.org/10.3390/su131910950.

Ma, Y., Lu, T., Yang, L., Wu, L., Li, P., Tang, J., Chen, Y., Gao, F., Cui, S., Qi, X., & Zhang, Z. (2022). Efficient adsorptive removal of fluoroquinolone antibiotics from water by alkali and bimetallic salts co-hydrothermally modified sludge biochar. *Environmental Pollution (Barking, Essex: 1987), 298*. https://doi.org/10.1016/j.envpol.2022.118833.

Macusi, E. D., Macusi, E. S., Jimenez, L. A., & Catam-isan, J. P. (2020). Climate change vulnerability and perceived impacts on small-scale fisheries in eastern Mindanao. *Ocean & Coastal Management, 189*, 105143. https://doi.org/10.1016/j.ocecoaman.2020.105143.

Mafi-Gholami, D., Jaafari, A., Zenner, E. K., Nouri Kamari, A., & Tien Bui, D. (2020). Spatial modeling of exposure of mangrove ecosystems to multiple environmental hazards. *The Science of the Total Environment, 740*. https://doi.org/10.1016/j.scitotenv.2020.140167.

Mallik, B. K., Panda, T., & Padhy, R. N. (2012). Traditional herbal practices by the ethnic people of Kalahandi District of Odisha, India. *Asian Pacific Journal of Tropical Biomedicine, 2*, S988–S994. https://doi.org/10.1016/S2221-1691(12)60349-9.

Marin-Monroy, E. A., Romero-Canyas, R., Fraire-Cervantes, J. A., Larson-Konar, D., & Fujita, R. (2020). Compliance with rights-based fisheries management is associated with fishermen's perceptions of peer compliance and experience: A case study in the Upper Gulf of California. *Ocean & Coastal Management, 189*, 105155. https://doi.org/10.1016/j.ocecoaman.2020.105155.

Mazda, Y., Magi, M., Nanao, H., Kogo, M., Miyagi, T., Kanazawa, N., & Kobashi, D. (2002). Coastal erosion due to long-term human impact on mangrove forests. *Wetlands Ecology and Management, 10*, 1–9. https://doi.org/10.1023/A:1014343017416.

Middleton, B. A., & David, J. L. (2022). Trends in vegetation and height of the topographic surface in a tidal freshwater swamp experiencing rooting zone saltwater intrusion. *Ecological Indicators, 145*, 109637. https://doi.org/10.1016/j.ecolind.2022.109637.

Miguel, L. L. A. J. (2018). Study on seasonal hydrology and biogeochemical variability in a tropical estuarine system, Central Mozambique Coast, Africa. *Marine Pollution Bulletin, 131*, 674–692. https://doi.org/10.1016/j.marpolbul.2018.04.071.

Morgenstern, Y., Storrs, K. R., Schmidt, F., Hartmann, F., Tiedemann, H., Wagemans, J., & Fleming, R. W. (2024). High-level aftereffects reveal the role of statistical features in visual shape encoding. *Current Biology: CB, 34*, 1098–1106.e5. https://doi.org/10.1016/j.cub.2023.12.039.

Moschetto, F. A., Ribeiro, R. B., & De Freitas, D. M. (2021). Urban expansion, regeneration and socioenvironmental vulnerability in a mangrove ecosystem at the southeast coastal of São Paulo, Brazil. *Ocean & Coastal Management, 200*, 105418. https://doi.org/10.1016/j.ocecoaman.2020.105418.

Oktariani, A. F., Sudaryatma, P. E., Ramona, Y., Wirasuta, I. M. G., Darmayasa, I. B. G., Wiradana, P. A., & Okabayashi, T. (2023). Heavy metals content in fresh tuna and swordfish caught from Hindian and Pacific Oceans: Health risk assessment of dietary exposure. *VeterinaryWorld, 16*, 858–868. https://doi.org/10.14202/vetworld.2023.858-868.

Omoregie, A. I., Ouahbi, T., Kan, F. K., Sirat, Q. A., Raheem, H. O., & Rajasekar, A. (2025). Advancing slope stability and hydrological solutions through biocementation: A bibliometric review. *Hydrology, 12*. https://doi.org/10.3390/hydrology12010014.

Onofri, L., & Maynou, F. (2020). Unwanted catches, quota systems and the EU landing obligation: An economic and econometric analysis. *Ocean & Coastal Management, 189*, 105159. https://doi.org/10.1016/j.ocecoaman.2020.105159.

Pateiro, M., Munekata, P. E. S., Wang, M., Barba, F. J., Berm, R., & Lorenzo, J. M. (2020). Nutritional profiling and value of processing by product from gilthead sea bream (*Spaurus aurata*). *Marine Drugs, 18*, 1–18.

Pathak, J., Priyadarshini, E., Rawat, K., & Bohidar, H. B. (2017). Complex coacervation in charge complementary biopolymers : Electrostatic versus surface patch binding. *Advances in Colloid and Interface Science, 250*, 40–53. https://doi.org/10.1016/j.cis.2017.10.006.

Pradhan, N., Nayak, R. R., Mishra, D. K., Priyadarshini, E., Sukla, L. B., & Mishra, B. K. (2012). Microbial treatment of lateritic Ni-ore for iron beneficiation and their characterization. 2, 110–115. https://doi.org/10.5923/j.env.20120206.01.

Pratiwi, W. H., Putri, G. L., Pratama, M. A., Zulkarnain, F., & Priadi, C. R. (2021). Health risk analysis of nitrite, nitrate, and heavy metal pollution in groundwater near landfill area: A case study of the Sumur Batu village in Bekasi, Indonesia. *IOP Conference Series: Earth and Environmental Science, 633*. https://doi.org/10.1088/1755-1315/633/1/012015.

Provencher, J. F., Liboiron, M., Borrelle, S. B., Bond, A. L., Rochman, C., Lavers, J. L., Avery-Gomm, S., Yamashita, R., Ryan, P. G., Lusher, A. L., Hammer, S., Bradshaw, H., Khan, J., & Mallory, M. L. (2020). A Horizon Scan of research priorities to inform policies aimed at reducing the harm of plastic pollution to biota. *The Science of the Total Environment, 733*, 139381. https://doi.org/10.1016/j.scitotenv.2020.139381.

Qi, C., Song, X., Wang, H., Yan, Y., & Liu, B. (2022). The role of exercise-induced myokines in promoting angiogenesis. *Frontiers in Physiology, 13*, 1–10. https://doi.org/10.3389/fphys.2022.981577.

Rangel-Buitrago, N., Neal, W. J., Bonetti, J., Anfuso, G., & de Jonge, V. N. (2020). Vulnerability assessments as a tool for the coastal and marine hazards management: An overview. *Ocean & Coastal Management, 189*, 105134. https://doi.org/10.1016/j.ocecoaman.2020.105134.

Rather, S. A., Hussain, P. R., & Suradkar, P. (2024). Evaluating the effects of chitosan incorporated with seabuckthorn leaf extract composite edible coatings on the shelf life of peach (*Prunus persica* L.) fruits. *Food Humanit, 2*, 100280. https://doi.org/10.1016/j.foohum.2024.100280.

Ricolfi, L., Barbieri, M., Muteto, P. V., Nigro, A., Sappa, G., & Vitale, S. (2020). Potential toxic elements in groundwater and their health risk assessment in drinking water of Limpopo National Park, Gaza Province, Southern Mozambique. *Environmental Geochemistry and Health, 42*, 2733–2745. https://doi.org/10.1007/s10653-019-00507-z.

Rodella, I., Madau, F., Mazzanti, M., Corbau, C., Carboni, D., Simeoni, U., & Parente, L. (2020). Carrying capacity as tool for beach economic value assessment (case studies of Italian beaches). *Ocean & Coastal Management, 189*, 105130. https://doi.org/10.1016/j.ocecoaman.2020.105130.

Rodríguez Guitián, M. A., Real, C., Ramil-Rego, P., Franco, R. R., & Castro, H. L. (2020). Characteristics, vulnerability and conservation value of active tufa-forming springs on coastal cliffs in the NW Iberian Peninsula. *Ocean & Coastal Management, 189*. https://doi.org/10.1016/j.ocecoaman.2020.105122.

Sadeghi, H., Fazlzadeh, M., Zarei, A., Mahvi, A. H., & Nazmara, S. (2022). Spatial distribution and contamination of heavy metals in surface water, groundwater and topsoil surrounding Moghan's tannery site in Ardabil, Iran. *International Journal of Environmental Analytical Chemistry, 102*, 1049–1059. https://doi.org/10.1080/03067319.2020.1730342.

Samal, P., Mohanty, A. K., Khaoash, S., Mishra, P., & Ramaswamy, K. (2022). Health risk assessment and hydrogeochemical modelling of groundwater due to heavy metals contaminants at Basundhara coal mining region, India. *International Journal of Environmental Analytical Chemistry, 00*, 1–20. https://doi.org/10.1080/03067319.2021.2023511.

Sarath, N. G., Shackira, A. M., El-Serehy, H. A., Hefft, D. I., & Puthur, J. T. (2022). Phytostabilization of arsenic and associated physio-anatomical changes in *Acanthus ilicifolius* L. *Environmental Pollution (Barking, Essex: 1987), 298*, 118828. https://doi.org/10.1016/j.envpol.2022.118828.

Shi, H., Zeng, M., Peng, H., Huang, C., Sun, H., Hou, Q., & Pi, P. (2022). Health risk assessment of heavy metals in groundwater of Hainan Island using the monte carlo simulation coupled with the APCS/MLR model. *International Journal of Environmental Research and Public Health, 19*, 7827. https://doi.org/10.3390/ijerph19137827.

Simha, P., Lalander, C., Nordin, A., & Vinnerås, B. (2020). Alkaline dehydration of source-separated fresh human urine: Preliminary insights into using different dehydration temperature and media. *The Science of the Total Environment, 733*, 139313. https://doi.org/10.1016/j.scitotenv.2020.139313.

Simpson, S. D., Jennings, S., Johnson, M. P., Blanchard, J. L., Schön, P. J., Sims, D. W., & Genner, M. J. (2011). Continental shelf-wide response of a fish assemblage to rapid warming of the sea. *Current Biology: CB, 21*, 1565–1570. https://doi.org/10.1016/j.cub.2011.08.016.

Smith, K. F., Sax, D. F., & Lafferty, K. D. (2006). Evidence for the role of infectious disease in species extinction and endangerment. *Conservation Biology: The Journal of the Society for Conservation Biology, 20*, 1349–1357. https://doi.org/10.1111/j.1523-1739.2006.00524.x.

Takal, S. U., Tahiru, A. W., Fattah, I. R., Cobbina, S. J., Asare, W., & Abanyie, S. K. (2025). Enhancing resilience to climate change: A comprehensive PRISMA review of agricultural and non-agricultural adaptation strategies for Ghana. *Cogent Soc. Sci. 11*. https://doi.org/10.1080/23311886.2025.2453905.

Wang, S., Gao, Y., Jia, J., Lu, Y., Wang, J., Ha, X., Li, Z., & Sun, K. (2022). Determining whether hydrological processes drive carbon source and sink conversion shifts in a large floodplain-lake system in China. *Water Research, 224*, 119105. https://doi.org/10.1016/j.watres.2022.119105.

Wang, X. M., Hou, X. Q., Zhang, Y. Q., Yang, R., Feng, S. F., Li, Y., & Ren, Y. (2012). Genetic diversity of the endemic and medicinally important plant rheum officinale as revealed by inter-simpe sequence repeat (ISSR) markers. *International Journal of Molecular Sciences, 13*, 3900–3915. https://doi.org/10.3390/ijms13033900.

Wang, Y., Chao, B., Dong, P., Zhang, D., Yu, W., Hu, W., Ma, Z., Chen, G., Liu, Z., & Chen, B. (2021). Simulating spatial change of mangrove habitat under the impact of coastal land use: Coupling MaxEnt and Dyna-CLUE models. *The Science of the Total Environment, 788*, 147914. https://doi.org/10.1016/j.scitotenv.2021.147914.

Wijgerde, T., van Ballegooijen, M., Nijland, R., van der Loos, L., Kwadijk, C., Osinga, R., Murk, A., & Slijkerman, D. (2020). Adding insult to injury: Effects of chronic oxybenzone exposure and elevated temperature on two reef-building corals. *The Science of the Total Environment, 733*, 139030. https://doi.org/10.1016/j.scitotenv.2020.139030.

Woldeyohannis, Y. S., Hiremath, S. S., Tola, S., & Wako, A. (2022). Investigation of soil physiochemical properties effects on soil compaction for a long year tilled farmland. *Applied and Environmental Soil Science, 2022*. https://doi.org/10.1155/2022/8626200.

Yu, J. K., Ma, J. Q., & Liu, D. (2020). Historical evolution of marine functional zoning in China since its reform and opening up in 1978. *Ocean & Coastal Management, 189*, 105157. https://doi.org/10.1016/j.ocecoaman.2020.105157.

Yu, M., Li, Y., Luo, G., Yu, L., & Chen, M. (2022). Agroecosystem composition and landscape ecological risk evolution of rice terraces in the southern mountains, China. *Ecological Indicators. 145*, 109625. https://doi.org/10.1016/j.ecolind.2022.109625.

Zhang, Y., Wang, P., Mao, H., Zhang, Y., Zheng, L., Yu, P., Guo, Z., Li, L., & Jiang, Q. (2021). PEGylated gold nanoparticles promote osteogenic differentiation in in vitro and in vivo systems. *Materials & Design, 197*, 109231. https://doi.org/10.1016/j.matdes.2020.109231.

Zhou, S., Tian, Q., Li, W., Li, J., Li, W., & Han, F. (2024). Corrigendum to "Functional study of Cygb in the immune response to Vibrio harveyi disease in yellow drum (*Nibea albiflora*)" *Fish & Shellfish Immunology, 143*(2023), 109217. https://doi.org/10.1016/j.fsi.2023.109217 Fish Shellfish Immunol. 150, 109587. https://doi.org/10.1016/j.fsi.2024.109587.

Chapter 19

Valuing wetland ecosystem services: mapping the state of the art of research in India

Arghya Chakrabarty[1,2] and Ritesh Kumar[1]
[1]*Wetlands International South Asia, New Delhi, India,* [2]*ICFRE—Rain Forest Research Institute, Jorhat, Assam, India*

19.1 Introduction

India is endowed with a diverse wetland regime that ranges from the high-altitude lakes of the Himalayas, the floodplains and marshes of the Ganga–Brahmaputra alluvial plains, the saline flats of the Great Indian Desert, to extensive mangroves and coral reefs bordering the country's eastern and western coasts. The most recent assessment of wetland extent in the country pegs the extent of these ecosystems at 16.89 million hectares, equating to about 5.12% of the country's total geographic area (SAC, 2024).

India's national framework for wetland conservation and management has undergone considerable evolution over the last four decades. An initial focus on conservation based on wildlife values graduated into a commitment to organized efforts for wetlands conservation as India ratified the Ramsar Convention in 1982, and a separate Ministry of Environment and Forest (presently, the Ministry of Environment, Forest and Climate Change (MoEFCC)) was established in 1985. The Ministry introduced a scheme for the conservation of wetlands in the Seventh Five-Year Plan, enabling state governments to develop and implement management plans for prioritized wetlands. The Ministry has continued support to the state governments for designing and implementing management plans for priority wetlands, under centrally sponsored schemes such as the National Plan for Conservation of Aquatic Ecosystems, Integrated Development of Wildlife Habitats covering wetlands located within the Protected Area network and the scheme for the Conservation of mangroves and coral reefs. The Ministry's flagship scheme aims at "mainstreaming the full range of ecosystem services and biodiversity values within developmental plans and programs at various levels." As India's network of Ramsar Sites has steadily increased to 100, currently the largest in Asia, a dedicated scheme, "Amrit Dharohar," has been launched to promote the unique conservation values of these sites. Thus, ecosystem services and wetland values have come to acquire a central role in wetland policymaking and programming.

Wetlands ecosystem services are further spotlighted through the endorsement of Ramsar Convention's wetland "wise use" as the central tenet for conservation action in the guidelines of the national program, as well as the regulatory framework for wetlands, the Wetlands (Conservation and Management) Rules, 2017. The Ramsar Convention defines wise use of wetlands as "the maintenance of their ecological character, achieved through the implementation of ecosystem approaches, within the context of sustainable development." Ecological character is "the combination of the ecosystem components, processes and services that characterize the wetland at any given point in time." The inclusion of ecosystem services within the definition of ecological character has tended to formally bridge conservation and sustainable use of wetlands (Braat & de Groot, 2012), implementation of which calls for their integrated management based on the full range of ecosystem services and biodiversity values while framing management objectives (Parrott & Quinn, 2016; Zsuffa et al., 2014).

The term "value of ecosystem services and biodiversity" can be interpreted in various ways, including importance, preference, or a measure thereof. However, value can also refer to a principle or core belief underlying one's preferences. The way nature is valued can relate to diverse ontologies and epistemologies, which influence the constitution and

Wetlands: Biodiversity, Health and their Role in Ecosystem Services. DOI: https://doi.org/10.1016/B978-0-443-34222-6.00006-1

conceptualization of value, as well as changes brought about in various decision-making contexts. Valuation is "the process of expressing a value for a particular good or service…in terms of something that can be counted, often money, but also through methods and measures from other disciplines (sociology, ecology and so on)" (Limburg et al., 2002). Valuation involves assigning relative weights to the various aspects of individual and social decision problems, with these weights reflecting the goals and worldviews of the community, society, and cultures to which individuals belong (Costanza & Folke, 1997). Economic valuation expresses these weights in monetary terms, making them comparable to alternative uses, which often have benefit and cost flows defined in similar units. It is essentially an anthropocentric way of looking at nature, wherein values are assigned to the extent that these fulfill and directly or indirectly contribute to human well-being (i.e., a positive change in well-being).

Functioning wetlands provide multiple provisioning, regulating, supporting and cultural services, which, being interlinked, are positively or adversely affected in response to wetland use within the context of broader developmental programming. Achieving wetland-wise use requires addressing trade-offs between the values people hold for these ecosystems and the benefits derived. Unfortunately, being public goods, a large category of wetland ecosystem services is not considered and internalized in public policymaking. Economic values make explicit the impact of public policy or private decisions on ecosystem service values, enabling the expression of these value changes in units that facilitate their incorporation into public decision-making (Mooney, 2005). Economic valuation improves the possibility of achieving sound decisions on wetland use and management by acting as a societal feedback mechanism alerting society to the consequences of consumption choices and behavior (Brown & Zavestoski, 2004).

The science and practice of valuing ecosystem services and their application in decision-making have undergone considerable evolution in recent years (Gómez-Baggethun et al., 2016; Constanza et al., 2017). While valuation is recognized as an important lever for sustainability transformation, a major emphasis is on the fact that generating knowledge of ecosystem services is an important starting point, but it is not by itself sufficient for integration into policy and management (Pascual et al., 2023; Primmer et al., 2015). Insights into decision-making, particularly in understanding how stakeholders make decisions, embedding research into institutional design and policy processes, and developing interventions that facilitate monitoring, evaluation, and adaptation over time are crucial for effective management (Daily et al., 2009).

The provision of ecosystem services depends on the functioning and complexity of ecosystems and landscape context (Raudsepp-Hearne et al., 2010). Seen from a biophysical perspective, services emerge from ecological production functions (Daily et al., 2009), which are supported by ecological structures and processes, often referred to as supporting services (de Groot et al., 2010). Ecosystem functions differ from components or processes, as functions capture not only their combination but also the capacity of ecosystems to deliver services (Naeem, 1998). The spatial and temporal scales at which services are produced, accessed, and used shape how societies value them and what tools are applied for management (Raudsepp-Hearne & Peterson, 2016).

Wetlands deliver multiple ecosystem services, which are best considered as bundles rather than isolated services. Some services covary positively (e.g., improving flood-buffering capacity of wetlands can support increased groundwater recharge and help maintain productive fisheries), whereas in several circumstances, ecosystem services may covary negatively (use of wetlands for permanent agriculture may reduce the capability of wetlands to moderate water regimes). Trade-offs among bundles are inevitable, since services rarely change in the same direction under different patterns of use and management (MEA, 2005; Raudsepp-Hearne et al., 2010). Management that maximizes a single service, typically a provisioning one, usually reduces the availability of regulating and cultural services (Bennett et al., 2009; Russi et al., 2013). Shifting management systems to recognize and reward regulating and cultural services, which are not traded in markets, remains a major challenge (Guerry et al., 2015). Key questions for wetland management are whether diverse service bundles are considered in setting objectives and whether the ecological functions that sustain them are adequately addressed in management actions.

Ecosystem services are also framed politically, often leading to the creation of new markets, property regimes, and power relations, with clear distributional consequences (Kull et al., 2015). The transformation of natural capital into ecosystem services is shaped by institutions that operate across levels (Duraiappah et al., 2014). These institutions govern access through various processes, including entitlements (Leach et al., 1999; Sen, 1984), power asymmetries (Robards et al., 2011), social differentiation, and poverty. Power relations, embedded within institutional and governance systems, shape the ability of ecosystems to provide ecosystem services (Felipe-Lucia et al., 2015; Ribot & Peluso, 2003). From a social perspective, it is essential to understand how human actions generate services, who benefits from them, and how values are articulated and integrated into decision-making (Ernstson, 2013). Thus, assigning value to services is inherently subjective, varying across individuals, groups, time, and location, and influenced by cultural and institutional norms (Spangenberg et al., 2014). Frameworks such as the cascade model have tended to capture the interlinkages between ecosystem services, biophysical processes, and social systems (Nassl & Löffler, 2015; Potschin-young et al., 2018; Boulton et al., 2016).

In India, economic valuation has received attention as a significant research area only in the last two and a half decades. One of the early structured attempts was undertaken under the Ministry of Environment and Forests' Ecodevelopment program, in which valuation techniques were applied to Keoladeo National Park (Rajasthan) to provide possible policy options for improving people-park relationships. Subsequently, the World Bank supported 'Environmental Management Capacity Building Technical Assistance (EMCaB)' Project, implemented during 1996–2004 by the Ministry of Environment and Forests with the Indira Gandhi Institute of Development Research, Institute of Economic Growth, Madras School of Economics, and other agencies, put significant focus on promoting research using economic valuation tools, of which wetlands were one of the priority areas. Since then, the subject matter has been accorded high priority within research programs of the MoEFCC and several universities. In 2010, the MoEFCC launched the Economics of Ecosystems and Biodiversity—India Initiative (TII) to highlight the economic consequences of the loss of biological diversity and the decline in ecosystem services. The initiative envisioned mainstreaming ecosystem services and biodiversity values into developmental programming, utilizing an evidence-building approach for three ecosystem types: inland wetlands, forests, and coastal and marine ecosystems.

While the initiatives mentioned previously and several others have been crucial in advancing the application of economic valuation tools, a synoptic view of research progress and its application in planning and decision-making, with specific reference to wetlands, has yet to be undertaken, to the best of our knowledge. In this chapter, we present an overview of progress in the economic valuation of wetland ecosystem services over the last 25 years and offer recommendations for strengthening the interface with planning and decision-making.

19.2 Methods

The wetland economic valuation research content was identified through searches in Google Scholar, Scopus, and ResearchGate, and supplemented with relevant reports, government documents, and other grey literature. Search terms combined keywords such as "wetland ecosystem services AND valuation AND India," "wetland economics AND India," and service- or method-specific terms (e.g., provisioning, regulating, contingent valuation method [CVM], travel cost method [TCM], benefit transfer). Reference lists of the retrieved papers were also screened.

The review followed PRISMA guidelines (Page et al., 2021) across three phases: identification, screening, and integration. From 64 records, three duplicates were removed. Studies were excluded if (1) their scope extended beyond India, (2) wetland services were aggregated with other ecosystems without disaggregation, or (3) empirical valuation methods were insufficiently rigorous. The final dataset comprised 46 publications.

Each study was coded for:

- Wetland type—inland, coastal/marine, or human-made (Gardner & Davidson, 2011).
- Ecosystem services—provisioning, regulating, cultural, and supporting (MEA, 2005).
- Valuation methods—market/production-based, revealed preference, stated preference, benefit transfer, and modeling/integrated approaches (TEEB, 2010; de Groot et al., 2012).
- Geographic scope—state-level and spatial coverage.
- Valuation results—standardized to Rs. (billion/million) for comparability.

Ambiguous approaches were coded to the closest-fit category, with dual coding where appropriate. Descriptive statistics were used to analyze distributions across wetland types, services, and methods, highlighting research gaps and biases.

To assess balance in research effort, diversity indices were applied: the Shannon–Wiener index (H') to capture overall diversity (Shannon, 1948) and the Simpson index ($1-D$) to reduce sensitivity to rare classes (Simpson, 1949). Research inequality was evaluated using the Lorenz curve and Gini coefficient (G), adapted from economics to scientometric analysis (Gini, 1912; Lorenz, 1905; Rousseau, 1993). G values range from 0 (equal distribution) to 1 (maximum inequality).

Of the 64 studies retrieved, 46 met the eligibility criteria and formed the evidence base for the review.

19.3 Results

19.3.1 Application of valuation frameworks and methods

Various conceptual frameworks have been employed, although not always consistently. The total economic value (TEV) framework is most often employed. One study on estuarine and lagoon systems in Kerala arrived at a figure of about Rs.

1.92 billion (Anoopa, 2012). In contrast, when the same framework is applied at the national level, the reported values are much larger, reaching several trillions of rupees (Gopal & Marothia, 2016). Despite its widespread use, the application of TEV was geographically concentrated in states such as West Bengal, Odisha, and Kerala, with inland, arid, and semiarid systems receiving less attention.

The application of frameworks varied considerably across wetland types. The distribution of studies was uneven across wetland types. Research on marine and coastal ecosystems was slightly more common, at just under half of the total (49%). Inland wetlands were examined almost as frequently (46%), whereas human-made systems appeared only rarely in the literature, at about 5%. Looking more closely at inland systems, freshwater marshes and swamps were the most frequently examined (17%), with smaller proportions devoted to floodplain wetlands (9%), rivers and streams (9%), and freshwater lakes (7%). Specialized inland systems such as peatlands (1%), high-altitude wetlands (1%), and springs (1%) received limited valuation attention. Among coastal and marine wetlands, estuaries accounted for 15%, mangroves 13%, mudflats 9%, coastal lagoons 6%, shallow marine waters 3%, and mixed coastal systems 3%. Human-made wetlands were represented exclusively by reservoirs and tanks (5%) (Fig. 19.1).

Indices of research distribution confirmed moderate diversity. The Shannon–Wiener index (H'=2.68) indicated a moderately high spread across wetland types, and the Simpson index (0.91) suggested a high likelihood that two randomly selected studies concerned different wetland types. The Gini coefficient (0.53) demonstrated a moderate degree of inequality, with estuaries, mangroves, freshwater marshes, and floodplains accounting for a disproportionate share of publications.

A wide range of valuation methods has been employed, although conventional techniques continue to dominate. Stated preference approaches were the most common, appearing in approximately one-third of the studies. Market- and production-based methods were applied almost as frequently, while revealed-preference techniques accounted for roughly one-fifth. The remaining work drew on benefit transfer (just over 8%) and on modeling or integrated approaches (around 9%). Within market-based applications, direct market valuation (16.28%) was most common, particularly for fisheries, fodder, and timber. Only a small share of studies, approximately 8%, have used replacement cost methods. These were mainly used to assess regulating functions, for instance, water purification or protection against storms. Production function approaches (4.65%) linked wetlands to agricultural outputs, while damage cost avoided methods (1.16%) were rarely used.

Revealed preference approaches focused mainly on recreational services. The TCM was employed in approximately one-fifth of the studies, primarily used to measure visitor spending and consumer surplus. Hedonic pricing, by contrast, was used only once, in a case where property values were linked to wetland proximity. Among stated-preference techniques, contingent valuation was by far the most common, accounting for nearly 28% of applications. It was generally used to gauge willingness to pay for conservation as well as for option and nonuse values. Choice modeling appeared in 3.49% of studies. Benefit transfer (8.14%) was mostly limited to unit-value transfers, used to scale local estimates up to regional or national levels. Integrated and modeling approaches, though less frequent (9.30%), provided dynamic perspectives. Simulation models (5.81%), InVEST (Integrated Valuation of Ecosystem Services and Trade-Offs) modeling (2.33%), and biophysical–economic linkages (1.16%) incorporated ecological processes and climate change scenarios into valuation exercises (Fig. 19.2).

The use of valuation frameworks varied by the type of ecosystem service studied. Provisioning services were most frequently assessed (72%), followed by regulation (33%), cultural (13%), and supporting services (2%), while nearly three-quarters of studies examined multiple services together. Standardized estimates show that wetlands deliver substantial annual benefits, ranging from Rs. 1315 to 157,569 ha^{-1} $year^{-1}$ for cultural and regulating functions to Rs. 2463 to 24,366 ha^{-1} $year^{-1}$ for provisioning flows, with integrated assessments reaching Rs. 40,000 to 60,000 ha^{-1} $year^{-1}$ (Table 19.1).

19.4 Valuation outcomes and their relevance

The range of outcomes reported through these valuation methods demonstrates the multiple dimensions of the contributions wetlands make to society and the economy. Provisioning services featured prominently in site-level studies. Estuarine fisheries in Kerala were valued at Rs. 7.72 million annually (Anoopa, 2012), while sand extraction was estimated at nearly Rs. 25.75 billion annually (Gopal & Marothia, 2016). Such results underline the importance of wetlands in sustaining livelihoods, even as they point to mounting stress from unsustainable extraction.

Regulating services were often valued in relation to risk reduction. In the Sundarbans, mangrove stands act as a first line of defense during cyclones by breaking the force of incoming waves and slowing down storm surges, and one study placed the annual value of this protection at about Rs. 4.2 billion (Kadaverugu et al., 2021). Earlier studies observed that this

FIGURE 19.1 Distribution of wetland valuation studies by wetland type and subtype.

natural protection translated into real savings for local households, which faced lower repair costs after major storms (Badola & Hussain, 2005). Such evidence illustrates the way wetlands themselves function as frontline defenses for human settlements in hazard-prone regions.

Cultural and recreational services were quantified using revealed preference methods. Recreation in selected wetlands was estimated at Rs. 150 million annually, with consumer surplus per visit of INR 555 (Guha & Ghosh, 2009). Hedonic pricing applied in peri-urban contexts demonstrated that wetland proximity contributed positively to property values. Supporting and nonuse values were elicited through stated preference methods. For example, Anoop and Suryaprakash (2008) reported option values of Rs. 0.84 million and aggregate nonuse values of Rs. 3.88 million. Subsequent willingness-to-pay surveys (Verma and Negandhi, 2011; Vijayan et al., 2019) extended such findings, directly informing conservation finance mechanisms and payment for ecosystem services schemes.

Scaling up through benefit transfer produced higher-order outcomes. Saline wetlands in Gujarat were valued at Rs. 1.9 billion annually (Dixit et al., 2016), while aggregated national estimates placed wetland values in trillions of rupees (Gopal & Marothia, 2016). Dynamic modeling highlighted the vulnerability of wetland services to

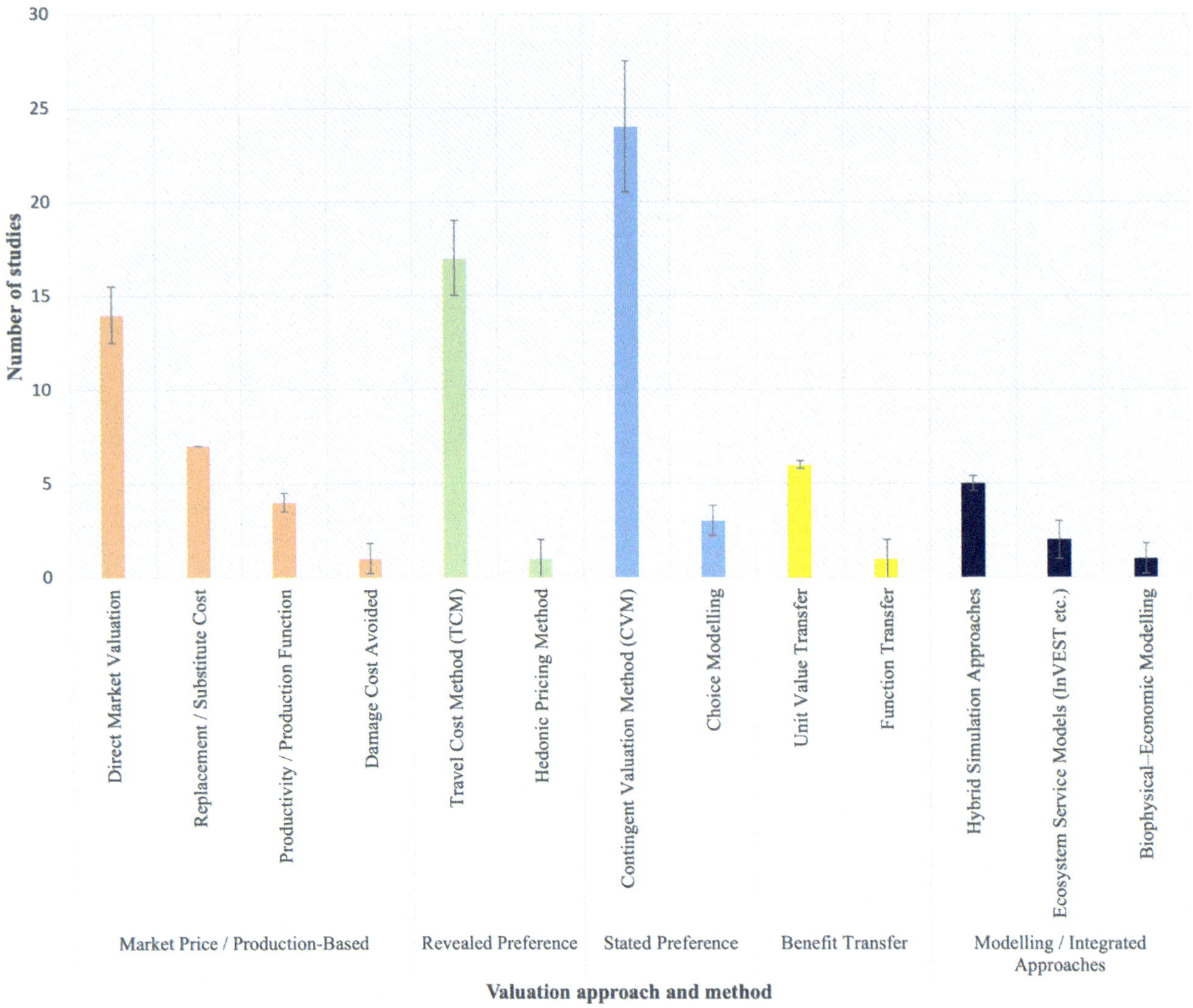

FIGURE 19.2 Methodological distribution of wetland valuation studies in India.

TABLE 19.1 Ecosystem services valued in reviewed wetland studies.

Ecosystem services category	Proportion of studies (%)	Representative range (Rs. ha^{-1} year^{-1}, 2025) (applying a social discount rate of 5% (Kumar, 2012))
Provisioning	71.74	2463–24,366
Regulating	32.61	10,000–157,569
Cultural	13.04	1315–15,000
Supporting	2.17	15,000–25,000
Multiple/Combined	69.57	40,000–60,000

TABLE 19.2 Examples of valuation results and their decision relevance.

Approach/method	Study (references)	Key result	Decision relevance
Market/Production	Anoopa (2012)	Estuarine TEV Rs. 1.92 billion; provisioning services Rs. 7.72 million annually	Basis for livelihood security
Market/Production	Gopal and Marothia (2016)	Sand extraction Rs. 25.75 billion annually	Evidence of unsustainable extraction
Regulating (damage cost avoided)	Das and Vincent(2009)	Mangrove storm protection provides household-level savings	Informing disaster risk reduction
Revealed preference (travel cost method)	Guha and Ghosh (2009)	Recreation value Rs. 150 million annually; consumer surplus INR 555/visit	Basis for ecotourism initiatives
Revealed preference (hedonic pricing method)	Everard et al. (2019)	Wetland proximity increases property values	Urban planning application
Stated preference (contingent valuation method)	Anoop and Suryaprakash (2008)	Option values Rs. 0.84 million; aggregate Rs. 3.88 million	Evidence for PES design
Stated preference and market	Kadaverugu et al. (2021)	Mangrove storm protection Rs. 4.2 billion	Application to climate resilience
Benefit transfer	Dixit et al. (2016)	Saline wetland TEV Rs. 1.9 billion annually	Cost–benefit analysis tool
Modeling	Khan et al. (2019)	Climate projections: 43% increase in water yield; 54.5% increase in sediment export	Adaptation planning
Modeling	Khatun and Das (2025)	Floodplain wetland decline valued at Rs. 127.6 billion (1990), significantly reduced by 2020	Long-term degradation evidence

PES, Payment for ecosystem services; *TEV*, total economic value.

environmental change. Reservoir modeling points to noticeable changes under climate stress. One study estimated that water yield could increase by over 40%, but at the same time, sediment export might rise by more than half (Khan et al., 2019). Floodplain wetlands show a different story: their estimated economic value, which was close to Rs. 127.6 billion in 1990, has dropped sharply by 2020 (Khatun & Das, 2025) (Table 19.2).

19.5 Spatial distribution of studies

The geographical distribution of valuation studies was uneven, with evidence concentrated in a few ecologically significant regions. Out of the 46 studies reviewed, about 46% (21 studies) were conducted in coastal and estuarine areas. The Western Ghats accounted for 7 studies (roughly 15%), while the North-East contributed 5 (around 11%). Four studies focused on the Himalayan states (about 9%), and three looked at arid or semiarid areas (about 7%). At a broader scale, two studies (4%) were carried out at the national level, and the Indo-Gangetic plains were represented by only one study (2%). The remaining 3 studies (approximately 7%) were based in central Indian regions.

The coastal belt, including the Sundarbans of West Bengal, Chilika Lagoon and Bhitarkanika mangroves in Odisha, and the backwaters of Kerala, was the most intensively studied. In the Sundarbans, cyclone buffering was valued at Rs. 4.2 billion annually and recreation at Rs. 28.3 million. Chilika and Bhitarkanika valuations ranged from fisheries in the tens of lakhs to ecosystem services valued at Rs. 59.25 billion. Kerala's Ashtamudi and Vembanad–Kol wetlands reported provisioning values of Rs. 7.72 million annually, TEV of Rs. 1.92 billion, and option values of Rs. 3.88 million. In Tamil Nadu and Puducherry, reservoirs and tanks generated provisioning values in millions, while a restoration study estimated an annual value of Rs. 27.38 billion.

In Himalayan states such as Himachal Pradesh, Uttarakhand, Sikkim, and Jammu and Kashmir, high-altitude wetlands were the focus. Valuations emphasized provisioning and cultural services, with household-level benefits in Himachal Pradesh estimated at Rs. 0.27 million±Rs. 11,000 annually. In the Western Ghats, studies in Karnataka and inland Kerala have focused on riverine and forest-linked wetlands, with wetland regulating services valued at Rs. 138.4 million annually. Additionally, cultural services linked to navigation and livelihoods have also been highlighted.

The North-East was represented by valuation studies on floodplains and lakes in Assam and Manipur. In Assam, fisheries were valued at Rs. 4.6 million annually, aggregate services at Rs. 4.65 billion, and broader assessments at Rs. 34.5 billion. In Manipur, Loktak Lake yielded provisioning values in the billions annually, with over 112 t of edible plants traded each year. Arid and semiarid regions contributed fewer studies but produced high valuations: wetlands in Gujarat were estimated at Rs. 18.98 billion annually, wetlands in Madhya Pradesh at Rs. 2.96 billion annually, and wetlands in Rajasthan in the range of millions to billions.

Two studies extended beyond individual states. One combined the Himalayan states into a regional assessment, while another produced national-level TEV estimates in the trillions of rupees. Across all regions, coastal belts generated the highest-value outcomes, Himalayan states reported household-scale contributions averaging Rs. 0.27 million annually, Western Ghats wetlands yielded mid-range values in the tens of crores, the North-East was dominated by provisioning services up to Rs. 34.5 billion annually, and arid and semiarid wetlands produced some of the highest per-site valuations, such as Rs. 18.98 billion annually in Gujarat.

Trade-offs arising from policy decisions constitute a useful application area for economic valuation tools. The study of the Yamuna floodplains assessed the opportunity cost of converting them for development and concluded that such conversion could not be justified on grounds of economic efficiency (Kumar et al., 2001). Economic valuation was used as a tool to assess the impacts of freshwater flow regulation on the ecosystem services of Chilika Lake. The assessment highlighted the benefits of floods to floodplain agriculture and downstream wetland fisheries. It also indicated that reducing the freshwater flows had negative economic consequences in terms of the values of fisheries, flooding, and waterlogging likely to be created due to policy decisions (WISA, 2004). An analysis of net and gross value added in fisheries of Chilika Lake has been linked to an analysis of livelihood systems to validate the distributional consequences of an increase in fish landing from Chilika (Kumar et al., 2020).

Only two studies have attempted extrapolation of economic values of wetlands or impacts of change in wetland extent to state or national-level aggregates. Pandey et al. (2004) have computed state-level aggregated values of wetland wealth using the data on wetland extent (from the Directory of Wetlands, 1990) and economic values from Costanza et al. (1997), and Mitsch and Gosselink (2000). The study ranks Karnataka, Gujarat, and Andhra Pradesh as the states having the highest wetland wealth, and Nagaland, Meghalaya, and Sikkim as having the lowest wealth. More recently, a framework for accounting for inland wetland ecosystems in selected Indian states has been proposed by Kumar (2012). The study uses the benefit transfer method to determine the impacts of physical area losses of wetlands in Gujarat, Jammu and Kashmir, Kerala, Rajasthan, and West Bengal. Value estimates from 18 wetlands have been used to develop a metaregression model to finally compute the loss of per capita wetland wealth for 1991–2001. The study concludes that the State of Jammu and Kashmir had the highest per capita wealth loss (US$ 211.83), and that the identified states had an average loss of US$ 11.57.

19.6 Discussion

There has been considerable progress in applying economic valuation tools to wetlands in the last 25 years. However, a critical review indicates substantive gaps in spatial coverage, ecosystem services coverage, and application to decision-making contexts. Studies show that larger wetlands and particular types are studied more often than smaller or less visible types, creating an imbalance in representation. Valuation studies have focused on ecologically prominent systems, such as estuaries, mangroves, floodplains, and freshwater marshes. By contrast, high-altitude wetlands, peatlands, saline wetlands, springs, and human-managed systems, including aquaculture ponds and rice paddies, remain largely unexamined.

A second limitation lies in the asymmetrical treatment of service categories. Provisioning services, including fisheries, agriculture, fuelwood, and tourism, dominate the valuation studies because they are readily quantified through market-based methods. Regulating services, such as storm protection and water purification, receive moderate attention, often through cost-based proxies. However, supporting services, including nutrient cycling and biodiversity maintenance, have been assessed in only about 5% of studies, while cultural services account for approximately 15%.

The methodological landscape is also constrained. Approaches, such as the CVM, direct market valuation, and the TCM, together constitute nearly two-thirds of all studies considered in this chapter. By comparison, techniques such as benefit transfer, which accounts for only 8% of studies, and integrated biophysical–economic modeling, which represents only 1%, remain underutilised.

Spatial and temporal mismatches further undermine the evidence base. Research is geographically concentrated in West Bengal, Odisha, and Kerala, particularly in the Sundarbans, Chilika Lagoon, and the Kerala backwaters. Inland arid and semiarid wetlands, high-altitude systems in the Himalayas, and human-managed wetlands are rarely valued. In addition, most studies remain site-specific and short-term, failing to reflect the larger basin-level and migratory flyway scales at which many wetland services are generated. This disconnect reduces the applicability of valuation results for integrated water resource management and transboundary ecological governance.

In India, most valuation studies have taken a technocratic route, leaving little room for local communities to take part or for more open, deliberative approaches. Some pilot efforts, such as those under the TEEB–India Initiative, involved communities in data collection. Concerns such as equity, access, and the cultural value of wetlands are seldom given serious attention. As a result, valuation findings often struggle to gain legitimacy and have a limited impact on policy decisions.

Finally, very few studies have been applied to decision-making contexts. A key barrier is the weak link between valuation results and formal policy processes. The Wetlands (Conservation and Management) Rules, 2017, recognize the ecological value of wetlands, but valuation evidence is rarely used in environmental impact assessments, land-use planning, or cost–benefit studies.

Political economy dynamics also shape the reception of valuation findings. Wetlands are often regarded as potential sites for agriculture, aquaculture, or urban expansion, and these uses typically receive greater political and economic support. Because of this, valuation studies that highlight cultural or regulatory services often carry less weight than those that support short-term development goals. Moreover, when valuation studies are introduced through technical frameworks that appear externally imposed, they are sometimes received with skepticism by local stakeholders.

19.7 Enhancing integration in wetland policy and programs

As India's policy and programming on wetlands advances, it is pertinent that research on the multiple values is advanced and integrated in planning and decision-making. Some of the ways in which this can be achieved are: (1) promoting interdisciplinary approaches to economic valuation, such as the use of biophysical and economic modeling tools, for example ARIES (artificial intelligence for environment and sustainability) and InVEST, (2) increased coverage of studies on hitherto underrepresented wetland types and landscapes, (3) broadening research on functional aspects of wetlands so as to understand combinations of ecosystem structures and processes under which service delivery can take place; (4) building ecosystem services values into inventory and assessment protocols to enable identification and prioritization of services; (5) integrating ecosystem services related objectives and implementation strategies in wetlands management planning processes. This would also include elucidating the spatial and temporal aspects of ecosystem services, as well as the interactions between ecosystem services and livelihood capitals, thereby allowing assessment of equity and social fairness related outcomes of wetland management.

Stronger mandates for incorporating ecosystem service values into environmental assessments, land use planning, and water allocation would improve integration. In addition, embedding wetland valuation in natural capital accounting and Green GDP initiatives would firmly place wetlands within national economic and planning frameworks.

References

Anoop, P., & Suryaprakash, S. (2008). Estimating the option value of Ashtamudi Estuary in South India: A contingent valuation approach.

Anoopa, P. (2012). Economic valuation of use benefits of Ashtamudi estuary in South India. *Indian Journal of Agricultural Economics, 67*(3), 351.

Badola, R., & Hussain, S. A. (2005). Valuing ecosystem functions: An empirical study on the storm protection function of Bhitarkanika mangrove ecosystem, India. *Environmental Conservation, 32*(1), 85–92.

Bennett, E. M., Peterson, G. D., & Gordon, L. J. (2009). Understanding relationships among multiple ecosystem services. *Ecology Letters, 12*(12), 1394–1404.

Boulton, A. J., Ekebom, J., & Gislason, G. M. (2016). Integrating ecosystem services into conservation strategies for freshwater and marine habitats: A review. *Aquatic Conservation: Marine and Freshwater Ecosystems, 26*(5), 963–985.

Braat, L. C., & De Groot, R. (2012). The ecosystem services agenda: Bridging the worlds of natural science and economics, conservation and development, and public and private policy. *Ecosystem Services, 1*(1), 4–15.

Brown, P., & Zavestoski, S. (2004). Social movements in health: An introduction. *Sociol Health Illn, 26*(6).

Costanza, R., d'Arge, R., de Groot, R., Farber, S., Grasso, M., Hannon, B., Limburg, K., Naeem, S., O'Neill, R. V., Paruelo, J., Raskin, R. J., Sutton, P., & van den Belt, M. (1997). The value of the world's ecosystem services and natural capital. *Nature, 387*, 253–260.

Costanza, R., & Folke, C. (1997). Valuing ecosystem services with efficiency, fairness and sustainability as goals. *In: Nature's services: Societal dependence on natural ecosystems,* 49–70.

Costanza, R., De Groot, R., Braat, L., Kubiszewski, I., Fioramonti, L., Sutton, P., et al. (2017). Twenty years of ecosystem services: How far have we come and how far do we still need to go? *Ecosystem Services, 28*, 1–16.

Daily, B. F., Bishop, J. W., & Govindarajulu, N. (2009). A conceptual model for organizational citizenship behavior directed toward the environment. *Business & Society, 48*(2), 243–256.

Das, S., & Vincent, J. R. (2009). Mangroves protected villages and reduced death toll during Indian super cyclone. *Proceedings of the National Academy of Sciences, 106*(18), 7357–7360.

De Groot, R. S., Alkemade, R., Braat, L., Hein, L., & Willemen, L. (2010). Challenges in integrating the concept of ecosystem services and values in landscape planning, management and decision making. *Ecological Complexity, 7*(3), 260–272.

De Groot, R., Brander, L., Van Der Ploeg, S., Costanza, R., Bernard, F., Braat, L., Christie, M., Crossman, N., Ghermandi, A., Hein, L., & Hussain, S. (2012). Global estimates of the value of ecosystems and their services in monetary units. *Ecosystem services, 1*(1), 50–61.

Dixit, A. M., Bandyopadhyaya, S., Kumar, L., & Bedamatta, S. (2016). Economic valuation of landscape level wetland ecosystem and its services in Little Rann of Kachchh, Gujarat. In *The economics of ecosystems and biodiversity India initiative*. GIZ India, 1–160.

Duraiappah, A. K., Asah, S. T., Brondizio, E. S., Kosoy, N., O'Farrell, P. J., Prieur-Richard, A. H., et al. (2014). Managing the mismatches to provide ecosystem services for human well-being: A conceptual framework for understanding the new commons. *Current Opinion in Environmental Sustainability, 7*, 94–100.

Ernstson, H. (2013). The social production of ecosystem services: A framework for studying environmental justice and ecological complexity in urbanized landscapes. *Landsc Urban Plan, 109*(1), 7–17.

Everard, M., Kangabam, R., Tiwari, M. K., McInnes, R., Kumar, R., Talukdar, G. H., & Das, L. (2019). Ecosystem service assessment of selected wetlands of Kolkata and the Indian Gangetic Delta: Multi-beneficial systems under differentiated management stress. *Wetlands Ecology and Management, 27*, 405–426.

Felipe-Lucia, M. R., Comín, F. A., & Escalera-Reyes, J. (2015). A framework for the social valuation of ecosystem services. *Ambio, 44*(4), 308–318.

Gardner, R. C., & Davidson, N. C. (2011). The ramsar convention. In *Wetlands: Integrating multidisciplinary concepts*. Dordrecht: Springer Netherlands, 189–203.

Gini, C. (1912). Variabilità e mutabilità. Studi economico-giuridici, Università di Cagliari, III(2a). Reprinted in: Gini, C., 1955. pp. 211–382.

Gómez-Baggethun, E., Barton, D. N., Berry, P., Dunford, R., & Harrison, P. A. (2016). Concepts and methods in ecosystem services valuation. *Routledge Handbook of Ecosystem Services*, 99–111.

Gopal, B., & Marothia, D. K. (2016). Economics of biodiversity and ecosystem services of rivers for sustainable management of water resources. In *The economics of ecosystems and biodiversity India initiative*. GIZ India.

Guerry, A. D., Polasky, S., Lubchenco, J., Chaplin-Kramer, R., Daily, G. C., Griffin, R., et al. (2015). Natural capital and ecosystem services informing decisions: From promise to practice. *Proceedings of the National Academy of Sciences, 112*(24), 7348–7355.

Guha, I., & Ghosh, S. (2009). *A glimpse of the tiger: How much are Indians willing to pay for it?* Institute of Development Studies and Partner Organisations.

Kadaverugu, R., Dhyani, S., Dasgupta, R., Kumar, P., Hashimoto, S., & Pujari, P. (2021). Multiple values of Bhitarkanika mangroves for human well-being: Synthesis of contemporary scientific knowledge for mainstreaming ecosystem services in policy planning. *Journal of Coastal Conservation, 25*, 1–15.

Khan, M., Sharma, A., & Goyal, M. K. (2019). Assessment of future water provisioning and sediment load under climate and LULC change scenarios in a peninsular river basin, India. *Hydrological Sciences Journal, 64*(4), 405–419.

Khatun, R., & Das, S. (2025). Valuing wetland ecosystem services in relation to the area and landscape dynamics of Rarh region of West Bengal, India. *Wetlands Ecology and Management, 33*(1), 7.

Kull, C.A., de Sartre X.A., Castro-Larrañaga M. (2015). The political ecology of ecosystem services. *Geoforum*, 1(61), 122–134.

Kumar, P., Babu, C., Sharma, R., & Love, A. (2001). Valuation of ecological services of wetland ecosystems: A case study of Yamuna floodplains in the corridors of Delhi. In *Mimeograph*. Institute of Economic Growth.

Kumar, P. (2012). Measuring natural capital: Accounting of inland wetland ecosystems from selected states of India. *Economic and Political Weekly*, 77–84.

Kumar, R., Pattnaik, A. K., & Finlayson, C. M. (2020). Ecosystem services: Implications for managing Chilika. In *Ecology, conservation, and restoration of Chilika Lagoon, India*. Springer International Publishing, 63–94.

Leach, M., Mearns, R., & Scoones, I. (1999). Environmental entitlements: Dynamics and institutions in community-based natural resource management. *World Development, 27*(2), 225–247.

Limburg, K. E., O'Neill, R. V., Costanza, R., & Farber, S. (2002). Complex systems and valuation. *Ecological Economics, 41*(3), 409–420.

Lorenz, M. O. (1905). Methods of measuring the concentration of wealth. *Publications of the American Statistical Association, 9*(70), 209–219.

MEA (Millennium Ecosystem Assessment). (2005). *Ecosystems and Human Well-Being: Synthesis*, Island Press, Washington, DC.

Mitsch, W. J., & Gosselink, J. G. (2000). *Wetlands* (3rd ed.). Wiley.

Mooney, G. (2005). Addictions and social compassion. *Drug Alcohol Rev, 4*(2), 137–141.

Naeem, S. (1998). Species redundancy and ecosystem reliability. *Conservation Biology, 12*(1), 39–45.

Nassl, M., & Löffler, J. (2015). Ecosystem services in coupled social–ecological systems: Closing the cycle of service provision and societal feedback. *Ambio, 44*(8), 737–749.

Page, M. J., McKenzie, J. E., Bossuyt, P. M., Boutron, I., Hoffmann, T. C., Mulrow, C. D., Shamseer, L., Tetzlaff, J. M., Akl, E. A., Brennan, S. E., & Chou, R. (2021). The PRISMA 2020 statement: An updated guideline for reporting systematic reviews. *BMJ, 372*.

Pandey, J. S., Joseph, V., & Kaul, S. N. (2004). A zone-wise ecological-economic analysis of Indian wetlands. *Environmental Monitoring and Assessment, 98*(1), 261–273.

Parrott, L., & Quinn, N. (2016). A complex systems approach for multiobjective water quality regulation on managed wetland landscapes. *Ecosphere, 7*(6), e01363.

Pascual, U., Balvanera, P., Anderson, C. B., Chaplin-Kramer, R., Christie, M., González-Jiménez, D., et al. (2023). Diverse values of nature for sustainability. *Nature, 620*(7975), 813–823.

Potschin-Young, M., Haines-Young, R., Görg, C., Heink, U., Jax, K., & Schleyer, C. (2018). Understanding the role of conceptual frameworks: Reading the ecosystem service cascade. *Ecosystem Services, 29*, 428–440.

Primmer, E., Jokinen, P., Blicharska, M., Barton, D. N., Bugter, R., & Potschin, M. (2015). Governance of ecosystem services: A framework for empirical analysis. *Ecosystem Services, 16*, 158–166.

Raudsepp-Hearne, C., & Peterson, G. D. (2016). Scale and ecosystem services: how do observation, management, and analysis shift with scale—Lessons from Québec. *Ecology and Society, 21*(3).

Raudsepp-Hearne, C., Peterson, G. D., Tengö, M., Bennett, E. M., Holland, T., Benessaiah, K., et al. (2010). Untangling the environmentalist's paradox: Why is human well-being increasing as ecosystem services degrade? *BioScience, 60*(8), 576–589.

Ribot, J. C., & Peluso, N. L. (2003). A theory of access. *Rural Sociol, 68*(2), 153–181.

Robards, M. D., Schoon, M. L., Meek, C. L., & Engle, N. L. (2011). The importance of social drivers in the resilient provision of ecosystem services. *Global Environmental Change, 21*(2), 522–529.

Rousseau, R. (1993). Measuring concentration: Sampling design issues, as illustrated by the case of perfectly stratified samples. *Scientometrics, 28*(1), 3–14.

Russi, D., ten Brink, P., Farmer, A., Badura, T., Coates, D., Förster, J., et al. (2013). *The economics of ecosystems and biodiversity for water and wetlands*. IEEP, 78–118.

SAC (Space Applications Centre). (2024). Indian Space Research Organisation (ISRO), National Wetlands Inventory and Assessment (NWIA) Atlas: 2024 Space Applications Centre Ahmedabad.

Sen, A. (1984). *Resources, values and development*. Blackwell.

Shannon, C. E. (1948). A mathematical theory of communication. *The Bell System Technical Journal, 27*(3), 379–423.

Simpson, E. H. (1949). Measurement of diversity. *Nature, 163*(4148), 688.

Spangenberg, J. H., Görg, C., Truong, D. T., Tekken, V., Bustamante, J. V., & Settele, J. (2014). Provision of ecosystem services is determined by human agency, not ecosystem functions. *Four Case Studies. International Journal of Biodiversity Science, Ecosystem Services & Management, 10*(1), 40–53.

TEEB, R.O. (2010). *Mainstreaming the economics of nature*. TEEB Geneva.

Verma, M., & Negandhi, D. (2011). Valuing ecosystem services of wetlands: A tool for effective policy formulation and poverty alleviation. *Hydrological Sciences Journal*, 56, 1622–1639.

Vijayan, A., Job, E., & Sathyan, A. R. (2019). Addressing sustainability of fresh water lake: The ecosystem valuation of Vellayani Lake in South India (Poster presentation). Tropentag 2019: International Research on Food Security, Natural Resource Management and Rural Development, Kassel, Germany. https://www.tropentag.de/2019/abstracts/links/Vijayan_302IW19J.pdf.

Wetlands International South Asia (WISA), 2004. Socio-economic assessment of environmental flow scenarios of Chilika. Final report. Wetlands International South Asia, New Delhi.

Zsuffa, I., Van Dam, A. A., Kaggwa, R. C., Namaalwa, S., Mahieu, M., Cools, J., & Johnston, R. (2014). Towards decision support-based integrated management planning of papyrus wetlands: A case study from Uganda. *Wetlands Ecology and Management, 22*(2), 199–213.

Index

Note: Page numbers followed by *f* indicate figures, *t* indicate tables.

Z

Printed and bound by CPI Group (UK) Ltd, Croydon, CR0 4YY
14/09/2026
02216414-0002